75th ANNIVERSARY OF THE TRANSISTOR

IEEE Press
445 Hoes Lane
Piscataway, NJ 08854

75th ANNIVERSARY OF THE TRANSISTOR

Edited by

Arokia Nathan
Darwin College, University of Cambridge, Cambridge, UK

Samar K. Saha
Prospicient Devices, Milpitas, CA, USA

Ravi M. Todi
President, IEEE EDS, USA

Published by John Wiley & Sons, Inc., Hoboken, New Jersey.
Published simultaneously in Canada.

Library of Congress Cataloging-in-Publication Data Applied for:

Hardback ISBN: 9781394202447

Cover Design: Wiley
Cover Image: © Ali Kahfi/Getty Images

Set in 10/12pt Times LT Std by Straive, Pondicherry, India

SKY10053117_081023

Contents

About the Editors

Arokia Nathan is a leading pioneer in the development and application of thin film transistor technologies to flexible electronics, display and sensor systems. His current research interests lie in integration of devices, circuits, and systems using a broad range of inorganic and organic thin film material systems on rigid and mechanically flexible substrates for large area systems. Following his PhD in Electrical Engineering, University of Alberta, Canada in 1988, he joined LSI Logic Corp., Santa Clara, CA, where he worked on advanced multi-chip packaging techniques and related issues. Subsequently he was at the Institute of Quantum Electronics, ETH Zürich, Switzerland, before joining the Electrical technology and was a recipient of the 2001 Natural Sciences and Engineering Research Council E.W.R. Steacie Fellowship. In 2004, he was awarded the Canada Research Chair in nano-scale flexible circuits. In 2005/2006, he was a visiting professor in the Engineering Department, University of Cambridge, UK. Later in 2006, he joined the London Centre for Nanotechnology, University College London as the Sumitomo Chair of Nanotechnology. He moved to Cambridge University in 2011 as the Chair of Photonic Systems and Displays, and he is currently a Bye-Fellow and Tutor at Darwin College. He has over 600 publications in the field of sensor technology and CAD, and thin film transistor electronics including 6 books, and more than 130 patents and four spin-off companies. He is a recipient of the Royal Society Wolfson Research Merit Award, the BOE Award for contributions to TFT CAD, and winner of the 2020 IEEE EDS JJ Ebers Award. He serves on technical committees of professional societies and conferences in various capacities. He is currently the vice president of Publications and Products in IEEE's Electron Devices Society. He is a fellow of IEEE, a Distinguished Lecturer of the IEEE Electron Devices Society and Sensor Council, a chartered engineer (UK), Fellow of the Institution of Engineering and Technology (UK), Fellow of the Royal Academy of Engineering, of the Society for Information Displays, and Fellow of the Canadian Academy of Engineering.

Samar K. Saha is the Chief Research Scientist at Prospicient Devices, Milpitas, California, USA. Since 1984, he has worked at various technical and management positions for SuVolta, Silterra USA, DSM Solutions, Synopsys, Silicon Storage Technology, Philips Semiconductors, LSI Logic, Texas Instruments, and National Semiconductor. He has, also, worked as an Electrical Engineering faculty at Santa Clara University, California; the University of Colorado at Colorado Springs, Colorado; the University of Nevada at Las Vegas, Nevada; Auburn University, Alabama; and Southern Illinois University at Carbondale, Illinois. He has authored over 100 research papers; two books entitled, *FinFET Devices for VLSI Circuits and Systems* (CRC Press, 2020) and *Compact Models for Integrated Circuit Design: Conventional Transistors and Beyond* (CRC Press, 2015); one book chapter, "Introduction to Technology Computer Aided Design," in *Technology computer Aided design: Simulation for VLSI MOSFET* (C.K. Sarkar, ed., CRC Press, 2013); and holds 13 US patents. His book on FinFET has been translated into Chinese language entitled, *Nanoscale Integrated Circuits FinFET Device Physics and Modeling* (CRC Press/China Machine Press, 2022).

Dr. Saha served as the 2016–2017 President of the *Institute of Electrical and Electronics Engineers* (IEEE) *Electron Devices Society* (EDS) and currently serving as a member of the IEEE Fellow Committee. He is a life fellow of IEEE, a fellow of the Institution of Engineering and Technology (IET), UK, and a distinguished lecturer of IEEE EDS. Previously, he has served EDS as senior past and junior past president; Awards chair; vice president of publications; an elected member of the Board of Governors; fellow evaluation committee chair; editor-in-chief of IEEE *QuestEDS*; chair of George Smith and Paul Rappaport awards; editor of Region-5 and 6 Newsletter; chair of Compact Modeling Technical Committee; and Chair of North America West Subcommittee for Regions/Chapters; IEEE as a member of the Conference Publications Committee and TAB Periodicals Committee; and Santa Clara Valley/San Francisco EDS chapter as the treasurer, vice chair, and chair.

In publications, he has served as a Guest Editor of six Special Issues (SIs) of IEEE EDS journals including head guest editor of the IEEE Transactions on Electron Devices SIs on *Advanced Compact Models and 45-nm Modeling Challenges* and *Compact Interconnect Models for Giga Scale Integration* and two IEEE Journal of Electron Devices Society SIs from the selected extended papers presented at conferences. Dr. Saha received the PhD degree in Physics from Gauhati University, India, and MS degree in Engineering Management from Stanford University, USA. He is the recipient of the 2021 IEEE EDS Distinguished Service Award.

Ravi M. Todi received his Masters and Doctoral degrees in Electrical Engineering from University of Central Florida (UCF), Orlando, Florida. His graduate research was focused on gate stack engineering, with emphasis on binary metal alloys as gate electrode and on high-mobility germanium channel devices. In his early career, as advisory engineer/scientist at semiconductor research and development center (SRDC) at IBM microelectronics division, his work was focused on high-performance embedded dynamic random-access memory (eDRAM) integration on 45 nm silicon on insulator (SOI) logic platform. For his many contributions to the success of the eDRAM program at IBM, Ravi was awarded IBM's prestigious outstanding technical achievement award. Over the past decade, Ravi has held several lead technical and management positions at Qualcomm, GlobalFoundries, and Western Digital. Most recently he is the silicon technologist responsible for foundry technology at an early-stage Silicon Valley startup. With over 60 US granted patents, over 30 peer-reviewed journal publications, over 40 international conference presentations, and over 50 invited distinguished lectures, Ravi is well known in the semiconductor industry as a technical/business leader.

Ravi served as an associate editor for IEEE Transactions on Electron Devices from 2014 to 2019, as IEEE-EDS treasurer from 2012 to 2015, as IEEE-EDS vice president for technical activities and conferences from 2016 to 2019, as EDS president-elect in 2020, and is currently serving as EDS president from 2021 to 2023. He is the recipient of 2011 IEEE EDS Early Career Award. He is an IEEE Fellow and is a Distinguished Lecturer for IEEE Electron Devices Society.

Preface

The invention of the point-contact transistor on 16 December 1947 and subsequently the bipolar junction transistor (BJT) on 23 January 1948 has revolutionized the semiconductor industry worldwide, bringing profound changes to humanity. Over the past 75 years, the device structure and fabrication technology have continuously evolved from the point-contact transistor to the current state-of-the art ultrathin-body field-effect transistor (FET). This continuous evolution of device architecture and disruptive innovation in IC fabrication processes led to the production of high performance, low power, high density, and low-cost very large scale integrated (VLSI) circuits and systems or integrated circuit (IC) chips, enabling smart environments and integrated ecosystems. The relentless pursuit of worldwide research and development (R&D) on electron devices continues to push transistor performance to new limits. This has served to continuously improve computational efficiency and computing power to support wireless communication technology, 5G and beyond, not to mention computationally intelligent devices to support machine learning and artificial intelligence (AI). While there is a large volume of published works in the literature, no comprehensive systematic narratives are available to offer students, young professionals, and experts to reflect on the invention and evolution of the *transistor*, which ignited the digital revolution. Thus, the *75th Anniversary of the Transistor* was conceived to showcase short stories told by the people who contributed to this digital revolution. The objective of this book is to reiterate the importance of the invention of the *transistor* that changed the world, and which continues to bear an overwhelming presence in the engineering community, and in particular, the new generation of engineering professionals and academics.

This *75th Anniversary of the Transistor* commemorative volume celebrates the 75th anniversary of the invention of the *transistor*, providing a comprehensive coverage of the underlying historical perspective since conception and subsequent evolution into a multitude of integration and manufacturing technologies and applications. The level of the book is introductory and aimed at pre-university and undergraduate students serving as a useful reference on transistors and ICs. The stories of the invention and evolution of transistor device technology are told in such a way that readers do not need to have prerequisite knowledge of semiconductor physics.

Before we dig deep into the invention of the *transistor* and its evolution over the last 75 years, the *75th Anniversary of the Transistor* commemorative volume presents two introductory articles. The first, a brief introduction by Leo Esaki of the Yokohama College of Pharmacy, Totsuka-ku, Yokohama, Kanagawa, Japan, corecipient of the 1973 Nobel Prize in Physics and an Institute of Electrical and Electronics Engineers (IEEE) Electron Devices Society (EDS) Celebrated Member, describes his 1957 invention, "The First Quantum Electron Device" – the *Esaki diode*. The second, "The IEEE Electron Devices Society: A Brief History," by Samar K. Saha of Prospicient Devices and 2016–2017 president of the IEEE EDS, describes the origins of the EDS and its growth over the past seven decades, as a volunteer-led and -driven global organization of the IEEE, with the evolution of transistor architecture and its manufacturing technology.

The first 12 articles which then follow narrate the stories of R&D efforts on semiconductors, with particular focus on the invention and post-invention evolution of device architecture and fabrication processes for VLSI chip manufacturing. The first of this series of articles, "Did Sir J. C. Bose Anticipate the Existence of p- and n-Type Semiconductors in His Coherer/Detector Experiments?," by Prasanta K. Basu of the Institute of Radio Physics and Electronics, Kolkata, India, presents a brief history of the understanding and development of semiconductor materials, devices, and doping during the 1890s and early 1900s as well as Bose's work on coherers or detectors. The second, "The Point-Contact Transistor: A Revolution Begins," by John Dallesasse and Robert B. Kaufman of the University of Illinois,

Urbana-Champaign, USA, presents the auspicious invention of the point-contact germanium (Ge) transistor that started the digital revolution leading to profound changes in the lives of people around the world. The third, "On the Shockley Diode Equation and Analytic Models for Modern Bipolar Transistors," by Tak H. Ning of IBM Thomas J. Watson Research Center, Yorktown Heights, New York, USA, presents the theory of the BJT mathematically formulated by William Shockley describing his invention of the junction diode and the BJT that started the practical manufacturing of transistors and VLSI chips. The fourth, "Lilienfeld's Concept on Junctionless Field Effect Transistors," by Mamidala Jagadesh Kumar and Shubham Sahay of the Indian Institute of Technology Delhi, India, recounts the original concept of the junctionless transistor device by Julius E. Lilienfeld in the early 1930s, forming the basis of the metal-oxide-semiconductor FET (MOSFET), the ubiquitous device of choice for manufacturing IC chips in the 1970s. This is followed by, "The First MOSFET and its Long Journey from Concept to Implementation," by Hiroshi Iwai of the Tokyo Institute of Technology, Tokyo, Japan. He discusses the evolution of device architecture and disruptive innovation in manufacturing technology over six decades for IC chip development leading to today's digital ecosystem.

The sixth and seventh articles, "The Invention of the Self-Aligned Silicon Gate Process," by Robert E. Kerwin of AT&T Bell Labs, USA, and "Application of Ion Implantation to Device Fabrication: The Early Days," by Alfred U. MacRae of MacTech Consulting, Seattle, Washington, USA, provide an up close and personal account of fabrication process development in Bell Laboratories. Kerwin's invention of self-aligned silicon gate process in 1966 enabled scaling of silicon devices to smaller dimension for manufacturing complementary MOS (CMOS) technology in the micron to nanometer regime. MacRae's successful implementation of the ion implantation process offered controlled amount of target impurities or dopants in fabricating transistors in a CMOS technology enabling continued scaling of MOSFETs to the nanometer regime.

The eighth article in the series describes the "Evolution of the MOSFET: From Microns to Nanometers," by Yuan Taur of the University of California, San Diego, California, USA, dwelling on evolutionary scaling of device architectures toward the nanometer regime. This is followed by, "The SOI Transistor: Its 60-Year Journey," by Sorin Cristoloveanu of Grenoble INP Minatec, France, gives an elegant history of silicon on insulator (SOI) transistor device technology and its apparent prospect to replace the conventional planar-CMOS technology. However, the SOI transistor lost to the fin-FET (FinFET) for deployment in nanometer-scale ICs. The story of the FinFET goes further with two dedicated articles, "FinFET – The 3D Thin-Body Transistor," by Chenming Hu of the University of California, Berkeley, California, USA, followed by "Historical Perspective of the Development of the FinFET and Process Architecture," by Digh Hisamoto of Hitachi, Ltd., Kokubunji, Tokyo, Japan. The former provides a chronological narrative on the development of the FinFET and its adoption by the microelectronics industry. The latter begins with the double-gate MOSFET and describes its evolution to the three-dimensional FinFET. Both clearly illustrate the FinFET as the alternative to planar CMOS technology, enabling continued miniaturization of FETs with high performance and low power dissipation to avert VLSI chip power crises.

The last of the series, "Origins of the Tunnel FET," by Gehan Amaratunga of Cambridge University, Cambridge, UK, elucidates the introduction of a tunnel junction within a MOSFET structure creating the transistor known as the Tunnel FET (TFET) from which emerged a whole new class of low-power electronics where the subthreshold current level operation of the TFET offer an advantage.

The next three articles provide an illuminating account of the evolution of transistor architecture to achieve solid-state memories. The first, "Floating-Gate Memory: A Prime Technology Driver of the Digital Age" by Simon M. Sze of National Yang Ming Chiao Tung University, Hsinchu, Taiwan and an EDS Celebrated Member, describes the historical events leading to the development of the floating-gate memory (FGM) as well as future technology trends to the year 2030. The second, "Development of ETOX NOR Flash Memory" by Stefan K. Lai of Intel Corporation, California, USA, recounts the development of Intel's Erasable Programmable Read Only Memory (EPROM) Tunnel Oxide (TOX) or ETOX NOR flash memory and onward to the in-system alterability function of an Electrically Erasable

Programmable Read Only Memory (EEPROM) circa mid-1980s. The third and final article in this series, "History of MOS Memory Evolution on DRAM and SRAM" by Mitsu Koyanagi of Tohoku University, Sendai, Japan, provides a historical perspective of the development of dynamic random-access memory (DRAM) and his invention of the three-dimensional stacked capacitor cell (STC) in 1976, overcoming the scaling limitations of conventional planar-type DRAM cells.

The next group of four articles describes derivatives of transistors continuing its evolution in the ensuing decades after its invention. The first, "Silicon-Germanium Heterojunction Bipolar Transistors: An IBM Retrospective," by Subramaniam S. Iyer of the University of California, Los Angeles, California, USA and John D. Cressler, the Georgia Institute of Technology, Georgia, USA, narrates the authors' personal endeavors to develop SiGe heterostructure BJT device technology against all odds including convincing IBM's senior management for R&D support for project completion, which as it turned out was highly lucrative for IBM. This is followed by, "The 25-Year Disruptive Path of InP/GaAsSb Double Heterojunction Bipolar Transistors," by Colombo Bolognesi of ETH Zürich, Zürich, Switzerland. The article describes the phenomenal progress of InP/GaAsSb double heterojunction bipolar transistors (DHBTs) over the last 25 years: from a laboratory oddity in 1997 to a commercial product by 2004. This evolved from an unknown heterojunction system to THz transistors and record-breaking mixed-mode ICs by 2022. The third article, "The High Electron Mobility Transistor: 40 Years of Excitement and Surprises," by Jesus del Alamo of the Massachusetts Institute of Technology, Cambridge, Massachusetts, USA, briefly reviews the development of high-electron mobility transistor (HEMT) based on III–V arsenides and phosphides in the last 40 years. Finally, comes the conception and evolution of the transistor for an entirely different suite of applications in: "The Thin Film Transistor and Emergence of Large Area, Flexible Electronics and Beyond," by Yue Kuo of Texas A&M University, College Station, Texas, USA, Jin Jang of Kyung Hee University, Seoul, Korea, and Arokia Nathan of Darwin College, Cambridge, UK. They discuss the general trend in thin film transistor (TFT) technology development to achieve high performance, low-power consumption, flexible electronics, biological systems, and low-cost solution processing.

The next four articles present application specific evolution of the transistor. The first, "Imaging Inventions: Charge Coupled Devices," by Michael F. Tompsett of AT&T Bell Labs, USA, recounts the invention of charge-coupled devices using MOS technology at Bell Telephone Laboratories in 1970s, and its subsequent evolution. The next article, "CMOS Image Sensors: A Camera in Every Pocket," by Eric R. Fossum of Dartmouth College, Hanover, New Hampshire, USA, describes the invention and underlying technology in CMOS image sensors. The third article, "From Transistors to Microsensors – a Memoir," by Henry Baltes of ETH Zürich, Zürich, Switzerland, narrates the author's personal historical perspective on the development of transistor-based microsensors and its evolution with new manufacturing process technologies including CMOS and micromachining. Finally, is the article, "The Insulated Gate Bipolar Transistor Power Device," by B. Jayant Baliga of North Carolina State University, Raleigh, North Carolina, USA, which describes the author's personal effort driving the invention and development of the insulated gate bipolar transistor (IGBT) technology at General Electric Company as well as its adoption by the consumer, industrial, lighting, and other corporate sectors leading to IGBT commercialization around the world.

The continuous evolution of the transistor architecture for IC chip manufacturing has been matched by advancements in development of tools to characterize critical performance and reliability as well as mathematical formulation to simulate device performance. These developments are described in the following three papers. The first, "History of Noise in Metal-Oxide-Semiconductor Field-Effect Transistors," by Renuka P. Jindal of the VanderZiel Institute of Science and Technology, LLC, Princeton, New Jersey, USA, describes the author's personal memoirs on the development of the fundamental understanding of noise in sub-micrometer MOS devices in the early 1980s at Bell Labs. The next article, "A Miraculously Reliable Transistor: A Short History," by Muhammad Alam and Ahmed Islam of Purdue University, West Lafayette, Indiana, USA, summarizes some of the key reliability challenges. They explain how resolution at the material, device, circuit, system, and algorithm levels made

Moore's Law possible. The final article, "Technology Computer-Aided Design: A Key Component of Microelectronics' Development," by Siegfried Selberherr and Viktor Sverdlov of the Institute for Microelectronics, TU Wien, Vienna, Austria, briefly reviews the evolution of technology computer-aided design (TCAD) and the key mathematical formulations to numerically characterize the performance of transistors.

The next four articles describe the evolution of ICs, along with design and emerging computational paradigms. The first, *"Early Integrated Circuits,"* by Willy Sansen of Katholieke University, Leuven, Belgium, discusses his personal experience of transitioning circuit design from vacuum tubes to solid-state transistors. The second, "A Path to the One-Chip Mixed-Signal SoC for Digital Video Systems," by Akira Matsuzawa of the Tokyo Institute of Technology, Tokyo, Japan, provides a narrative on the progress of IC technology for digital television and video systems over last five decades. The next article, "Historical Perspective of the Nonvolatile Memory and Emerging Computing Paradigms," by Ming Liu of Fudan University, Shanghai, China, reviews the historical perspective of the emergence and development of nonvolatile memory and R&D efforts on emerging computing paradigm. Finally, the article, "CMOS Enabling Quantum Computing," by Edoardo Charbon of EPFL, Lausanne, Switzerland, describes cryo-CMOS as being the technology of choice for the core of processors in a scalable quantum machine for computing.

The final three articles discuss the advances in materials and materials science enabling the development of the transistor architecture for manufacturing IC chips. The first, "Materials and Interfaces – How They Contributed to Transistor Development," by Bruce Gnade of the University of Texas at Dallas, Richardson, Texas, USA, briefly reviews the significance of the continuous advances in materials and materials engineering pertinent to the development of advanced silicon-based transistor technology. Then comes, "The Magic of MOSFET Manufacturing," by Kelin J. Kuhn of Cornell University, Ithaca, USA, who describes how the MOS device architecture interlaces with the semiconductor process flow to create IC chips and predicts the evolution of transistor device technology in the next 75 years. The third article, "Materials Innovation – Key to Past and Future Transistor Scaling," by Tsu-Jae King Liu and Lars P. Tatum of the University of California at Berkeley, California, USA, discusses the importance of materials innovations and advanced transistor structures enabling transistor scaling to below 10 nm minimum feature size. They provide a prospective on materials innovation to enable continuous scaling of the transistor to 1 nm to sustain Moore's Law into the next decade.

The final narrative in this *75th Anniversary of the Transistor* commemorative volume, "Germanium: Back to the Future," by Krishna Saraswat of Stanford University, California, USA, describes the historical progress of the Ge-MOSFET as well as progress in Ge-based devices for optical interconnects. He outlines the potential rebirth of the Ge-transistor, where the revolution started 75 years ago with Brattain and Bardeen's Ge point-contact transistor and Shockley's Ge-diode and Ge-BJT, outlining Ge-based MOSFETs and CMOS technology of the future to overcome the scaling limitations of present-day IC technologies.

We take great pride and honor in commemorating the 75th anniversary of the birth of the transistor, the undertaking of which would not have been possible if not for the passionate narrative of the authors whom with startling rapidity and shared enthusiasm lent their invaluable contributions to this epic story. Shepherding this timely compilation has been an inspiring, once-in-a-lifetime journey. We hope that the reader will be equally inspired by the key historical milestones made possible by the humble yet life-changing transistor whose pioneers are owed a huge debt of gratitude by humanity. We look forward to what the new pioneers can achieve in the next 75 years!

Arokia Nathan
Samar K. Saha
Ravi M. Todi
June 2023

Chapter 1

The First Quantum Electron Device

Leo Esaki

Yokohama College of Pharmacy, Yokohama, Kanagawa, Japan

It was the advent of the transistor that inspired me to study semiconductors at an early stage. Around 1950, the technology of Ge p–n junctions had been developed; so efforts were being made to understand the junction properties, including the breakdown in the reverse direction. McAfee, Ryder, Shockley, and Sparks at Bell Telephone Laboratories all asserted that the low-voltage breakdown in Ge junctions resulted from the Zener mechanism, that is to say, interband electron tunneling from the valence band in the p-region to the empty conduction band in the n-region.

However, later studies did not support the Zener mechanism and it was therefore concluded that the breakdown in Ge junctions was always due to an avalanche mechanism. Although Zener had proposed quantum mechanical interbank electron tunneling as an explanation for dielectric breakdown in 1934, this sophisticated electron tunneling mechanism had never been proved to be important in reality. We may call this proposal "a creative failure."

Although that was still the situation in 1956, somehow I had become convinced that electron tunneling must be observable under suitable conditions, so I risked investigating the tunneling mechanism in narrow Ge p–n junctions.

The figure shows the current–voltage characteristics for diodes of three kinds. I first obtained a backward diode which was more conductive in the reverse direction than in the forward direction, as shown in the figure. The junction width of that diode was 20 nm or less; the possibility of an avalanche was completely excluded because the breakdown occurred well below the threshold voltage for electron–hole pair production. When the junction width narrowed down to about 10 nm, the current-flow mechanism was convincingly tunneling not only in the reverse direction but also in the

75th Anniversary of the Transistor, First Edition. Edited by Arokia Nathan, Samar K. Saha, and Ravi M. Todi.

low-voltage range in the forward direction, giving rise to a prominent current-peak, as shown in the figure. Since a current-peak associated with a negative resistance had never predicted, the Esaki diode – the very first quantum electron device – came as a total surprise in 1957 [1].

Leo Esaki

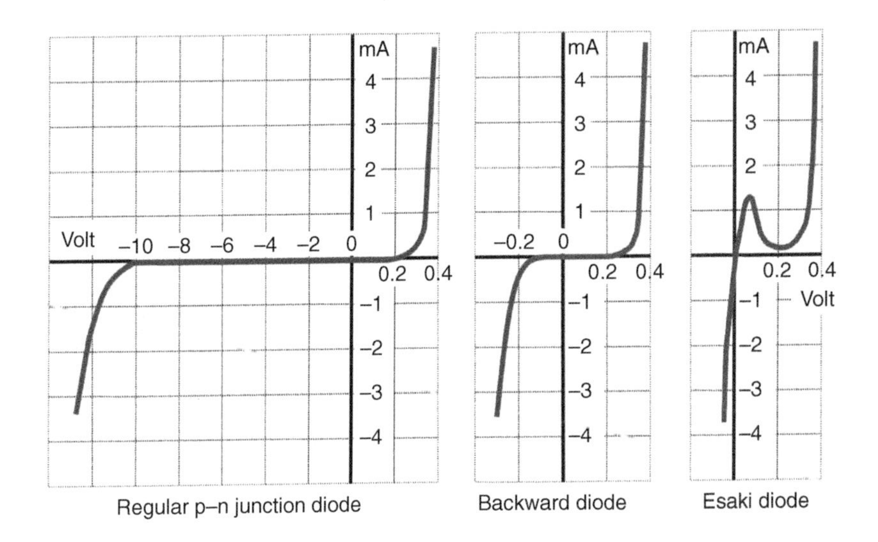

Reference

[1] Esaki, L. (1958). New phenomenon in narrow germanium *p-n* junctions. *Physical Review* 109 (2): 603–604. https://doi.org/10.1103/PhysRev.109.603.

Chapter 2

IEEE Electron Devices Society

A Brief History

Samar K. Saha

Department of Process, Device, and Compact Modeling, Prospicient Devices, Milpitas, CA, USA

2.1 Introduction

Since the invention of the *Transistor* [1–4], the solid-state electronics has made profound impact on humanity. The unprecedented progress in solid-state device technology over the past seven decades is the result of pioneering contributions and dedicated efforts of the people of electrical and electronics engineering, specifically electron devices community. The *Electron Devices Society* (EDS) [5] is such a community of the professional association, the *Institute of Electrical and Electronics Engineers* (IEEE) [6]. Currently, in the hierarchy of IEEE, the EDS is one of the 39 technical societies and seven technical councils under the IEEE Technical Activities Board (TAB) [7]. The growth of EDS over the last seven decades is inherently related to the evolution of transistor and transistor manufacturing technology for very large-scale integrated (VLSI) circuits and systems enabling digital ecosystem. Thus, the history of the IEEE EDS is the story of the global engineering and academic community dedicated to advancing electron and ion devices related to technology for the benefit of humanity.

In this article, we reflect the origins of the IEEE EDS and its growth becoming a true volunteer-led volunteer-driven global professional organization along with its diverse portfolio of publications as well as meetings and conferences on topics of interests to its technical community. Furthermore, the article also outlines the strategic initiatives including educational program and recognition of members' and luminaries of the electron devices community.

75th Anniversary of the Transistor, First Edition. Edited by Arokia Nathan, Samar K. Saha, and Ravi M. Todi.

2.2 Origins of EDS

The origins of the IEEE EDS lie in the year 1952 as a committee of the then professional association, the *Institute of Radio Engineers* (IRE), established in the year 1912 [8–11]. However, the *EDS* can trace its origins back to the 1930s, when the IRE *Technical Committee on Electronics* used to coordinate Institute's technical activities in the field of electronics; e.g., first three conferences on Electron Tubes starting in 1938 were held under auspices of the *IRE Committee on Electronics* [8–11]. In the meanwhile, with the increasing demands for electrical engineering (EE) professionals and the solid-state electronics in the 1940s, there had been a huge growth of IRE membership [8–11]. In order to address this growth, the IRE committee on electronics coordinating activities on Electron Tubes was extended in 1949 to include *Solid-State Electron Devices*, and renamed the entity to *IRE Committee on Electron Tubes and Solid-State Devices*, often referred to as to as the "Committee 7" [5, 11]. The committee's first Chairman was *Leon S. Nergaard* of *Radio Corporation of America* (RCA), who served in that position until 1951 [8–11].

Note that the "Committee 7" formed the IRE Professional Group on Electron Devices which would become the IEEE Electron Devices Society in ensuing decades as shown in Figure 2.1 and described below.

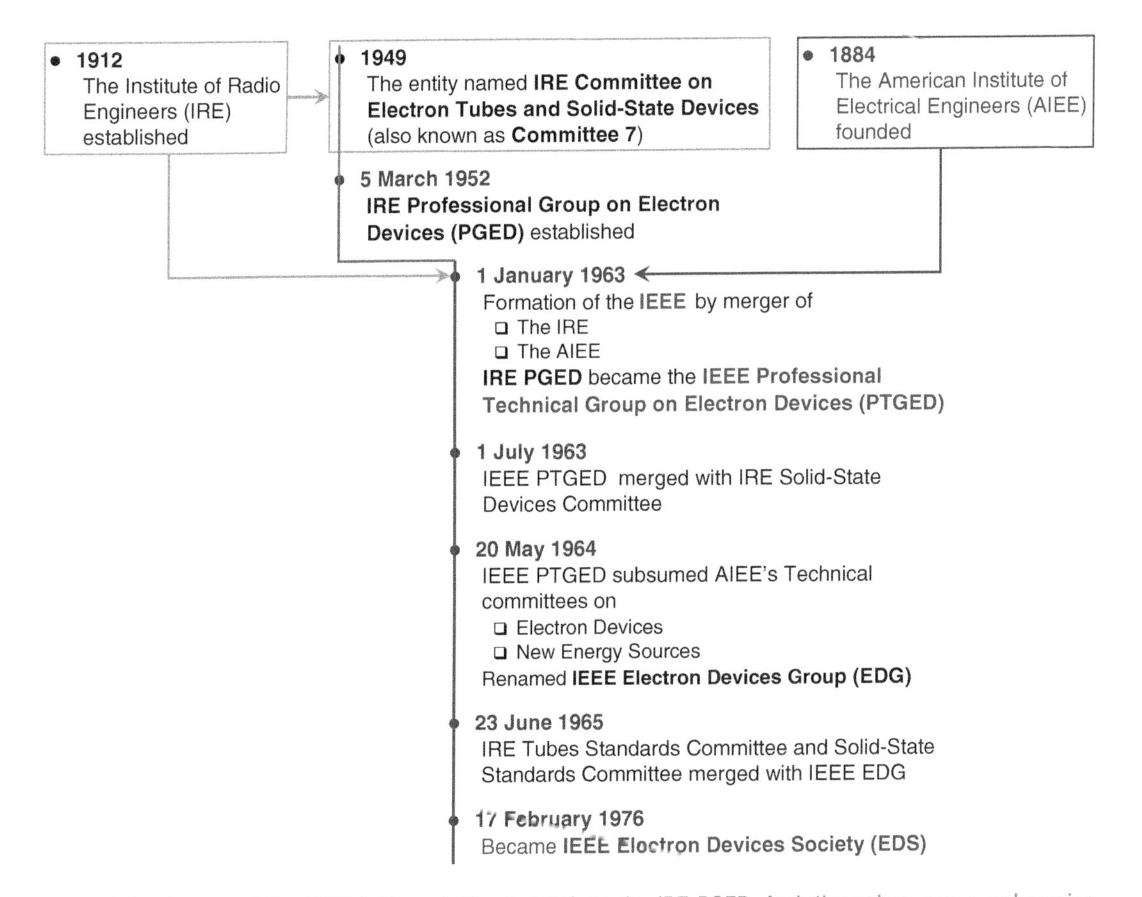

Figure 2.1 Origins of the IEEE EDS on 5 March 1952 as the IRE PGED. And, through mergers, subsuming different functionalities, and name changes in ensuing decades, IRE PGED became the IEEE EDS on 17 February 1976.

2.2.1 IRE Professional Group on Electron Devices

After the invention of the *Transistor* at Bell Telephone Laboratories by John Bardeen and Walter Brattain in December 1947 [1, 2] and William Shockley in January 1948 [3, 4], the solid-state electron devices started to attract major attention. Thus, in order to interact more directly to rapidly changing electron devices technical community, the *IRE Committee on Electron Tubes and Solid-State Devices* proposed to form a new group under the IRE's Professional Group system in the year 1951 [8–11]. By the end of 1951, the proposal was approved by the *IRE Executive Committee* (ExCom). And, on 5 March 1952, the *IRE Professional Group on Electron Devices* (PGED) was established. The *Administrative Committee* (AdCom) of the PGED held its first meeting on 5 March 1952 at IRE headquarters in New York City [11]. The first AdCom volunteers include *George D. O'Neill* of Sylvania as the first Chairman, Leon Nergaard as the founding Vice Chairman, and *John Saby* of General Electric (GE) as the first Secretary [11]. *Thus, on 5 March 1952, the antecedent of the IEEE Electron Devices Society was launched.*

The IRE PGED became the IEEE Professional Technical Group on Electron Devices (PTGED) after the amalgamation of the IRE and American Institute of Electrical Engineers (AIEE) in 1963 to become the IEEE [9–12].

2.2.2 IEEE Professional Technical Group on Electron Devices

In the 1940s, the dynamic growth of radio technology and the emergence of the new discipline of electronics led to stiff competition between the IRE and AIEE [9–12], though the AIEE had been the standard of excellence for EE professionals longer than the then rival association IRE. The AIEE was founded in 1884 by some of the most prominent inventors and innovators including *Nikola Tesla*, *Thomas Alva Edison*, *Elihu Thomson*, *Edwin J. Houston*, and *Edward Weston* [9–12]. In the early 1960s, the IRE and AIEE recognized that their constituencies had been increasingly overlapped with needless duplication of staff, publications, and activities [9–12]. Thus, on 1 January 1963, the two institutes formally became one, *the Institute of Electrical and Electronic Engineers* or simply referred to by the letters "I.E.E.E. (pronounced I-triple-E)," in the year after the IRE celebrated its 50th (and the last) anniversary [9–13]. On the very same day (1 January 1963), the *IRE PGED* became the *IEEE* PTGED [5, 11].

After the formation of the IEEE, the duplicate functionalities and committees of the IRE and AIEE related to electron devices were consolidated to form the IEEE Electron Devices Group (EDG).

2.2.3 IEEE Electron Devices Group

The IEEE PTGED merged with the *Solid-State Devices Committee* on 1 July 1963 and subsumed the functionalities of the AIEE's technical committees (TCs) on *Electron Devices* and *New Energy Sources* on 20 May 1964, and was renamed the *IEEE* EDG. On 23 June 1965, the original *IRE Tubes Standards Committee* and *Solid-State Standards Committee* also fell under the new *IEEE EDG* [9–11] which after a decade, became today's *IEEE Electron Devices Society* [5].

2.2.4 IEEE Electron Devices Society

In 1975, the *IEEE EDG* AdCom petitioned the IEEE TAB to ordain it as a *full-fledged Society* describing its maturity and increasing stability as an entity of the IEEE [11]. Under the Institute's general practices, such a designation indicates that a Group's financial resources are deemed strong enough to survive on its own as a completely independent organization. The petition cited the group's *field-of-interest* (FoI) as *electron and ion devices including electron tubes, solid-state and quantum devices, energy sources, and other devices which are related to technology*. The request was quickly granted. And, on 17 February 1976, the EDG formally became the *IEEE Electron Devices Society, EDS* [5, 11].

As shown in Figure 2.1, on 5 March 1952, the IRE PGED was established to interact directly with *electron devices community* and therefore, the origins of the IEEE EDS lie on 5 March 1952 as the *IRE PGED*; which on 1 January 1963 became the *IEEE Professional Technical Group on Electron Devices*; on 20 May 1964 was renamed the *IEEE EDG*; and on 17 February 1976 became the *IEEE Electron Devices Society*, or EDS.

Throughout the twentieth century (second millennium) to-date in the third millennium, the Society has grown to be one of the world's largest association of electron devices professionals with about 10,000 members in over 160 countries worldwide.

2.3 Growth of EDS

After the establishment of the *IEEE PTGED* on 1 January 1963, the Group's major efforts had been consolidation of its technical activities, membership drive, and formation of chapters in the *United States and beyond its borders* as well as globalization, sponsor meetings and conferences, launch new journals and a Newsletter, and so on, as described below.

2.3.1 Consolidation of Technical Activities

The responsibility for consolidating all professional activities of the merged IRE and AIEE groups under the new *IEEE PTGED* was primarily on *Ray Sears* of Bell Com, AdCom Chairman from July 1963 to June 1964 and *Earl Thomas*, Chairman from July 1964 to June 1966 [11]. Thomas setup the new merged organization and its procedures to ensure that the major technical activities previously carried on by the groups and committees being merged are not lost. By the time Thomas stepped down in June 1966, the IEEE EDG activities had been reorganized into *five* associated subfields: (i) *Electron Tubes*, (ii) *Solid-State Devices,* (iii) *Energy Source Devices*, (iv) *Integrated Electronics*, and (v) *Quantum Electronics* each under its own TC [11]. Note that the EDS TCs continue to grow to diversify its technical activities within its FoI counting 16 at the end of the year 2022 [5].

2.3.2 Membership Drive

The membership drive started as early as in the 1950s by the IRE PGED. Through Group's concentrated efforts to form local chapters around the United States (US), there were active chapters in Boston, Los Angeles, New York City, Philadelphia, San Francisco, and Washington, D.C. as well as lightly active groups in other areas of the country [11]. By December 1953, paid membership of the IRE PGED exceeded 1000 engineers and scientists [11]. The Group continued its steady growth, and by December 1959, its membership exceeded 5000 including student members [11]. In the early 1970s, the total (IEEE EDG) membership grew over 9300 including 32% student members and another 16% from overseas [11]. Consequently, the EDG was among the largest of the IEEE's professional groups.

2.3.3 Electron Devices Group Beyond US Borders

During the 1970s, the IEEE EDG expanded its activities beyond US borders. These efforts to expand had actually begun back in the mid-1960s, when the IRE PGED's annual *Fall Washington* D.C. *Meeting* was renamed the *International Electron Devices Meeting* (IEDM) in 1965 [14]. During the late 1960s, non-US membership grew from less than 1000 to just over 1500, representing about 16% of the total membership [11]. The largest foreign contingents were in Canada, Europe, and Japan; the Tokyo Section alone recorded nearly 300 members in 1969. European members took increasing part in the activities of the IEEE EDG occasionally cosponsoring meetings such as the *European Microwave Conference* and the *Symposium on Solid-State Device Technology*. There were also frequent exchanges of information with the *Japan Society of Applied Physics* [11]. And, in 1972, the annual *Device Research Conference* (DRC) was held for the first time beyond the US borders at the University of Alberta in Edmonton, Canada [11].

In the continued efforts to grow EDS beyond US boarders, AdCom began to include Japanese representatives on its roster of elected members during the 1980s. In 1984, for the first time Asian member Takuo Sugano was elected to serve at AdCom followed by Yoshiyuki Takeishi in 1987 [11]. Previously, there had been *two* European members, Adolf Goetzberger and Cyril Hilsum, on the committee. Thus, electing Japanese members, the IEEE EDS AdCom became a truly global committee. Besides, the Society's policy to reimburse travel expenses of AdCom members, non-US members have been a regular and continuing feature of today's EDS AdCom (renamed *BoG* in 2013) [5, 11, 15].

1. *Managing Growth and Diversification*: In the late 1980s, the EDS membership grew over 10,000 and it published *two* professional journals of its own and cosponsored *four* others as well as sponsored almost 20 conferences a year and cosponsored nearly as many more [5, 11]. Thus, the Society's operations had become too complex and far-reaching for an all-volunteer organization. Therefore, during the year 1987, discussions began on the need to hire permanent professional staff to manage the day-to-day operations of the Society [5, 11].

   In the late 1980s, with finances in much better condition, the day before June 1989 AdCom meeting, EDS President *Craig Casey* of Duke University, Meetings Committee chair *Michael S. Adler* of GE, Treasurer *Lu A. Kasprzak* of IBM, and several past presidents met to discuss hiring management staff. They concluded that such management was indeed required. Subsequently, in October 1989, Adler, Casey, Kasprzak, *Friedolf M. Smits* (past Treasurer, 1980–1987), and the incoming EDS President *Lewis Terman* of IBM recommended to create a new full-time position of EDS *Executive Officer* [5, 11]. And, in December 1990 meeting, AdCom unanimously endorsed their recommendation *to having Society's sprawling activities managed by an Executive Office with full-time staff of salaried personnel.*
2. *EDS Executive Office Established*: Upon AdCom approval, the search committee including Casey, Smits, and Terman interviewed candidates and hired *William (Bill) F. Van Der Vort* who had been working at IEEE since 1977 [5, 11]. Bill started as the *EDS Executive Director* in August 1990. He led a growing team at the Society's Executive Office in Piscataway, New Jersey, to manage the Society's business and finances, coordinate its myriad meetings, and support the editing and publishing of its *Newsletter* and professional *Journals*. After Bill's retirement in the year 2009, Christopher Jannuzzi was hired as the EDS Executive Director in 2010 who served in full capacity through 2015 and part-time in the same position as the supervisor of the EDS Executive office Operations Director, James Skowrenski, during 2016–2019. Since 2020, the EDS Executive office is managed by Operations Manager, *Ms. Laura Riello*, who has been working at IEEE since 1987 [5].

2.3.4 Globalization of EDS

In the 1990s, one of the major initiatives was to complete the globalization efforts to *convert the Society into a truly international organization* [5, 11]. In this endeavor, *Roger Van Overstraeten* of Belgium was elected to AdCom, its first European member in more than a decade [5, 11]. Shortly after, it became the standard EDS policy to have *at least two elected AdCom* members from IEEE Region-8

(Africa, Europe, and Middle East) and another *two* from IEEE Region-10 (Asia and Pacific) [5]. By implementing this policy, AdCom's total membership grew from 18 to 22. In order to make Society's growing activities more effective, an *EDS Executive Committee* (ExCom) was established consisting of the *Elected Officers*, *Junior and Senior Past Presidents*, *Chairs* (renamed as the Vice-Presidents, VPs, from the mid-2004) of key committees, and the *Executive Director* [5, 16].

During the 1990s, the Society's efforts to globalization had been a top priority. Under the leadership of EDS Presidents Terman (1990–1991) and Adler (1992–1993), new chapters were established in Australia, Canada, China, Egypt, France, and Germany as well as in other countries [5, 11]. In order to sustain this rapid global expansion, the Society created a new *Regions/Chapters Committee* (SRC) appointing *Cary Yang* of Santa Clara University as its first Chairman. ExCom members occasionally visited these chapters and regions to help fostering better communications and membership services [17]. And, the Society started the *Distinguished Lecturer* (DL) *Program* to present leading-edge and exciting technical research by quality lecturers to EDS chapters and members as well as facilitate communications among members, chapters, EDS, and IEEE [17]. Currently, the DL program is extended to providing single or multiple DLs by a single or multiple lecturers in a Session/Day and referred to as the "EDS Distinguished Lecturer/ Mini-Colloquia (DL-MQ) Program" [5]. By the time Adler stepped down in 1993, there were 59 EDS chapters in all, with 26 *of them located beyond US* borders; by the end of 1996, more than half of the chapters, 48 out of 85, were outside the US. The new EDS logo, *sporting a lone electron, represented by a spin vector, circling the globe*, was redesigned by Terman in 1992, reflects the Society's broad international character [5, 11]. Thus, at the closing of the 1990s, the IEEE EDS, as the true global association for the electron devices community, found itself in an enviable intellectual and financial position. The total number of 99 EDS chapters spread widely across the globe and the total membership exceeded 13,000 including more than 5000 outside the US [5, 11].

In the third millennium, the Society's global activities continue to grow. At the end of 2022, the EDS chapter grew to 240 including 88 joint chapters and 97 student branch chapters spread across the world with a total number of EDS membership of 10,769 [5]. Note that 37.7% of total members are from IEEE Region 10 (Asia and Pacific) [5].

In the new millennium, the EDS leadership implemented new initiatives on educational as well as humanitarian activities to better serve the electron devices community.

2.3.5 Educational Initiatives

The Society continues to launch new strategic initiatives to enhancing the value of EDS membership and chapters worldwide including student fellowship, *Science, Technology, Engineering, and Mathematics* (STEM) education program, and humanitarian activities.

1. *STEM Education:* For high school students' education, the EDS-ETC (*Engineers Demonstrating Science – Engineer Teacher Connection*) program was developed by Mid-Hudson region EDS chapter, New York, USA, in the year 2010 through the leadership of *Fernando Guarin* of IBM, East Fishkill, New York. The EDS-ETC program is a highly successful and sought out program by EDS chapters worldwide [5].
2. *EDS Webinar*: In the year 2011, the EDS (2010–2011) President *Renuka P. Jindal* of the University of Louisiana at Lafayette, Louisiana, launched the EDS *webinar series* to deliver live lectures by luminaries within the EDS FoI [5, 18]. The first webinar entitled, "The FinFET 3D Transistor and the Concept Behind It," was offered by *Chenming Hu* of the University of California, Berkeley, on 27 July 2011. The online repository of past webinars provides EDS members with on-demand access to streaming videos of the past events [5, 18].
3. *EDS Center of Excellence*: Another educational program to engage high school students including under-represented girls, an *EDS Center of Excellence* was established at the Heritage Institute of

Technology, Kolkata, India, in 2017 through the effort of (2016–2017) President *Samar Saha* of Prospicient Devices, Milpitas, California [19]. This is the first of its kind educational center in EDS and IEEE and is aimed to encourage high school and undergraduate boy and girl students in EE and specifically device engineering, and choose device engineering as their professional career.

4. *EDS Podcast*: Similar to EDS webinar, the Society under the leadership of 2020 President *Meyya Meyyappan* of National Aeronautics and Space Administration (NASA), launched EDS Podcast Series in 2020 to host interviews with some of the most successful EDS luminaries sharing their lives and careers to inspire students and young professionals. On 15 January 2021, the first podcast with Chenming Hu was hosted by *Muhammad Mustafa Hussain* of the University of California, Berkeley (now Purdue University, West Lafayette, Indiana) [5].

With increasing efforts to globalization and diversification of Society's activities, it became important to rename the Society's governing body to appropriately represent its true Mission.

2.3.6 AdCom Became BoG

At the June 2012 AdCom meeting in Leuven, Belgium, Renuka motioned to renaming AdCom to *Board of Governors* to more accurately reflect the volunteer-led volunteer-driven spirit of the Society. The EDS AdCom unanimously voted in favor of the Motion [15]. Subsequently, the change was approved by IEEE TAB. Thus, from January 2013, *the EDS AdCom became the EDS Board of Governors or BoG*. All roles and functions of the BoG remain the same as AdCom. However, "BoG" represents the true governing body of the true global Society to carry out its envisioned Mission [5].

In order to financially support the growing educational and humanitarian initiatives, the Society established an EDS Mission Fund.

2.3.7 EDS Mission Fund

In the year 2013, under the leadership of (2012–2013) President *Paul Yu* of the University of California, San Diego, California, the EDS partnered with the *IEEE Foundation* to establish the *IEEE EDS Mission Fund* of the *IEEE Foundation* [5, 20]. The fund is aimed to greatly enhance the *humanitarian, educational*, and *research initiatives* within the EDS FoI by providing members and other constituents of the EDS community with the ability to contribute directly to its mission-driven initiatives [5].

With continued changes in the electron devices technical area, the EDS leadership created a dedicated Future Directions Committee.

2.3.8 Future Directions

Since the formative years, the Group/Society continuously diversified its technical activities in emerging areas to strategically position itself in the dynamic technical fields. With recent rapidly changing device technology, it has become crucial to move forward EDS in the emerging device technology areas and beyond. In this effort, Samar initiated the EDS first five-year strategic planning with his goal to *building EDS on the foundation of the past to meet the challenges of the future* [19]. In continuation, (2018–2019) President Fernando appointed an ad hoc committee with Samar as the Chairman to continue strategic planning. In order have concentrated efforts on EDS future directions, Meyya created a Standing *Future Directions Committee* in 2020 with *Paul Berger* of Ohio State University as the VP. In 2022, Paul was replaced by *Douglas P. Verret* as the VP of the EDS *Future Directions Committee* [5].

Thus, the EDS has grown worldwide over the past seven decades and continue to grow promoting excellence in the field of electron devices for the benefit of humanity [5].

The Society's growth has been culminated by a growing publication portfolio of top-tiered professional journals and conferences on emerging topical areas.

2.4 Publications

In pursuant of the effort to publish a journal and a monthly newsletter, the *IRE PGED* launched its first journal in 1952 and many more in ensuing decades; some of which are described below.

2.4.1 Transactions on Electron Devices

1. *Transactions of the IRE PGED:* Through the dedicated effort of the AdCom subcommittee on publications successive Chairmen *Herbert J. Reich* and John Saby, the *Transactions of the IRE PGED* was published in November 1952 [21]. For the first few years, the *Transactions* was published on an *irregular* basis depending on the availability of high-quality papers, often from conferences or symposia that did not publish a proceedings [22]. There was only a *single issue* in 1952, *three* in 1953, and *four* in 1954 [22, 23], when AdCom agreed to begin publishing the *Transactions* on a regular *quarterly* basis. Also, from the year 1954, the *Transactions* volume started as ED-1 [23]. In 1955, *Earl L. Steele* of GE was appointed as the first formal Editor of the *Transactions* and the publication was renamed the *IRE Transactions on Electron Devices* [24]. Starting from *IRE Transactions on Electron Devices*, volume ED-2, 1955, the quarterly publications continued till 1960, volume ED-7 [22]. From 1961, the *Transactions* became *bimonthly* [22]. After Steele's resignation publishing ED-8, issue 5, *Glen Wade* of Raytheon, Burlington, Massachusetts, took over as the Editor bringing out *Transactions*, volume ED-8, issue 6, 1961, and all six issues of ED-9, 1962 [25], before renaming the publication to the *IEEE Transactions on Electron Devices*.
2. *IEEE Transactions on Electron Devices*: In 1963, the principal publication of the *IEEE PTGED* was renamed the *IEEE Transactions on Electron Devices* (T-ED) [26]. Wade continued to serve as the Editor through 1960s, bringing on Associate Editors and incorporating numerous changes. Under his leadership, the *T-ED* became *a monthly journa*l in 1964, ED-11 [22].

In the 1970s, the IEEE T-ED continued to expand its editorial board with growing number of high-quality submissions. In 1971, Wade stepped down as Editor and was quickly replaced by *John Copeland* of Bell Labs, who served until 1974. There was an extensive *Editorial Board*, with Associate Editors within the EDS FoI including *bipolar devices*, *display devices*, *energy sources*, *electron tubes*, and *solid-state power devices*. Thus, a new Editor could be readily drafted from these ranks whenever the existing one decided to retire without any disruption in the quality and efficiency of publication [5, 22]. In 2023, Patrick Fay of the University of Notre Dame is appointed as the EiC. *Note that since February 1996, the Society's Editor and Associated Editor are renamed as the Editor-in-Chief (EiC) and Editor, respectively* [22].

2.4.2 EDG Newsletter

In the year 1966, the first issue of the *Newsletter of the IEEE EDG* was published in June, edited by *Jan M. Engel* of IBM Research in San Jose, California, USA [11]. The first issue provided an envisioned convenient forum for recent news of the organization and its day-to-day operation, beyond the scope of the *Transactions*. It included the highlights of AdCom meetings, reports from conferences and other professional gatherings, and information of the forthcoming meetings of interest to the membership. In 1970, Engel stepped down and the EDG Newsletter continued with *John Szedon* of Westinghouse as the editor [11]. However, in the year 1984, the quarterly EDS Newsletter was replaced by the new *Division I Circuits and Devices Magazine*. Though EDS AdCom reinstated the *Newsletter* in 1994 with *Krishna Shenai* of the University of Wisconsin, Madison, as the EiC [5, 11, 17].

2.4.3 IEEE Electron Device Letters

In January 1980, a quick-turnaround journal, the *IEEE Electron Device Letters* (EDL), was launched with *George E. Smith* of Bell Labs as the founding Editor [27]. Note that Smith *won the 2009 Nobel Prize in physics*. With Smith's efforts, the time to publication dropped to 10–13 weeks. Currently, *in the year* 2022, *the average time to online-publication of EDL is about four weeks, the fastest in the IEEE publications* [5]. From January 2023, *Sayeef Salahuddin* of the University of California, Berkeley, is serving as the EiC of EDL [5]. Again, *since February 1996, the EDL Editor and Associated Editor are renamed as the EiC and Editor, respectively* [22].

2.4.4 IEEE Journal of Microelectromechanical Systems

In the year 1990, AdCom approved a new publication, *IEEE/ASME Journal of Microelectromechanical Systems* (JMEMS), with the *American Society of Mechanical Engineers (ASME).* In March 1992, JMEMS began publication with the *IEEE Robotics and Automation Society* (RAS) and *Industrial Electronics Society* (IES) collaborating along with the effort. *Richard Muller* of the University of California, Berkeley, was the founding Editor along with a number of Associated Editors [28]. Since 2013, the JMEMS is only sponsored by IEEE, renamed *IEEE JMEMS* with EDS, RAS, and IES as the cosponsors using subscription-based sponsorship model [28]. After Richard stepped down, *Christofer Hierold* of ETH Zurich was appointed as the EiC in 2013; Christofer was replaced by *Gianluca Piazza* of the Carnegie Mellon University, Pittsburgh, Pennsylvania, in 2019 [5, 28].

2.4.5 IEEE Journal of Photovoltaics

In 2011, through the efforts of (2010–2011) President Renuka Jindal along with Samar, VP publications, and *Timothy* (Tim) *Anderson* of the University of Florida, the Society launched the multi-society publication, the *IEEE Journal of Photovoltaics* (J-PV) [5]. The inaugural issue was published in July 2011 with Tim as the EiC along with editorial boards of Editors [29]. After Tim stepped down, *Angus Rockett* of Colorado School of Mines was appointed as the EiC of J-PV in January 2021.

2.4.6 Journal of Electron Devices Society

Through the efforts of EDS ExCom including Renuka, (2008–2009) President *Cor Claeys*, Paul Yu, and Samar, the Open Access (OA) publication, the *IEEE Journal of Electron Devices Society* (J-EDS) was launched. In January 2013, the first issue of the J-EDS appeared with Renuka as the founding EiC along with *five* Editors [30]. After Renuka stepped down, *Mikael Ostling* was appointed in 2017 and he was replaced by *Enrico Sangiorgi* in 2020 without interruption of publication cycle and processes.

2.4.7 IEEE EDS Magazine

Through the efforts of 2020 President Meyya, 2021–2023 President *Ravi Todi* of Rivos, and the VP of Publications and Products Committee (PPC) *Joachim N. Burghartz* of the Institut für Mikroelektronik Stuttgart, Germany, the publication of an EDS Magazine was approved by the IEEE Periodicals

Committee (PerCom) in 2022. The first issue of the *IEEE Electron Devices Magazine* (EDM) is scheduled to be published in July 2023 with Joachim as the founding EiC [5].

2.4.8 Open Journal on Immersive Displays

Under the leadership of 2021–2023 President Ravi and PPC VP *Arokia Nathan* of Cambridge University, UK, the EDS proposal for an *Open Journal on Immersive Displays* was approved by the IEEE PerCom in November 2022. The inaugural issue is scheduled to be published on 1 January 2024 with Arokia as the founding EiC [5].

Similar to archival publications, the EDS continues to diversify its technical activities through sponsored or cosponsored conferences within the EDS FoI. Some of these meetings and conferences are highlighted below.

2.5 Conferences

During the formative years, the IRE PGED actively *sponsored* and *cosponsored* meetings and conferences. It sponsored parallel sessions at the annual IRE meetings in New York and the *Pacific Coast Council* or WESCCON conference [11]. The PGED often *cosponsored* more specialized meetings of interest to its members with other IRE professional groups including *Symposium on Microwave Radio Relay Systems* held in November 1953 [31], and a *Symposium on Fluctuation Phenomena in Microwave Sources* in 1954 [32].

2.5.1 Device Research Conferences

In the early 1950s, the IRE "Committee 7" controlled the sponsorship of the annual *Electron Device Research* conferences. However, this gathering became two separate annual meetings in 1952: the *Conference on Electron Device Research* (CEDR) and the *Solid-State Device Research Conference* (SSDRC) [11]. The major features of CEDR and SSDR were *by-invitation-only* and off-the-record without proceedings facilitating open discussions of advanced cutting-edge research without violating the proprietary concerns of employers. However, interested authors could publish in the *PGED Transactions* [11, 21–23]. In 1969, under the leadership of *Herbert Kroemer* of the University of Colorado at Boulder (now, at the University of California, Santa Barbara, California) and *Calvin F. Quate* of Stanford University, CEDR and SSDR merged back into a single DRC which continued to be held annually around June [33].

Although the IRE PGED actively involved cosponsoring meetings, membership strongly supported an idea of sponsoring a technical meeting of their own originating annual IRE Fall Electron Devices Meetings at Washington, D.C. [5, 11].

2.5.2 Annual Electron Devices Meeting

1. *IRE Fall Electron Devices Meeting:* The overwhelming support for holding a technical meeting every fall led PGED AdCom approved such a meeting at the January 1955 meeting. George O'Neill agreed to serve as the Chairman of the first annual meeting in Washington, D.C. George made the necessary arrangements with support from the local *PGED Chapter* holding the first *IRE Fall Electron Devices Meeting* in Washington, D.C., in the Fall of 1955 [5, 11, 34]. It occurred during 24–25 October 1955 at the *Shoreham Hotel* near Rock Creek Park, with over 600 in attendance [34]. Following 25–26 October 1956 meeting attracted more than 1000 attendees, *Shockley* delivered the luncheon speech [35]. By the mid-1960s, sessions of this annual meetings were typically devoted to

electron tubes, *solid-state devices*, *energy-conversion devices*, *integrated circuits*, and *quantum devices* as well as emerging areas such as *imaging displays* and *sensors* [36].

In the mid-1960s, as the IEEE EDG began expanding its activities beyond the US borders, the annual Fall Washington Meeting was renamed the International Electron Devices Meeting (IEDM) in 1965.

2. *International Electron Devices Meeting*: The first IEDM was held at the Sheraton-Park Hotel and Motor Inn, Washington, D.C., during 20–22 October 1965, under the leadership of the General Chairman *Clarence G. Thornton* of Philco Corporation, Lansdale, Pennsylvania [36].

In the year 1982, the IEDM was first held outside the Washington, D.C., in San Francisco Hilton, San Francisco, California, during 13–15 December 1982, with *Al F. Tasch, Jr.* of Motorola, Austin, Texas, as the Conference Chairman. In IEDM, sessions on complementary metal-oxide-semiconductor (CMOS) technology for VLSI circuits attracted large crowds [36]. The success of the IEDM in San Francisco in terms of both attendance and finances, convinced AdCom to hold the IEDM on the West Coast every other year. In 1984, when the IEDM again occurred in San Francisco, a record 2900 attended, and the Society received a $70,000 surplus, a major part of the total EDS surplus of $175,000 that year [11]. Except for 1986, when it was held in *Los Angeles*, the IEDM has ever since returned to San Francisco in the *even years*. Since 2016, the IEDM is held annually in San Francisco only [5, 36].

As the MOS devices became the pervasive technology for VLSI circuits in the 1980s, a concentrated effort started to support the emerging area with a new Symposium on VLSI Technology.

2.5.3 Symposium on VLSI Technology

In support of rapid changes in technology, the Society cosponsored a *VLSI Workshop* in 1980, and in 1981 cosponsored the first *Symposium on VLSI Technology* in Maui, with the *Japan Society of Applied Physics* [11].

Historically, the Electron Devices Group/Society strategically position itself in the frontiers of emerging technology through collaboration in organizing meetings and conferences within its technical FoI as in the case of a conference on photovoltaic devices.

2.5.4 Photovoltaic Specialists Conference

The first meeting of *photovoltaic device specialists* was organized by the *Institute for Defense Analysis* (IDA) at NASA Head Quarter (HQ) in Washington, D.C. on 14 April 1961 [37]. The subsequent *Solar Working Group Conferences* in 1962 and 1963 were held in Washington, D.C. From the June 1964, the conference began using a numbering system along with the title *Photovoltaic Specialists Conference* (PVSC) [37]. Thus, the 1964 annual meeting was formally called the 4th PVSC. The fourth and the fifth PVSC on 2–3 June 1964 and 18–20 October 1965, respectively, were *jointly* sponsored by IEEE, *American Institute of Aeronautics and Astronautics* (AIAA), and NASA. Since the sixth PVSC, on 28–30 March 1967 at Cocoa Beach, Florida, the PVSC has been solely sponsored by the *IEEE EDG/EDS* [5, 37].

The list of professional meetings supported by the EDS continues to grow and diversify. Where EDS had supported 39 meetings at the beginning of the 1990s, the total grew to 68 by 1996 and continued to climb for the rest of the decade with the rapid changes of VLSI manufacturing landscape.

2.5.5 Electron Devices Technology and Manufacturing Conference

Through the efforts of EDS VP of Conferences and Technical Activities, Ravi Todi, and *VLSI Technology and Circuits TC* chairman, *Shuji Ikeda* of Tei Solutions Co. Ltd., Tsukuba, Japan, and the President Samar, the 1st *IEEE Electron Devices Technology and Manufacturing* (EDTM) Conference was launched in Toyama, Japan, during 12–16 March 2017 with Shuji as the General chair [38]. The EDTM continues to grow and the seventh EDTM held in 2023 in-person in Seoul, Korea [38].

In mid-2010s, the EDS BoG felt to execute new initiatives to revitalize Society activities in the IEEE Region 7 (Canada). This finds the origins of the IEEE International Flexible Electronics Technology Conference (IFETC).

2.5.6 International Flexible Electronics Technology Conference

In late 2017, through the effort of *Gaozhi (George) Xiao* and *Ta-Ya Chu* of National Research Council Canada, Canada, and EDS (2016–2017) President Samar, the first IEEE IFETC was held during 7–9 August 2018 at the Delta Hotel City Centre in Ottawa, Ontario, Canada, with George and Samar as the General Co-Chairs and Ta-Ya and *Ye Tao* of Advance Electronics and Photonics, Canada, as the Technical Program Co-Chairs [39]. The 5th IEEE IFETC 2023 is scheduled during 14–16 August in San Jose, California, USA, with Samar as the General Chair.

Similarly, to support and growth of EDS membership in IEEE Region 9 (Latin America), a dedicated EDS flagship conference Latin American Electron Devices Conference (LADEC) was established by the effort of EDS (2018–2019) President Fernando Guarin of Globalfoundries, New York.

2.5.7 Latin American Electron Devices Conference

The inaugural LADEC was held during 24–27 February 2019 in Armenia, Colombia, with General Chairs, *Johan Sebastián Eslava* of National University of Columbia (UNAL), Colombia, and *Andrei Vladimirescu* of Paris Institute of Digital Technology (ISEP), France, and the University of California, Berkeley, USA. The LADEC 2023 is scheduled during 3–5 July 2023 in Pueble, Mexico, with *Esteban Arias* of Instituto Tecnológico de Costa Rica Cartago, Costa Rica and Cor Claeys of KU Leuven, Belgium as the Co-General Chairs [40].

Along with the growth in membership, supporting members needs for publications and meetings and conferences, the Society established different awards to recognize technical and professional accomplishments its members and individuals within the EDS FoI as outlined below.

2.6 Awards and Recognition

Since 1970s, AdCom/BoG leadership continues to establish a series of special awards to honor and recognize major accomplishments of individuals in the *field of electron devices* [5]. Some of awards include *J.J. Ebers Award* (1971); *Jack A. Morton Award* (1974) – Institute-level technical field award replaced by *Andrew S. Grove Award* in 2000; *William R. Cherry Award* (1980); *Paul Rappaport Award* (1984); *Distinguished Service Award* (1993); *Chapter of the Year Award* (1997); *Millennium Medals* (2000); *George E. Smith Award* (2002); *Region 9 Biennial Outstanding Student Paper Award* (2002); *Education Award* (2005); *Early Career Award* (2009); *Robert Bosch Micro and Nano Electro Mechanical Systems Award* (2014); *Lester F. Eastman Award* (2019); *Leo Esaki Award* (2019); and student Fellowships for PhD, masters, and undergraduate-level research and studies within the EDS FoI [5, 11].

2.7 Conclusion

The IEEE EDS is a true volunteer-led volunteer-driven global association of electron devices community. Since 1950s, the EDS AdCom/BoG successfully expanding its activities within the US – coast-to-coast, beyond US borders, and globally. Along with globalization, the Society continues to launching sponsored and cosponsored journals and conferences, implementing new initiatives to support educational activities of members and students, and recognizing the accomplishments of its members as well as supporting student education. Though a new horizon coming into view of the electron and ion devices related to technology, the long glorious history of EDS over seven decades tells us that the Society is strategically well-positioned to navigate through the technological changes by timely strategic planning for *Future Directions*. Thus, BoG is on the right path to *building EDS on the past to meet the challenges of the future* and EDS will continue to enjoy an enviable intellectual and financial position.

References

[1] Bardeen, J. and Brattain, W.H. (1950). Three-electrode circuit element utilizing semiconductive materials. US Patent 2,524,035, 3 October 1950.

[2] Bardeen, J. and Brattain, W.H. (1948). The transistor, a semi-conductor triode. *Physical Review* 74 (2): 230–231.

[3] Shockley, W. (1951). Circuit element utilizing semiconductive material. US Patent 2,569,347, 25 September 1951.

[4] Shockley, W. (1949). The theory of p–n junctions in semiconductors and p–n junction transistors. *Bell System Technical Journal* 28 (3): 435–489.

[5] IEEE EDS (2023). About. https://eds.ieee.org/about-eds.

[6] IEEE (2021). At a glance. https://www.ieee.org/about/at-a-glance.html.

[7] IEEE (2023). IEEE Technical Activities Board (TAB). https://www.ieee.org/about/volunteers/tab.html.

[8] IRE (2023). History of the Institute of Radio Engineers 1912–1963. https://ethw.org/IRE_History_1912-1963.

[9] McMahan, A.M. (1984). *The Making of a Profession: A Century of Electrical Engineering in America*. New York: IEEE Press. https://ethw.org/Archives:The_Making_of_a_Profession:_A_Century_of_Electrical_Engineering_in_America.

[10] Ryder, J.D. and Fink, D.G. (1984). *Engineers & Electrons: A Century of Electrical Progress*. New York: IEEE Press. https://ethw.org/Archives:Engineers_%26_Electrons:_A_Century_of_Electrical_Progress, 2023.

[11] IEEE EDS (2002). 50 years of electron devices: the IEEE electron devices society and its technologies, 1952–2002. https://eds.ieee.org/images/files/About/eds_anniversarybooklet.pdf.

[12] AIEE (2023). History of the American Institute of Electrical Engineers 1884–1963. https://ethw.org/AIEE_History_1884-1963.

[13] IEEE (2023). History of IEEE. https://www.ieee.org/about/ieee-history.html.

[14] (1965). International Electron Devices Meeting (20–22 October 1965). https://ieeexplore.ieee.org/stamp/stamp.jsp?tp=&arnumber=1474089.

[15] IEEE EDS (ed.) (2013). EDS announces Administrative Committee now Board of Governors. *The Electron Devices Newsletter* 20 (1): 28–28. https://eds.ieee.org/images/files/newsletters/newsletter_jan13.pdf.

[16] IEEE EDS (2004). Electron Devices Society. *IEEE Electron Devices Newsletter* 11 (3): 2. https://eds.ieee.org/images/files/newsletters/Newsletter_July04.pdf.

[17] Adler, M.S. (1994). Message from the outgoing president. *IEEE Electron Devices Newsletter* 1 (1): 3–5. https://eds.ieee.org/images/files/newsletter/Newsletter_July94_color.pdf.

[18] IEEE EDS (2023). Webinar archive. https://eds.ieee.org/education/webinars/webinar-archive.

[19] Saha, S.K. (2018). Message from EDS president. *IEEE Electron Devices Newsletter* 25 (1): 11–12. https://eds.ieee.org/images/files/newsletters/newsletter_jan18.pdf.

[20] Yu, P. (2013). Message from President Paul Yu. *Electron Devices Newsletter* 20 (4): 7–8. https://eds.ieee.org/images/files/newsletters/newsletter_oct13.pdf.

[21] Front Cover (1952). Papers on electron devices. *Transactions of the IRE Professional Group on Electron Devices* PGED-1: c1–c1. https://ieeexplore.ieee.org/stamp/stamp.jsp?tp=&arnumber=6811049.

[22] IEEE Xplore (1963–2023). All issues. *IEEE Transactions on Electron Devices*. https://ieeexplore.ieee.org/xpl/issues?punumber=16&isnumber=10025571.

[23] IEEE Xplore (1954). Table of Contents. *Transactions of the IRE Professional Group on Electron Devices* ED-1 (1): c1–c1. https://ieeexplore.ieee.org/stamp/stamp.jsp?tp=&arnumber=1471805.

[24] IEEE Xplore (1955). Table of Contents. *IRE Transactions on Electron Devices* ED-2 (1): c1–c1. https://ieeexplore.ieee.org/stamp/stamp.jsp?tp=&arnumber=1471914.

[25] IEEE Xplore (1962). Table of Contents. *IRE Transactions on Electron Devices* ED-9 (6): c1–c1. https://ieeexplore.ieee.org/stamp/stamp.jsp?tp=&arnumber=1473242.
[26] IEEE Xplore (1963). Table of Contents. *IEEE Transactions on Electron Devices* ED-10 (1): c1–c1. https://ieeexplore.ieec.org/stamp/stamp.jsp?tp=&arnumber=1473374.
[27] Smith, G.E. and Dixon, R.W. (1980). Editorial. *IEEE Electron Device Letters* 1: 1–1, 1. https://ieeexplore.ieee.org/stamp/stamp.jsp?tp=&arnumber=1481066.
[28] IEEE Xplore (2013). Table of Contents. *Journal of Microelectromechanical Systems* 22 (2). https://ieeexplore.ieee.org/stamp/stamp.jsp?tp=&arnumber=6490356.
[29] Jindal, R.P. (2011). Capturing the solar wind. *IEEE Journal of Photovoltaics* 1 (1): 1–2. https://ieeexplore.ieee.org/stamp/stamp.jsp?tp=&arnumber=6055242.
[30] Jindal, R.P. (2013). Editorial. *IEEE Journal of Electron Devices Society* 1 (1): 1–8. https://ieeexplore.ieee.org/stamp/stamp.jsp?tp=&arnumber=6471856.
[31] Saad, T.S. and Wiltse, J.C. (2002). 50 years of the IEEE microwave theory and techniques society. *IEEE Transactions on Microwave Theory and Techniques* 50 (3): 612–624. https://doi.org/10.1109/22.989946.
[32] Symposium on Fluctuation Phenomena in Microwave Sources (1954). *1954 Symposium on Fluctuation Phenomena in Microwave Sources* (18–19 November 1954). Professional Group on Electron Devices. https://openlibrary.org/books/OL17127945M/1954 Symposium_on_Fluctuation_Phonomena_in_Microwave_Sources_held_at_Western_Union_Autidorium_New_York.
[33] Franklin, A., Jena, D., and Akinwande, D. (2018). 75 years of the device research conference – a history worth repeating. *IEEE Journal of the Electron Devices Society* 6: 116–120. https://doi.org/10.1109/JEDS.2017.2780778.
[34] IEEE. (1955). First Annual Technical Meeting on Electron Devices by IRE, Washington, DC (24–25 October 1955), pp. 1–6. https://ieeexplore.ieee.org/stamp/stamp.jsp?tp=&arnumber=1471950.
[35] IEEE. (1956). Second annual technical meeting on electron devices by IRE, Washington, DC (25–26 October 1956), pp. 1–7. https://ieeexplore.ieee.org/stamp/stamp.jsp?tp=&arnumber=1472109.
[36] IEEE Xplore (1965–2022). Browse conferences. International electron devices meeting, Washington, DC/San Francisco. https://ieeexplore.ieee.org/browse/conferences/title?selectedValue=TitleRange:E&queryText=International%20Electron%20Devices%20Meeting.
[37] PVSC History (1961–2020). USA/Japan/Spain. https://ieee-pvsc.org/PVSC49/about-history.php.
[38] IEEE Xplore (2023). *IEEE Electron Devices Technology and Manufacturing Conference*, Toyama, Japan (2017); Seoul, Korea (2023). https://ieeexplore.ieee.org/browse/conferences/title?queryText=Electron%20Devices%20Technology%20and%20Manufacturing%20Conference.
[39] IEEE Xplore (2023). *International Flexible Electronics Technology Conference*, Ottawa, Canada (7–9 August 2018); San Jose, USA (14–16 August 2023). https://ieeexplore.ieee.org/browse/conferences/title?queryText=International%20Flexible%20Electronics%20Technology%20Conference.
[40] IEEE Xplore (2023). *Latin American Electron Devices Conference (LAEDC)*, Pueble, Mexico (3–5 July 2023). https://ieeexplore.ieee.org/xpl/conhome/1831384/all-proceedings.

Chapter 3

Did Sir J.C. Bose Anticipate the Existence of p- and n-Type Semiconductors in His Coherer/Detector Experiments?

Prasanta Kumar Basu

Institute of Radio Physics and Electronics, University of Calcutta, Kolkata, India

3.1 Introduction

On 23 December 1947, two scientists of the Bell Labs, John Bardeen and Walter Brattain, demonstrated before other workers the amplifying property of their point-contact transistor [1]. Their team leader, William Shockley, in the next few months, developed the theory of p–n junction and came out with the idea of a more robust device, the junction transistor [2]. The trio became the inventor of a solid-state device, which transformed the world in the decades to come, and started a race, still unabated, of announcing newer devices and systems outperforming the earlier ones.

In their work, the three inventors either used a point contact between a metal and a doped semiconductor or considered junctions between semiconductor layers alternately doped to be p-type or n-type. Although the term semiconductor was well known in their time or even earlier, this special class of solid materials received the attention of many workers starting from Volta [3–5] including the rectification and other properties of metal–semiconductor junction. However, it was Sir Jagadish Chunder (Chandra) Bose, a Calcutta-based physicist working as a professor in the Presidency College, under the then British Indian Government, who first demonstrated the use of metal–semiconductor junction as a detector of electromagnetic (EM) waves. The detector was termed as coherer at that time. Bose's application for US patent, filed in 1901, was finally approved in 2004 [6]. Bose's Galena detector is now universally recognized as the first semiconductor device [3, 5, 7–11]. Walter H. Brattain, the coinventor of the transistor, and coauthor of 1955 paper "History of Semiconductor Research" acknowledged Bose's priority in using a semiconductor crystal to detect radio waves [3].

75th Anniversary of the Transistor, First Edition. Edited by Arokia Nathan, Samar K. Saha, and Ravi M. Todi.

While Bose's work on detectors/coherers is well documented, there is less discussion in the literature or printed media about an oft-quoted comment on Bose. Sir Neville Mott, Nobel laureate in physics in 1977 for his work on semiconductor and solid-state physics, remarked that *Bose was at least 60 years ahead of his time* and *in fact he anticipated the existence of n-type and p-type semiconductors*. This is recorded in the paper by Mitra [12] and thereafter the statement is mentioned in the title page of the website of the Bose Institute [13], an Institute founded by Bose, by Johnson [14] and by Emerson [15], and subsequently in many popular articles, journal publications, and books, the number of such is too many to be cited here. In order to avoid repetition of the words quoted by Mott, we shall henceforth use the word *Remark*.

It may be mentioned that Bose never used the term *semiconductor* in all his papers and in his time the physics of semiconductors and the effect of doping was unknown. His experiments on coherers however classified different metals and other materials as positive, negative, or neutral, showing, respectively, decreasing, increasing, and no change of resistance with increase in electro motive force (EMF).

The question then arises when, where, in what context, to whom, and on what scientific basis the *Remark* in relation to Bose's anticipation about conductivity types of semiconductor was made. Some answers to these questions were given by Mitra [12]; however, the explanation he presented needs closer examination.

The present paper makes an attempt to give some more information about the time and source and the interpretation of the *Remark* by a few workers, adding the author's own views. Since the story dates back to almost 125 years from today, it will not be out of place to briefly introduce Bose, his work environment, and his work to the present readers and narrate the background leading to the *Remark*.

The paper is organized as follows. Section 3.2 gives a brief bio-sketch of Bose. Section 3.3 describes Bose's work on coherer or detector and semiconductors. Section 3.4 narrates what is known about the time, place, and the person meeting Mott. Section 3.5 gives a brief history of the understanding and development of semiconductor materials, devices, and doping. Section 3.6 examines Mott's remark. Section 3.7 concludes the paper.

3.2 J.C. Bose: A Brief Biography

Sir Jagadish Chandra Bose was born on 30 November 1858 in Munshiganj, Bengal Presidency, now in Bangladesh. He studied First Arts (FA) at St. Xavier's College, Calcutta, and obtained the FA and the BA from Calcutta University in 1877 and 1880, respectively. He then went to England to study medicine, but gave it up due to ill health. This led him to enter the Christ's College, Cambridge, UK, in 1881, and he came out successfully from Cambridge with a Natural Tripos in 1884 and in the same year received the BSc degree from London University (Figure 3.1).

Bose joined the Presidency College as an officiating professor after his return. However, his salary was much less than the British Professors there and in protest he did not accept his salary for nearly three years. Finally his struggle ended and he became a regular professor.

Bose started research at his 36th birthday in a makeshift laboratory in his college. He demonstrated how electric waves could propagate over a distance of 75 ft through the intervening rooms, which could make a bell ring and fire a pistol. The demonstration was made to the public in the Town Hall Calcutta, as well as at the Presidency College. In 1896, Bose could send a radio signal covering a distance of nearly 5 km between two colleges in Calcutta University. To detect the signal, he used one of his inventions: "a mercury coherer with a telephone detector." The development was reported by him in a paper presented to the Royal Society in London in 1899. It may be mentioned here that Marconi used exactly the same principle to receive the first transatlantic wireless signal in 1901, but he said that he had received the design from an Italian colleague [10].

Bose published his first scientific paper, "On Polarization of Electric Rays by Double Refracting Crystals," in *The Electrician* in 1895. He conducted extensive research on microwave wireless systems

Figure 3.1 Prof. Bose along with his experimental setup at the Friday evening discourse on electric waves before the Royal Institution (1896). Source: Pictorial Press Ltd/Alamy Stock Photo.

between 1894 and 1901. Of 128 papers he published on his entire research activities, 27 were on wireless propagation studies. He communicated his research findings to the Royal Society of London through Lord Rayleigh, who taught Bose physics at Cambridge and later acted as his mentor during his research in physics. The research of Bose after 1901 focused primarily on plant physiology. In this area also, he made significant contributions.

After his retirement from the Indian Educational Service in November 1915, he served as an Emeritus Professor of Presidency College. He was conferred with knighthood in 1917 and in the same year he founded the Bose Research Institute (now the Bose Institute) in Calcutta. It included a large lecture theater whose purpose, he said, was to disseminate knowledge of scientific advances to the widest possible public "without any academic limitations, henceforth to all races and languages, to both men and women alike, and for all time coming."

In 1920, he was the first Indian scientist to become a Fellow of the Royal Society. He left a lasting legacy for science in Asia – and the world.

He breathed his last on 23 November 1937, at the age of 79.

3.3 Bose's Work on Detectors

When Bose started his work, Hertzian waves were detected with a "coherer," invented in about 1890 by the Frenchman Edouard Branly [16]. It was constructed by placing metal fillings between two electrodes in a glass tube. The impressed radio-frequency caused the loose metal filings to clump together, or cohere, thereby increasing the resistance. They had to be shaken loose again before another signal could be detected. Cuff [17] gives a detailed account of coherers.

Figure 3.2 View of the spiral spring receiver developed by Bose. Source: Wikipaedia.

Bose improved the performance of the coherers by designing and fabricating his spiral spring coherer, which used fine wires of metal, mainly steel, shaped in the form of springs and encased in a box (see Figure 3.2) [18]. The system irradiated by EM waves produced current in the presence of EMF applied between two electrodes at the ends. Bose could find ways to avoid tapping of the coherers, used by earlier workers, to bring the devices back to their original state and to detect new EM waves. He observed hysteresis in the *I–V* curves, but later could obtain self-cohering action [6, 18–23].

Bose produced another new coherer using a metal cup containing mercury that was covered in a very thin insulating film of oil. An iron disc suspended above touched the film of oil without breaking it. The presence of a radio signal broke the film; an electrical current then passed through the device and operated a telephone receiver. The system restored itself automatically, making it a self-cohering detector.

Bose then investigated the effects of contacts between metal and different kinds of other metals, metalloids, nonmetals, amalgams, and compounds. He found that in some of the materials, the resistance decreased when irradiated but in others it increased. He named the first category as the positive-type coherer substances. Galena, Te, and Mg belong to this class showing self-recovery, but the other members, Cr, Mn, and Zr lack the self-recovery character. On the other hand, the negative-type substances, like K, Na, Pb, Sn, allotropic Ag, etc., show increasing resistance with increasing EMF. He also observed that the sensitivity and other attributes of the coherer depended on environmental conditions and the important role of oxidation. A photograph of the Galena detector is shown in Figure 3.3.

Bose also carried out experiments with both single and multiple contacts of the receiver setup. For a single contact device, the pressure of contact was adjusted by means of a fine micrometer screw. In some cases, the contact ends were both rounded; in others, a pointed end was pressed against a flat piece.

Figure 3.3 Photograph of the Galena detector. Source: D.T. Emerson.

While studying the optical properties of EM waves, Bose discovered that polarizing crystals have selective conductivity [18, 23]. The conductivity was found larger in one direction, while it was lower in the perpendicular direction. One of these crystals was galena, the mineral form of lead sulfide. Bose made a pair of point contacts from galena and linked them in series with a voltage source and a galvanometer.

In his patent disclosure [6], Bose mentioned the existence of both positive- and negative-type substances, self-recovery, hysteresis in the *I–V* curves, and the use of substances like galena, tellurium, etc., which were later identified as semiconductors. In his letter patent, he claimed to have made "a coherer or detector of electrical disturbances, Hertzian waves, light waves or other radiations." Bose called his device "a universal radiometer," one of whose uses could be to detect "signals in wireless or other telegraphy."

The physical processes occurring in the substances used in coherer are not clear and such studies led to controversies. However, coherers remained a subject of interest and according to Cuff [17], the investigations led to metal-oxide-metal devices and STM.

3.4 Mott's Remark

We now mention in what context, Mott made the *Remark*. Emerson [15] mentioned that it appeared first in the paper by Mitra [12], a Scientist at the Bose Institute. Mitra wrote that Shyamadas Chatterjee who worked at the Bose Institute, and thereafter at the Physics departments of the University of Calcutta and Jadavpur University, met Prof. Mott at Cambridge and showed him the papers of Bose.

Professor S.D. Chatterjee later wrote (The Oracle, 1982) "The Nobel Prize winner Sir Neville F. Mott, Cavendish Professor of Physics, Cambridge University, once remarked to the (present) writer that J. C. Bose was at least · sixty years ahead of his time." "In fact, he had anticipated the existence of P-type and N-type semiconductors" (see Mitra [12]).

Somewhat later, Dwaraka (Nath) Bose (DNB) remembered the conversation between Mott and Chatterjee in more detail [24]. "Shyamadas Chatterjee who worked at the Bose Institute in the early forties discussed these (Bose's) results with Neville Mott when he visited Cavendish Lab in mid-fifties. Mott was struck by these observations which led him to make the oft quoted *Remark*. He added, "this incident was narrated to me by Chatterjee shortly after his visit and is a mark of recognition by one of the doyens in the field."

Unfortunately, DNB, an expert in Materials Science and Semiconductors, breathed his last more than a year ago. The present author is unable to seek further details of the meetings and clarification of the remark. He gave his own interpretation of the *Remark* which will be discussed in the next section.

3.5 Understanding Semiconductors and Doping

We now need to answer the questions: (i) what was the status of semiconductor, knowledge about it and the devices, and (ii) what was known about doping before and after Bose. To get answers to these primary questions raised in this paper, we mention a few of the important milestones in this area, derived mainly from [3–5, 25] in Table 3.1. Neither all the works are included in the list, nor the table is complete, due to limited space.

It may now be relevant to discuss the scientific knowledge available and developed during the period 1895–1901 when Bose published his papers related to coherers and Galena detectors. As is clear from Table 3.1, most of the papers published by all workers including Bose from the earlier period to even

Table 3.1 Important milestones in semiconductor research.

Year	Author	Observation (source)
1782	Volta	First used the term materials of semiconducting nature [26]
1821	Davy	Conductivity varies with temperature [27]
1833	Faraday	Decrease of resistance with increasing temperature [28]
1839	Becquerelle	Photovoltaic effect [29]
1851	Hittorf	First measurement of conductivity of semiconductor with temperature [30]
1873	Smith	Photoconductivity [31]
1874	Braun	Rectification by metal–galena contact [32]
1878	Hall	Hall effect [33]
1895–1901	Bose	Coherer, detector, Galena detector [6, 19–23]
1899	Riecke	Concept of positive and negative charge carriers with different mobilities [34]
1900	Drude	Free electron theory of metals [35]
1906	Pierce	Cat's whisker diode with Si [36]
1907	Koeniksberger	Classification as metals, insulators, and variable conductors [37]
1908	Baedeker	Hall measurement of CuI showing the presence of positive charges [38]
1910	Weiss	First concept of semiconductor [39]
1924	Gudden	Impurity hypothesis [40]
1928	Bloch	Electron theory in periodic lattice [41]
1929	Peierls	Concept of holes [42]
1931	Heissenberg	Concept of holes [43]
1931	Wilson	Band theory: filled and empty bands [44]
1930–1939	Schottky and coworkers	Theory of metal–semiconductor junction [45–47]
1939	Mott	Theory of rectifiers [48]
1938–1939	Davydov	Theory of rectification in Cu oxide considering the presence of p–n junction [49, 50]
1941	Ohl	Identification of impurities creating p–n junction [51]
1941	Schaff	Designation of n and p types [52]
WWII, 1950	Woodyard	Methods for adding tiny amounts of solid elements from the nitrogen column of the periodic table to germanium to produce rectifying devices [53] (legal fight for patent, see. [3])
1947	Bardeen, Brattain	Ge point contact transistor [1]
1949	Shockley	Junction transistor, p–n junction theory [54]
1949	Schaff and coworkers	Formation of barrier in p–n junction of Si [55]
1950	Sparks and Teal	Growth of single crystal Ge by Czochralsky method [56]
1951	Schokley, Sparks and Teal	First Ge n–p–n transistor [57]
1952	Pfann	Zone refining [58]

Table 3.2 Comparison between coherer and current metal–semiconductor diode.

Coherer	M–S diode
Purely experimental	Experimental and theoretical
Acceptable theory unavailable	Sound theoretical model available
Used polycrystalline material	Pure single crystal
Contacts between two metals and between metal and other substances studied	Metal semiconductor contacts used
Unknown doping	Controlled doping
Unstable/partially stable structure	Stable structure
Characteristics susceptible to environment, risk for oxidization	Characteristics not susceptible to environment
I–V curve of positive coherer concave	*I–V* curve of metal-n Ge cat's whisker concave
I–V curve of negative coherer convex	*I–V* (considering magnitudes of both) curve of metal-p Ge cat's whisker concave

the first two decades of the twentieth century were purely experimental. The samples and materials used, including metals, were mostly amorphous or at best made of very small polycrystals, containing impurities and susceptible to environmental conditions with surfaces tarnished by oxides. The structures used were unstable or at most partially stable. The only available theory was Maxwell's equations and known laws due to Faraday, Ampere, Gauss et al. Electron was discovered only at that time and quantum theory initiated by Planck was due after four to five years. Another point to note is that Bose never used the word semiconductor, but found the existence of *sensitive materials for* positive and negative coherers and the substances exhibiting the behavior.

On the basis of such observations, we may now compare the characteristics of the materials used in coherers and current metal–semiconductor diode using doping. Table 3.2 summarizes the comparison.

The table clearly indicates that concrete picture of properties of semiconductors, the effect of doping, and the necessity of using very pure semiconductors with precisely controlled amount of dopants to get reliable operation of materials and devices was obtained about four to five decades after the work by Bose and his predecessors and contemporaries. The entries in the last two rows are obtained from Figures 3.4 and 3.5 in the next section, in which details are given.

3.6 Interpretation of Mott's Remark

We may now examine Bose's work in the light of modern developments in the understanding of physics and technology of semiconductors.

In particular, we may seek the answer to the question posed in the title. As is clear from the reports of Mitra [12] and DNB [24], Mott was unaware of the pioneering contributions of Bose. After reading his papers, Mott was justified to declare that Bose was at least 60 years ahead of his time. Doubts however arise about the second part of his remark. Bose conducted his experiments with amorphous materials, or at best with tiny polycrystalline aggregates. His samples might have defects to give rise to p-type behavior.

In this connection, Mitra [12] thought that some of the results of Bose could be attributed to the formation of oxide layers transforming them into p-type and n-type semiconductors. DNB [60] observed

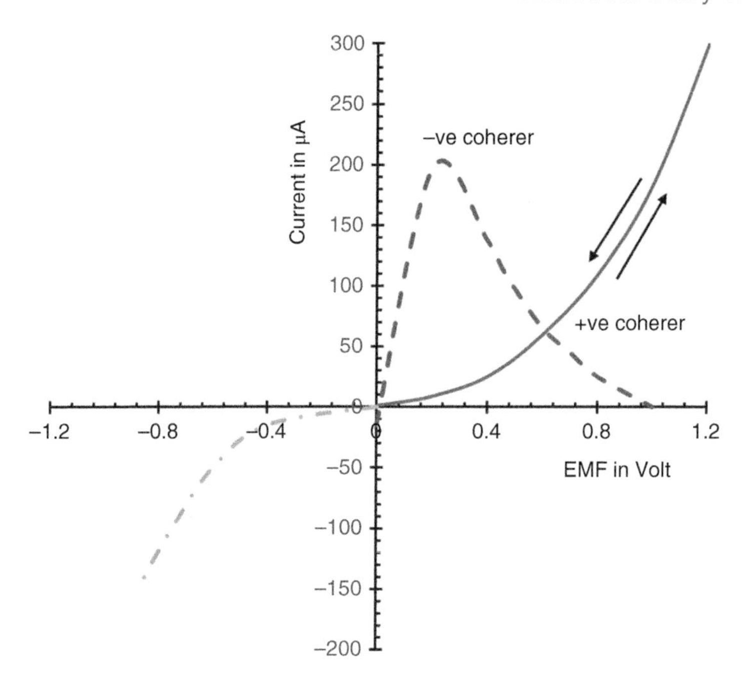

Figure 3.4 *I–V* characteristics of self-recovering positive and negative coherers of Bose. Characteristics of +ve coherer is redrawn using figure 57 of Collected Physical Papers, that of −ve coherer drawn according to Bose's observation. The *I–V* curve in the fourth quadrant corresponds to metal-p-type semiconductor diode. The arrows indicate the variation with increasing and decreasing EMF.

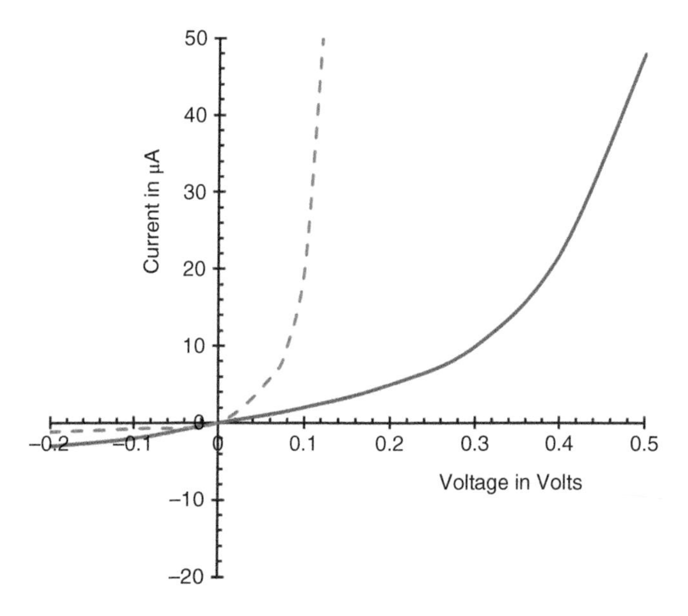

Figure 3.5 *I–V* curves of the replica of Bose detector (solid line in red) and of a typical point contact Ge diode like 1N34 or OA 81 (light blue dotted line). Source: Adapted from Groenhaug [59].

that "he found the direction of the rectification depended on the type of crystal. He thus called these positive or negative coherers. It is now known that these were due to n-type or p-type PbS."

The possibility of doping of semiconductors near a point contact was not ruled out by Shockley himself, while explaining the operation of point-contact transistor of Bardeen and Brattain. As noted by Warner [61], "he (Shockley) postulated the presence of two or more curved junctions (PN or Schottky) under the collector point, a kind of speculation stemming partly from the forming process, because some of the copper and phosphorus in the collector wire could have entered the germanium as doping impurities."

It is highly likely that Mott too postulated the presence of doping in the vicinity of contact in Bose's Galena detector.

However, attempts by DNB [60] to identify positive and negative coherers, respectively, with n-type and p-type semiconductors, remain unacceptable on scientific basis. The *I–V* characteristics of self-recovering positive coherer as measured by Bose, and as shown by figure 57 in [22], is redrawn as a solid curve in Figure 3.4. The arrows indicate the variation with increasing and decreasing EMF. The two variations are identical for a self-cohering detector and the concave nature of the curve is characteristic of positive-type material. Bose did not give any plot of negative type, but only mentioned that the curve had a convex shape. Bose stated in [22] "an increase of EMF is attended by a diminution of current, so that at a critical EMF the current disappears altogether." We illustrate in Figure 3.4 the variation for negative type accordingly by a dashed curve showing zero currents at both no EMF and at a large critical EMF and convex nature. The maximum current is chosen arbitrarily, as no such value was given by Bose [18–23].

It is to be noted that the magnitude of current increases with increasing magnitude of EMF for both types of metal–semiconductor diodes. However, *I–V* characteristics of negative coherer show different behavior.

The *I–V* characteristic of metal–Galena detector, a positive coherer, is similar in nature as in Figure 3.4. Improvements in design, structure, and material were made and the product cat's whisker diode was in extensive use during World War II [4]. As will be detailed in the next paragraph and Figure 3.5, the plot for +ve coherer is of the same nature exhibited by metal-n-type semiconductor. Based on this, the qualitative nature of the *I–V* curve for metal/p-type semiconductor is shown in Figure 3.4. It shows concave nature considering the magnitudes of I and V, unlike that exhibited by the −ve coherer.

Recently, there have been a few attempts to reproduce the results of Bose [59, 62]. The *I–V* curves obtained by Groenhaug [59] for a replica of Bose detector and for a 1N34 cat's whisker diode, a contact between a metal and n-type Ge, are shown in Figure 3.5. The solid curve is actually a replica of Marconi's coherer used in radio communication, and as clarified in [10], is actually developed by Bose. The plot for 1N34 diode is more or less as sensitive as Bose's Galena detector [59]. The plot for metal-p-type Ge for forward bias should be similar to the curves shown in the 4th quadrant of Figures 3.4 and 3.5.

Bose made several pioneering experiments that are relevant in the perspectives of today's electronics and photonics. Apart from detectors and microwave components, Bose was the first to demonstrate infrared detector, light tunneling, chiral metamaterial, and even memristors [63]. Mott truly remarked Bose was at least 60 years ahead of his time. However, it is difficult to accept his remark that Bose anticipated the existence of two conductivity types of semiconductors.

3.7 Conclusion

The aim of the paper is to critically examine Mott's statement that Bose was able to anticipate the existence of p- and n-type semiconductors. An attempt has been made to specify the time and context for Mott's remark. The scientific basis behind the *Remark* is then assessed by considering the views of past workers. The two types of coherers: positive and negative, were thought to be the precursors of p-type and n-type semiconductors. The *I–V* characteristics of the two types of coherers and of metal-n-type and metal-p-type semiconductors are compared next in this paper. No similarities are found however.

In conclusion, the remark by Mott on Bose's anticipation, and the mystery, as posed in the title of this paper, remains unsolved.

Acknowledgments

The author is thankful to Dr. Anutosh Chatterjee and Dr. Suprakash Roy, formerly of the Bose Institute, Prof. B.M. Arora, ex-TIFR and IITB, Prof. Baidya Nath Basu, ex IT BHU, and Prof. Akhlesh Lakhtakia, Penn State Univ., for useful comments and clarifications. Thanks are also due to Dr. Atanu Kundu of Heritage Institute of Technology Kolkata, for kindly introducing the author to the Editors of this special issue. Help by Dr. Bratati Mukhopadhyay and Mr. Shyamal Mukhopadhyay of Institute of Radio Physics and Electronics, University of Calcutta, in the preparation of the manuscript is gratefully acknowledged.

References

[1] Bardeen, J. and Brattain, W.H. (1948). The transistor, a semiconductor triode. *Phys. Rev.* 74: 230–231. (1949). Physical principles involved in transistor action. *Phys. Rev*. 75: 203–231.
[2] Shockley, W. (1976). The path to the conception of junction transistors. *IEEE Trans. Electron. Dev.* ED-23(7): 597–620.
[3] Pearson, G.L. and Brattain, W.H. (1955). History of semiconductor research. *Proc. IRE* 43 (12): 1794–1806.
[4] Busch, G. (1989). Early history of the physics and chemistry of semiconductors- from doubts to fact in a hundred year. *Euro. J. Phys.* 10: 254–264.
[5] Lukasiak, L. and Zakubowski, A. (2010). History of semiconductors. *J. Telecomm. Info. Technol.* 1: 3–9.
[6] Bose, J.C. (2004). US Patent 755,840, granted on 29 March 2004. See Bose, J.C. (1998). Detector for electrical disturbances patent. *Proc. IEEE* 86(1): 229–234.
[7] (1901). Semiconductor rectifiers patented as cat's whisker detectors. https://www.computerhistory.org/siliconengine/semiconductor-rectifiers-patented-as-cats-whisker-detectors.
[8] Sengupta, D.P., Engineer, M.H., and Shepherd, V.A. (2009). *Remembering Sir J. C. Bose*. Singapore: IISc Press-World Scientific Publications. ISBN: 13-978-981-4271-61-5.
[9] Bondyopadhyay, P.K. (1998). Under the glare of a thousand suns-the pioneering works of sir J.C. Bose. *Proc. IEEE* 86 (1): 218–224. https://doi.org/10.1109/5.658772.
[10] Bondyopadhyay, P.K. (1998). Sir J. C. Bose's diode detector received Marconi's first transatlantic wireless signal of December 1901. *Proc. IEEE* 86 (1): 259–285.
[11] Sengupta, D.L., Sarkar, T.L., and Sen, D. (1998). Centennial of the semiconductor diode detector. *Proc. IEEE* 86 (1): 235–243. 10.1109/5.658772.
[12] Mitra, B. (1984). Early microwave engineering: J C Bose's physical researches during 1895–1900. *Sci. Cult.* 50 (5): 147–154.
[13] https://www.boseinstitute.
[14] Johnson, M. J C Bose (1863–1937). https://www.christs.cam.ac.uk/jagadis-chandra-bose-1858-1937 (accessed 5 May 2023).
[15] Emerson, D.T. (1997). The work of Jagadish Chandra Bose: 100 years of millimetre-wave research. *IEEE Trans. MTT* 45 (12): 2267–2273.
[16] Branly, E. (1900). Accroiissments de Resistance des Radioconducteurs (Increase of resistance of radio-conductors). *Comptes Rendus Hebomadaries des Seances de l'Academie des Science [The weekly Review Session of the Academy of Science (Paris)]* 130: 1068–1071.
[17] Cuff, T.M. (1993). Coherers, a review. MS Engg thesis, at Temple University, pp. 1–378. http://doi.org/10.13140.RG.2.1.1757.8079) Cuff (p. 11) writes: Note, at the time Branly wrote this article, the term'coherers' was not yet in wide use and hence Branly referred to them as 'radioconductors'.

[18] Bose, J.C. (1895). *On Polarization of Electric Rays by Double Refracting Crystals*, 1–10. Asiatic Society of Bengal; Collected Physical Papers. Longmans Green, and Co. London, 1927.
[19] Bose, J.C. (1899). *On a Self-Recovering Coherer and the Study of the Cohering Action of Different Metals*, vol. A65, 166–172. Proc. Royal Soc.; Collected Physical Papers. Longmans Green, and Co. London, 1927.
[20] Bose, J.C. (1900). *On Electric Touch and Molecular Changes Produced in Matter by Electric Waves*, vol. A66, 452–474. Proc. Roy Soc; pp. 127–162 in Collected Physical Papers.
[21] Bose, J.C. (1901). *On the Continuity of Effect of Light and Electric Radiation on Matter*. Proc. Roy Soc; pp. 163–191 in Collected Physical Papers.
[22] Bose, J.C. (1901). *On the Change in Conductivity of Metallic Particles Under Cyclic Electromotive Variation, Sec. A*. Bose Inst. Trans., 223–252. British Association of Glasgow; Collected Physical Papers.
[23] Bose, J.C. (1987). *On the Selective Conductivity Exhibited by Certain Polarizing Substances*, 71–76. Proc. Roy. Soc.; Collected Physical Papers.
[24] Bose, D. (2010). Physics and India (correspondence). *Current Science* 99 (1): 12.
[25] https://en.wikipedia.org/wiki/Doping_(semiconductor).
[26] Volta, A. (1782). Del modo di render sensibilissimala piu debole electtricita sia naturale, sia artiliciale. *Phil. Trans. R. Soc.* 72: 237–281.
[27] Davy, H. (1821). Further researches on the magnetic phenomena produced by electricity with some new experiments on the properties of electrified bodies in their relation to conductivity powers and temperature. *Phil. Trans. R. Soc.* 111: 236–244.
[28] Faraday, M. (1839). *Experimental Researches in Electricity*, vol. I, 122–124. London: Bernard Quaritch.
[29] Becquerel, A.E. (1839). On electric effects under the influence of solar radiation. *Comptes Rendus de l'Academie des Sciences* 9: 711–714.
[30] Hittorf, J.W. (1851). Ueber das elektrische Leitungsvermogen des Schwefelsilbers und des Halbschwefelkupfers. *Ann. Phys. Lpz.* 84: 1–28.
[31] Smith, W. (1873). The action of light on selenium. *J. Soc. Telegr. Eng.* 2 (1): 31–33.
[32] Braun, F. (1874). Ueber die Stromleitung durch Schwefelmetalle. *Annalen der Physikt und Chemie* 153 (4): 556–563.
[33] Hall, E.H. (1879). On a new action of the magnet on electric currents. *Am. J. Math.* 2: 287–291.
[34] Drude, P. (1900). Zur Elektronentheorie der Metalle. *Ann. Phys. Lpz.* 1: 566–613.
[35] Riecke, V.E. (1901). Ist die metallische Leitung verbunden mit einem Transport von Metallionen? *Phys. Z.* 2: 639.
[36] Pierce, G.W. (1907). Crystal rectifiers for electric currents and electrical oscillators. *Physical Review* 25: 31–60. (1909) 28: 153–187; (1909) 29: 478–484.
[37] Koenigsberger, J. and Reichenheim, O. (1906). Ueber ein Temperaturgesetz der elektrischen Leigfahigkeit fester einheitlicher Substanzen und einige Folgerungen daraus. *Phys. Z.* 7: 570–578.
[38] Baedeker, K. (1907). Ueber die elektrische Leitfahigkeit und die thermoelektrische Kraft einiger Schwermetall-Verbindungen. *Ann. Phys. Lpz.* 22: 749–766.
[39] Weiss, J. (1910). Experimentelle Beitrage zur Elektronentheorie aus dem Gebiet der Thermoelektrizitat. Inaugural Dissertation, Albert-Ludwigs Universitat. Freiburg im Breisgau (Emmendingen: Dolter), pp. 110–114.
[40] Gudden, B. (1924). Elektrizitatsleitung in Kristallisierten Stoffen unter Ausschluss der Metalle. *Ergeb. Exakten Naturwiss.* 3: 116–159.
[41] Bloch, F. (1928). Ueber die Quantenmechanik der Elektronen in Kristallgittern. *Z. Phys.* 52: 555–560.
[42] Peierls, R. (1929). Zur Theorie derGalvanomagnetischen Effekte. *Z. Phys.* 53: 255–266.
[43] Heisenberg, W. (1931). Zum Paulischen Ausschliessungsprinzip. *Ann. Phys. Lpz.* 10: 888–904.
[44] Wilson, A.H. (1931). The theory of electronic semiconductors. *Proc. R. Soc. Lond.* A133: 458–491. (1931). A134: 277–287.
[45] Wagner, C. and Schottky, W. (1930). Theorie der Geordneten Mischphases. *Zeitschrift fur Physikalische Chemie* BIt: 163–210.
[46] Schottky, W. and Waibel, F. (1933). Die Elektronenleitung des Kupperoxyduls. *Physikalische Zeitschrift* 34: 858–864.
[47] Schottky, W. (1939). Zur Halbleitertheorie der Sperrschichtrichter und Spitzengleichrichter. *Zeitschrift fur Physik* 113: 367–414.
[48] Mott, N.F. (1939). The theory of crystal rectifiers. In: *Proc. Royal Soc*, vol. 171, 27–38. London.

[49] Davydov, B. (1938). The rectifying action of semiconductors. *Tech. Phys. USSR* 5 (2): 87–95.
[50] Davydov, B. (1939). On the contact resistance of semiconductors. *J. Phys. (USSR)* 1 (2): 167–174.
[51] Ohl, R.S. (1941). US Patent 2,402,662, filed 27 May 1941.
[52] Scaff, J.H. (1941). US Patent 2,402,582, filed 4 April 1941.
[53] Woodyard, J.R. (1950). Nonlinear circuit device utilizing germanium. US Patent 2,530,110, filed 1944, granted 1950.
[54] Shockley, W. (1949). The theory of p–n junctions and p–n junction transistors. *Bell. Syst. Tech. J.* 28: 435–489.
[55] Scaff, J.H., Theurer, H.C., and Schumacher, E.E. (1949). p-type and n-type silicon and the formation of the photovoltaic barrier in silicon ingots. *Trans. Am. Inst. Min. Metall. Eng.* 185: 383–388. Comment in [4] Although a major portion of the work reported in this reference was done before the war, publication was held up for security reasons. Scaff, J.H. and Theuerer, H.C. US Patent 2,567,970 (18 September 1951).
[56] Sparks, M. and Teal, G.K. (1953). Method of making p–n junctions in semiconductor materials. US Patent 2,631,356, filed 15 June 1950, issued 17 March 1953.
[57] Shockley, W., Sparks, M., and Teal, G.K. (1951). p–n junction transistors. *Phys. Rev.* 83: 151. 10.1103/PhysRev.83.151.
[58] Pfann, W.G. (1952). Principles of zone refining. *Trans. Am. Inst. Min. Metall. Eng.* 194: 747–753.
[59] Groenhaug, K.-L. Experiments with a replica of the Bose detector. www.kgroenha.net. Marcon (accessed 5 May 2023).
[60] Bose, D.N. (1997). Transistors: from point contact to single-electron. *Resonance* 2: 39–54.
[61] Warner, R.M. (2001). Microelectronics: its unusual origin and personality. *IEEE Trans. ED* 48 (11): 2457–2467.
[62] Gandhi, G., Aggarwal, V., and Chua, L.O. (2013). The first radios were made using memristors! *IEEE Circ. Syst. Mag.* 13: 8–16, 2nd Qtr.
[63] Basu, P.K. (2022). *Acharya Jagadish Chandra's Four Pioneering Endeavours and Their Relevance in the Perspectives of Current Electronics and Photonics, Keynote Address at 5th RCRS 2022, Organized by Indian Radio Science Society*. IIT Indore (30 November–4 December 2022).

Chapter 4

The Point-Contact Transistor

A Revolution Begins

John M. Dallesasse and Robert B. Kaufman

Holonyak Micro and Nanotechnology Lab, University of Illinois, Urbana, IL, USA

4.1 Introduction

Few would disagree that the revolution in electronics that began with the invention of the transistor has had world-changing impact. More may debate the exact event or series of events that define when that revolution began. There are many accounts from those involved, and in-depth examinations from historians of technology such as Michael Riordan and Lillian Hoddeson [1, 2] who have written the most definitive accounts of transistor history. While ideas regarding semiconductor-based switching devices may have existed prior to the discovery of transistor action in point-contact transistors by John Bardeen and Walter Brattain in December of 1947, it was this demonstration that finally validated the ongoing investment in semiconductor work at Bell Labs and was ultimately followed by the invention of the junction transistor by William Shockley in the year 1948. And, the trio were awarded the 1956 Nobel Prize in Physics. This enabled semiconductor device innovations at Bell Labs and elsewhere, and ultimately to the development of the building blocks of silicon technology. The point-contact transistor was not a false start, it was not isolated work without forward continuity, it was a true beginning. Like many revolutionary innovations, the discovery of transistor action was enabled by a convergence of the right people working on the right things in the right place at the right time. There was an organization that had a need for better switching devices and consequently was willing to support work on new material and device technologies. There was a shortage of office space that put an exceptional experimental physicist (Brattain) in the same office as a person, with his eventual two Nobel Prizes in Physics, who could be argued to be one of the greatest theorists of

75th Anniversary of the Transistor, First Edition. Edited by Arokia Nathan, Samar K. Saha, and Ravi M. Todi.

(a)

(b)

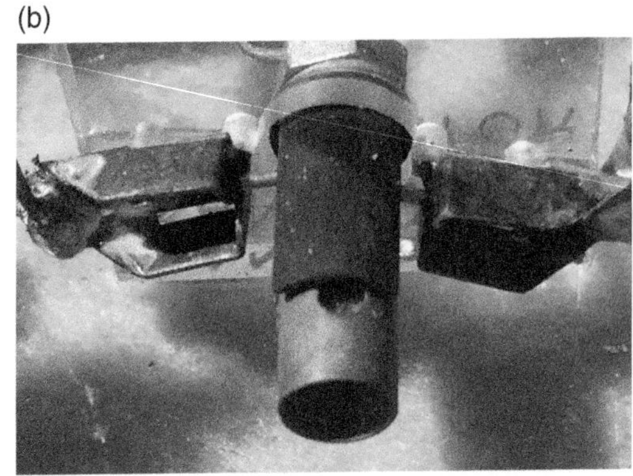

Figure 4.1 (a) Type A point-contact transistor. Source: Spurlock Museum/University of Illinois Board of Trustees. (b) Leads and housing of the transistor are clamped into connectors that provide mechanical support and electrical connection to the emitter, base, and collector. Contact to the base is made through the can, and the two leads provide connections to the emitter and collector points. This transistor (b) is in what may be the oldest transistor circuit still in existence, the "Oscillator-Amplifier Box" or "Bardeen Music Box" presently housed in the collection at the Spurlock Museum at the University of Illinois at Urbana-Champaign.

the last century (Bardeen). The right material was also needed. Brattain's experimental work guided by Bardeen's insights led them to move away from silicon onto pieces of "high back-voltage" germanium – a rectifier material developed by Purdue during WWII with higher resistance under reverse bias. A material other than germanium with a doping type or level other than the sample Brattain had would likely not have shown transistor action, or not have shown the amplification needed for applications in telephony. Finally, there needed to be an appreciation of potential impact. While the scale of the microelectronic revolution was impossible to predict at that time, it was appreciated that the discovery of transistor action was important and that transistors would be useful electronic devices. Patent filings, publications, and public announcements led to significant interest in sample devices which were delivered by a point-contact transistor pilot line set up at Mervin Kelly's direction by Jack Morton. "Type A" transistors like the one shown in Figure 4.1 were the first commercial devices used in exploring circuit applications of the emerging transistor technology.

While point-contact transistors were only in use for a few years before they were replaced by the junction transistor, which was proposed by Shockley, it was the demonstration of gain in the point-contact transistor, Bardeen's understanding of band-bending to produce an inversion region, and Shockley's description of minority carrier injection which enabled the next step, and the steps that followed.

4.2 Background and Motivation

Ideas do not magically transform into reality. The art of creating a thing is often underestimated in value by the thinking world but without knowledge and skill in rendering, ideas would remain solely in the world of thought. Today's integrated circuit cannot be made without an infrastructure that costs billions of dollars and the know-how associated with that infrastructure. There is an art even to the small things. When it was decided to recreate the discovery of the point-contact transistor, the first thought might be "how hard can it be?" Answering that question requires an appreciation of the time and environment that Brattain and Bardeen were working in, and insight into the art that existed at that time. What we know, in the art as well as the science, changes with time. Going backward can be as challenging as going forward, even with the benefit of historical accounts.

Before the transistor, there was the rectifier. In 1874, Braun discovered that pressing a wire onto a piece of galena (lead sulfide) would create a device that would allow current to flow freely in one direction but not in the other, and later patented the use of "crystal rectifiers" in wireless receivers. These discoveries would lead to him being awarded the 1908 Nobel Prize with Marconi for their "contributions to the development of wireless telegraphy." [3] Silicon crystals with tungsten whisker contacts became widely used for rectifiers and were eventually discovered to have advantages in high-frequency applications where tubes were not fast enough. While these devices were widely used, they were not understood. Physics had not yet caught up to phenomenon – science had not caught up to art – and there was no basis from which an operating mechanism could be postulated. Things changed in the early twentieth century. Schrodinger and de Broglie postulated the wave nature of electrons, and Davisson and Germer experimentally demonstrated wave properties. Sommerfeld then proposed that a metal resembled a sea of electrons following Fermi–Dirac statistics surrounding a background of fixed positive metal ions. The physics that would allow an explanation of rectification was beginning to emerge, and Mott and Schottky were able to pierce the veil and postulate theories of rectification at metal–semiconductor contacts.

Before his work on the transistor, Brattain worked on copper-oxide rectifiers with Becker (1930s). They, and later Shockley, postulated that these devices might be made into amplifiers with a third "grid" electrode. This proved not to be possible or practical, but thinking was underway on how a third terminal might be used to switch or amplify current in a semiconductor. It is worth noting that Lilienfeld proposed a semiconductor device that used a third terminal to control current several years earlier (1930), but this device was never reduced to practice nor were the underlying physics or technology that would enable such a device to be made understood at that time. The value of crystal rectifiers as high-speed devices for high-frequency applications motivated further experimental work, and the "art" of producing high-performance devices followed the science. Techniques such as "contact forming" and crystal purification processes were developed to improve crystal rectifier performance, and silicon crystal detectors became a key component in radar systems. The need for such systems driven by WWII led to significant advances in the processes for making silicon and germanium, and in the understanding of conduction in these materials. It was with this backdrop that Bardeen and Brattain worked on semiconductor devices trying to create an improved switch, and discovered the transistor effect – point-contact transistors were born and the electronics revolution began.

Seventy-five years later, to commemorate the 75th Anniversary of this discovery, the IEEE Electron Devices Society encouraged an effort to recreate Bardeen and Brattain's work. Thus would begin the journey to make our own point-contact transistor guided by the early work and studies of devices from the late 1940s and early 1950s. This process started with the belief that we could simply press contacts onto a piece of correctly doped n-type germanium and that would be it, *transistor accompli*. This presumption was overly optimistic as it would take months of spotted effort to realize a device that can even begin to approach the performance of Bardeen and Brattain's first published transistor results. A key statement in the original Bardeen–Brattain paper proved to be the stumbling block: "The collector point may be electrically formed by passing large currents in the reverse direction" [5]. Best practices for contact preparation, material types, and operating conditions had to be relearned through the old literature and accounts of the point-contact transistor discovery and first and foremost, a better understanding of the physics behind the transistor action that arose from a properly fabricated device.

4.3 Inventors' Understanding How a Point-Contact Transistor Operates

A key physical effect behind Bardeen and Brattain's discovery of the first transistor is the formation of a thin p-type inversion layer at the free surface of the germanium crystal. Early thought after the transistor discovery suggested that this inversion layer was created by the high work functions of some metals coupled with Fermi-level alignment within the n-type semiconductor. The resulting band bending and creation of a space-charge region would lead to a p-type inversion layer at the semiconductor–metal interface. Various experiments with different metals later indicated that this effect is actually a feature of surface-state Fermi-level pinning at the free surface of germanium [6]. How Bardeen and Brattain depicted this band-bending effect is shown in Figure 4.2, from their publication immediately following the first transistor paper. This article investigated the physics behind point-contact formation that enabled the observed transistor effect [4]. States at the surface of the germanium pin the Fermi level at some barrier energy, ϕ_s, from the conduction band edge. Due to the electron affinity and bandgap of germanium, this places the Fermi level closer to the valence band for a thin region near the surface of the device, thus creating a p-type inversion layer around 1 μm thick [4, 6].

For the point-contact transistor, two top-side point contacts are made to this layer, minimally affecting the surface condition and allowing the inversion layer to be maintained, while an ohmic contact is formed across the bottom of the block of germanium to create the third contact. Bardeen coined these top contacts the emitter and collector and the back-side contact was termed the base. On first look, the point-contact transistor and a p–n–p bipolar junction transistor (BJT) appear similar as both devices consist of emitter and collector contacts incident to p-type regions and a base contact residing an n-type region. However, for the point-contact transistor, the emitter and collector terminals contact the same, continuous p-type region and thus electrical behaviour is quite different. The emitter operates in much the same way as a BJT emitter, providing minority carrier injection in the form of holes, but the base and collector terminals operate differently. Usually for a BJT the injected minority carriers from the emitter flow through the base, with some of the minority carriers recombining, and the rest exiting the collector such that the *alpha*, collector current divided by emitter current, is less than one. The small base current that compensates the recombination process is instead amplified in the collector current output leading to a large *beta*, collector current divided by base current.

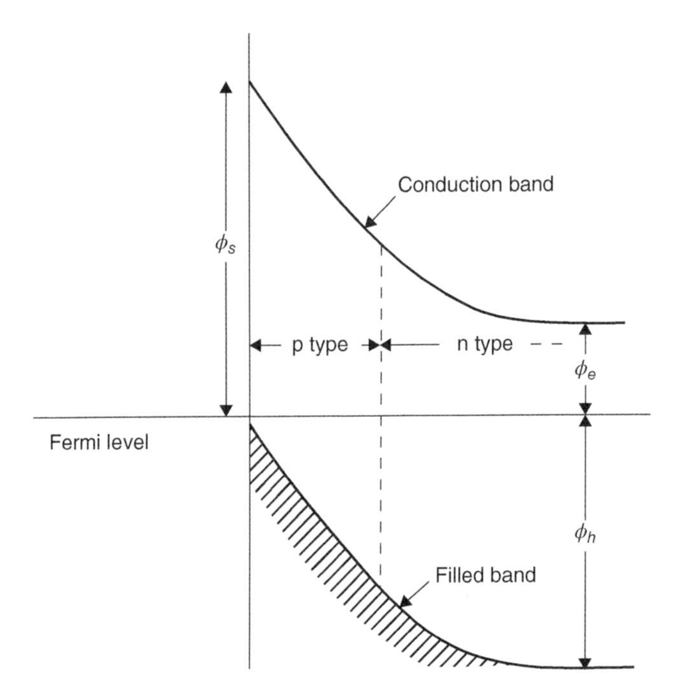

Figure 4.2 Bardeen and Brattain's explanation of the formation of a p-type inversion layer on the surface of a germanium crystal under the top-side contacts in the point-contact transistor. Source: Reprinted with permission from Brattain and Bardeen [4]. Copyright 2023 by the American Physical Society.

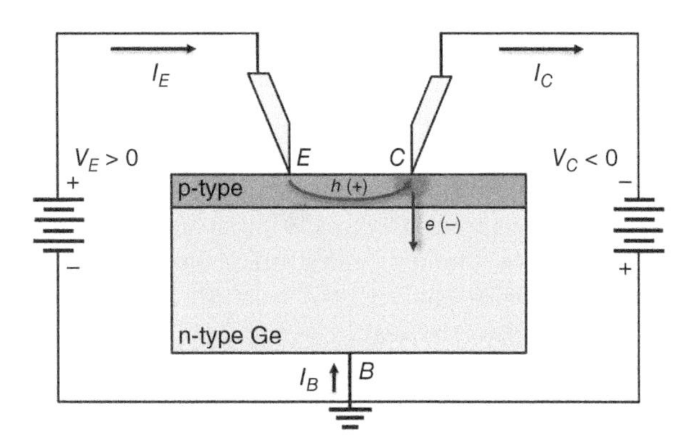

Figure 4.3 Illustration of an active-mode point-contact transistor with a positive emitter bias and negative collector bias. Current directions of each terminal are indicated and the rough propagation of carriers contributing to the collector current are shown.

For a point-contact transistor, the *beta* is now less than one while the emitter current gain, *alpha*, is the high-gain output. The *alpha* gain metric that Bardeen and Brattain first defined was a differential current gain defined as $\alpha = \partial I_c/\partial I_e$ [5]. A standard operating configuration for a point-contact transistor is to apply a positive bias to the emitter terminal, ground the base-terminal, and apply a negative bias to the collector terminal; akin to common-base operation of a BJT. In this way, a net field exists between the emitter terminal and the collector contact in their shared p-type region, allowing the holes injected in the emitter to travel through to the collector. The additional current traveling through the collector to provide the *alpha* gain then must come from the base, and this is roughly explained as electrons flowing from the collector contact to the n-type base. Figure 4.3 shows an illustration of a point-contact transistor in an active operating mode. The difference in operation compared to a BJT is clear as both the emitter and base current are directed into the device whereas a p–n–p BJT would feature base current directed out of the device.

4.4 Recreating the Point-Contact Transistor

For the original transistor, Brattain obtained his germanium from Purdue where it was being used for high back-voltage rectifiers [2]. This sample was described as a lightly n-type germanium crystal doped with antimony having a resistivity on the order of 10 Ω-cm [5, 7]. For our attempt at recreating the point-contact transistor, we started by procuring a lightly-doped n-type germanium wafer with a resistivity specified in the range from 1 to 10 Ω-cm. This is somewhat lower than the resistivity used by Bardeen and Brattain in their first paper, leading to the need to perform our work on a sample more highly doped than desired [5]. Our choice was constrained by material availability – with the COVID pandemic, supply chain issues limiting options and the cost and lead time of substrates with custom doping levels prohibitive, a compromise was needed. A consequence of higher doping is reduced thickness of the inversion layer that is key to point-contact transistor operation; it would eventually be determined that our point-contact transistor needed closer contacts likely due to this effect. That said, unlike for Bardeen and Brattain, the preparation of our germanium's surface was not as critical since the surface was already as expected. At the time of the first transistor, the as-grown germanium crystals often had to be etched back to expose the desired n-type region [5, 7]. Our examination of using a NO_3-acid-based etch on the germanium sample to remove material from the surface prior to point-contact formation showed no change in the electrical properties of the device.

As with the original transistor, the first step for our device was preparing a small "block" of material; in our case a 1 cm × 1 cm piece diced from the 2″ diameter wafer. From there, a back-side metal of Ni/Au was evaporated to form the ohmic contact to the base and two tungsten probe trips positioned by individual micromanipulators were used to form the top-side emitter and collector contacts. Three-terminal measurements were taken with a HP4155C semiconductor parameter analyzer using the common-base configuration described previously. In the initial attempts to create a point-contact transistor, the importance of the "contact forming" step mentioned in the original Bardeen–Brattain paper [5] was not fully appreciated, and it was omitted. The result was a device that showed a very weak transistor effect. Reverse-biased collector current was controlled by emitter injection, but with no gain of any sort.

Despite how simple it superficially appeared to be to recreate this historic device, the poor initial result led to the realization that something was missing from our setup or process that led to this sub-par performance. The lack of the expected *alpha* gain was due to a collector current nearly 10× smaller than the emitter injection. Not only were the holes injected in the emitter not making it to the collector contact, but any base current into the collector was also negligible. This result would lead us to investigate more deeply the history and physics of the point-contact transistor to determine how we could improve the performance of our own device. This investigation led us to determine that the most critical element we were missing from our process was the collector-forming step. While not present in the original gold-wedge transistor demonstration, it was a key part of the first published device results and commercial point-contact transistors to follow [2, 5]. Other accounts of earlier devices that lead Bardeen and Brattain to the point-contact transistor also demonstrated a transistor effect with different contact types, including that original gold-wedge transistor. Looking through these accounts, published papers, and Bardeen's Nobel lecture [2, 5, 8], we focused our experiments on a variety of collector-forming methods, contact metal types, and contact separation.

4.4.1 Rediscovering Collector Forming

The collector-forming process is done to improve the current flow between the base and collector terminals. Collector forming is accomplished by passing short pulses of high current between the contacts, typically in a reverse-bias configuration and thus requiring a high voltage. One method reported to achieve this was to charge a capacitor to a high voltage, typically 75–300 V, and then quickly discharge it across the contact delivering a high-current pulse [9]. The exact charging voltage would have to be experimented with as the process was not well defined and prone to yield issues, but when it did work, it improved the emitter gain *alpha* of the point-contact transistor tremendously. Besides just the high reverse-bias voltage pulse, another method used by Bell Labs for rectifiers, and likely Bardeen and Brattain's first transistor, consisted of repeated pulses of alternating current, meaning both a forward and reverse current was being passed between the base–collector contacts [7]. Due to the forward-bias component of the pulses, a lower voltage was used, around 30 V, since this would deliver a high current as is. A 50-Ω resistor was placed in series with the collector–base connections to limit the current below 1 A [7]. Trying both methods, strictly reverse-biased pulses and AC pulses, using either a pulsed voltage source (a HP 214B) or a curve-tracer with a 300 µs pulsed output (a Tektronic 577), different collector-forming parameters were tested on our point-contact transistor samples. By far the most consistent method was found to be achieved when using lower-voltage AC pulses. The best results were achieved with a 70 V pulse with current limited to 1.4 A using a tungsten probe as the collector contact. The collector current (I_C) vs. emitter current (I_E) plot, which follows the format of the original transistor curve plot shown by Bardeen and Brattain, is measured for this device before and after the collector-forming process. This chart enables quick identification of the *alpha* gain of the output as one can look at the slope of the lines – the more positive the slope, the higher the emitter gain. Depicted in Figure 4.4 is the comparison of the pre and post forming *alpha* curves which shows a noticeable improvement of

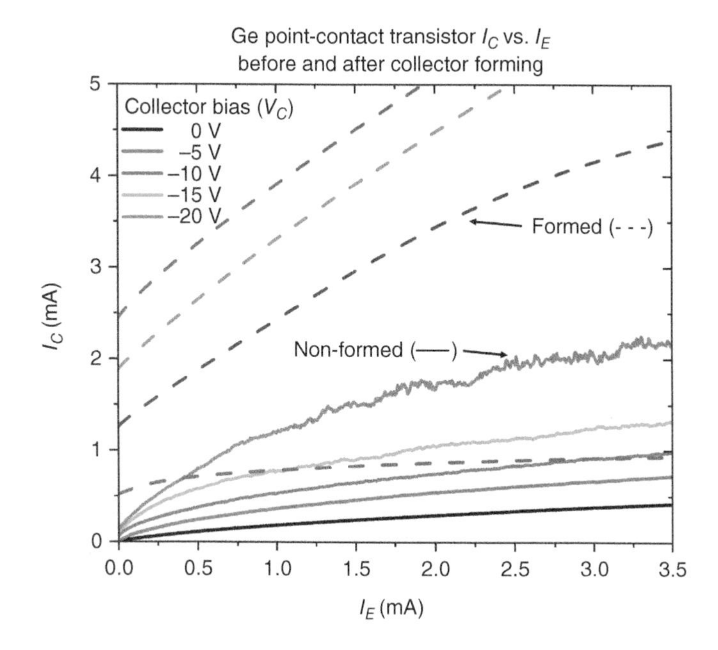

Figure 4.4 I_C vs. I_E plot comparing the *alpha* gain characteristics between formed and non-formed contacts. Contacts are tungsten probe tips and the forming was done with repeated 300 μs, 70 V AC pulses.

the device gain for the formed collector contact. Not only that, but the stability of the output is improved too, especially for larger current injection, which was a trend seen for all forming attempts.

By just passing this high current through our collector contact, we were able to take our no-gain transistor to one with gain greater than 1. The exact reason contact forming greatly enhances the performance of the point-contact transistor has been debated with a few different theories proposed but no definitive answer. Bardeen and Pfann initially suggested that the effect was caused by a change in the surface states and possibly n-type doping introduced into the germanium from impurities in the point contacts [10]. These would induce a change in the Fermi level at the surface and thus change the contact barrier. In their initial testing, they found only a reverse-bias current flow to be beneficial in collector forming. By decreasing the barrier layer and thus raising the Fermi level, they enabled better electron current flow out of the contact [10]. Shockley sought to describe the forming action more physically and attributed the increased gain to one of two outcomes: either a formation of hole trap states or what he coined the p–n hook theory [11]. For the trap theory, if there were additional hole states there would be a decrease in hole mobility and an enhanced positive space charge attracting electrons from the collector contact [11]. The p–n hook theory posited that the forming process creates a larger, localized p-type region with an n-type inversion layer at the contact point leading to an n–p–n BJT across the base–collector terminals [11]. This p–n hook theory was also supported by measurements by Valdes of a presumed extended p-type region with higher conductivity centered from the point where the contact was formed [12] combined with knowledge that high heat could transform n-type germanium to p-type [7]. In this theory, the p-type emitter injection of the point-contact transistor would essentially form the base-current of the n–p–n BJT and facilitate electron flow from the point-contact transistor collector to base with the *beta* gain of a BJT translating to the *alpha* gain of the point-contact transistor [11]. The general theme of these explanations is that the forming process enhances the current that can flow between the base and collector which is the source of the gain for the emitter current.

4.4.2 Exploring Contact Type Effects

The tests leading up to our final point-contact transistor result involved exploring different contact variations, and given the undesirable performance of our non-formed tungsten contacts, a variety of Bardeen and Brattain's reported contact schemes were examined. Their first experiment to demonstrate a hole-based transistor effect occurred when they deposited small gold pads on the surface of an n-type germanium piece as one contact and used a tungsten point as the other top contact [2, 8]. Initially, they had intended to evaporate this gold contact on top of an oxide layer in an attempt to make something more akin to a field-effect transistor, but as part of the preparation of the germanium block, they had accidentally washed off the oxide [2]. Instead, the gold pad was contacted directly to the p-type inversion layer on the germanium surface. The reason this device did not produce high gain was due to the large spacing between the two top-side contacts, which would lead Bardeen to suggest the gold-covered wedge design [2]. By wrapping gold around the corner of an insulating wedge and cutting a slit at the point, two top-side gold contacts were made that were spaced only 40 μm apart – close enough to enable gain and demonstrate the first transistor [2]. Sometime between this discovery and the publishing of their first paper on the point-contact transistor in June the following year, Bardeen and Brattain had refined the design using two tungsten or bronze-phosphor contacts spaced closely together with the collector contact appropriately formed [5].

Taking inspiration from this progression, we looked at implementing various gold contact schemes to see if we could observe a stronger transistor effect, despite the choice of metal supposedly not changing the Fermi-level pinning and p-type inversion layer (see discussion above). This first took the form of a crude recreation of the wedge-contact (as seen in the top-right of Figure 4.5b), though one that lacked the slit precision of Brattain's and thus was not practical in testing. Attempts at gold-wrapped probe-tips or scalpels were also used (Figure 4.5c), but the pressure of the contact to the surface was poor and it was difficult to closely align the emitter and collector contacts together. That left us to attempt the first nearly functional method that Bardeen and Brattain had done – the evaporated gold pad. To do this, 200 nm thick squares of gold measuring 20 μm × 20 μm, 30 μm × 30 μm, and 50 μm × 50 μm

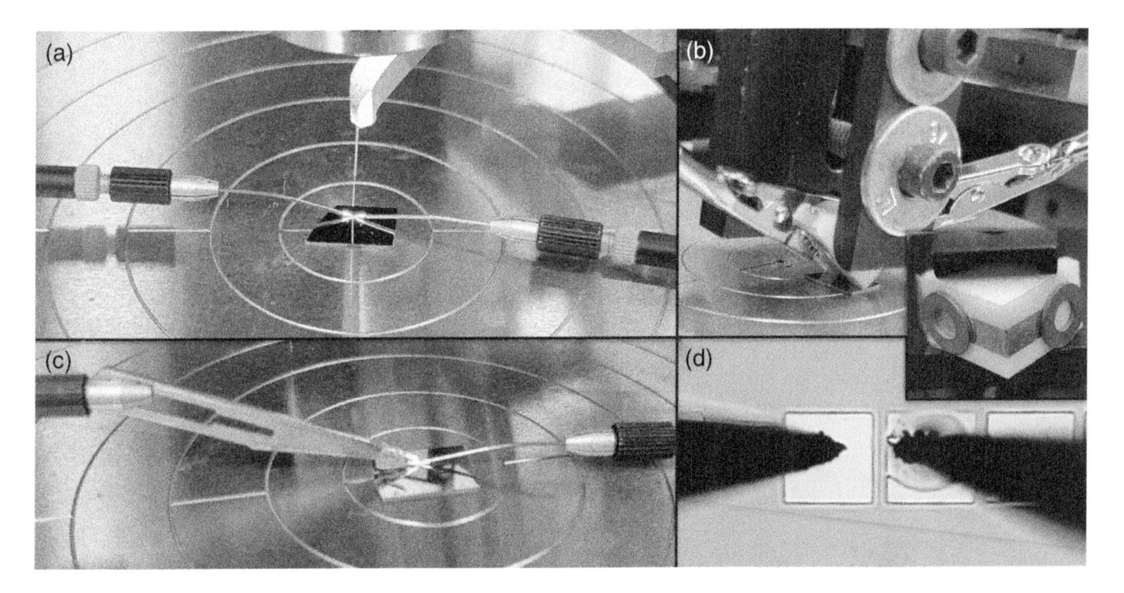

Figure 4.5 Collage of various point-contact transistor setups we attempted: (a) two tungsten probe setup as emitter and collector contacts for point-contact transistor. (b) Gold-wedge point-contact transistor setup. (c) Gold-wrapped scalpel as emitter and tungsten probe as collector setup. (d) Evaporated gold pads setup with a formed collector contact over a degraded pad.

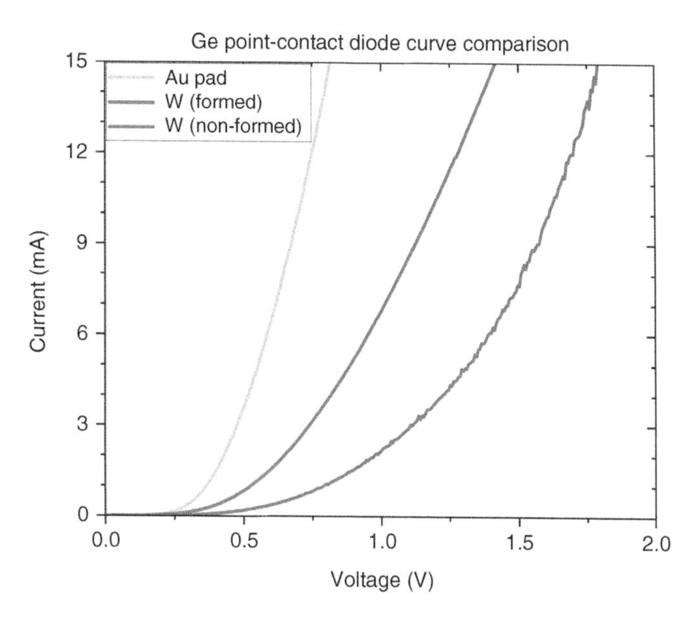

Figure 4.6 Forward-bias diode curves between different top-side contact types and back-side base contact on n-type germanium piece used for point-contact transistor attempts; gold pad was 20 μm × 20 μm, tungsten was a probe tip, and forming was done with multiple 300 μs, 70 V AC pulses.

were evaporated on pieces of n-type germanium (Figure 4.5d). One of the ways that the differences of these contact types could be observed is by taking the diode curve between the top-side contact and the back-side, n-type base contact. Figure 4.6 depicts the diode curves measured for the three different contact types that were tested: a 20 μm × 20 μm gold pad, a formed tungsten contact, and a non-formed tungsten probe contact. As can be seen, the gold contact produced the lowest turn-on voltage and the least resistance with the formed and non-formed tungsten contact having even larger turn-on voltages in that order. In theory, there should not be much of a difference between these curves and the most likely reason for these differences has to do with the size of the contact region with the gold pads being larger than the points of the tungsten probes and thus less resistive. This would then make sense as to why the formed tungsten contact performed better than the non-formed if the theory of a larger p-type region formation is correct. That said, the main advantage of the formed contact occurs during reverse-bias operation, so this forward-bias improvement means less for the final point-contact transistor performance. Whatever the exact cause for these discrepancies, between this simple test and the various full transistor measurements, it was determined that the gold contact was ideal for the emitter as it provided the best current injection. The formed tungsten contact was the best solution for the collector contact due to the improved base current to enhance the gain, as learned in the prior forming experiments. Perhaps gold may have worked well here as well, but the pads deposited on the sample are severely degraded or destroyed during the forming process.

4.4.3 Optimizing Contact Separation

One of the critical design elements of a point-contact transistor is the closeness of the contacts. For Bardeen and Brattain's first demonstrated transistor, they used a plastic wedge covered by a piece of gold foil with a slit at the end creating a separation of about 40 μm, which is on the order of a width of human hair [2]. It can be understood from the operation of the device as to why – a short distance between the contacts reduces recombination losses that the holes encounter when traversing between the emitter and collector contacts. With contact material determined, the last parameter to optimize for our point-contact

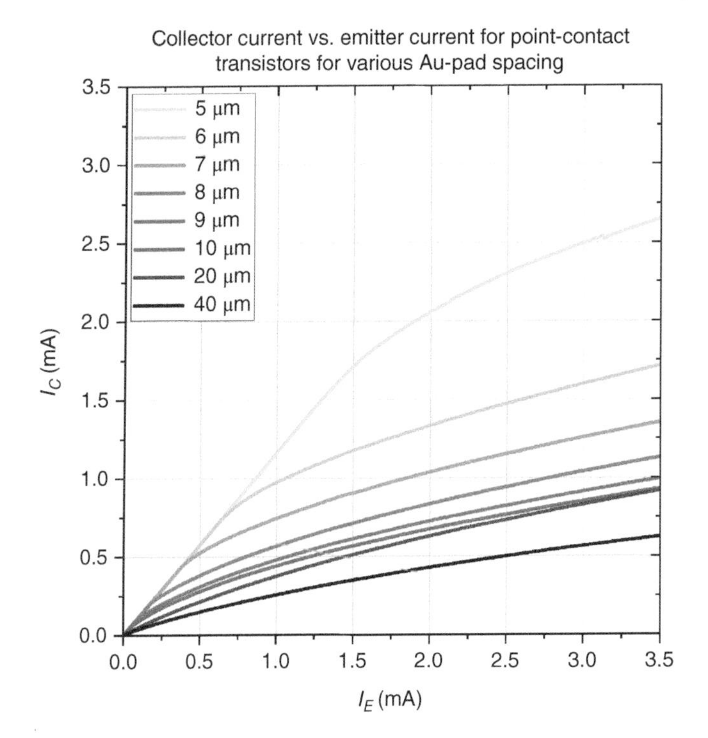

Figure 4.7 I_C vs. I_E between two 50 μm × 50 μm pads of various separations on a germanium point-contact transistor sample.

transistor was the distance between contacts. The gold pads deposited on the surface of our germanium piece were spaced at various distances from 2 to 100 μm. To demonstrate the effect pad spacing had on the collector current and any potential gain out of the device, point-contact transistor measurements were taken with both the emitter and collector being composed of gold pads spaced at different intervals. Figure 4.7 shows the *alpha* curve results which indicate a clear trend that the closer the contacts, the higher the collector current and gain. These measurements were done with a collector voltage of −20 V. Without the formed collector contact, there is limited gain available in the device, though for low injection current, there exists positive gain for the closer contacts. However, once you get to a spacing of 8 μm and further, all the emitter current is not making it through the collector, even for low injections. At 40 μm, the distance of the contacts in Bardeen's and Brattain's first wedge transistor, the performance is extremely poor. Effects of the emitter gain on contact spacing was something that Bardeen and Brattain tested as well and found an approximately exponential decrease as separation increased which aligns well with our data [6]. For our final device attempt, this told us that we had to make the contacts as close as possible if we wanted to see a larger gain. Unfortunately, lateral surface defects formed during collector forming would propagate between the top-side contacts and short them for close separation. This meant that the closest separation we could use for a formed device was ~10 μm.

4.4.4 Our Own Point-Contact Transistor: How Did it Do?

Just as Bardeen and Brattain went through many different experiments and iterations before finally arriving at the point-contact transistor, we too cycled through many different attempts at recreating it. Our best device, combining the superior injection qualities of a 50 μm × 50 μm gold emitter contact and the gain-provisions of a formed tungsten collector contact spaced 10 μm apart, was able to produce a differential *alpha* gain peaking around 1.6 for constant emitter voltage measurements and around 1.2 for

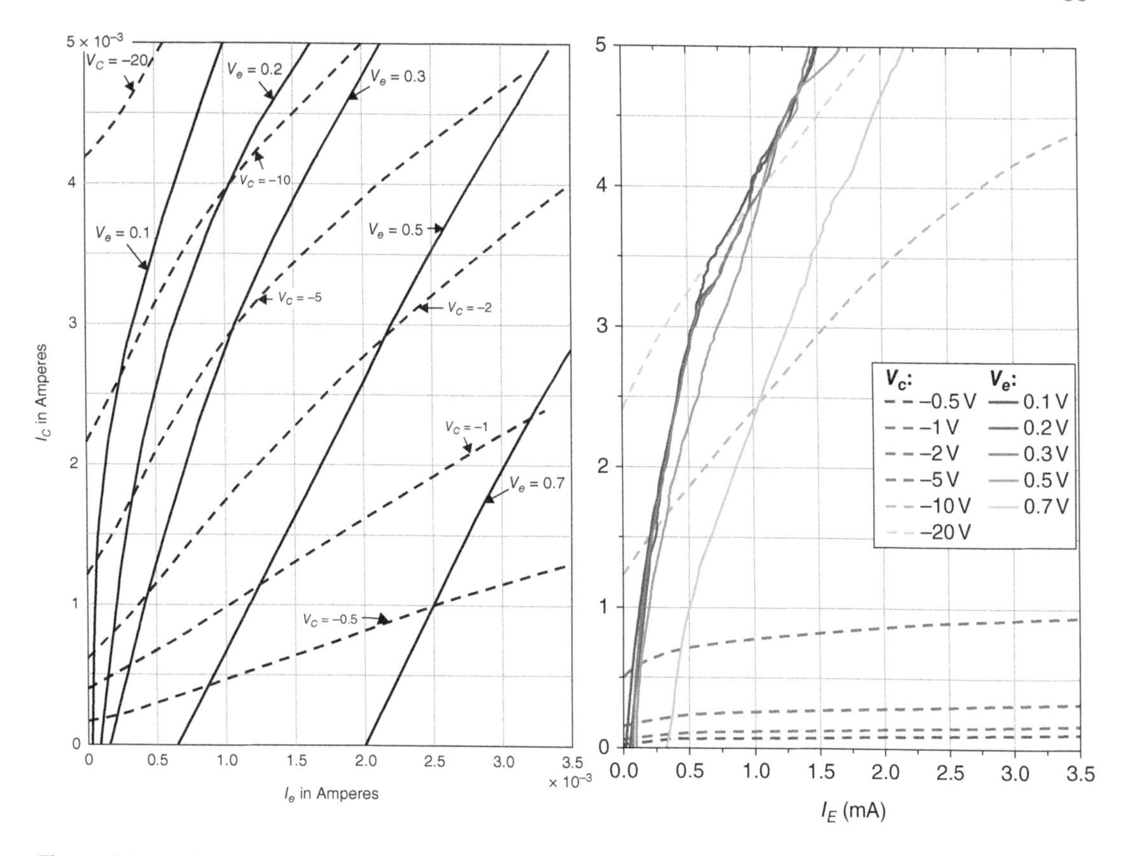

Figure 4.8 *Left*: I_C vs. I_E plot of Bardeen and Brattain's first published point-contact transistor demonstrating the *alpha* gain of the device for either fixed emitter or fixed collector voltages. Source: Reprinted with permission from Bardeen and Brattain [5]. Copyright 2023 by the American Physical Society. *Right*: The same plot and parameters measured for our own point-contact transistor attempt using a 50 μm × 50 μm gold pad as the emitter and a formed tungsten probe as the collector.

constant collector voltage measurements. It is compared side-by-side to the Bardeen and Brattain's original published transistor's *alpha* curves in Figure 4.8 [5]. The low-injection emitter curves (solid line) look very similar to that of the original transistor, though more unstable and producing less overall current for the same voltages. The collector voltage curves (dashed line) for our transistor is lacking compared to the originals, especially for low collector voltages below −10 V. That said, the above unity slopes of the −10 and −20 V curves mirror some of the curves in the original transistor showing that comparable performance can be achieved, but at a high applied bias. While there is an array of differences between the two devices, the most critical culprit for the performance results come down to the piece of germanium used. The higher doping of our sample hinders the physics that enabled the point-contact transistor to work in the first place as higher doping leads to a thinner p-type inversion layer.

After the demonstration of the first transistor, Bardeen, Brattain, and others at Bell Labs continued experiments to better understand its operation and improve the device. Refinement of the contacts, forming process, and packaging would lead to the Type A transistor, their first commercial version of the device. Instead of having an *alpha* gain in the 1–2 range, these more mature devices had emitter gain exceeding 4 and were used in a variety of circuits requiring signal gain. Bardeen had a famous demonstration of the usefulness of the point-contact transistor with two of them acting as critical components in a portable music box. One transistor would provide gain to sustain the oscillations of an LC tank while the other would amplify the oscillating signal to drive a speaker. To see how our point-contact

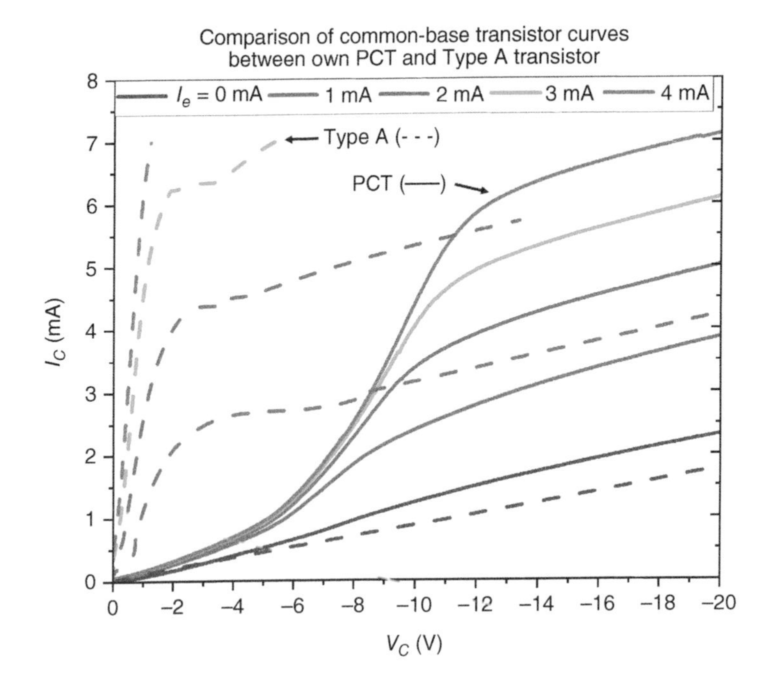

Figure 4.9 Common-base transistor curves for comparing a Type A point-contact transistor to our point-contact transistor attempt.

transistor compares to the commercialized Type A transistor, common-base measurements were taken while sweeping the reverse-bias collector voltage and injecting different emitter current. Figure 4.9 shows this measurement for both our transistor (solid lines) as well as a Type A transistor we had access to (dashed lines). While our own attempt at the point-contact transistor may have approached the performance of the original published device, it clearly trails even further when compared to the refined Type A transistor. Here, the figure of merit is the magnitude of the collector current as it relates to the direct emitter gain (I_C/I_E). The Type A transistor not only reaches its maximum current level at lower collector biases, but emits collector currents consistently higher than our point-contact transistor did for the same emitter injection. This is as expected given our device was only able to show a gain less than 2 whereas the Type A transistor can go much higher. The interesting effect for ours is how it does not reach its saturation point until around −10V on the collector meaning that our contact, even while formed, struggles to attract the holes injected in the emitter.

4.5 Concluding Remarks

Despite performance that was not as good as that in Bardeen and Brattain's seminal publication, we were able to recreate a point-contact transistor that had a positive gain. This gain was sufficient to drive oscillations in a recreation of the oscillator circuit used in Bardeen's Oscillator-Amplifier Box, an apparatus assembled for early transistor seminars and demonstrations. The time and effort to get to function what is by today's standards a relatively simple device reinforced our appreciation for the work done on the discovery of the original transistor 75 years ago. Without the conviction and dedication of Bardeen and Brattain as well as their unique combination of practical and theoretical knowledge, it is difficult to say how many years would have elapsed before the genesis of transistor electronics. Even with the luxury of knowing how the device was supposed to operate, and with publications and historical accounts to guide us, the process of relearning the art and developing a more subtle understanding of the physics was needed to produce working devices. Not all things are obvious in retrospect, and the unique skills of those who lived and created history need to be appreciated and acknowledged by those who come after and benefit from their achievements.

References

[1] Riordan, M. and Hoddeson, L. (1997). *Crystal Fire*. New York, NY: W.W. Norton & Company, Inc.
[2] Hoddeson, L. and Daitch, V. (2002). *True Genius: The Life and Science of John Bardeen*. Washington, DC: Joseph Henry Press.
[3] The Nobel Prize in Physics 1909 (2023) Nobel Prize Outreach AB 2023. http://NobelPrize.org. https://www.nobelprize.org/prizes/physics/1909/summary (accessed 17 January 2023).
[4] Brattain, W.H. and Bardeen, J. (1948). Nature of the forward current in germanium point contacts. *Physical Review* 74 (2): 231–232.
[5] Bardeen, J. and Brattain, W.H. (1948). The transistor, a semi-conductor triode. *Physical Review* 74 (2): 230–231.
[6] Bardeen, J. and Brattain, W.H. (1949). Physical principles involved in transistor action. *Physical Review* 75 (8): 1208–1225.
[7] Torrey, H.C. and Whitmer, C.A. (1948). Chapter 12: high-inverse-voltage rectifiers. In: *Crystal Rectifiers*, 361–397. New York, NY: McGraw-Hill.
[8] Bardeen, J. (1964). Semiconductor research leading to the point contact transistor. Nobel lecture, December 11, 1956. In: *Physics 1942–1962*, 318–341. Amsterdam: Elsevier Publishing Company.
[9] Slade, B.N. (1953). *Factors in the Design of Point-Contact Transistors*. Harrison, NJ: Radio Corporation of America. Tube Dept.
[10] Bardeen, J. and Pfann, W.G. (1950). Effects of electrical forming on the rectifying barriers of n- and p-germanium transistors. *Physical Review* 77 (3): 401–402.
[11] Shockley, W. (1950). Theories of high values of alpha for collector contacts on germanium. *Physical Review* 78 (3): 294–295.
[12] Valdes, L.B. (1952). Transistor forming effects in n-type germanium. *Proceedings of the IRE* 40 (4): 445–448.

Chapter 5

On the Shockley Diode Equation and Analytic Models for Modern Bipolar Transistors

Tak H. Ning*

IBM Thomas J. Watson Research Center, Yorktown Heights, NY, USA

*Retired

5.1 Introduction

Structuarally, an n–p–n bipolar transistor is formed with two p–n diodes having a common p-region which is the base of the transistor. Similarly, a p–n–p bipolar transistor consists of two p–n diodes with a common n-region. For a modern n–p–n transistor, the applied voltages in normal operation and the corresponding energy-band diagram are illustrated schematically in Figure 5.1. In a typical operation, the base–emitter diode is forward biased, with electrons injected from the emitter into the base and holes injected from the base into the emitter. The electrons flow in the base by diffusion, recombining with holes as they travel toward the collector. Those electrons reaching the end of the quasineutral base layer, located at $x = W_B$, are collected by the adjacent n-type region as collector current.

The transport of electrons in the p-type base layer is governed by the diffusion equation, namely,

$$\frac{d^2 n_p(x)}{dx^2} - \frac{n_p(x) - n_{p0}}{L_n^2} = 0, \tag{5.1}$$

where n_p is the electron density in the p-type region, n_{p0} is the electron density in the p-type region at thermal equilibrium, and L_n is the electron diffusion length. Equation (5.1) is a second-order differential equation, and its solution requires two boundary conditions for $n_p(x)$. The boundary condition at $x = W_B$ is intuitively obvious, and is usually taken as $n_p(x = W_B) = n_{p0}$. What is needed is the boundary condition at $x = 0$, i.e. we need to know $n_p(x = 0)$.

75th Anniversary of the Transistor, First Edition. Edited by Arokia Nathan, Samar K. Saha, and Ravi M. Todi.

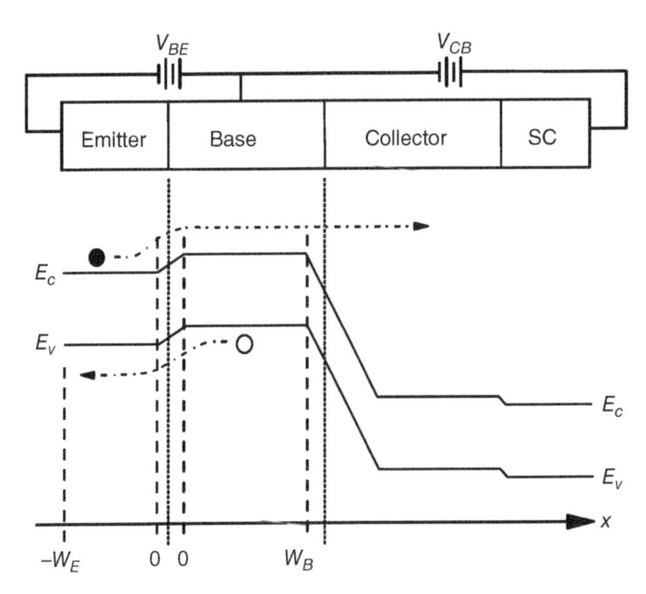

Figure 5.1 Schematics illustrating the applied voltages in normal operation of an n–p–n bipolar transistor (top), and the corresponding energy-band diagram, carrier flows, and locations of the boundaries of the quasineutral emitter and base regions (bottom).

The bipolar transistor was invented in 1948. Not long after that, in a paper published in 1949 [1] and again in a book published in 1950 [2], Shockley showed that $n_p(x = 0)$ is given by the equation (see eq. (8) in Ref. [1]).

$$p_p(x=0)n_p(x=0)=n_i^2\exp\left(\frac{qV_{BE}}{KT}\right), \tag{5.2}$$

where $p_p(0)$ and $n_p(0)$ are the hole density and electron density, respectively, at $x = 0$ where the injected electrons first enter the quasineutral p-side of the diode, n_i is the intrinsic-carrier density, and V_{BE} is the base–emitter forward bias voltage. Shockley emphasized that Eq. (5.2) is "the key equation of the rectification theory." Equation (5.2) is often referred to as the Shockley diode equation.

In the 1949 paper [1], using Eq. (5.2) as the boundary condition for solving the minority-hole diffusion equation, Shockley derived an analytic model for the collector current of a p–n–p transistor and discussed its properties. In the discussion section, he said the following: "The p–n–p transistor has the interesting feature of being calculable to a high degree. One can consider such questions as the relative ratios of width to length of the n-region and the effect of altering impurity contents and scaling the structure to operate in different frequency ranges. …"

Since the publication of Shockley's classic paper in 1949 and book in 1950, bipolar transistor structures and the associated device fabrication technology have evolved tremendously, and analytic device models have been developed to explain the characteristics of these modern bipolar devices as well as their performance benefits. An objective of this paper is to show that, indeed, many of these analytic models can be formulated, quite simply and elegantly, by adapting the Shockley diode equation to account for the new physics governing the modern transistors. As examples, we discuss the models for the collector current, Early voltage, and base transit time of vertical SiGe-base bipolar transistors and symmetric lateral bipolar transistors on SOI.

5.2 Adaptation of Shockley Diode Equation to Modern Bipolar Transistors

It should be noted that the factor n_i^2 in Eq. (5.2) is a property of the semiconductor used for fabrication of the bipolar transistor. It is given by

$$n_i^2 = N_c N_v \exp\left(\frac{-E_g}{kT}\right), \tag{5.3}$$

where N_c and N_v are the effective density of states of the conduction band and the valence band, respectively, and E_g is the bandgap energy. That is, the $p_p(x=0)n_p(x=0)$ product varies exponentially with the bandgap of the quasineutral p-region.

As the technology of bipolar transistors evolved from 1948 to the present time, the physics behind the increase in the collector-current density, at a fixed V_{BE}, for the advanced transistors can be traced to a reduction of the bandgap in the base region, as will be explained later. The Shockley diode equation can be adapted to model the collector current of these modern transistors simply by replacing E_g with an effective base-region bandgap that is consistent with the physics of the corresponding transistor.

5.2.1 Heavy-Doping Effect

Prior to the 1970s, areal dimensions of transistor were measured in mils (0.1 mil = 2.54 μm). The emitter area of a typical bipolar transistor was large, usually larger than 10 μm^2, and the base doping concentration for these large-emitter transistors was typically less than 1×10^{17} cm^{-3}. At such low doping levels, the dopant impurities have little effect on the bandgap, and the Shockley diode equation was used without any modification. The value of n_i for Si is about 1×10^{10} cm^{-3} at room temperature.

However, modern bipolar transistors all have thin but heavily doped base regions, with base doping concentration, N_B, typically in the 1×10^{18}–1×10^{19} cm^{-3} range. In such heavily doped regions, the orbitals associated with the impurity atoms overlap to form a band of finite width, causing the bandgap to appear narrower than in lightly doped regions. This heavy-doping effect is included in modeling the minority-carrier transport in modern bipolar transistors by defining an effective intrinsic-carrier density, n_{ie}, and lumping all the heavy-doping effects into a parameter called apparent bandgap narrowing, ΔE_g, given by

$$p_0(\Delta E_g, x)n_0(\Delta E_g, x) \equiv n_{ie}^2(x)$$
$$= n_i^2 \exp\left[\frac{\Delta E_g(x)}{kT}\right], \tag{5.4}$$

where $p_0(\Delta E_g, x)n_0(\Delta E_g, x)$ is the $p_0 n_0$ product at location x where the bandgap is smaller by an amount ΔE_g due to heavy-doping effect [3]. Heavy-doping effect increases with doping concentration. For $N_B = 10^{17}$ cm^{-3}, ΔE_g is negligible, usually taken as 0 meV. For $N_B = 10^{19}$ cm^{-3}, ΔE_g is about 65 meV. The model for the collector current in these bipolar transistors is the same as that obtained by Shockley, except that n_i in the Shockley diode equation is replaced by n_{ie}.

5.2.2 SiGe-Base Bipolar Transistors

The bandgap energy of Ge (≈0.66 eV) is significantly smaller than that of Si (≈1.12 eV). When Ge is incorporated into Si, the Si bandgap becomes smaller. By incorporating Ge into the base region of a Si bipolar transistor, the bandgap of the base region, and hence the accompanied device characteristics, can be modified. The presence of Ge in the base region can be accounted for simply by extending the parameter n_{ie} in Eq. (5.4) to include bandgap narrowing caused by the Ge distribution [4], i.e.

$$n_{ie}^2(SiGe,x) = n_{ie}^2(x)\exp\left[\frac{\Delta E_{g,SiGe}(x)}{KT}\right], \tag{5.5}$$

where $n_{ie}(x)$ is the effective intrinsic-carrier density in Eq. (5.4), due to heavy-doping effect alone, and $\Delta E_{g,SiGe}(x)$ is the local bandgap narrowing in the base due to the presence of Ge. The model for the collector current in a SiGe-base bipolar transistors is the same as that obtained by Shockley, except that n_i in the Shockley diode equation is replaced by $n_{ie}(SiGe, x)$.

5.3 Modern Bipolar Transistors Structures

Figure 5.2 shows schematically two modern bipolar transistor structures. Figure 5.2a is the schematic of a vertical n–p–n bipolar transistor. A typical vertical bipolar transistor has a collector, the n pedestal collector directly beneath the emitter in the figure, doping concentration that is significantly lower than the base-region doping concentration, i.e. $N_C \ll N_B$. The n$^+$ subcollector, which has a doping concentration much larger than N_C, acts as an extension of the reach-through connecting the collector to the contact C; it does not determine the maximum speed of the transistor. The most advanced vertical bipolar transistors today have a SiGe base, with a graded Ge distribution causing a drift field in the base to greatly enhance the device performance.

Figure 5.2b is the schematic of symmetric lateral bipolar transistors on SOI, suggesting the possibility of integrating n–p–n and p–n–p on the same chip. A symmetrical lateral transistor, by symmetry, has $N_C = N_E \gg N_B$. As of today, there has been no report of demonstration of lateral bipolar transistors having a graded Ge distribution in the base region.

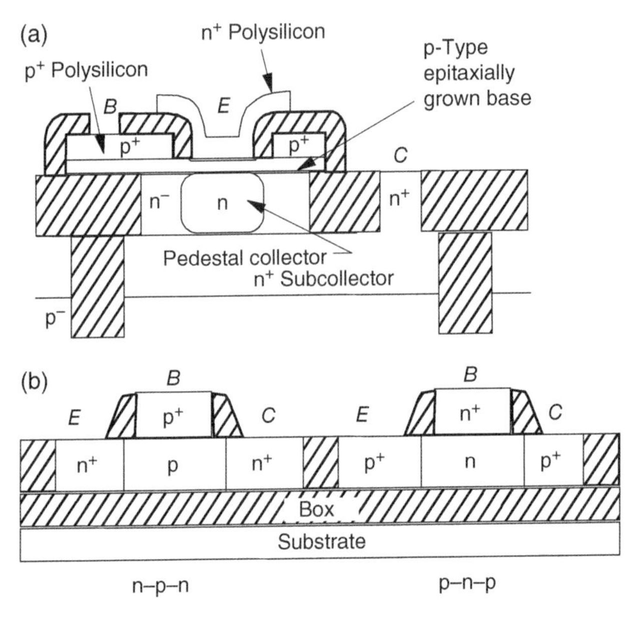

Figure 5.2 (a) Schematic of a modern vertical n–p–n bipolar transistor structure. The epitaxially grown p-type base could be a SiGe alloy. (b) Schematic of symmetric lateral bipolar transistors on SOI. n–p–n and p–n–p transistors can be integrated in a manner analogous to SOI CMOS.

Let us first focus on the vertical transistor. In operation, the device speed increases with collector current, reaching a peak, and then falls off rapidly as the collector current is increased further. The falloff of speed at high collector current is due to base-widening effect [5]. It occurs when the n-type collector doping concentration is not high enough to support the electron current injected from the n^+ emitter into the p-type base. To avoid base-widening effect, bipolar circuits using vertical transistors are designed to operate at V_{BE} values where the injected electron density in the base is comparable to or smaller than the collector doping concentration. With $N_C \ll N_B$, vertical bipolar transistors are therefore designed to operate at minority-carrier densities in the base region much smaller than N_B to avoid speed degradation due to base-widening effect. That is, the operation of a vertical bipolar transistor is limited to low-injection levels of minority carriers in the base.

For a symmetric lateral bipolar transistor, with $N_C = N_E \gg N_B$, there is no base-widening effect. As a result, symmetric lateral bipolar transistors can be operated at arbitrarily large minority-carrier densities in the base region without speed degradation. That is, the operation of a symmetric lateral bipolar transistor is not limited to low-injection levels [6].

5.3.1 Shockley Diode Equation for Arbitrary Injection Levels

The left hand side (LHS) of Eq. (5.2) can be expanded as

$$p_p(x=0)n_p(x=0) = \left[p_{p0}(0) + \Delta p_p(0)\right]\left[n_{p0}(0) + \Delta n_p(0)\right]$$
$$\approx \left[p_{p0}(0) + \Delta n_p(0)\right]\Delta n_p(0), \tag{5.6}$$

where in the second line we have used the fact that n_{p0} is negligible compared to Δn_p when the emitter–base diode is forward biased in device operation, and quasineutrality means $\Delta p_p = \Delta n_p$. Therefore, the Shockley diode equation for a modern bipolar transistor, from Eq. (5.2), is

$$\left[p_{p0}(0) + \Delta n_p(0)\right]\Delta n_p(0) = n_i^2 \exp\left(\frac{qV_{BE}}{kT}\right), \tag{5.7}$$

which is a quadratic equation in $\Delta n_p(0)$, with a solution [6]

$$\Delta n_p(0) = \frac{p_{p0}(0)}{2}\left[\sqrt{1 + \frac{4n_i^2}{p_{p0}^2(0)}\exp\left(\frac{qV_{BE}}{kT}\right)} - 1\right]. \tag{5.8}$$

Since no assumption about the magnitude of $\Delta n_p(0)$ is made, Eq. (5.8) is valid for arbitrary levels of Δn_p. At low V_{BE}, $\Delta n_p(0)$ increases as $\exp(qV_{BE}/kT)$. However, at large V_{BE}, it increases as $\exp(qV_{BE}/2kT)$. Equation (5.8), with n_i replaced by n_{ie} given by Eq. (5.4), should be used for modeling symmetric lateral bipolar transistors.

5.3.2 Shockley Diode Equation for Vertical Bipolar Transistors

As explained above, vertical bipolar transistors are designed to operate at injection levels where Δn_p is comparable to or smaller than $N_C \ll N_B$, which means the second term within the square root in Eq. (5.8) is small compared to unity. In this case, (5.8) is reduced to

$$\Delta n_p(0) \approx \frac{n_i^2}{p_{p0}(0)}\exp\left(\frac{qV_{BE}}{kT}\right). \tag{5.9}$$

This is the familiar approximation of the Shockley diode equation in the literature, valid for low-injection levels, before base-widening effect kicks in. It leads to a collector current that increases with V_{BE} as $\exp(qV_{BE}/kT)$. To model a SiGe-base vertical bipolar transistor, one simply replaces n_i by $n_{ie}(SiGe, x)$ given by Eq. (5.5).

5.4 Analytic Models for Modern Bipolar Transistors

As discussed in Section 5.1, the Shockley diode equation gives the boundary condition needed for determining $n_p(x)$, the electron density in the p-type base region. This in turn enables the hole density in the p-type base region, $p_p(x)$, to be determined through the fact that $p_p(x) = p_{p0}(x) + \Delta n_p(x)$. Since Shockley's 1949 paper [1], analytic models for many key parameters of a bipolar transistor have been developed. These analytic models typically involve the integration across the quasineutral base region, from $x = 0$ to $x = W_B$, of some function of $p_p(x)$ and $n_{ie}^2(x)$ for a thin-base bipolar transistor, or some function of $p_p(x)$ and $n_{ie}^2(SiGe, x)$ for a SiGe-base vertical bipolar transistor. For some base doping profiles, e.g. a boxlike doping profile or a Gaussian doping profile, and some Ge distributions, e.g. a triangular Ge distribution or a trapezoidal Ge distribution, the integrals involved can be evaluated explicitly, enabling easy device design optimization. Here, we illustrate several of these analytic models.

5.4.1 Analytic Model for Collector Current Density

Moll and Ross [7] showed that the collector current density J_C of an n–p–n bipolar transistor can be calculated from the expression

$$J_C = \frac{q\exp(qV_{BE}/kT)}{\int_0^{W_B}\left[p_p(x)/D_n(x)n_{ie}^2(x)\right]dx}, \tag{5.10}$$

where $D_n(x)$ is the electron diffusion coefficient in the p-type base region, and $n_{ie}(x)$ is the effective intrinsic-carrier density given by Eq. (5.4). For a uniformly doped base region, such as the base region of a symmetric lateral bipolar transistor on SOI, illustrated in Figure 5.2b, both D_n and n_{ie} are independent of x.

For a vertical SiGe-base bipolar transistor, $p_p(x)$ can be replaced by the base doping distribution $N_B(x)$ since the transistor is operated at low-injection level where $p_p(x)$ is essentially the same as $N_B(x)$. However, $n_{ie}(x)$ should be replaced by $n_{ie}(SiGe, x)$ given by Eq. (5.5).

5.4.2 Analytic Model for Early Voltage

When a bipolar transistor is operated at a fixed V_{BE}, it has a fixed base current that is independent of the collector-emitter voltage V_{CE}. However, the collector current I_C increases as V_{CE} is increased. These output characteristics are illustrated in Figure 5.3. For circuit modeling purposes, I_C in the nonsaturation region, where $V_{CE} > V_{BE}$, is often assumed to depend linearly on V_{CE}. The collector voltage at which the linearly extrapolated I_C reaches zero is denoted by $-V_A$. Early [8] showed that V_A is given by the expression

$$V_A + V_{CE} = \frac{qD_n(W_B)n_{ie}^2(W_B)}{C_{dBC}}\int_0^{W_B}\frac{p_p(x)}{D_n(x)n_{ie}^2(x)}dx, \tag{5.11}$$

where C_{dBC} is base–collector junction capacitance per unit area. V_A is referred to as the Early voltage.

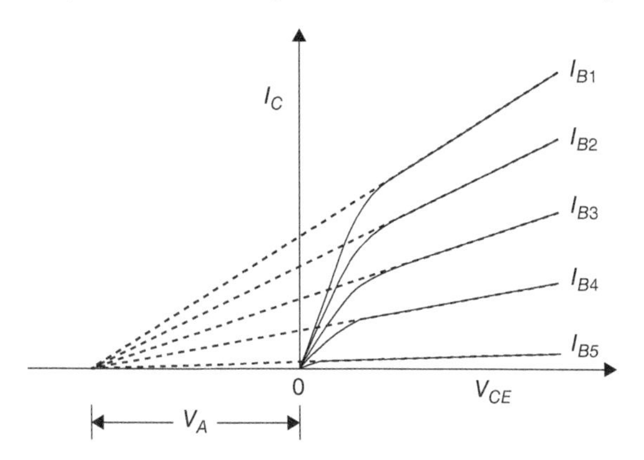

Figure 5.3 Schematic illustrating the approximately linear dependence of I_C on V_{CE}. The linearly extrapolated I_C intersects the V_{CE}-axis at $-V_A$.

To calculate the Early voltage of a vertical SiGe-base bipolar transistor from Eq. (5.11), $p_p(x)$ can be replaced by $N_B(x)$ and $n_{ie}(x)$ should be replaced by $n_{ie}(SiGe, x)$ given by Eq. (5.5).

5.4.3 Analytic Model for Base Transit Time

Kroemer [4] showed that the base transit time t_B of a bipolar transistor is given by the general expression

$$t_B = \int_0^{W_B} \frac{n_{ie}^2(x)}{p_p(x)} \int_x^{W_B} \frac{p_p(x')}{D_n(x')n_{ie}^2(x')} dx'dx. \tag{5.12}$$

For a transistor having a constant base doping concentration and bandgap, it gives the familiar result of $t_B = W_B^2 / 2D_n$.

For a vertical SiGe-base bipolar transistor, $p_p(x)$ can be replaced by $N_B(x)$ and $n_{ie}(x)$ should be replaced by $n_{ie}(SiGe, x)$ given by Eq. (5.5). The result is

$$t_B(SiGe) = \int_0^{W_B} \frac{n_{ie}^2(SiGe,x)}{N_B(x)} \int_x^{W_B} \frac{N_B(x')}{D_n(SiGe,x')n_{ie}^2(SiGe,x')} dx'dx. \tag{5.13}$$

Equation (5.13) can be used to study the sensitivity of the base transit time of a vertical SiGe-base bipolar transistor to the details of the Ge distribution profile.

5.5 Discussion

As Shockley stated in his 1949 paper, "The p–n–p transistor has the interesting feature of being **calculable** to a high degree. ..." The reason the transistor features being calculable to a high degree is because the minority-carrier current flow in the base is one-dimensional, which makes the mathematics involved relatively simple. In the case of an n–p–n bipolar transistor, the starting point for developing an analytic model is solving for $n_p(x)$ from the minority-carrier diffusion equation, using the Shockley diode equation, Eq. (5.2), as a boundary condition. The n_i^2 factor in the Shockley diode equation is adapted to account for the new physics, namely heavy-doping effect and base-bandgap engineering using Ge, in modern bipolar transistors. The key device parameters, such as collector current, Early voltage, and base transit time, are calculable from analytic equations, provided that the base dopant distribution, and the Ge distribution in the case of a SiGe-base transistor, are known.

It is a CMOS world. I spent much time in my career dealing with both CMOS device models and bipolar device models. I must say the bipolar device models are more elegant and mathematically much simpler than the CMOS models.

References

[1] Shockley, W. (1949). The theory of p–n junctions in semiconductors and p–n junction transistors. *Bell Syst. Tech. J.* 28: 435–489.

[2] Shockley, W. (1950). *Electrons and Holes in Semiconductors*. Princeton, NJ: D. Van Nostrand.

[3] Slotboom, J.W. and de Graaff, H.D. (1976). Measurements of bandgap narrowing in Si bipolar transistors. *Solid State Electron.* 19: 857–862.

[4] Kroemer, H. (1985). Two integral relations pertaining to the electron transport through a bipolar transistor with a non-uniform energy gap in the base region. *Solid State Electron.* 28: 1101–1103.

[5] Kirk, C.T. Jr. (1962). A theory of transistor cutoff frequency (*fT*) falloff at high current densities. *IEEE Trans. Electron Devices* ED-9: 164–174.

[6] Taur, Y. and Ning, T.H. (2022). *Fundamentals of Modern VLSI Device*, 3e. Cambridge: Cambridge University Press.

[7] Moll, J.L. and Ross, I.M. (1956). The dependence of transistor parameters on base resistivity. *Proc. IRE* 44: 72–78.

[8] Early, J.M. (1952). Effects of space-charge layer widening in junction transistors. *Proc. IRE* 40: 1401–1406.

Chapter 6

Junction-Less Field Effect Transistors

The First Transistor to be Conceptualized

Mamidala Jagadesh Kumar[1] and Shubham Sahay[2]

[1] Department of Electrical Engineering, Indian Institute of Technology, Delhi, India
[2] Department of Electrical Engineering, Indian Institute of Technology, Kanpur, India

6.1 Introduction

In this era of big data and internet-of-things (IoT), the smart devices have revolutionized every facet of human endeavor. The rapid advancements in the fundamental component of the functional circuit of these smart devices, known as transistors, has driven the smart revolution and influenced our everyday lives. Downscaling the transistor dimensions enabled the device designers to realize a significantly improved speed, reduced dynamic power dissipation, and more functionality per chip for several generations. However, the scaling process reached its fundamental limits as the transistors were scaled to the short-channel regime owing to the significantly increased static power dissipation and the complex fabrication process. Furthermore, scaling the dimensions of the workhorse of the semiconductor industry, the metal-oxide-semiconductor field-effect transistor (MOSFET), requires fabrication techniques for realizing ultra-steep doping profiles at the metallurgical junction along with a high dopant activation [1]. However, this requirement presents a complex constraint on the thermal budget: realizing ultra-steep junctions with complementary doping requires a low thermal exposure while high-temperature annealing process is indispensable for dopant activation [2]. Therefore, the design of incessantly scaled transistors with ultrahigh performance would be simplified if the origin of this conflicting requirement, i.e. the metallurgical junctions, is somehow eliminated from the transistors.

Interestingly, the concept of a transistor without any metallurgical junction was formulated and patented by Julius Edgar Lilienfeld in 1930 (Figure 6.1) [3] even before the bipolar junction transistors

75th Anniversary of the Transistor, First Edition. Edited by Arokia Nathan, Samar K. Saha, and Ravi M. Todi.

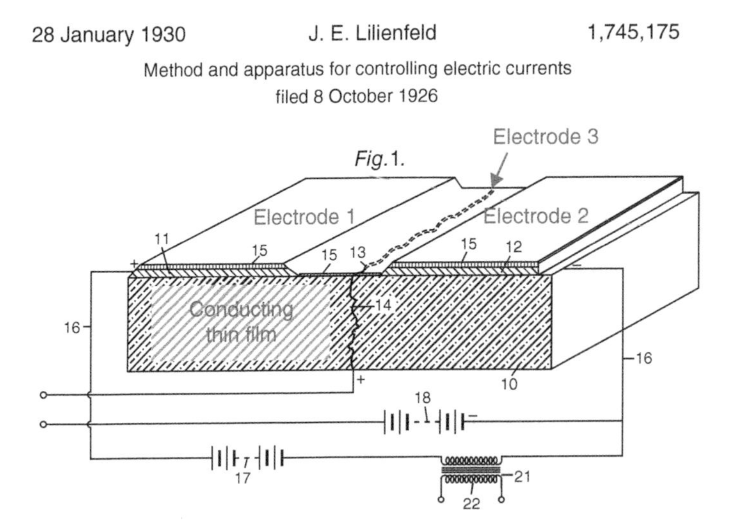

Figure 6.1 The device proposed by Lilienfeld to modulate current between two electrodes with the aid of voltage applied on the third electrode. Source: Adapted from [3].

(BJTs) were invented by Shockley [4]. As shown in Figure 6.1, Lilienfeld conceptualized an apparatus consisting of a thin conducting film of a compound of copper and sulfur deposited over three metal electrodes. Lilienfeld proposed that the electric field generated by the application of a voltage on electrode 3 may penetrate the ultrathin conducting film and modulate its conductivity and the current flowing between electrode 1 and electrode 2 similar to the modern junction (J) FETs.

Furthermore, in 1933, Lilienfeld also conceptualized and patented the transistor action in a metal-oxide-semiconductor (MOS) configuration consisting of an ultrathin conducting film of copper sulfide (semiconductor) deposited over an ultrathin insulating compound (aluminum oxide) of the metal electrode (aluminum) [5]. Lilienfeld postulated that the electric field through the insulator may be used to modulate the conductivity of the semiconducting thin film like the modern-day MOSFETs. However, the Lilienfeld transistor did not contain any metallurgical junction and may be perceived as a gate-controlled resistor. In the quest for realizing ultra-scaled transistors with improved performance using simplified fabrication process by eliminating the metallurgical junctions, Lilienfeld's seminal works were revisited and nanowire transistors without any junction, popularly known as the junctionless FETs were experimentally demonstrated in 2010 [6]. Now, we will discuss the structure of the JLFETs and how its unique architecture leads to an altogether different operating mechanism as compared to the MOSFETs.

6.2 Structure and Operation

The 3D view of an n-type double gate JLFET is shown in Figure 6.2. The JLFET consists of a uniformly doped semiconductor film with a gate stack to modulate the conductivity and the current between the source and drain electrodes. Since the source, channel, and drain regions are uniformly doped with same type and same concentration of dopant atoms in JLFETs, it does not contain any metallurgical junctions unlike the MOSFETs [7]. The absence of metallurgical junctions not only reduces the fabrication cost [8] but also relaxes the constraint on the thermal budget. This facilitates the fabrication of JLFETs even in the back-end-of-line (BEOL) [9] and provides flexibility to the designers while selecting materials for the gate stack of JLFETs. Furthermore, the semiconductor is heavily doped to realize Ohmic contacts at the source and drain electrodes.

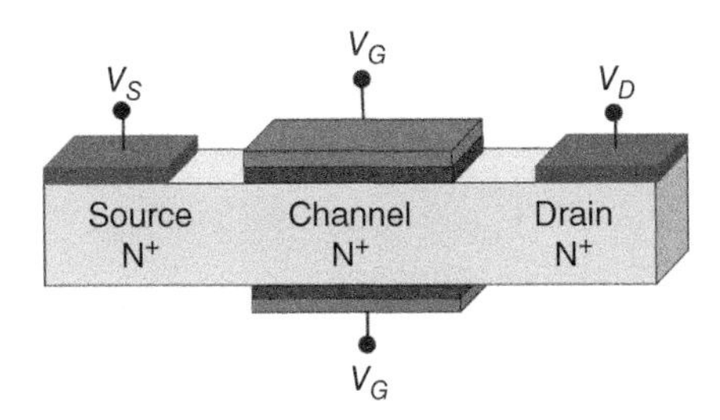

Figure 6.2 The structure of a n-type double-gate junction-less FET. Source: Sahay and Kumar [1]/ John Wiley & Sons.

Application of a gate voltage changes the majority carrier concentration in the channel region and modulates the drain current. To realize a significantly reduced leakage current between the drain and source electrodes in the OFF-state ($V_{GS} = 0\,V$, $V_{DS} = V_{DD}$) of JLFETs, the channel region must be completely depleted of the carriers. This volume depletion of the channel region in the OFF-state of JLFETs (Figure 6.3a) is achieved via a proper design of the gate electrode and the channel region: (i) a gate electrode with a high work function (≥5.1 eV) is chosen for n-type JLFETs while a gate metal with a low work function (≤3.9 eV) is selected for p-type JLFETs [7]. A large work function difference between the gate metal and the semiconductor (Φ_{MS}) results in a high inbuilt electric field emanating from the semiconductor to the gate electrode even when no external voltage is applied on the gate. (ii) an ultrathin (~6–10 nm) channel is used to ensure that this inbuilt electric field penetrates the entire semiconductor thickness and depletes it of the majority carriers in the OFF-state [6], and (iii) a moderately high doping (~$10^{19}\,cm^{-3}$) is chosen for the semiconductor film. While a lightly doped semiconductor leads to Schottky contacts at the source and drain electrodes and degrades the performance, a heavily doped semiconductor exhibits a large work function, minimizes Φ_{MS}, and leads to a significant reduction in the inbuilt electric field responsible for realizing volume depletion in the OFF-state of JLFETs [1].

As the gate voltage increases, the inbuilt electric field decreases leading to a reduction in the gate-induced depletion region widths. The gate voltage at which the depletion regions induced by the front gate and the back gate just contact each other at the center of the channel as shown in Figure 6.3b is defined as the threshold voltage (V_{th}) of JLFETs [6]. It may be noted that this definition of threshold voltage is significantly different from MOSFETs where the threshold voltage is defined as the gate voltage at which appreciable minority carriers accumulate in the channel region and the concentration of the inversion layer becomes same as the bulk concentration [1].

As the gate voltage is increased beyond V_{th}, the gate-induced depletion region widths shrink further and an undepleted (neutral) region is uncovered at the center of the channel as shown in Figure 6.3c. The heavily doped neutral region connects the source and drain regions and leads to a large drain current in this partial depletion mode of operation [6]. It may be noted that the drain current flows through the neutral regions in the bulk (center) of the channel region in the JLFETs unlike MOSFETs where the current flows through the inversion layer at the surface.

The neutral region continues to expand while the depletion regions shrink as the gate voltage is increased further. The gate voltage at which the depletion regions vanish, and the entire channel region becomes neutral (Figure 6.3d) is known as the flat band voltage [6]. In this flat-band mode of operation, the entire channel conducts current between the source and drain regions and the electric field along the channel thickness (longitudinal direction) is minimum as shown in Figure 6.4. Further increasing the gate voltage leads to accumulation of majority carriers at the channel surface and the surface concentration becomes larger than the bulk concentration (Figure 6.3e). In this accumulation mode of operation, the drain current is dominated by the majority carriers flowing at the surface. However, the JLFETs are typically designed such that they operate in the flat-band condition in the ON-state ($V_{GS} = V_{DS} = V_{DD}$) to

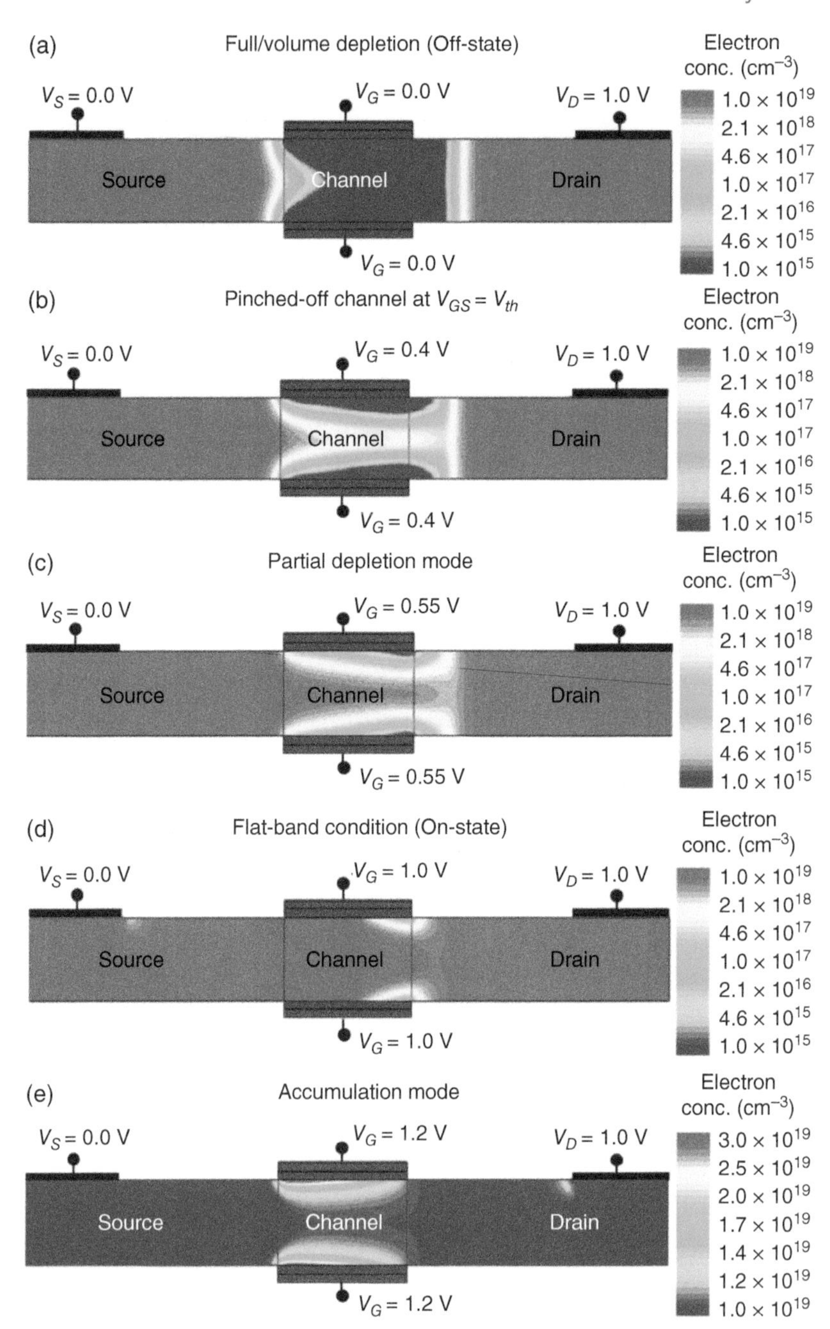

Figure 6.3 Majority carrier (electron) concentration for different modes of operation of a n-type double-gate junction-less FET: (a) volume depletion mode, (b) at threshold voltage, (c) partial depletion mode, (d) at flat-band voltage, and (e) accumulation mode. Source: Sahay and Kumar [1]/John Wiley & Sons.

minimize the surface scattering-induced mobility degradation and to reduce the influence of the interface traps and defects present at the surface [1].

We observe two major differences between the operating mechanism of the JLFETs and the MOSFETs: (i) while the accumulation and flat band conditions are achieved in the MOSFETs below the threshold voltage, the JLFETs are biased in the flat band and accumulation mode of operation above the threshold

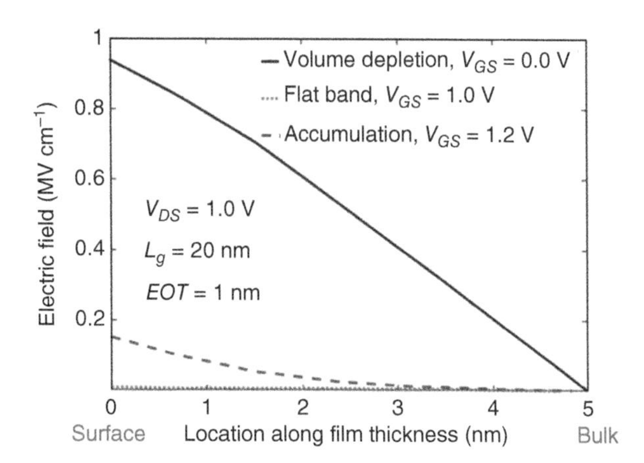

Figure 6.4 Electric field profile for different modes of operation of a JLFET.

voltage. Moreover, (ii) while the drain current of the MOSFET is dominated by the minority carriers in the inversion layer at the surface, the majority carriers flowing in the bulk of the channel region are responsible for current conduction in JLFETs. This bulk conduction property of JLFETs in the ON-state (flat band condition with minimum longitudinal electric field) renders several advantages with respect to the factors that affect the performance metrics which we shall discuss in the next section.

6.3 Salient Features of JLFETs

6.3.1 Reliability

The incessant scaling of the gate oxide in MOSFETs has led to significantly increased gate stress resulting in reliability issues such as increased threshold voltage instability [10]. The extended stress-measure-stress tests indicate that the JLFETs exhibit a significantly low bias temperature instability (BTI), transconductance degradation and hot carrier degradation as compared to the MOSFETs [11]. This is attributed to the bulk conduction and the low electric field in the gate oxide in the ON-state of JLFETs due to the flat-band regime of operation as shown in Figure 6.4. Therefore, as compared to the MOSFETs, the JLFETs exhibit more resiliency against reliability issues.

6.3.2 Carrier Ballisticity

The ultra-short channel MOSFETs exhibit ballistic transport mechanism as their channel lengths are scaled below the mean-free path of the carriers [12]. Under these conditions, the carriers move from the source to the drain region with an extremely high velocity without encountering any scattering event. The finite source-to-channel barrier height in MOSFETs in the ON-state (Figure 6.5a) results in back-scattering of electrons via the optical phonons [13]. However, the smooth transition of energy bands from source to drain region in JLFETs as shown in Figure 6.5a eliminates the possibility of backscattering and increases their ON-state current in the ballistic regime.

6.3.3 Non-Silicon High Mobility Channel Materials

The high mobility materials such as germanium, III–V materials, etc., do not have a stable native oxide and exhibit a significantly high trap density and interfacial defects with the CMOS-compatible insulators. Therefore, the MOSFETs based on high mobility materials exhibit a degraded performance and a poor reliability [1]. The bulk conduction reduces the influence of the traps and interface defects on the

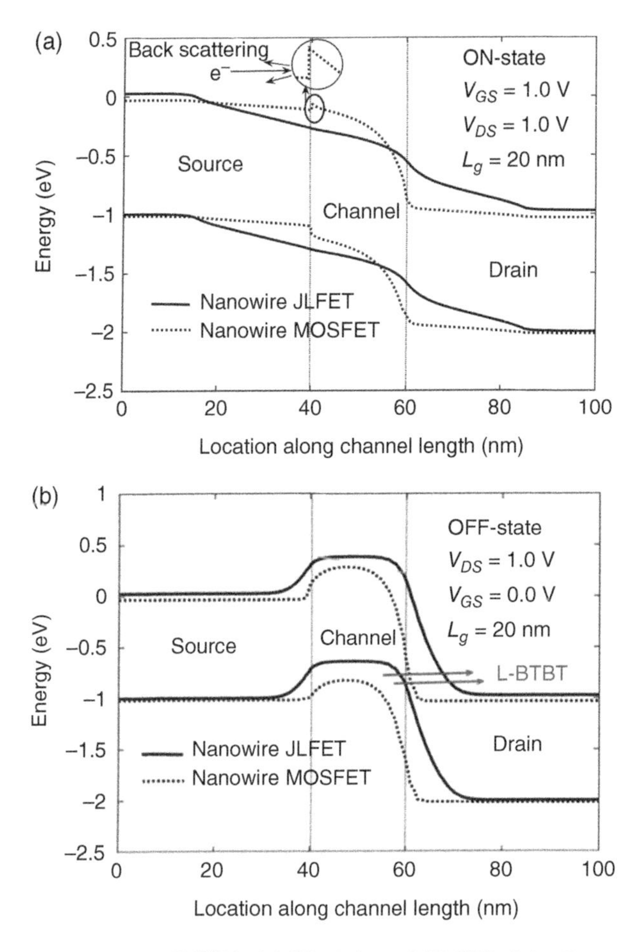

Figure 6.5 Energy band profiles of a JLFET in (a) ON-state and (b) OFF-state. Source: Sahay and Kumar [1]/ John Wiley & Sons.

carriers and the low electric field in the ON-state (flat band as shown in Figure 6.4) enhances the reliability facilitating the application of the junction-less architecture for transistors based on non-silicon high mobility materials. Therefore, JLFETs based on germanium, InGaAs, GaNs, etc., were extensively investigated [14–18].

6.3.4 Low Frequency Noise

The low frequency noise or 1/*f* noise originates due to the random trapping and de-trapping of the carriers by the traps present at the semiconductor/oxide interface and degrades the radio frequency (RF) performance and the signal-to-noise ratio (SNR) of sensors [1]. The bulk conduction mechanism in JLFETs ensures a limited interaction of the majority carriers with the interface traps resulting in a significantly reduced noise spectral density as compared to the MOSFETs [19, 20]. Therefore, the JLFETs exhibit an improved SNR and a better RF performance [1].

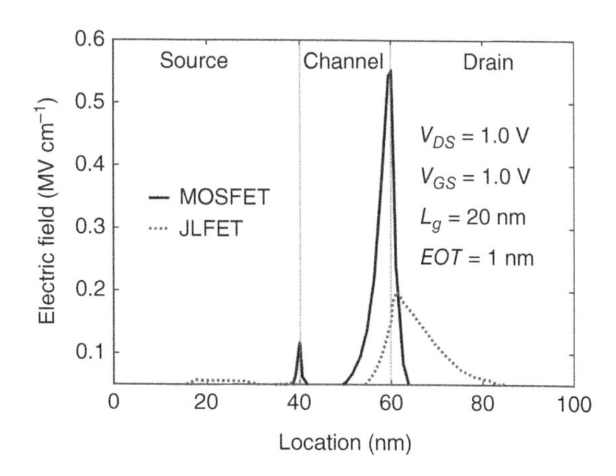

Figure 6.6 Lateral electric field profile of the JLFETs and the MOSFETs. Source: Sahay and Kumar [1]/ John Wiley & Sons.

6.3.5 Short-Channel Effects

The short-channel effects such as threshold voltage roll-off, drain-induced barrier lowering (DIBL), hot carrier, etc., significantly degrade the performance of the MOSFETs and restrict their aggressive scaling. While the absence of metallurgical junctions in JLFETs eliminates the possibility of channel charge sharing by the source-channel/channel-drain depletion regions mitigating the threshold voltage roll-off, the lower lateral electric field at the channel–drain interface in JLFETs (Figure 6.6) leads to a reduced coupling of the drain field at the source–channel interface resulting in a significantly reduced DIBL in JLFETs as compared to the MOSFETs [21, 22]. Moreover, JLFETs also exhibit a significantly improved reliability against the hot-carrier effects due to reduction in longitudinal (Figure 6.4) and lateral (Figure 6.6) electric field.

6.3.6 Mobility

The inversion layer carriers in the MOSFETs exhibit a significant mobility degradation due to the high longitudinal electric field and surface scattering [1]. Although the bulk conduction reduces surface scattering in the volume/partial depletion and flat band regime of operation in JLFETs, the majority carriers still suffer from Coulomb scattering due to the ionized impurities [23, 24]. Moreover, the carriers are also subjugated to surface scattering in the accumulation mode of operation. However, the JLFETs exhibit a high mobility in the weak inversion mode of operation due to the shielding effect [24]. The ionized impurities which form the scattering centers in JLFETs are of opposite polarity to the majority carriers and attract them. The majority carriers accumulated around the ionized impurities shield the other carriers from getting scattered and improve the mobility of JLFETs.

6.3.7 Temperature Dependence and Cryogenic Operation

An increase in the operating temperature of JLFETs leads to a larger subthreshold leakage current. However, the carrier mobility also decreases due to increased phonon scattering and nullifies the increment in the subthreshold current [25]. Therefore, the short-channel JLFETs exhibit a zero temperature coefficient (ZTC) point in the transfer characteristics which represents the gate voltage at which the impact of high thermal energy of the carriers is compensated by the reduced mobility. Furthermore, due to the presence of the flat band capacitance in the ON-state of JLFETs, they also exhibit a ZTC point in the capacitance-voltage characteristics unlike MOSFETs [26].

The need to operate the control circuitry for quantum computers at sub-77 K has attracted significant interest in cryogenic operation of CMOS devices. The JLFETs exhibit an anomalous trend for mobility at reduced temperatures: while the mobility degrades till 15 K due to the Coulomb scattering, it starts to increase due to dopant inactivation as the temperature is reduced below 15 K [27]. Furthermore, the drain conductance of JLFETs also shows an unusual behavior: it increases with the drain voltage and then starts decreasing [28]. This is attributed to the formation of Schottky contacts due to dopant freezing at 4 K.

Moreover, the JLFETs also exhibit transconductance oscillations at low temperatures for low drain voltages. The ionized dopant atoms may act like quantum dots and provide a tunneling path to the carriers leading to these oscillations. Furthermore, the band edge states may also lead to similar oscillations in the transconductance [29].

6.3.8 Impact-Ionization at Lower Voltages

The high semiconductor film doping in JLFETs leads to a reduced band gap which results in an increased impact ionization rate. Moreover, the impact-ionization takes place within the bulk of JLFETs which further increases the impact ionization rates. Furthermore, as shown in Figure 6.6, the lateral electric field responsible for impact ionization in JLFETs peaks inside the drain region and remains high for a larger portion of the drain region accelerating the carriers to attain sufficient kinetic energy for inducing impact ionization unlike MOSFETs where the electric field peaks the channel–drain junction. Therefore, the JLFETs exhibit impact ionization-induced steep slope in the output characteristics at a smaller drain voltage as compared to the MOSFETs [30, 31].

The voltage required for achieving the steep slope may be further reduced by employing independent gate technique or misaligned gates [32, 33]. Furthermore, the JLFETs do not show a sharp "kink" effect due to the dynamic nature of impact ionization owing to the large recombination of the impact ionization-generated carriers with the majority carriers flowing in the bulk. The JLFETs also exhibit single transistor latch-up phenomena and fail to turn OFF if not designed properly [34].

6.4 Challenges for JLFETs

In this section, we will discuss about the major challenges which inhibit widespread application of JLFETs.

6.4.1 Source/Drain Series Resistance and Fabrication Issue

The moderately highly doped source/drain regions offer a high source/drain series resistance and reduce the ON-state current of JLFETs as compared to the MOSFETs as shown in Figure 6.7. Furthermore, it is also difficult to realize Ohmic contacts on such source/drain regions [1]. Moreover, realizing uniform ultrathin semiconductor films across wafers is a technological challenge. Also, the gate electrodes with high work function such as platinum are expensive and difficult to pattern [1].

6.4.2 Increased Sensitivity to Variations

Since the JLFETs employ a heavily doped semiconductor film, they exhibit a larger variation in the threshold voltage and drain current due to random dopant fluctuation as compared to the MOSFETs [35, 36]. Moreover, volume depletion in JLFETs depends critically on the semiconductor

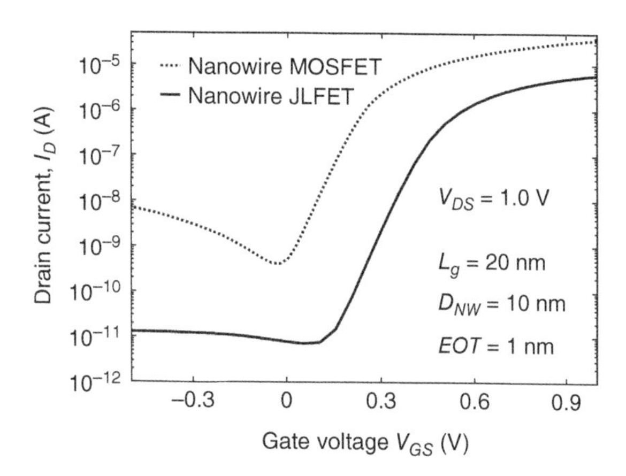

Figure 6.7 Transfer characteristics of the JLFETs and the MOSFETs. Source: Sahay and Kumar [1]/ John Wiley & Sons.

film thickness and the work function of the gate electrode. Therefore, JLFETs also show a higher sensitivity to the variations in the channel thickness, work function variations of polycrystalline metals such as TiN, and line edge roughness [36–39]. Therefore, mismatch effects are larger in JLFETs which is detrimental for analog circuit and memory design.

6.4.3 Parasitic BJT Action in the OFF-State

Although an efficient design and gate control in JLFETs leads to volume depletion, it also results in overlap of the valence band of the channel region with the conduction band of the drain region leading to lateral band-to-band tunneling (L-BTBT) as shown in Figure 6.5b [40–42]. This L-BTBT leads to hole accumulation in the channel and transforms it into a p-type region leading to the formation of a parasitic BJT with the p-type channel region as the base and source/drain regions as the emitter/collector, respectively, of the parasitic BJT. The gate loses control over the channel region once the parasitic BJT is triggered (V_{GS}<0 as shown in Figure 6.7) [40–42]. Moreover, the tunneling current which is essentially the base current of the parasitic BJT gets amplified by the large current gain (owing to the ultra-scaled channel length/base width) to yield the collector (drain) current. This results in a significantly high leakage current in JLFETs increasing their static power dissipation and restricting their aggressive scaling.

Although these challenges hinder the adoption of JLFETs as the mainstream transistor technology, the unique attributes of JLFETs such as low temperature processing, bulk conduction, etc., also open up a new horizon for efficient design of memories, biosensors, 3D thin film transistors (TFTs), and BEOL-compatible transistors which we shall discuss next.

6.5 Unconventional Applications of JL Architecture

6.5.1 Memory

The presence of external capacitor in conventional 1T-1C DRAMs limits their scalability. Therefore, 1T capacitorless DRAMs based on impact ionization-generated carrier storage in the floating body were proposed [43]. However, MOSFET-based 1T DRAMs consume a large energy since impact ionization is triggered at high drain voltages. The snapback characteristics observed in the impact ionization-dominated regime of JLFETs at low drain voltages may be harnessed for low power capacitorless DRAM application [44].

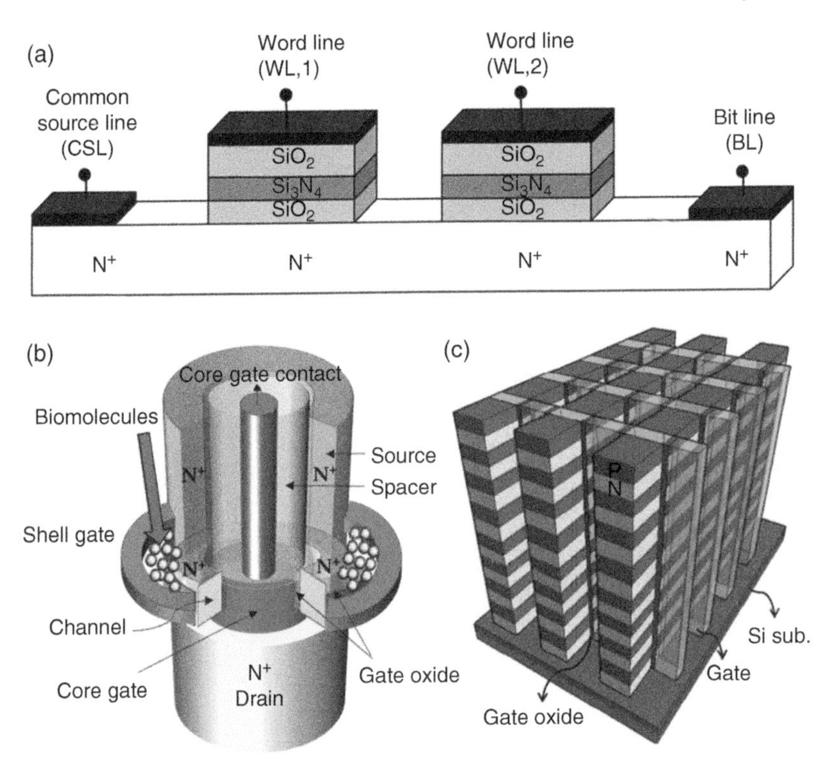

Figure 6.8 Unconventional application of junction-less architecture: (a) JL-NAND flash memory, (b) JL-biosensor, and (c) 3D stacked JL-TFTs. Source: Sahay and Kumar [1]/John Wiley & Sons.

Furthermore, the conventional 3D NAND flash memories essentially utilize the JL architecture with electrostatically induced virtual source/drain regions to reduce the cost and fabrication complexity [45, 46]. However, the undoped polysilicon string also leads to a small read current and reduced memory window. The read current can be increased and the endurance, memory window, and the distribution of the program/erase threshold voltages could be further improved by utilizing a uniformly doped polysilicon string in the 3D NAND flash architecture as shown in Figure 6.8a [47, 48].

6.5.2 Biosensors

FET-based biosensors have attracted significant attention owing to their low cost, portability, high sensitivity, and reliability. However, the MOSFET-based biosensors suffer from a reduced selectivity, limited scalability, and poor SNR due to their inherent conduction mechanism and short-channel effects [49]. The sensitivity, selectivity, and the response time of the FET-based biosensors could be significantly improved by utilizing JL architecture in the impact ionization-induced steep slope regime [50]. JL-biosensors were extensively explored for efficient detection of biomolecules (Figure 6.8b).

6.5.3 3D Thin Film Transistors

The absence of metallurgical junctions facilitates low temperature processing of JL devices and opens the possibility of realizing these devices on single crystal silicon-on-glass substrate for display devices and optoelectronic applications. Therefore, JL TFTs with hybrid P/N channel (Figure 6.8c) which exhibit high performance as well as improved reliability and low noise behavior were demonstrated experimentally and studied extensively [51, 52].

6.5.4 Back-End-of-Line Transistors

The 3D integrated circuits require vertical stacking of transistors over the BEOL interconnects. Since the thermal stability of interconnects may be degraded if the processing temperature in the subsequent steps increases above 430 °C, BEOL-compatible transistors are needed for 3D ICs. Owing to their low temperature processing capability, JLFETs offer a new paradigm for BEOL-compatible transistors [9, 53].

6.6 Conclusions

In this chapter, we have provided a comprehensive review of the first transistor to be conceptualized: JLFETs. We discuss the unique attributes such as low temperature processing, bulk conduction, and low-field operation in the ON-state which leads to several advantages including enhanced reliability, immunity against short-channel effects, high carrier ballisticity, impact ionization-induced steep slope at low voltages, etc. Furthermore, we also provide a detailed overview of the challenges faced by the JLFETs which hinders their adoption as the mainstream transistor technology. Moreover, we also discuss the emerging avenues which get significant benefits from the junction-less architecture.

References

[1] Sahay, S. and Kumar, M.J. (2019). *Junctionless Field-Effect Transistors: Design, Modeling, and Simulation*. Wiley.
[2] Gibbons, J.F. (1972). Ion implantation in semiconductors – part II: damage production and annealing. *Proc. IEEE* 60 (9): 1062–1096.
[3] Lilienfeld, J.E. (1930). Method and apparatus for controlling electric currents. US Patent 1,745,175, January 1930.
[4] Shockley, W. (1951). Circuit element utilizing semiconductive material. US Patent 2,569,347, 25, September 1951.
[5] Lilienfeld, J.E. (1933). Device for controlling electric current. US Patent 1,900,018, March 1933.
[6] Colinge, J.-P., Lee, C.-W., Afzalian, A. et al. (2010). Nanowire transistors without junctions. *Nat. Nanotechnol.* 5 (3): 225–229.
[7] Lee, C.-W., Afzalian, A., Akhavan, N.D. et al. (2009). Junctionless multigate field-effect transistor. *Appl. Phys. Lett.* 94 (5): 053511–053512.
[8] Wen, S.M. and Chui, C.O. (2013). CMOS junctionless field-effect transistors manufacturing cost evaluation. *IEEE Trans. Semi. Manuf.* 26 (1): 162–168.
[9] Vandooren, A., Franco, J., Parvais, B. et al. (2018). 3D sequential stacked planar devices on 300mm wafers featuring replacement metal gate junctionless top devices processed at 525°c with improved reliability. *IEEE Symposium on VLSI Technology*, Honolulu, HI (18–22 June 2018). IEEE.
[10] Cho, M., Lee, J.-D., Aoulaiche, M. et al. (2012). Insight into N/PBTI mechanisms in sub-1-nm-EOT devices. *IEEE Trans. Electron Devices* 59 (8): 2042–2048.
[11] Toledano-Luque, M., Matagne, P., Sibaja-Hernández, A. et al. (2014). Superior reliability of junctionless pFinFETs by reduced oxide electric field. *IEEE Electron Device Lett.* 35 (12): 1179–1181.
[12] Datta, S. (2005). *Quantum Transport: Atom to Transistor*. Cambridge University Press.
[13] Akhavan, N.D., Ferain, I., Razavi, P. et al. (2011). Improvement of carrier ballisticity in junctionless nanowire transistors. *Appl. Phys. Lett.* 98 (10): 103510.
[14] Wong, I.H., Chen, Y.T., Huang, S.H. et al. (2014). In-situ doped and tensile stained Ge junctionless gate-all-around n-FETs on SOI featuring Ion= 828 μA/μm, Ion/Ioff ~ 1× 105, DIBL= 16 – 54 mV/V, and 1.4 X external strain enhancement. *IEEE International Electron Devices Meeting*, San Francisco, CA (15–17 December 2014), pp. 6–9. IEEE.

[15] Wong, I.H., Chen, Y.T., Huang, S.H. et al. (2015). Junctionless gate-all-around pFETs using in-situ boron-doped Ge channel on Si. *IEEE Trans. Nano.* 14 (5): 878–882.
[16] Djara, V., Czornomaz, L., Deshpande, V. et al. (2016). Tri-gate InGaAs-OI junctionless FETs with PE-ALD Al2O3 gate dielectric and H2/Ar anneal. *Solid State Electron.* 115: 103–108.
[17] Goh, K.H., Yadav, S., Low, K.I. et al. (2016). Gate-all-around In0.53Ga0.47As junctionless nanowire FET with tapered source/drain structure. *IEEE Trans. Electron Devices* 63 (3): 1027–1033.
[18] Im, K.S., Seo, J.H., Yoon, Y.J. et al. (2014). GaN junctionless trigate field-effect transistor with deep-submicron gate length: characterization and modeling in RF regime. *Jap. J. Appl. Phys.* 53 (11): 118001.
[19] Singh, P., Singh, N., Miao, J. et al. (2011). Gate-all-around junctionless nanowire MOSFET with improved low-frequency noise behaviour. *IEEE Electron Device Lett.* 32 (12): 1752–1754.
[20] Jeon, D.Y., Park, S.J., Mouis, M. et al. (2013). Low-frequency noise behavior of junctionless transistors compared to inversion-mode transistors. *Solid State Electron.* 81: 101–104.
[21] Rios, R., Cappellani, A., Armstrong, M. et al. (2011). Comparison of junctionless and conventional trigate transistors with Lg down to 26 nm. *IEEE Electron Device Lett.* 32 (9): 1170–1172.
[22] Colinge, J.P., Lee, C.-W., Ferain, I. et al. (2010). Reduced electric field in junctionless transistors. *Appl. Phys. Lett.* 96 (7): 073510.
[23] Goto, K.I., Yu, T.H., Wu, J. et al. (2012). Mobility and screening effect in heavily doped accumulation-mode metal-oxide-semiconductor field-effect transistors. *Appl. Phys. Lett.* 101 (7): 073503.
[24] Rudenko, T., Nazarov, A., Ferain, I. et al. (2012). Mobility enhancement effect in heavily doped junctionless nanowire silicon-on-insulator metal-oxide-semiconductor field-effect transistors. *Appl. Phys. Lett.* 101 (21): 213502.
[25] Doria, R.T., Trevisoli, R.D., de Souza, M. et al. (2012). The zero temperature coefficient in junctionless nanowire transistors. *Appl. Phys. Lett.* 101 (6): 062101.
[26] Han, M.H., Chen, H.B., Yen, S.S. et al. (2013). Temperature-dependent characteristics of junctionless bulk transistor. *Appl. Phys. Lett.* 103 (13): 133503.
[27] Trevisoli, R., de Souza, M., Doria, R.T. et al. (2016). Junctionless nanowire transistors operation at temperatures down to 4.2 K. *Semicond. Sci. Technol.* 31 (11): 114001.
[28] Ma, L., Han, W., Wang, H. et al. (2013). Temperature dependence of electronic behaviors in n-type multiple-channel junctionless transistors. *J. Appl. Phys.* 114 (12): 124507.
[29] de Souza, M., Pavanello, M.A., Trevisoli, R.D. et al. (2011). Cryogenic operation of junctionless nanowire transistors. *IEEE Electron Device Lett.* 32 (10): 1322–1324.
[30] Lee, C.W., Nazarov, A.N., Ferain, I. et al. (2010). Low subthreshold slope in junctionless multigate transistors. *Appl. Phys. Lett.* 96 (10): 102106.
[31] Yu, R., Nazarov, A.N., Lysenko, V.S. et al. (2013). Impact ionization induced dynamic floating body effect in junctionless transistors. *Solid State Electron.* 90: 28–33.
[32] Parihar, M.S. and Kranti, A. (2014). Back bias induced dynamic and steep subthreshold swing in junctionless transistors. *Appl. Phys. Lett.* 105 (3): 033503.
[33] Gupta, M. and Kranti, A. (2012). Transforming gate misalignment into a unique opportunity to facilitate steep switching in junctionless nanotransistors. *Nanotechnology* 27 (5): 455204.
[34] Parihar, M.S., Ghosh, D., and Kranti, A. (2013). Single transistor latch phenomenon in junctionless transistors. *J. Appl. Phys.* 113 (18): 184503.
[35] Leung, G. and Chui, C.O. (2012). Variability impact of random dopant fluctuation on nanoscale junctionless FinFETs. *IEEE Electron Device Lett.* 33 (6): 767–769.
[36] Nawaz, S.M. and Mallik, A. (2016). Effects of device scaling on the performance of junctionless FinFETs due to gate-metal work function variability and random dopant fluctuations. *IEEE Electron Device Lett.* 37 (8): 958–961.
[37] Leung, G. and Chui, C.O. (2011). Variability of inversion-mode and junctionless FinFETs due to line edge roughness. *IEEE Electron Device Lett.* 32 (11): 1489–1491.
[38] Choi, S.J., Moon, D.I., Kim, S. et al. (2011). Sensitivity of threshold voltage to nanowire width variation in junctionless transistors. *IEEE Electron Device Lett.* 32 (2): 125–127.
[39] Nawaz, S.M., Dutta, S., Chattopadhyay, A., and Mallik, A. (2014). Comparison of random dopant and gate-metal workfunction variability between junctionless and conventional FinFETs. *IEEE Electron Device Lett.* 35 (6): 663–665.

[40] Gundapaneni, S., Bajaj, M., Pandey, R.K. et al. (2012). Effect of band-to-band tunneling on junctionless transistors. *IEEE Trans. Electron Devices* 59 (4): 1023–1029.

[41] Hur, J., Lee, B.H., Kang, M.H. et al. (2016). Comprehensive analysis of gate-induced drain leakage in vertically stacked nanowire FETs: inversion-mode vs. junctionless mode. *IEEE Electron Device Lett.* 37 (5): 541–544.

[42] Sahay, S. and Jagadesh Kumar, M. (2017). Physical insights into the nature of gate induced drain leakage in ultra-short channel nanowire field effect transistors. *IEEE Trans. Electron Devices* 64 (6): 2604–2610.

[43] Hu, C., King, T.-J., and Hu, C. (2002). A capacitorless double-gate DRAM cell. *IEEE Electron Device Lett.* 23 (6): 345–347.

[44] Parihar, M.S., Ghosh, D., Armstrong, G.A., and Kranti, A. (2012). Bipolar snapback in junctionless transistors for capacitorless dynamic random access memory. *Appl. Phys. Lett.* 101 (26): 263503.

[45] Tanaka, H., Kido, M., Yahashi, K. et al. (2007). Bit cost scalable technology with punch and plug process for ultra high density flash memory. *IEEE Symposium on VLSI Technology*, Kyoto (12–14 June 2007), pp. 14–15. IEEE.

[46] Jang, J., Kim, H.-S., Cho, W. et al. (2009). Vertical cell array using TCAT (terabit cell array transistor) technology for ultrahigh density NAND Flash memory. *Symposium on VLSI Technology*, Kyoto (15–17 June 2009), pp. 192–193. IEEE.

[47] Choi, S.J., Moon, D.I., Duarte, J.P. et al. (2011). A novel junctionless all-around-gate SONOS device with a quantum nanowire on a bulk substrate for 3D stack NAND flash memory. *Symposium on VLSI Technology – Digest of Technical Papers*, Kyoto (14–16 June 2011), pp. 74–75. IEEE.

[48] Chen, M.C., Yu, H.Y., Singh, N. et al. (2009). Vertical Si nanowire SONOS memory for ultra-high density application. *IEEE Electron Device Lett.* 30 (8): 879–881.

[49] Sen, D., Patel, S.D., and Sahay, S. Dielectric modulated nanotube tunnel field-effect transistor as a label free biosensor: proposal and investigation. *IEEE Trans. NanoBioscience* https://doi.org/10.1109/TNB.2022.3172553.

[50] Parihar, M.S. and Kranti, A. (2015). Enhanced sensitivity of double gate junctionless transistor architecture for biosensing applications. *Nanotechnology* 26 (14): 145201.

[51] Cheng, Y.C., Chen, H.B., Su, J.J. et al. (2015). Characteristics of a novel poly-Si P-channel junctionless transistor with hybrid P/N-substrate. *IEEE Electron Device Lett.* 36 (2): 159–161.

[52] Cheng, Y.C., Chen, H.B., Chang, C.Y. et al. (2016). A highly scalable poly-Si junctionless FETs featuring a novel multi-stacking hybrid P/N layer and vertical gate with very high Ion/Ioff for 3D stacked ICs. *IEEE Symposium on VLSI Technology*, Honolulu, HI (14–16 June 2016), pp. 188–189. IEEE.

[53] Vandooren, A., Witters, L., Vecchio, E. et al. (2017). Double-gate Si junction-less n-type transistor for high performance Cu-BEOL compatible applications using 3D sequential integration. *IEEE* S3S: 1–2.

Chapter 7

The First MOSFET Design by J. Lilienfeld and its Long Journey to Implementation

Hiroshi Iwai

National Yang Ming Chiao Tung University, Hsinchu, Taiwan

7.1 Introduction

Nearly 100 years ago, on 22 October 1925, Julius Edgar Lilienfeld submitted a patent, entitled "Method and apparatus for controlling electric current," to the Canadian Patent Office [1]. This was the world's first idea of solid-state amplifiers as well as field effect transistors (FETs) and actually the first MESFETs (MEtal Semiconductor FET) [2]. On 28 March 1928, he filed another patent, "Device for controlling electric current," to the United States Patent Office [3]. This was the first idea of MISFETs (Metal Insulator Semiconductor FET)/MOSFETs (Metal-Oxide-Semiconductor FET) that have been the main devices used for micro/nanoelectronics for 50 years since the beginning of the 1970s.

Julius Edgar Lilienfeld was born in Lemberg (or Lwów in Poland, now Liviv in Ukraine), as an Austro-Hungarian citizen, on 18 April 1882, received PhD degree from Berlin University on 18 February 1905, and worked as a professor at the Physics Institute of Leipzig University from 1916 to 1927 [4–6]. He carried out research on conduction in extreme vacuum and contributed to the development of X-ray tubes. He also worked with Count Ferdinand von Zeppelin on the design of hydrogen-filled dirigibles. He moved to the United States in 1927 to fight for his X-ray tube patent right there, and became director of research in Ergon Research Laboratories in Malden, MA, where he was interested in the research of the interfaces between electrolytes and salt and other materials, and semiconductor devices. He retired and moved to St. Thomas, the Virgin Islands, in 1935 to escape from an allergy associated with the growing of wheat, but continued his research there for some time. He died at home there on 28 August 1963 at the age of 81.

75th Anniversary of the Transistor, First Edition. Edited by Arokia Nathan, Samar K. Saha, and Ravi M. Todi.

He maintained communication with physicists such as Albert Einstein even after moving to the Virgin Islands [6], and his short obituary [4] appeared on "Physics Today" – a magazine published by the American Institute of Physics – on November 1963, immediately after his death. While his contributions on the X-ray tubes and dirigibles were mentioned in the obituary, those on his FET inventions were not. His name and his semiconductor patents had not been known in semiconductor device research communities until the end of 1940s, except for some Europeans, such as Heinrich J. Welker who filed an important patent of a MISFET in Germany in 1945 [7]. It is a famous story [8] that William Shockey was not aware of Lilienfeld's FET patents until he requested Bell Laboratories attorneys to submit his FET patent to the US Patent Office in early 1948. The filing was declined by the attorneys because of their discovery of the Lilienfeld's patents at that time.

One of the Lilienfeld's ideas described in the patent illustrated a prototype of today's MOSFETs more than 30 years before the first successful MOSFET realization by Dawon Kahng and Martin M. Atalla of Bell Laboratories in 1960 [9–13]. Probably, Lilienfeld did not think of the monolithic large-scale integration of his MOSFET, but it happened to be the most suitable device for the integration because of its planar structure. Unsurprisingly because of poor theories and technologies on the semiconductor in the 1920s, **his ideas were not sufficient for the practical MOSFET implementation, and further new ideas and experimental works were necessary for the realization of his idea. Because MOSFETs are the devices that utilize the semiconductor surface conduction, controlling the semiconductor surface had been the most critical issue that had not been solved until the development of semiconductor surface physics and technologies, and the progress in the purity and quality of the semiconductor materials to an extremely high level.**

In this paper, I will describe the background to the creation of Lilienfeld's idea, its importance and problems, and the path to implementing practical MOSFETs toward commercialization.

7.2 Demand for the Development of the Solid-State Amplifier and Its Difficulty

At the beginning of the twentieth century, wireless telecommunication started with coherers [14, 15], and then, with point-contact rectifiers (or crystal detectors) [16–18] as the receivers of signals. However, these devices could not amplify the signals in principle, and therefore, the development of radio wave amplifiers had been strongly required for detecting weak wireless signals from long distance.

On 27 January 1907, Lee de Forest filed a patent on the first triode vacuum tube as shown in Figure 7.1 [19]. This was the first three-terminal electron device, and the world's first analog and digital amplifier. The thermo-electron beam emitted from the heated cathode into the vacuum tube toward the

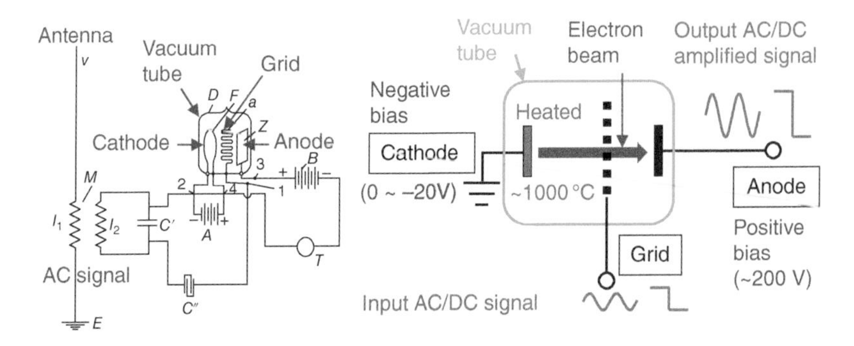

Figure 7.1 Triode vacuum tube invented by Lee de Forest. Source: Adapted from de Forest [19].

anode was modulated by the grid potential that was inserted in the beam. The triode can manipulate electron movement by the grid potential as we wish, and this was the beginning of "electronics" or "electronic engineering." With the electronics, we became able to treat information electrically, performing not only telecommunication (wireless, radio/TV broadcast, *etc.*) but also information/data processing (automatic machine-control, computer, *etc.*). Later at the beginning of the twenty-first century, even artificial intelligence (AI) had been realized.

However, the relatively long electron transit time due to the difficulty in decreasing the vacuum-tube size made it impossible to operate at microwave frequencies, until the end of the 1930s when klystrons (high-power microwave vacuum tube) were invented by Russell Varian and Sigurd Varian of Stanford University in 1937~1939 [20, 21]. Also, the need to heat the cathode of the vacuum tube above 1000 °C for emitting the electrons caused inconveniences for power saving and heat insulation of the systems. **Many people dreamed of replacing the triode vacuum tubes with some kind of three-terminal solid-state amplifiers to solve these problems.**

Controlling the electron current in the vacuum through the grid potential was easy. However, the control of the electron current in a solid-state conductor (or metal) was almost impossible because of the huge density of the free carriers existing as shown in Figure 7.2. On the other hand, in the case of insulators, there is no current available. Thus, semiconductors that have a lower electron density were thought to be the candidate.

As described, **the use of semiconductors had already started for commercial point-contact rectifiers (or crystal detectors) from the beginning of the twentieth century.** Semiconductor rectifiers were used popularly even in the 1930s and 1940s for low-priced radio receivers and also for high-frequency microwave receivers for which the triode vacuum tubes were not suitable in that period. **However, the rectifiers had limited ability to treat the signals resulting from the nature of the 2-terminal devices, and basically, signal amplification was impossible. Therefore, a demand for 3-terminal semiconductor rectifiers was necessitated.**

Probably, an easy idea for the 3-terminal solid-state amplifiers would be the similar structure as the triode vacuum tubes. The electron (or hole) current could be controlled by the potential of the grid electrode inserted into the depletion region where there are few free carriers as shown in Figure 7.3a, b. The concept of the depletion region in a semiconductor became recognized by the rectifying theories developed by Walter Schottky [22, 23], Nevil F. Mott [24], and Boris Davydov in 1938 [25] based on the energy band model created by Alan H. Wilson in 1931 [26, 27] using quantum mechanics.

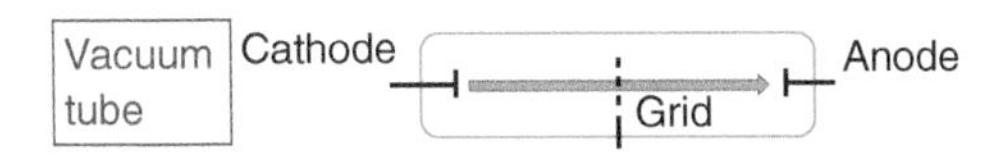

(a) Easy to control the current in vacuum by grid potential.

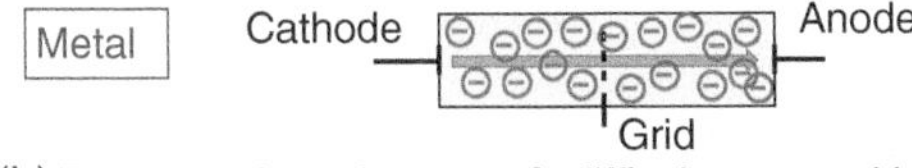

(b) Too many free electrons → difficult to control by grid potential.

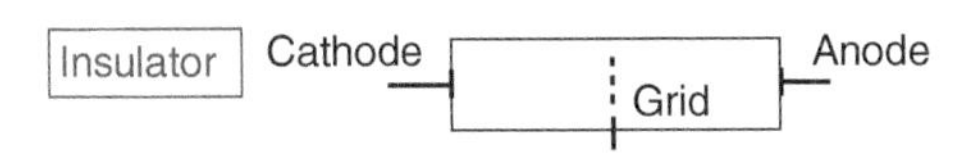

(c) No current available for conduction.

Figure 7.2 Difficulty in controlling current by grid potential inserted in the solid-state materials.

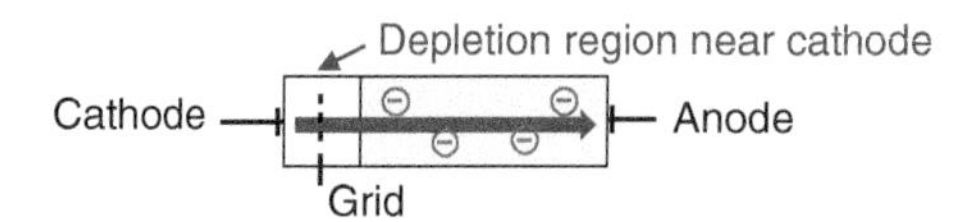

(a) Grid inserted type 1: Grid inserted in the depletion region near cathode.

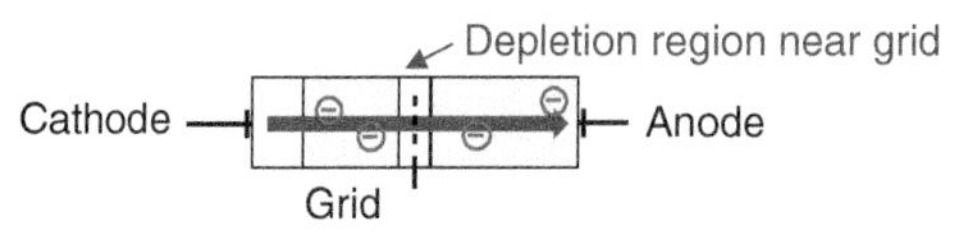

(b) Grid inserted type 2: Depletion region is created near mesh-type grid.

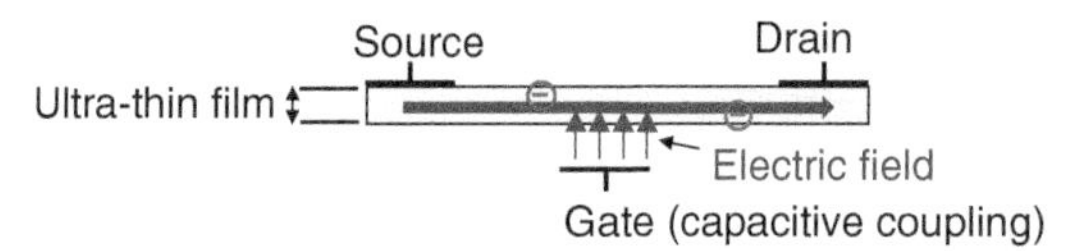

(c) Thin film channel and capacitive coupling gate electrode type.

Figure 7.3 Solid-state amplifier ideas using semiconductor: (a, b) MESFETs, (c) MESFETs, MISFETs, and MOSFETs.

Another approach for the semiconductor amplifier was to use an ultrathin semiconductor film and to control lateral current in the ultrathin film by vertical electric field. It was assumed that the vertical electric field easily penetrates the ultrathin film to control the lateral current. The vertical field was applied by the gate electrode voltage remotely with capacitive coupling as shown in Figure 7.3c. This is a field effect transistor (FET), although this type of the structure was conceived by Lilienfeld already in the 1920s, earlier than the grid-inserted structures (Figure 7.3a,b).

7.3 Grid-Inserted MESFETs

The grid-inserted MESFET device was proposed by Allgemeine Elektricitäts-Gesellschaft (AEG) in 1929 (a German company whose inventor names were not written in the patent) [28] as shown in Figure 7.4a. W. Shockley also thought of the structure in 1939 [29] as shown in Figure 7.4b.

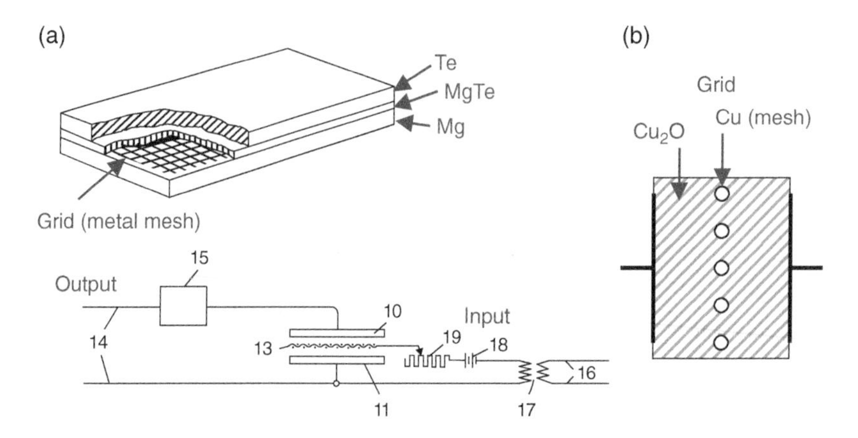

Figure 7.4 Ideas of grid-inserted type of solid-state amplifiers (MESFETs): (a) AEG's patent in 1929. Source: Adapted from Allgemeine Elekricitäcts-Gesellschaft [28]. Te and Mg films were separated by oxidized layers, respectively. (b) W. Shockley's idea in 1940. Source: Adapted from Shockley [29].

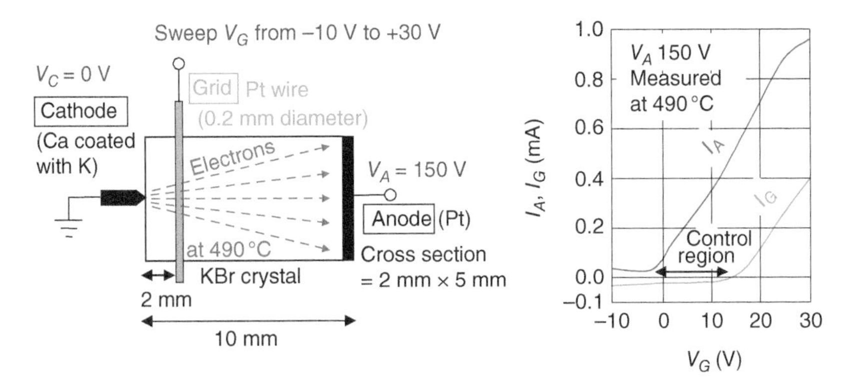

Figure 7.5 First record of signal amplification by a solid-state device (MESFET) in 1938. Source: Adapted from Hilsch and Pohl [30].

However, no successful results were reported for the grid-inserted MESFETs, except the case by Rudolf Hilsch and Robert Wichard Pohl of Göttingen Universität in 1938 [30] as shown in Figure 7.5. They used a KBr (potassium bromide) crystal. KBr is an ionic crystal with a wide bandgap (7 ~ 8 eV) and is an insulator rather than a semiconductor. Electrons were injected from the point-contact cathode to the KBr crystal (2 × 5 mm^2 cross section and 10 mm length) at high temperature, 490 °C. There were almost no carriers other than the injected electrons except ions of KBr whose density was 30 times smaller than that of the injected electrons at 490 °C. The injected current was modulated by the potential of a Pt (platinum) grid electrode that was inserted into KBr at a distance of 2 mm from the cathode. The Pt to KBr contact was a Schottky contact and there was insignificant leakage current at V_G (grid voltage) = 0 ~ 13 V. A maximum current amplitude (I_A/I_G) of 20 was observed by the experiment, where I_A is the anode current, and I_G is the grid current. The mobility of electrons in KBr was very low and the transit time of electrons in the KBr crystal of 10 mm was 10 seconds. Thus, the operating frequency was estimated to be only sub-Hz.

This was the first experimental recording of a solid-state amplifier. The performance of this MESFET was not satisfactory at all, essentially because KBr is an insulator. They also investigated the possibility of Cu-Cu_2O (semiconductor) rectifiers to operate at higher frequency at room temperature, but the depletion layer width was found to be 0.1 μm – that was too narrow to insert the grid, and they **concluded that it was impossible to fabricate the solid-state amplifier by inserting the grid into the semiconductor.**

7.4 Lilienfeld Patents for the MESFET and MOSFET

In 1925, the first idea of FET was filed as a patent by Julius E. Lilienfeld [1] as shown in Figure 7.6. This was a kind of Schottky metal gate FET and was the first idea of the solid-state amplifier in the world, even earlier than the grid-inserted type proposed by AEG in 1929 [28] as already described.

He proposed to use a SiO_2 glass substrate. The glass substrate was fractured (or broken) into two parts at first. Then, the fractured parts were put together to form a one substrate with an Al foil between them. After polishing the glass surface, Cu electrodes for the source/drain were deposited on the glass and a thin cuprous sulfide (Cu_2S) film was deposited on the structure as shown in Figure 7.7. An AC input signal was applied to both the Cu source and the Al gate electrodes in his patent. The current flowing in the Cu_2S film was supposed to be modulated by the Al foil film voltage change. The Al foil film thickness was as small as 2.5 μm to minimize the parasitic capacitance. By biasing the Al foil film positively, there is an insignificant leakage between the Al and Cu_2S films due to the rectifying effect.

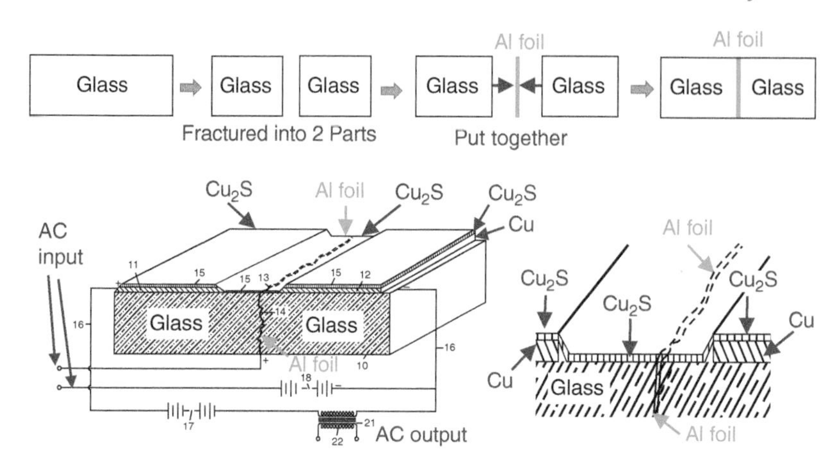

Figure 7.6 First idea of solid-state amplifier (MESFET) proposed by J. Lilienfeld in 1925. Source: Adapted from Lilienfeld [1].

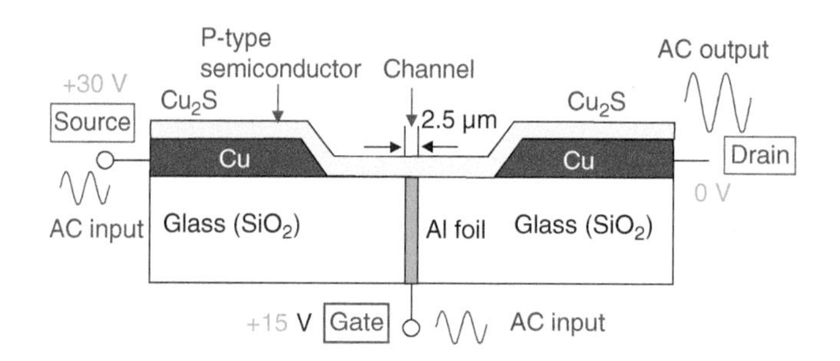

Figure 7.7 Schematic structure and operation of the Lilienfeld's MESFET.

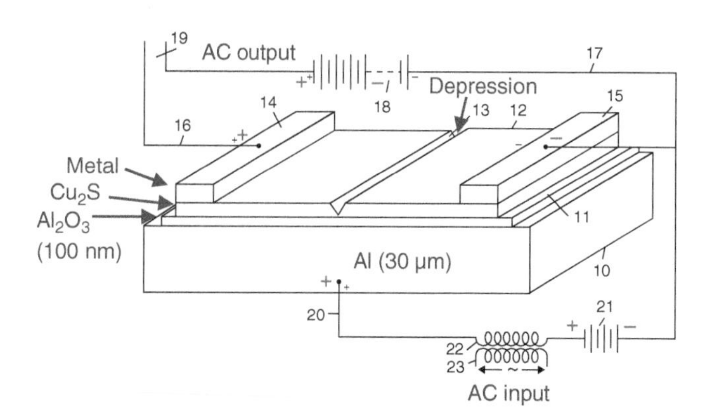

Figure 7.8 First idea of MISFET/MOSFET proposed by J. Lilienfeld in 1928. Source: Adapted from Lilienfeld [3].

In 1928, another FET, i.e. MOSFET, was proposed also by J. Lilienfeld [3] as shown in Figure 7.8. Here, a 100 nm thick Al_2O_3 was formed on a 30 μm thick Al substrate. Then, a thin Cu_2S film was deposited on the Al_2O_3 film. Al substrate, Al_2O_3, and Cu_2S were used as the gate electrode, gate oxide, and channel of the MOSFET, respectively. Different from the MESFET case, there was almost no leakage current between the Al gate and Cu_2S channel even under very high gate bias. There is a triangle-shaped depression at the center of the Cu_2S film. The thickness of the film at the depression was set to a molecular size so that the vertical electric field at the depression is enhanced significantly to control the channel current.

As described, J. Lilienfeld's patents were not recognized by Bell Laboratories until 1948. In Motorola, the patents were not recognized until 1953 when Virgil E. Bottom, director of research, was engaged in a patent search in connection with the development of the germanium transistor [5]. Unfortunately, there were no records of experimental results of J. Lilienfeld's MESFETs and MOSFETs, and it was uncertain if those devices worked successfully as amplifiers. According to J.B. Johnson of Thomas A. Edison Industries [31], **J. Johnson tried conscientiously to reproduce J. Lilienfeld's structure according to J. Lilienfeld's specification in the early 1960s but could observe no amplification or even modulation.** The reason was probably the very low mobility of holes in Cu_2S (~1 $cm^2 V^{-1} s^{-1}$), and the effect of surface states on the free surface of the film. While cupric sulfide, CuS, is a metallic conductor and could not have been used by Lilienfeld, the cuprous sulfide, Cu_2S, in its pure state is an insulator but is made a p-type semiconductor by an almost inevitable trace of CuS [31]. **J. Lilienfeld's FETs were normally-on type p-channel FETs controlled by a positively biased gate electrode, and this is a thin-film transistor (TFT)** [32] **as well (Figure 7.9).**

Regarding the TFT-type FETs, Oskar Heil filed a patent without experimental results in 1934 as shown in Figure 7.10 [33]. In his patent, the assumed semiconductors were Te, I, Cu_2O, V_2O_5, and the assumed gate insulators were air-gap, vacuum-gap, and dielectric films. Therefore, his patent describes

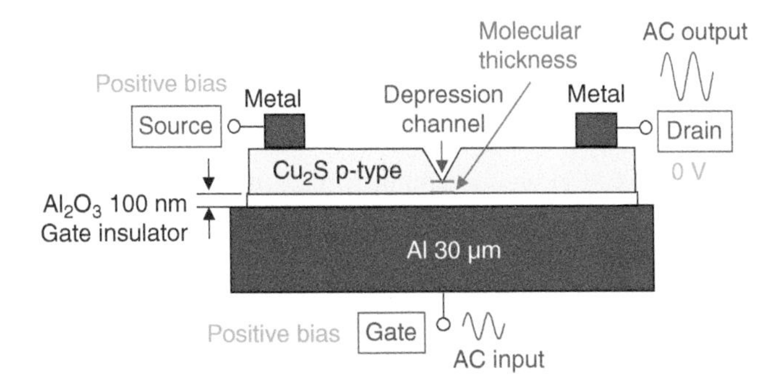

Figure 7.9 Schematic structure and operation of the Lilienfeld's MOSFET. Source: Adapted from Lilienfeld [3].

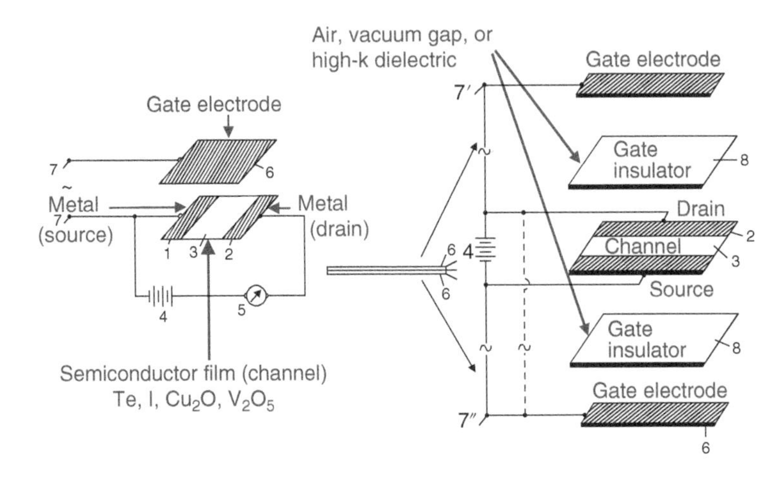

Figure 7.10 Thin-film MISFET idea by Oskar Heil in 1934. Source: Adapted from Heil [33].

about the MISFETs where the gate insulator was not limited to oxides which are used for MOSFETs. He proposed to put the channel sheet (semiconductor film) between the two gate electrodes vertically. Later, in 1957, John Toekel Wallmark also filed a similar structure without experimental results [34]. These are the double gate MOSFETs [35] that evolved to recent nano-sheet MOSFETs and CFET [36–40] that are now going to be used in production. By the way, Oskar Heil and his wife, Agnesa Arsenjewa-Heil, published velocity modulation vacuum tubes [41, 42] in 1935, which were the first practical high-power linear beam microwave electron tubes that predated the invention of klystron by the Varian brothers [20, 21] in 1937–1939.

7.5 Necessary Conditions for Successful MOSFET Operation, and MOSFET Development Chronology

Figure 7.11 shows the necessary conditions for the successful operation of bulk MOSFETs. They are **(i) Inversion surface channel:** There should be an inverted channel layer – at the surface of the semiconductor – that is electrically separated from the substrate via a depletion region. **(ii) pn Junction for source/drain connected to the surface inverted channel layer:** There should be semiconductor source and drain regions insulated by pn junctions from the substrate, and furthermore, the source should be connected to the inverted channel layer. **(iii) Dielectric gate insulator with interfacial dipoles:** The gate insulator should consist of a dielectric with certainly high dielectric constant so that there is a sufficiently high density of dipoles at the semiconductor interface to enhance the local electric field to control the surface carriers. **(iv) High-quality Si substrate:** The purity and quality of the semiconductor substrate should be good enough so that the mobility and lifetime of the carriers are high. Si and Ge were found to be good candidates. However, Ge is not suitable because of the bad property of Ge oxides. **(v) Gate SiO_2 film grown by high-temperature oxidation of Si:** The semiconductor interface with the gate insulator should be good enough so that the surface (interface) state density and surface roughness are small in order for the carrier mobilities to be high.

Table 7.1 shows the timeline for the completion of the MOSFET. J. Lilienfeld's ideas provided us with a prototype structure of today's MOSFET. However, semiconductor theories, materials, and technologies were in the primitive level in the 1920s, and J. Lilienfeld seemed to have recognized none of the above five necessary conditions. It took 30 years for the ideas, theories, materials, and technologies to reach a level sufficient to realize successful MOSFETs.

All of the three ideas of (i) surface inversion channel layer that is insulated from the substrate, (ii) the field enhancement effects of the gate dielectric dipoles, and (iii) Si substrate for high-carrier mobility were proposed by H.J. Welker in his patent in 1945 [7, 43, 44]. **The concept of**

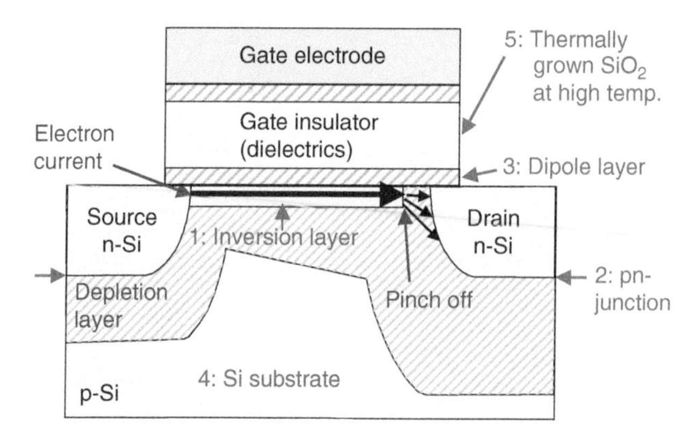

Figure 7.11 The five necessary conditions for bulk MOSFETs.

Table 7.1 Timeline for the FET development and related theories and technologies for the successful MOSFET operation.

1901: Point-contact semiconductor rectifier
1906: Triode vacuum tube
1925, 1928: Ideas of MESFET and MOSFET by J. Lilienfeld
1931: Concept of "hole"
1938: Schottky junction theory
1938: Amplification with KBr MESFET by R. Hilsh and R. Pohl
1930s–1940s: Improvement of Si and Ge semiconductor purity
1940: Discovery of the pn junction, concept of p- and n-type semiconductor
1945: Ideas of inversion channel, interfacial dipoles, Si and Ge substrate by H. Welker
1947: Amplification with electrolyte gate insulator by J. Bardeen and W. Brattain
1947, 48: Invention of point-contact transistor ('47) and junction transistor ('48)
1952, 53: Idea of JFET and concept of pinch-off by W. Shockley ('52), First operation of practical FET (JFET) ('53)
1953: Source/drain with pn junctions and connected to inversion layer at "one-sided" JFET by W. Brown and W. Shockley
1955: Proposal of practical MOSFET structure by I. Ross
1950s: Significant improvement of Si quality and semiconductor device fabrication technologies
1958: Stabilization of Si surface by high-temperature oxidation
1960: Successful MOSFET operation with thermally grown gate SiO_2 by D. Kahng and M. Atalla

"pinch-off" was given by Shockley in his junction FET (JFET) idea in 1952 [45]. **The concept of semiconductor source/drain insulated from the substrate by pn junctions and connected to the surface inversion channel layer was given by W.L. Brown and W. Shockley in 1953** [46, 47], although the structure was not a MISFET/MOSFET, but a kind of a "one-sided" JFET. The surface inversion channel conduction was controlled by the substrate p-Ge (or back-gate) potential given from the back electrode, as will be explained later in Section 7.10. A practical MISFET/MOSFET structure with gate electrode/insulator, surface inversion channel layer, and source/drain with pn junctions was proposed by Ian M. Ross in 1955 in his patent for a ferroelectric memory [48]. **The stabilization of the Si surface by high-temperature oxidation was found by Martin M. Atalla and others in 1958 and 1959** [49–52] and **the successful operation of MOSFET was obtained by Dawon Kahng and M. Atalla in 1960** [9–13].

7.6 Status of the Semiconductor Physics at the Lilienfeld Period (in the 1920s) and Thereafter

In the 1920s when Lilienfeld submitted the patents, the knowledge about semiconductor physics [53–55] was very poor. The mechanism or physics of the point-contact rectifier operation was not known at all, and furthermore, there were no concepts of the pn junction and holes.

The Hall effect [56] was discovered by Edwin Hall in 1879 before the discovery of electrons by Joseph J. Thomson in 1897 [57, 58], and the sign of the carriers flowing in conductors and semiconductors was able to be detected by the sign of the "Hall voltage" caused by the Lorenz force that is applied to the carriers under the vertical magnetic field. By the Hall measurement, the conduction of both negative carriers (electrons) and positive carriers were reported for semiconductor materials in the early

1930s [59–61]. The classic electron conduction models in conductors were proposed by Eduard Riecke [62, 63] and Paul Drude [64, 65] in the early twentieth century. However, the conduction of positive carriers had remained as a mystery.

In 1929, W. Schottky and W. Deutschmann suggested that the increase in the occupation of the states by electrons decreases the conduction of Cu_2O and that some non-occupied states are necessary for the conduction [66]. **The concept of the "hole" became clarified by Werner K. Heisenberg, Wolfgang E. Pauli, and Rudolf Peierls, when R. Peierls studied the "positive Hall" effect [67–70] in 1929.** Heisenberg stated that the lack of electrons in the closed cell of an atom is treated with the "electrons which have a positive charge" based on W. Pauli's suggestion in 1931 [67, 70]. **Thus, probably, J. Lilienfeld did not recognize the hole conduction of Cu_2S film in the 1920s.** In the case of Heil's patent in 1934 [33], both n- and p-type semiconductors were recognized, although there was not yet the terminology of n- and p-type semiconductors at that time.

7.7 Improvement of Si and Ge Material Quality and Discovery of the pn Junction in the 1940s

In the late 1930s, Si became the most popular semiconductor for the high-frequency (above 3 GHz) crystal detectors [71], and the purity and quality of Si crystals started to be improved [72]. Ge had also drawn attention because of its higher carrier mobility and lower melting temperature that made the crystal production easier [73]. **In February 1940, the pn junction was found by Russel S. Ohl's group in Bell Laboratories** [74–82]. They were conducting research on improving the purity of Si by melting it at high temperature and then cooling it for recrystallization. During the process, impurities included in Si precipitated at the crystal surface. They found black boundary regions in the recrystallized Si ingot. Si slabs including the boundary region showed photovoltaic and current-rectifying effects as shown in Figure 7.12. The boundary region was found to be the border between the p- and n-type Si with the acceptor (B), and donor (P), respectively. Impurities B, Al, and Ga had been recognized as forming p-type silicon and impurities P, Al, and Sb as forming n-type Si. **The terminology of "p-type" and "n-type" were named at that time** [81]. The discovery of the pn junction was a big step for the semiconductor technology development.

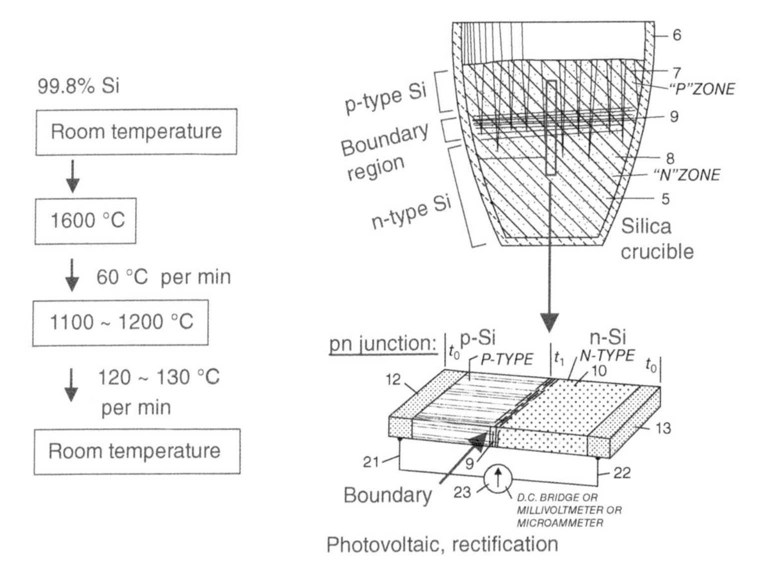

Figure 7.12 The pn junction discovered by R. Ohl in 1940. Source: Adapted from Ohl [75–77].

7.8 H. Welker's MISFET with Inversion Channel in 1945

With a priority date of 7 April 1945, a MISFET patent was filed with H.J. Welker as an inventor on 4 June 1945, as shown in Figure 7.13 [7]. At the beginning of 1945, Welker thought of forming a depletion layer at the surface of a semiconductor by applying a bias with an appropriate sign to the gate electrode. (For example, plus gate bias for p-type substrate semiconductor.) This depletion layer excludes the majority carriers (holes for this example) existing in the substrate. He thought that the injected minority carriers (electrons) from the source are attracted to the surface by the gate electric field and can be easily modulated by the gate field without any disturbance by the majority carriers. In his patent, the surface current flow layer was included in the depletion layer and he did not name the surface minority carrier layer. But it is obvious that the surface minority carrier layer is the inversion layer. The sign of the minority carriers (electrons) should be opposite to that of the majority carriers (holes).

He stated that the current injected from the source electrode to the semiconductor was divided into two parts. One part flows in the semiconductor substrate and the other part flows at the semiconductor surface. He mentioned that only the surface current was controlled by the gate field. He also noted that it is necessary to choose the appropriate semiconductor material in which the carrier density is small enough to suppress the substrate current. The substrate current problem was completely solved later by the introduction of the impurity-diffused semiconductor source/drain. Hence, the pn junction between the source/drain and substrate, in 1953 by W. Brown and W. Shockley [46, 47], as will be described later in Section 7.10.

The dielectric insulator was assumed as the gate insulator in the patent. The dielectric forms dipoles under electric field at the dielectric surface (or the interface to the semiconductor in this case). The dipoles enhance the controllability of the surface carriers by the increase in the local electric field (or more precisely, electric flux density) generated by the dipoles. He mentioned the dipole contribution in the patent. This effect was suggested by the research by Peter Brauer published in 1936 [43, 83]. P. Brauer showed that the Cu_2O conductivity decreased with the increase in the partial vapor pressure of H_2O at the surface, which can be attributed to the decrease in the hole concentration. H. Welker interpreted the result as the hole depletion at the surface of Cu_2O caused by water molecule dipole field at the surface. From this interpretation, he thought that the dipole field at the dielectrics/semiconductor interface would enhance the formation of the depletion layer in the semiconductor, and that a higher dielectric constant value is desirable for the effect.

He realized that Ge and Si should have higher upper frequency limits than Cu_2O due to their higher electron mobilities [43], and Ge and Si were proposed as the primary candidates for the substrate materials in the patent. Other candidates were Cu_2O, Se, CdS, and PbS.

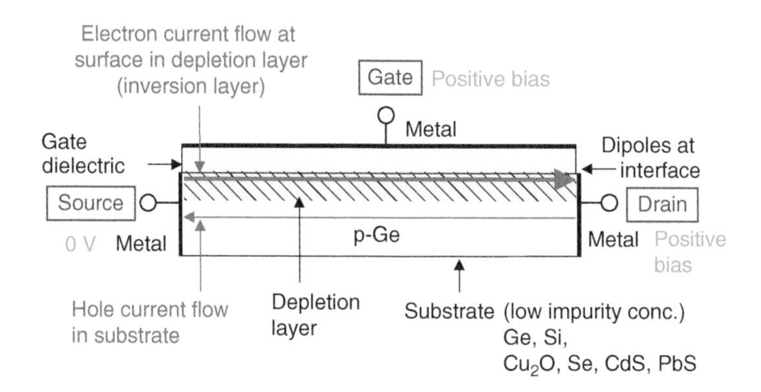

Figure 7.13 MISFET proposed by H. Welker in April 1945. Based on the description in the patent, n-MOSFET case is explained in this figure. Source: Adapted from Welker [7].

He calculated the depletion layer thickness, the inverted carrier density, and the current–voltage characteristics. **He formulated the MOSFET characteristics in an unpublished document written in March 1945 [44]**. However, as a result of the end of the war in 1945, the research on cm-waves and related devices were forbidden, and Welker's works were no longer published [44].

His formula of the MOSFET electrical characteristics was

$$I_D = kL/W\left\{U_{DS}\left(U_{GS} - U_h\right) - U_{GS}^2/2\right\}$$

using gradual channel model with the constant depletion layer depth through the channel, where U_{DS}, U_{GS}, U_h, L, W, k are drain voltage, gate voltage, threshold voltage, gate length, gate width, and a constant, respectively. Because of the assumption of the constant depletion width throughout the channel, this formula could not predict the channel "pinch-off," and no-saturation characteristics of MOSFETs were obtained. But this formula was popularly used as the model for the linear region ($V_G > V_D - V_{TH}$). The U_h is "a voltage characterized by the substrate impurity concentration" [44] and is the threshold voltage.

According to Ref. [44], in the spring of 1945, in the laboratory of the Steinheil Company, under the direction of Dr. Rollwagen, test structures were manufactured. At the suggestion of the work of P. Brauer, Cu_2O was used as the semiconductor that was produced by oxidizing a 10 μm thick copper layer. A thin glass plate and a silver electrode were placed on the Cu_2O, and "a control effect" could be observed at the measurement [44]. According to a paper [84], it is written "But tests he performed in March 1945 revealed no such amplification. In his logbook he recorded only small effects, orders of magnitude less than what was predicted by Schottky's theory." H. Welker also wrote in his paper [43], "Initial experiments with Cu_2O were begun, but were interrupted in May 1945 by circumstances over which I had no control."

The documents for H. Welker's patent application with a priority date, 6 April 1945, were destroyed or lost by the war. It was only in 1952 that the procedures at the German Patent Office were resumed after sufficient evidence had been collected in an "adventurous manner" [43]. The patent was granted in 1973 [7]. In his patent, he cited all the related patents and publications, by E. Lilienfeld [1, 3], AEG [28], and O. Heil [33], and R. Hilsch and R. Pohl [30].

H. Welker's work was not known in the 1940s and the early 1950s, but his proposed MISFET was almost the same as that of today's MOSFET, except that the pn junction was not used for the insulation between the source/drain and the substrate, although we have to be careful about the fact that his patent was resumed based on evidence in 1952. The existence of the inversion layer itself was presumed at the semiconductor part of Schottky junctions [85, 86] at the end of the 1930s [23], but **H. Welker's patent was the first one that assumed the use of the surface inversion layer for the FET conduction.**

7.9 Shockley's Group Study for MOSFET from 1945 to 1947

On 16 April 1945, W. Shockley thought of an FET idea, independently of J. Lilienfeld, O. Heil, and H. Welker as shown in Figure 7.14 [29].

In his idea, the concentration of the electrons or holes at the surface of a semiconductor is modulated by the voltage of the gate electrode separated from the semiconductor by an air gap. In his idea, there was no inversion layer, and the conduction was only an accumulation type. The change of the surface carrier concentration can be calculated by the energy band bending caused by the gate electric field.

In May and June, until 23 June 1945, various experiments were conducted using thin Si films deposited on ceramic [29]. It was reported that 90 V was applied laterally across the Si film, and 1000 V was applied to the gate electrode that was separated by an air gap with a distance of less than 1 mm, but that there were no current modulation at all [8, 29]. The effect observed was at least 1500 times smaller than that theoretically expected [29].

On 19 March 1946, John Bardeen explained the cause of no current-modulation for Shockley's FET as shown in Figure 7.15 [87]. The cause was surface states [88, 89]. Under a positive gate bias,

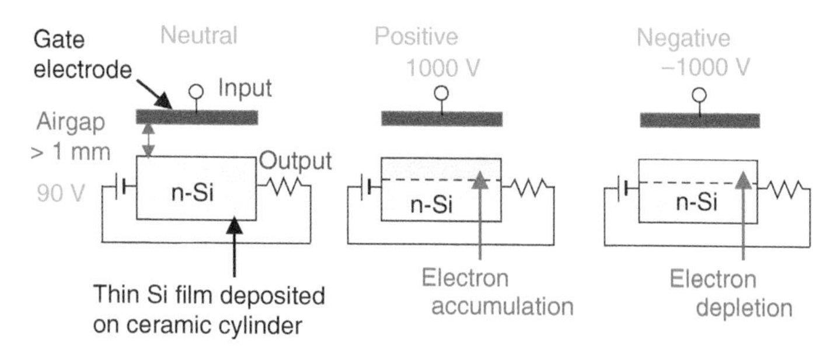

Figure 7.14 MISFET proposed by W. Shockley in April 1945. N-channel FET case. Accumulation conduction type. Source: Adapted from Riordan and Hoddeson [8] and Shockley [29].

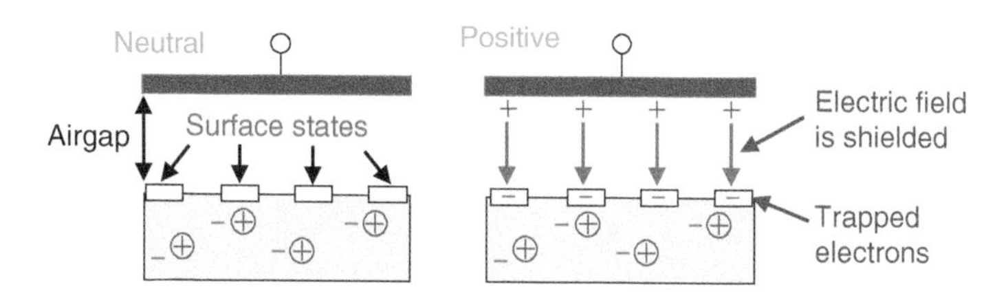

Figure 7.15 Electric field shielding by trapped electrons at surface states. Source: Adapted from Shockley [29].

for example, electrons are trapped by the surface states, and shield the electric field from the gate electrode. The electric field shield prevents the modulation of electron carriers at the surface.

Walter H. Brattain, Robert B. Gibney, and John Bardeen solved the problem by using electrolyte (distilled water in this case) as the gate insulator in November 1947 [90–93] as shown in Figure 7.16.

The resistivity of the distilled water is sufficiently high so that it can be used as the gate insulator in this case, although there was some leakage. The water molecules formed dipoles at the semiconductor surface to enhance the local electric field. The dielectric constant of the water is about 80, and the electric flux density in the water is about 80 times larger than that of air. This is a kind of a vertical MISFET. The current, which was injected from the point-contact source electrode and flowed laterally at the

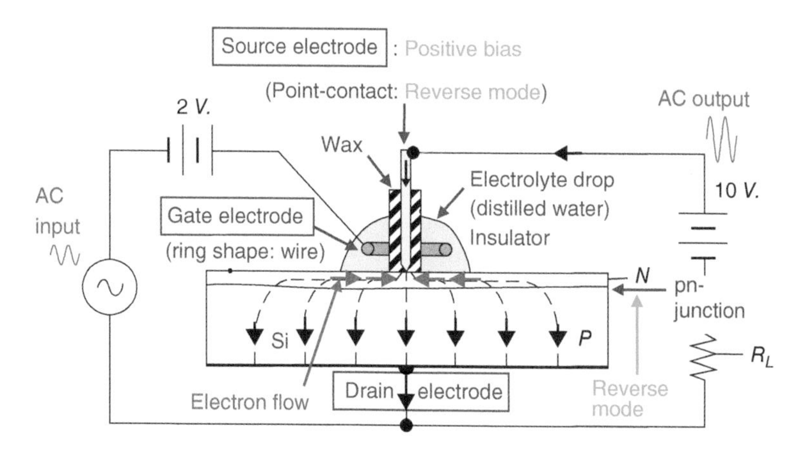

Figure 7.16 J. Bardeen's vertical-type MISFET using electrolyte (distilled water) as the gate insulator in November 1947. Source: Adapted from Bardeen [92].

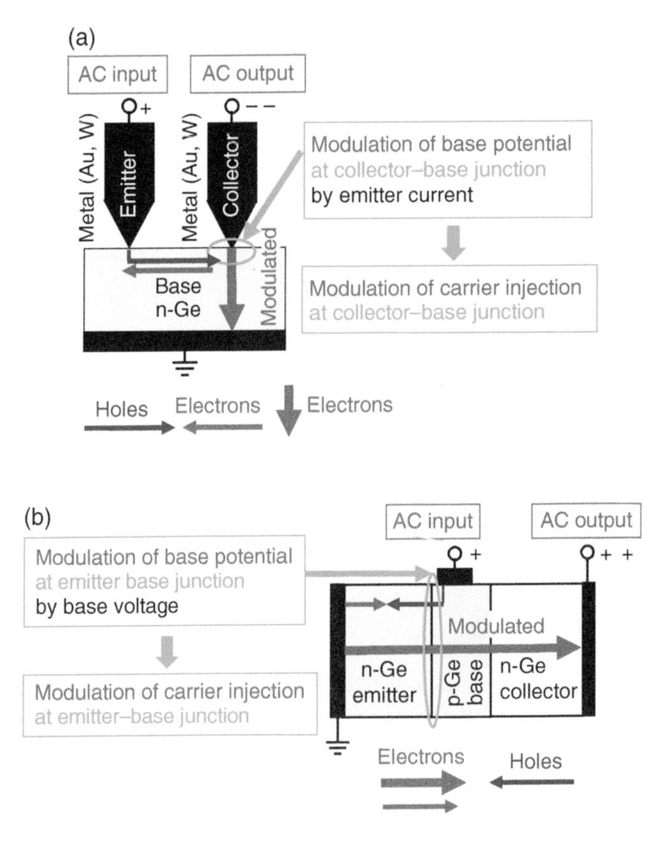

Figure 7.17 Junction transistor mechanism: (a) Point-contact transistor (PCT). (b) Bipolar junction transistor (BJT).

semiconductor surface, was modulated by the voltage of a ring-shaped gate electrode immersed in the water. Sufficiently high amplifications were confirmed by these types of structures in December 1947, but the maximum operating frequency was only 8 Hz, because of the low response time of the electrolytes.

Because of the low operating frequency, they replaced the electrolyte gate insulator with Ge oxide grown by anodic-oxidation on a Ge semiconductor [94]. However, it happened that the Ge oxide serendipitously did not exist and the gate electrode (AC input) directly touched the Ge surface [8, 29, 90–93, 95]. Then, they found AC signal amplification at the source terminal (AC output) at 1 kHz [96]. **This was the first transistor (point-contact transistor: PCT) (Figure 7.17a) that was the first practical solid-state amplifier and that became commercial devices soon after. W. Shockley further invented the bipolar junction transistor: BJT (Figure 7.17b) [97, 98] in January 1948 that became the most popular transistor in the 1950s and 1960s. Usually 'junction transistor' means only BJT, but PCT and BJT are junction transistors and also bipolar transistors and not the FETs that many people had in mind in the development. The operation mechanism was quite different as shown in Figure 7.17 [99, 100].**

Gerald L. Pearson and W. Shockley continued the MOSFET experiment using a quartz (SiO_2) film as the gate insulator as shown in Figure 7.18. They deposited semiconductor films on one side of the quartz that were used as the channel, and gold films on the other side of the quartz that were used as the gate electrode. They measured the conductance modulation by the change in the gate voltage on 26 December 1947 [101]. However, the modulation was only 10% of the expected value [100], and they did not continue the MOSFET development.

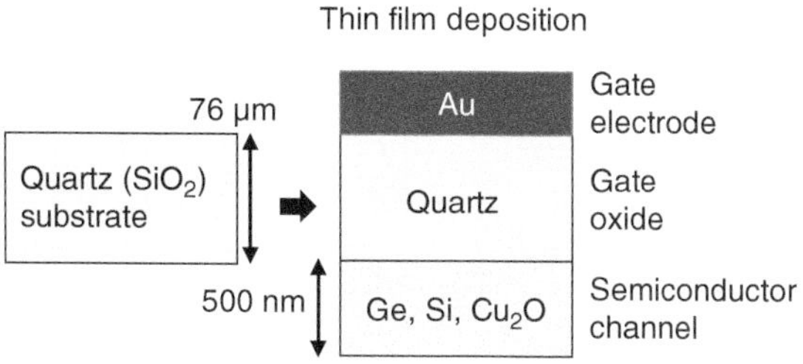

Thin film deposition

76 μm

Quartz (SiO_2) substrate

Au | Gate electrode

Quartz | Gate oxide

500 nm

Ge, Si, Cu_2O | Semiconductor channel

Figure 7.18 MOSFET experiment with SiO_2 gate insulator conducted by W. Shockley and G.L. Pearson on 26 December 1947. Source: Adapted from Shockley [100] and Lojek [104].

7.10 Technology Development in the 1950s Until the First Successful MOSFET Operation in 1960

Until the end of 1950s, there were no technologies to suppress the semiconductor surface state density for use as MISFETs, and thus, W. Shockley thought of a different type of thin-film field effect transistor (TFT) that does not use the semiconductor surface for the conduction in 1952 [45] as shown in Figure 7.19.

This was called the junction FET (JFET) later. The figure shows a p-channel JFET case, where the hole current flows at the center of the p-Ge thin film. There are positively biased n^+-Ge regions on both the sides of the p-Ge film. By changing the bias of the n^+-Ge regions (or gate regions), the channel potential of the p-Ge adjacent to the source region changes, resulting in a change in the hole injection from the source to the channel. W. Shockley formulated the channel current as a function of drain and gate voltages, taking account of the depletion width change across the channel accurately. He found a "pinch off" phenomenon in which the upper and lower inversion layers touch each other and there is no more increase in the drain current by further increasing the drain voltage. This operation region was called the saturation or pentode region. Shockley's formulation was the first accurate formulation of the FET characteristics in the full range of the drain and gate voltages. The formulation distinguished the linear (triode) and saturation (pentode) operation regions for FETs.

The experimental results of n-channel Ge JFETs were published by G.C. Dacey and Ian M. Ross of Bell Laboratories in 1953 [102, 103], after Shockley left Bell Laboratories. **This was the world-first successful FET operation for practical use, and is even now used for commercial products because of its nature of low noise due to no effects of the surface states.** However, the thin-film transistors with upper and bottom gates were not suitable for the monolithic integration at that period, and the bulk substrate FETs with surface conduction had been desired.

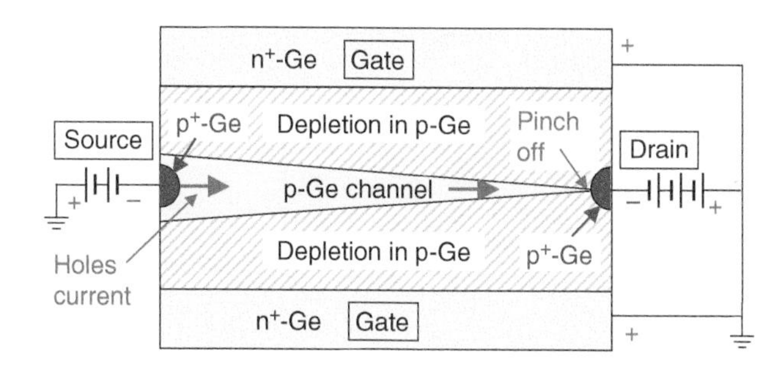

Figure 7.19 Idea of junction FET (JFET) proposed by W. Shockley 1952. Source: Adapted from Shockley [45].

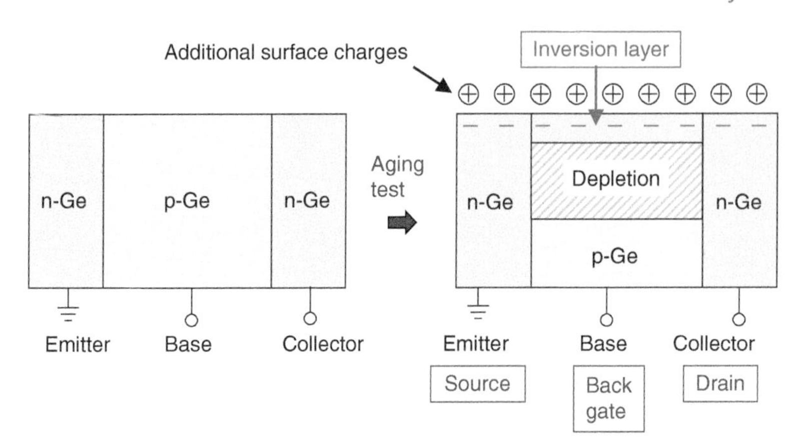

Figure 7.20 Concept of inversion layer connected to source/drain with pn junctions by W. Brown and W. Shockley 1953. The surface conduction In the inversion layer was modulated by base bias (or back gate bias). This structure is not MISFET/MOSFET, but a kind of "one-sided" JFET with only the back side gate. Source: Adapted from Brown and Shockley [46] and Brown [47].

The concept of the inversion layer channel FET with the pn junction source/drain electrically connected was given briefly by W.L. Brown and W. Shockley in January 1953 [46] before Shockley left Bell Laboratories in that year. They conducted aging tests of Ge npn bipolar junction transistors by applying reverse bias to emitter-base and base-collector junctions. They found that a surface inversion conduction layer was created at the base region and that the created inversion layer was electrically connected to the emitter and collector (Figure 7.20).

W.L. Brown published a detailed paper in 1953 [47] after W. Shockley left. He showed that the conduction of the inversion layer changed by changing the base potential via the back base electrode. This is not a MISFET/MOSFET, but a kind of "one-sided" JFET with only a back side gate that is the base. The **surface inversion channel layer** is electrically **connected with** the emitter (or **the source** of the "one-sided" JFET) as shown in the figure. The emitter (or **source** of the JFET)/collector (or **drain** of the JFET) **was insulated** from the base substrate (or **gate substrate** of the JFET) by **pn junctions**. The mechanism of the conduction modulation is similar to that in the previous W. Shockley's JFET [45] case, and the same formula can be used for the electrical characteristics. The measured channel conductance was 1/300 of the theoretical calculation probably due to the low carrier mobility at the surface.

In 1955, I.M. Ross showed a prototype design of the practical MOSFET in his ferroelectric nonvolatile memory device [48] as shown in Figure 7.21. The structure was the first one with the combination of inversion channel at the surface with the insulated front-gate electrode and the source/drain with pn

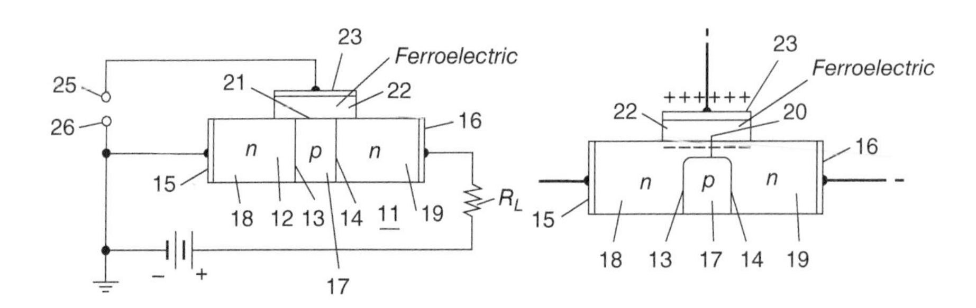

Figure 7.21 Practical MOSFET structure except the gate insulator (ferroelectric material) proposed for memory by I. Ross in 1955. Source: Adapted from Ross [48].

junctions, and it was the same as today's basic MOSFET structure, except for the fact that the gate insulator was a ferroelectric material ($SiTiO_3$).

The last condition having remained for realizing the successful MOSFET was to establish a good gate-insulator/semiconductor interface with a small density of surface-states and fixed-charges. In 1958, M. Attala, E. Tannenbaum, and E.J. Scheibner of Bell Laboratories found that SiO_2 gate films grown by high-temperature oxidation (such as 900 °C) of Si stabilize the Si interface [49, 50] with a small slow state density. In 1959, Joseph R. Ligenza found that high-pressure-steam oxidation ("steam bomb oxidation technique") at lower temperature such as 650 °C in 45 atm gave a good Si/SiO_2 interface without changing the impurity distribution [51, 52].

7.11 Success of MOSFET Operation by D. Kahng and M. Attala in 1960

In 1960, Dawon Kahng and M. Atalla of Bell Laboratories published the first successful operation of MOSFETs [9–12] as shown in Figure 7.22. Mechanically polished n- and p-type Si slices with [111] orientation were used as the substrates. The slices were 1.5 mm^2 and 250 μm thick. The p-type substrates were doped with boron with the concentration of $10^{13} cm^{-3}$ and the resistivity of the substrate was 1000 Ω cm. There was no information on the n-type substrate in the literatures. The substrates were oxidized in 1 atm steam for two hours at 1200 °C to form a 1 μm thick mask-oxide for the source/drain impurity diffusion. The vapor-phase diffusion was carried out at 1250 °C for phosphorus and 1300 °C for boron, respectively, by means of the "closed box technique." The junction depth was 12–25 μm. After etching the mask-oxide, Si surface was etched about 3 μm in a 10 to 1 solution of nitric and hydrofluoric acids for 20 ~ 30 seconds except for the contact portion of the source/drain region. 100 nm gate oxides were grown by "steam bomb oxidation technique" [51, 52] with 45 or 55 atm at 650 °C. Then, 150 nm thick Al gate electrode was evaporated. High-pressure oxidation was used in the fabrication of the first MOSFETs, but normal atmospheric-pressure oxidation at higher temperature such as above 1000 °C was used later by other companies as the standard. Contact holes were drilled through the oxide to the source/drain portions and a gold lead with diameter 38 μm was bonded to the Ge surface in the contact holes. A spring-loaded bronze point-contact was used to make contact with the Al gate electrode. The ratio of the channel width to the channel length was 14. The channel width was 250 μm according to Ref. [104], and thus the channel length was estimated to be 18 μm (or about 20 μm). P-channel MOSFETs showed good characteristics, while n-channel MOSFET characteristics were not good because of the large density of surface states.

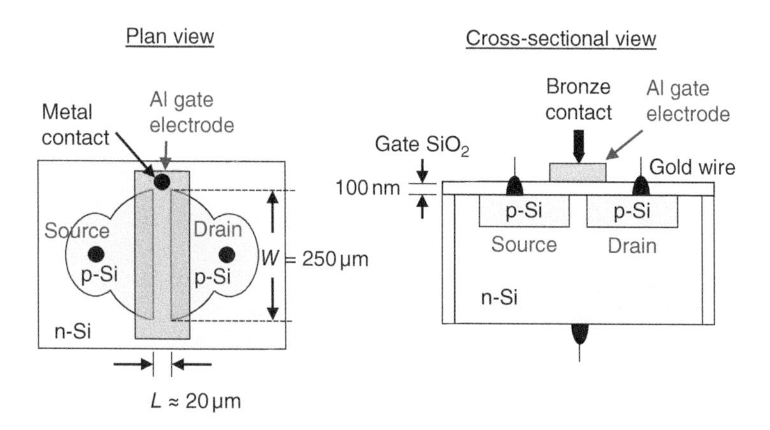

Figure 7.22 The structure of the first successful MOSFETs by D. Kahng and M. Atalla in 1960. Source: Adapted from Kahng [10].

The reason for the success of the MOSFET was choosing the silicon substrate and thermally grown SiO_2 as the gate insulator, as well as the progress in the Si material and various process technologies developed for bipolar junction transistors in the 1950s. They were Ge/Si crystal growth [105–107] and purification [108–110], photolithography [111–113], vapor-phase impurity diffusion [114–117], an oxide mask for protection and impurity diffusion [118, 119], Al-Si Ohmic contact [120, 121], cleaning technology [122–124], and planar technology [125, 126].

The implementation of the Lilienfeld's initial idea for MOSFET had to wait 30 years until those technologies were developed.

7.12 After the First Successful Operation of the MOSFET

Bell Laboratories manager, J. Morton, was not interested in developing large-scale MOS integrated circuits because of the inferior performance of the MOSFET (surface conduction device) to that of bipolar transistors (bulk pn junction device) and also because of the anticipated yield degradation problem caused by defects in the case of large number device integration [127]. However, MOSFETs are the most suitable devices for large-scale integration because of their two-dimensional flat structure, and thus, other companies such as Fairchild, IBM, RCA, and TI started to develop MOSFETs for integrated circuits. **Although high-temperature thermally grown SiO_2 stabilized the Si surface sufficiently for the initial performance to a certain degree, significant instability problems were found in the early 1960s** [128–134] **and it took about 10 years for the MOSFET to be used popularly in the integrated circuit**. By the beginning of the 1970s, ion-implantation [135–137], silicon-gate [138–140], LOCOS isolation [141–143], and PSG reflow [144] technologies were developed, and MOS LSI became the mainstream for large-scale integration [145–148], ultimately resulting in the recent nano-CMOS technology.

7.13 Summary and Conclusions

J. Lilienfeld's MOSFET in 1928 was the first and probably the grand design of the MOSFET, whose structure was found to be the most suitable for miniaturization and large-scale monolithic integration, resulting in today's nano-CMOS VLSI circuits. However, unsurprisingly because of the poor understanding of the semiconductor surface physics, poor technologies to control the semiconductor surface, and the poor purity/quality of semiconductor materials in the 1920s, further new ideas and experimental works were necessary for the first practical implementation by D. Kahng and M. Atalla in 1960.

One of the key technologies that contributed to the implementation was the stabilization of the Si surface by high-temperature oxidation of Si that was applied to the gate insulator. In the 1950s, purity and quality of Si made great progress through the development of bipolar junction transistor products. Also, many important process technologies were developed during that period. The developments of these technologies for the bipolar transistors were inevitable for implementation of MOSFETs.

However, control of the semiconductor surface property was not easy and the Si surface had not really been stabilized only by choosing high-purity Si substrates and the high-temperature gate oxidation. Various reliability instabilities occurred in the 1960s. It took another decade to resolve the reliability issue before the large-scale integration of MOSFET appeared in the market.

Acknowledgment

The old reference documents were collected with the help of the Library of Tokyo Institute Technology. I appreciate the library staff for their kind and great support.

References

[1] Lilienfeld, J.E. (1926). Method and apparatus for controlling electric current. US Patent 1,745,175, filed 8 October 1926 (priority CA272437TA·1925-10-22).

[2] Lepselter, M.P. and Sze, S.M. (1968). SB-IGFET: an insulated-gate field-effect transistor using Schottky barrier contacts for source and drain. *Proceedings of the IEEE* 56 (8): 1400–1402.

[3] Lilienfeld, J.E. (1928). Device for controlling electric current. US Patent 1,900,018, filed 28 March 1928.

[4] (1963). Obituaries: Julius E. Lilienfeld. *Physics Today* 16 (11): 104.

[5] Bottom, V.E. (1964). Invention of the solid-state amplifier. *Physics Today* 17 (2): 24–26.

[6] Kleint, C. (1998). Julius Edgar Lilienfeld: life and profession. *Progress in Surface Science* 57 (4): 251–328.

[7] Welker, H.J. (1945). Halbleiteranordnung zur kapazitiven Steuerung von Strömen in einem Halbleiterkristall (semiconductor arrangement for capacitive control of currents in a semiconductor crystal). Deutshe Patentamt DE 980,084, filed 4 June 1945, priority 7 April 1945.

[8] Riordan, M. and Hoddeson, L. (1997). *Crystal Fire: The Invention of the Transistor and the Birth of the Information Age*. W. W. Norton & Company.

[9] Kahng, D. and Attala, M.M. (1960). Silicon-silicon dioxide field induced surface devices. In: *IRE-AIEEE Solid-State Device Research Conference*. Pittsburgh, PA: Carnegie Institute of Technology.

[10] Kahng, D. (1991). Silicon-Silicon Dioxide Field Induced Surface Devices. *Technical Memorandum Issued by Bell Labs., January 16, 1961 Reprinted in S. M. Sze. Semiconductor Devices: Pioneering Papers*, 583–596. World Scientific Publishing Co.

[11] Kahng, D. (1960). Electric field controlled semiconductor device. US Patent 3,102,230, filed 31 May 1960.

[12] Attala, M.M. (1960). Semiconductor triode. US Patent 3,056,888, filed 17 August 1960.

[13] Kahng, D. (1976). A historical perspective on the development of MOS transistors and related devices. *IEEE Transactions on Electron Devices* 23 (7): 655–657.

[14] Branly, E. (1890). Variations de conductibilité sous diverses influences électriques (variations in conductivity under various electrical influences). *Comptes Rendus l'Académie des Sciences* 111: 785–787.

[15] Falcon, E. and Castaing, B. (2005). Electrical conductivity in granular media and Branly's coherer: a simple experiment. *American Journal of Physics* 73 (4): 302–307.

[16] Bose, J.C. (1901). Detector for electrical disturbances. US Patent 755,840, filed 30 September 1901.

[17] Pickard, G.W. (1906). Means for receiving intelligence communicated by electric waves. US Patent 836,531, filed 30 August 1906.

[18] Henry, H.C. (1906). Dunwoody. Wireless telegraphy system. US Patent 837,616, filed 23 March 1906.

[19] de Forest, L. (1907). Space telegraphy. U.S. Patent 879,532, filed 29 January 1907.

[20] Varian, R.H. and Varian, S.F. (1939). A high frequency oscillator and amplifier. *Journal of Applied Physics* 10: 321–327.

[21] Varian, D. (1984). From the inventor and the Pilot Russell and Sigurd Varian. *IEEE Transactions on Microwave Theory and Techniques* 32 (9): 1248–1263.

[22] Schottky, W. (1938). Halbleitertheoerie der sperrschicht. (semiconductor theory of the junction). *Naturwissenschaften* 26: 843.

[23] Schottky, W. and Spenke, E. (1939). Quantitative treatment of the space charge and boundary-layer theory of the crystal rectifier. *Wiss. Veroff. Siemens-Werken* 18: 225–291.

[24] Mott, N.F. (1938). Note on the contact between a metal and an insulator or semi-conductor. *Mathematical Proceedings of the Cambridge Philosophical Society* 34 (4): 568–572.

[25] Davydov, B. (1938). On the rectification of current at the boundary between two semiconductors. *Comptes Rendns (Doklady) de l'Académie des Sciences de l'URSS* 20 (4): 279–282.

[26] Wilson, A.H. (1931). The theory of electronic semiconductors I. *Proceedings of the Royal Society of London* A133: 458–491.

[27] Wilson, A.H. (1931). The theory of electronic semiconductors II. *Proceedings of the Royal Society of London* A134: 277–287.

[28] Allgemeine Elektricitäts-Gesellschaft (1930). Kontaktgleiehrichter mit zwei durch eine Sperrschicht getrennten Metallelektroden. (Contact rectifier with two metal electrodes separated by a barrier layer). Österreichisches Patentamt, AT13102, filed 11 July 1930, priority 11 July 1929.

[29] Shockley, W. (1976). The path to the conception of the junction transistor. *IEEE Transactions on Electron Devices* 23 (7): 597–620.

[30] Hilseh, R. and Pohl, R.W. (1938). Steuerung yon Elektronenströmen mit einem Dreielektrodenkristall und ein Modell einer Sperrschicht (Control of electron currents with a three-electrode crystal and a model of a barrier layer). *Zeitschrift für Physik* 111: 399–408.
[31] Johnson, J.B. (1964). More on the solid-state amplifier and Dr. Lilienfeld. *Physics Today* 17 (5): 60–62.
[32] Weimer, P.K. (1962). The TFT a new thin-film transistor. *Proceedings of the IRE* 50 (6): 1462–1469.
[33] Heil, O. (1935). Improvements in or relating to electrical amplifiers and other control arrangements and devices. British Patent GB439457 (A), convention date 2 March 1934 (Germany patent application: DEX439457), Application date 4 March 1935, Complete specification accepted 6 December 1935.
[34] Wallmark, J.T. (1957). Field effect transistor. US Patent 2,900,531, filed 28 February 1957.
[35] Sekigawa, T. and Hayashi, Y. (1984). Calculated threshold-voltage characteristics of an XMOS transistor having and additional bottom gate. *Solid State Eletronics* 27 (8/9): 827–828.
[36] Frank, D.J., Laux, S.E., and Fischetti, M.V. (1992). Monte carlo simulation of a 30 nm dual-gate MOSFET: how short can Si go? *International Technical Digest on Electron Devices Meeting* 1992: 553–556.
[37] Fiegna, C., Iwai, H., Wada, T. et al. (1993). A new scaling methodology for the 0.1 ~ 0.025 μm MOSFET. *Symposium on VLSI Technology*, pp. 33–34 (May 1993).
[38] Bidal, G., Boeuf, F., Denorme, S. et al. (2009). High velocity Si-nanodot: a candidate for SRAM applications at 16nm node and below. *2009 IEEE Symposium on VLSI Technology*, pp. 240–241 (June 2009).
[39] Barraud, S., Vizioz, C., Hartmann, J.M. et al. (2020). 7-levels-stacked nanosheet GAA transistors for high performance computing. *2020 IEEE Symposium on VLSI Technology*, pp. 1–2 (June 2020).
[40] Huang, C.-Y., Dewey, G., Mannebach, E. et al. (2000). 3-D self-aligned stacked NMOS-on-PMOS nanoribbon transistors for continued Moore's law scaling. *2020 IEEE International Electron Devices Meeting (IEDM)*, pp. 425–428 (December 2000).
[41] Arsenjewa-Heil, A. and Hell, O. (1935). Eine neue Methode zur Erzeugung kurzer, ungedämpfter, elektromagnetischer Wellen großer Intensität (a new method for the generation of short, undamped electromagnetic waves of high intensity). *Zeitschrift für Physik* 95: 752–762.
[42] Thumm, M. (2001). Historical German contributions to physics and applications of electromagnetic oscillations and waves. *Proc. Int. Conf. on Progress in Nonlinear Science*, Nizhny Novgorod, Russia, pp. 623–643 (January 2001).
[43] Welker, H.J. (1979). From solid state research to semi-conductor electronics. *Annual Review of Materials Research* 9: 1–22.
[44] Weiß, H. (1975). Steuerung von Elektronenströmen im Festkörper (Teil I) (Controlling electron currents in solids (part I)). *Physikalische Blätter* 31 (4): 156–165.
[45] Shockley, W. (1952). A unipolar "field-effect" transistor. *Proceedings of the IRE* 40: 1365–1376.
[46] Brown, W.L. and Shockley, W. (1953). N-type conduction on p-type germanium surfaces. *Physical Review* 90 (2): 336.
[47] Brown, W.L. (1953). N-type surface conductivity on p-type germanium. *Physical Review* 91 (3): 518–527.
[48] Ross, I.M. (1955). Semiconductive translating device. US Patent 2,791,760, filed 18 February 1955.
[49] Atalla, M.M., Scheibner, E.J., and Tannenbaum, E. (1958). Fabrication of semiconductor devices having stable surface characteristics. US Patent 2,899,344, filed 30 April 1958.
[50] Atalla, M.M., Tannenbaum, E., and Scheibner, E.J. (1959). Stabilization of silicon surfaces by thermally grown oxides. *The Bell System Technical Journal* 38 (3): 749–783.
[51] Ligenza, J.R. (1959). Method of treating silicon. US patent 2,930,722, filed 3 February 1959.
[52] Ligenza, J.R. (1962). Oxidation of silicon by high-pressure steam. *Journal of The Electrochemical Society* 109 (2): 73–76.
[53] Pearson, G.L. and Brattain, W.H. (1955). History of semiconductor research. *Proceedings of the IRE* 43 (12): 1794–1806.
[54] Busch, G. (1989). Early history chemistry of the physics and semiconductors-from doubts to fact in a hundred year. *European Journal of Physics* 10: 254–264.
[55] Hoddeson, L., Braun, E., Teichmann, J., and Weart, S. (1992). *Out of the Crystal Maze: Chapters from the History of Solid State Physics*. Oxford University Press.
[56] Hall, E.H. (1879). On a new action of the magnet on electric currents. *American Journal of Mathematics* 2: 287–292.
[57] Thomson, J.J. (1897). Cathode rays. *The Electrician* 40: 104–109.

[58] Thomson, J.J. (1897). Cathode rays. *Philosophical Magazine and Journal of Science, Series 5* 44 (269): 293–316.
[59] Voigt, W. (1930). Elektrisches und Optisches Verhalten von Halbleitern III (Electrical and optical behavior of semiconductors III). *Annalen der Physik* 7: 183–204.
[60] Englehard, E. (1933). Elektrisches und Optisches Verhalten von Halbleitern IX (Electrical and optical behavior of semiconductors IX). *Annalen der Physik* 17: 501–542.
[61] Schottky, W. and Waibel, F. (1933). Die Elektronenleitung des Kupperoxyduls (The electron conduction of the copper oxide). *Physikalische Zeitschrift* 34: 858–864.
[62] Riecke, E. (1898). Zur Theorie des Galvanismus und der Wärme (On the theory of galvanism and heat). *Annalen der Physik* 66 (11): 545–581.
[63] Riecke, E. (1906). Über die Elektronentheorie des Galvanismus und der Wärme (On the electron theory of galvanism and heat). *Jahrbuch der Radioaktivität und Elektronik Hrsg. von Joh. Stark* 3: 24–47.
[64] Drude, P. (1900). Zur Elektronentheorie der Metalle (On the electron theory of metals). *Annallen der Physik* 306 (3): 566–613.
[65] Drude, P. (1900). Zur Elektronentheorie der Metalle; II. Teil. Galvanomagnetische und thermomagnetische Effecte (On the electron theory of metals; part II. Galvanomagnetic and thermomagnetic effects). *Annallen der Physik* 308 (11): 369–402.
[66] Schottky, W. and Deutschmann, W. (1929). Zum Mechanismus der Richtwirkung in Kupferoxydulgleichrichtern (On the mechanism of directivity in copper oxide rectifiers). *Physikalische Zeitschrift* 30: 839–846.
[67] Peierls, R.E. (1980). Recollections of early solid state physics. *Proceedings of the Royal Society of London. Series A, Mathematical and Physical Sciences. The Beginnings of Solid State Physics* 371 (1744): 28–38.
[68] Peierls, R. (1929). Zur Theorie der Galvanomagnetischen Effekte (On the theory of galvanomagnetic effects). *Zeitschrift für Physik* 53: 255–266.
[69] Peierls, R. (1929). Zur Theorie des Hall-Effekts (On the theory of the Hall effect). *Physikalische Zeitschrift* 30: 273–274.
[70] Heisenberg, W. (1931). Zum Paulischen Ausschließungsprinzip (On Pauli's exclusion principle). *Annalen der Physik* 402 (7): 888–904.
[71] Scaff, J.H. and Ohl, R.S. (1947). Development of silicon crystal rectifiers for microwave radar receivers. *Bell System Technical Journal* 26 (1): 1–30.
[72] Seitz, F. and Einspruch, N.G. (1998). *Electronic Genie: The Tangled History of Silicon.* University of Illinoi Press.
[73] Clusiusr, K., Holz, E., and Welker, H. (1942). Elektrische Gleichrichteranordnung mit Germanium als Halbleiter und Verfahren zur Herstellung von Germanium für eine solche Gleichrichteranordnung (Electrical rectifier arrangement with germanium as a semiconductor and method for producing germanium for such a rectifier arrangement). Deutshe Patentamt 966,387 Patented October 3.
[74] Ohl, R.S. (1941). Alternating current rectifier US Patent 2,402,661, filed 1 March 1941.
[75] Ohl, R.S. (1941). Light-sensitive electric device. US Patent 2,402,662, filed 27 May 1941.
[76] Ohl, R.S. (1942). Thermoelectric device. US Patent 2,402,663, filed 11 April 1942.
[77] Scaff, J.H. (1941). Preparation of silicon materials. US Patent 2,402,582, filed 4 April 1941.
[78] Riordan, M. and Hoddeson, L. (1997). The origins of the pn junction. *IEEE Spectrum* 34 (6): 46–51.
[79] Pfann, W.G. and Scaff, J.H. (1949). Microstructures of silicon ingots. *Society of the AIME Metals Transactions* 185: 389–392.
[80] Scaff, J.H., Theuerer, H.C., and Schumacher, E.E. (1949). P-type and N-type silicon and the formation of the photovoltaic barrier in silicon Indots. *Society of the AIME Metals Transactions* 185: 383–388.
[81] Scaff, J.H. (1970). The role of metallurgy in the technology of electronic materials. *Metallurgical and Materials Transactions B* 1: 561–573.
[82] Pfann, W.G. (1974). The semiconductor revolution. *Journal of The Electrochemical Society*, vol. 121, no. 1, pp. 9–15.
[83] Brauer, P. (1936). Zum elektrischen Verhalten von Cupritkristallen (On the electrical behavior of cuprite crystals). *Annalen der Physik, Series 1* 24 (5): 609–624.
[84] Riordan, M. (2005). How Europe missed the transistor. *IEEE Spectrum* 42 (11): 52–57.
[85] Green, M.A. and Shewchun, J. (1973). Minority carrier effects upon the small signal and steady-state properties of Schottky diodes. *Solid-State Electronics* 16 (10): 1141–1150.
[86] Demoulin, E. and van de Wiele, F. (1974). Inversion layer at the interface of Schottky diodes. *Solid-State Electronics* 17 (8): 825–833.

[87] Bardeen, J. (1947). Surface states and rectification at a metal semi-conductor contact. *Physical Review* 71 (10): 717–727.
[88] Tamm, I. (1932). Über eine mögliche art der Elektronenbindung an Kristalloberflächen (On the possible bound states of electrons on a crystal surface). *Physikalische Zeitschrift der Sowjetunion* 1: 733–746.
[89] Shockley, W. (1939). On the surface states associated with a periodic potential. *Physical Review* 56 (4): 317–323.
[90] Brattain, W.H. (1968). Genesis of the transistor. *The Physics Teacher* 6 (3): 109–114.
[91] Brattain, W. (1976). The discovery of the transistor effect: one researcher's personal account. *Adventures in Experimental Physics ε* 5: 1–31.
[92] Bardeen, J. (1948). Three-electrode circuit element utilizing semiconductive materials. US Patent 2,524,033, filed 26 February 1948.
[93] Bardeen, J. (1957). Research leading to point contact transistor. *Science* 126 (3264): 105–112.
[94] Gibney, R.B. (1948). Electrolytic surface treatment of germanium. US Patent 2,560,792, filed 26 February 1948.
[95] Hoddeson, L. (1981). The discovery of the point-contact transistor. *Historical Studies in the Physical Sciences* 12 (1): 41–76.
[96] Bardeen, J. and Brattain, W.H. (1948). Three-electrode circuit element utilizing semiconductive materials. US Patent 2,524,034, filed on 17 June 1948 (a continuation-in-part of application Serial No. 11,165, filed 26 February 1948).
[97] Shockley, W. (1948). Circuit elements utilizing semiconductor material. US Patent 2,569,347, filed 26 June 1948.
[98] Shockley, W. (1949). The theory of p–n junctions in semiconductors and p–n junction transistors. *Bell System Technical Journal* 28 (3): 435–489.
[99] Bardeen, J. and Brattain, W.H. (1949). Physical principles involved in transistor action. *Physical Review* 75 (8): 1208–1225.
[100] Shockley, W. (1950). *Electrons and Holes in Semiconductors: With Applications to Transistor Electronics*. van Nostrand.
[101] Shockley, W. and Pearson, G.L. (1948). Modulation of conductance of thin films of semi-conductors by surface charges. *Physical Review* 74: 232–233.
[102] Dacey, G.C. and Ross, I.M. (1953). Unipolar field-effect transistor. *Proceedings of the IRE* 41: 970–979.
[103] Dacey, G.C. and Ross, I.M. (1955). The field effect transistor. *Bell System Technical Journal* 34 (6): 1149–1189.
[104] Lojek, B. (2007). *History of Semiconductor Engineering*. Springer.
[105] Teal, G.K. (1950). Methods of producing semiconductive bodies. US Patent 2,727,840, filed 15 June 1950.
[106] Sparks, M. and Teal, G.E. (1950). Method of making p–n junctions in semiconductor materials. US Patent 2,631,356, filed 15 June 1950.
[107] Teal, G.K. (1976). Single crystals of germanium and silicon – basic to the transistor and integrated circuit. *IEEE Transactions on Electron Devices* 23 (7): 621–639.
[108] William, G.P. (1952). Principles of zone melting. *Transactions of the American Institute of Mining and Metallurgical Engineers* 194: 747–753.
[109] Pfann, W. (1962). Zone melting. *Science* 135 (3509): 1101–1109.
[110] Theurer, H.C. (1952). Method of processing semiconductive materials. US Patent 3,060,123, filed 17 December 1952.
[111] Lathrop, J.W. and Nall, J.R. (1957). Semiconductor construction. US Patent 2,890,395, filed 3 October 1957.
[112] Lathrop, J.W. (2013). The Diamond Ordnance Fuze Laboratory's photolithographic approach to microcircuits. *IEEE Annals of the History of Computing* 35: 48–55.
[113] Andrus, J. (1957). Fabrication of semiconductor devices. US Patent 3,122,817, filed 15 August 1957.
[114] Fuller, C.S. (1954). Method of forming semiconductive bodies. US Patent 3,015,590, filed 5 March 1954.
[115] Derick, L. and Frosch, C.J. (1954). Manufacture of silicon devices. US Patent 2,804,405, filed 24 December 1954.
[116] Crishal, J.M., Wilcox, W.R., and Sandstrom, J.P. (1962). Impurity diffusion method. US Patent 3,244,567, filed 10 September 1962.

[117] Goldsmith, N. (1965). Method of forming an pn junction by vaporization. US Patent 3,374,125, filed 10 May 1965.
[118] Derick, L. and Frosch, C.J. (1956). Oxidation of semiconductive surfaces for controlled diffusion. US Patent 2,802,760, filed 2 December 1956.
[119] Frosch, C.J. and Derick, L. (1957). Surface protection and selective masking during diffusion in silicon. *Journal of the Electrochemical Society* 104 (9): 547–552.
[120] Matlow, S.L. and Ralph, E.L. (1958). Ruggedized solar cell and process for making the same and the like. US Patent 2,984,775, filed 9 July 1958.
[121] Moore, G.E. and Noyce, R.N. (1959). Method for fabricating transistors. US Patent 3,108,359, filed 30 June 1959.
[122] Lang, G.A. and Stavish, T. (1963). Chemical polishing of silicon with anhydrous hydrogen chloride. *RCA Review* 24 (4): 488–498.
[123] Kern, W. and Puotinen, D.A. (1970). Cleaning solutions based on hydrogen peroxide for use in silicon semiconductor technology. *RCA Review* 31: 187–206.
[124] Kriegler, R.J., Cheng, Y.C., and Colton, D.R. (1972). The effect of HCl and Cl_2 on the thermal oxidation of silicon. *Journal of The Electrochemical Society* 119: 388–392.
[125] Hoerni, J.A. (1959). Method of manufacturing semiconductor devices. US Patent 3,025,589, filed 1 May 1959.
[126] Hoerni, J.A. (1961). Planar silicon diodes and transistors. *1960 IEEE International Electron Devices Meeting (IEDM). IRE Transactions on Electron Devices* 8 (2): 178.
[127] Ross Knox Bassett (2002). *To the Digital Age, Research Labs. Start-Up Companies and the Rise of MOS Technology*. Johns Hopkins University Press.
[128] Deal, B.E., Sklar, M., Grove, A.S., and Snow, E.H. (1967). Characteristics of the surface surface-state(Q_{ss}) of thermally oxidized silicon. *Journal of the Electrochemical Society* 114: 266–274.
[129] Kooi, E. (1967). *The Surface Properties of Oxidized Silicon*. Technische Hogeschool Eindhoven.
[130] Hofstein, R.C. (1967). Stabilization of MOS devices. *Solid-State Electronics* 10: 657–670.
[131] Deal, B.E. (1980). Standardized terminology for oxide charges associated with thermally oxidized silicon. *IEEE Transactions on Electron Devices* 27 (3): 606–608.
[132] Snow, E.H., Grove, A.S., and Fitzgerald, D.J. (1967). Effects of ionizing radiation on oxidized silicon surfaces and planar devices. *Proceedings of the IEEE* 55 (7): 1168–1185.
[133] Snow, E.H., Grove, A.S., Deal, B.E., and Sah, C.T. (1965). Ion transport phenomena in insulating films. *Journal of Applied Physics* 36 (5): 1664–1673.
[134] Balk, P. (1965). Effects of hydrogen annealing on silicon surfaces. *Electrochemical Society Spring Meeting*, vol. 14, no. 1, pp. 237–240.
[135] Ohl, R.S. (1950). Semiconductor translating device. US Patent 2,750,541, filed 31 January 1950.
[136] Moyer, J.W. (1954). Method of making p-n junction semiconductor unit. US Patent 2,842,466, filed 15 June 1954.
[137] Shockley, W. (1954). Forming semiconductor devices by ion bombardment. US Patent 2,787,564, filed 28 October 1954.
[138] Dill, H.G. (1966). Insulated gate field effect transistor (IGFET) with semiconductor gate electrode. US Patent 3,544,399, filed 26 October 1966.
[139] Kerwin, R.E., Klein, D.L., and Sarace, J.C. (1967). Method for making MIS structure. US Patent 3,475,234, filed 27 March 1967.
[140] Watkins, B.G. (1966). IGBT comprising of n-type silicon substrate, silicon oxide gate insulator and p-type polycrystalline silicon gate electrode. US Patent 3,576,478, filed 22 July 1969 (continuation-in-part of Ser. No. 582,053, 26 September 1966).
[141] Kooi, E. and Appels, J.A. (1970). Method of making semiconductor devices with selective doping and selective oxidation. US Patent 3,755,001, filed 8 July 1971, priority 10 July 1970 NL 700206.
[142] Appels, J.A., Kooi, E., Paffen, M.M. et al. (1970). Local oxidation of silicon and its application in semiconductor technology. *Philips Research Report* 25: 118–132.
[143] Kooi, E. (1966). Method of producing a semiconductor device and a semiconductor device produced by said method. US Patent 3,970,486, filed 14 February 1974, priority NL 6614016, filed 5 October, 1966.
[144] Moore, G.E. (1970). Method of semiconductor device wherein film cracking is prevented by formation of a glass layer. US Patent 3,825,442, filed 22 January 1970.

[145] Sah, C.T. (1988). Evolution of the MOS transistor, from concept to VLSI. *Proceedings of the IEEE* 76 (10): 1208–1326.

[146] Iwai, H., Sze, S.M., Taur, Y., and Wong, H. (2013). MOSFETs. In: *Guide to State-of-the-Art Electron Devices* (ed. J.N. Burghartz), 21–36. Wiley-IEEE Press.

[147] Iwai, H. (2023). My LSI development experience at Toshiba (1973–1999). Hard working, exciting, and splendid days. *IEEE EDS Magazine* 1 (1): 41–57.

[148] Iwai, H. (2023). History of junction technologies, Commemorative talk for the 75th anniversary of the transistor. International Workshop on Junction Technology, pp. 1–77 (June 2023).

Chapter 8

The Invention of the Self-Aligned Silicon Gate Process

Robert E. Kerwin*

AT&T Bell Laboratories Fellow, Murray Hill, New Jersey, USA
*Retired

In early February of 1966, Donald Klein, supervisor of a small group of chemists in the Semiconductor Device Development Area of Bell Labs, Murray Hill, NJ, was presented with evidence of yield-limiting problems in the manufacturing of transistor memory arrays at Western Electric, Allentown, PA. Failure analysis at the plant had pinpointed the presence of impurities at the gate interface with the silicon substrate, and misalignments of the multi micron wide gates with respect to the underlying edges of the diffused sources and drains across the 2.54 cm silicon wafers.

On a Monday morning, Klein called together his team, several of whom had extensive experience in the material deposition, etching, and diffusion processes in use at that time. Klein had over the weekend thought that the use of a dual dielectric with appropriate self-limiting etchants under the gate would help preserve cleanliness of the silicon surface through all processing steps. The group set to lively discussion of possible dielectrics and their etchants for the proposed gate insulator sandwich. He assigned me to consider potential improvements in the photolithographic alignment process.

I strode around the small office staring at a well-drawn cross section of the metal-oxide-semiconductor (MOS) gate structure on the blackboard. At that time, I had been with Bell Telephone Laboratories (BTL) for 14 months studying the limits of each step in the photolithographic process including mask making, photoresist light sensitivities, adhesion, and etching control limits, what we might today call the modulation transfer function of each step. I had quite early discovered the oxygen sensitivity of the ultraviolet light-induced polymer cross-linking of negative photoresists that led me to recommend that the alignment and exposure tools be modified to include a nitrogen flood during the ultraviolet (UV) exposure, which led me to become the designated photolithography specialist on our team.

75th Anniversary of the Transistor, First Edition. Edited by Arokia Nathan, Samar K. Saha, and Ravi M. Todi.

While each new generation of smaller transistors was referred to in terms of the gate widths, this was not the critical dimension for alignment control. The critical dimension was the much narrower overlap of the gate edges with the underlying sources and drains to prevent mismatches in capacitance. Such mismatches would lead to slower performance of the transistors. Staring at the blackboard, it occurred to me that if we could reverse the process by putting in the gate first and allowing it to mask the diffusion process, it would provide perfect alignment by defining the edges and overlap with the sources and drains, i.e. *self-alignment*, independent of tool or operator error.

I thought first of suggesting high melting point metallic gates that could stand up to the 930 °C diffusion process. Two of my colleagues, Irving Amron and Bob Lieberman, were experts in solid-state diffusion and were unwilling to risk contamination of their furnaces by extraneous metals. After much thought, I asked them if I might put deposited silicon on the silicon wafer into their equipment. They reluctantly agreed to allow that, but why? I said that deposited silicon could handle the heat and might itself be doped sufficiently by the diffusion to carry enough current to establish the necessary field effect under the gate. Therefore, we went ahead. Jack Sarace devoted full time to establish optimum conditions for depositing polycrystalline silicon uniformly across the wafers and testing conductivity after diffusion. We prepared our first devices with silicon gates at the beginning of May 1966. These were tested by Roger Edwards and Joe Kleimack and found to be functional transistors. On 31 May 1966, I summarized the experiments on two pages of my lab notebook shown here and attached a photograph of the current and voltage behavior under varying gate voltages which show transistor action (see Figure 8.1).

We continued to hone the process through the Fall of 1966 and wrote up a patent application [1] and a paper for external publication [2]. This information was presented by Jack Sarace at the Metallurgical Society (AIME) meeting in New York City in August 1967. This self-alignment process was licensed and has been used broadly through the semiconductor industry for more than five decades during which other substantive changes in materials and processes have contributed as well to continuous improvements in device speed, reliability, and cost (Figure 8.2).

I should note that one member of our team, Joe Kleimack, had also tested the first transistors made by John Bardeen and Walter Brattain in 1947.

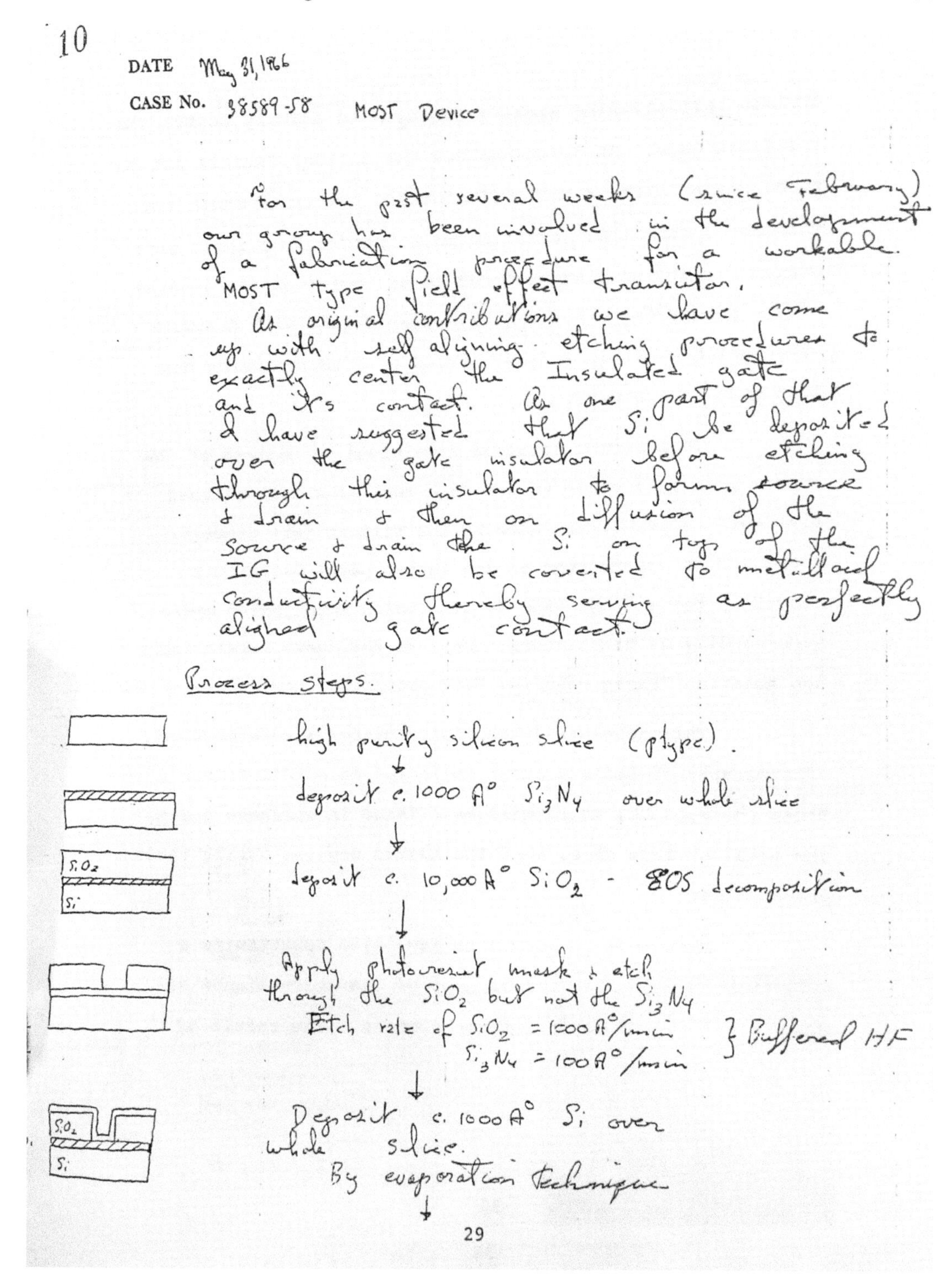

10

DATE May 31, 1966

CASE No. 38589-58 MOST Device

For the past several weeks (since February) our group has been involved in the development of a fabrication procedure for a workable MOST type field effect transistor.

As original contributions we have come up with self aligning etching procedures to exactly center the Insulated gate and its contact. As one part of that I have suggested that Si be deposited over the gate insulator before etching through this insulator to form source & drain & then on diffusion of the source & drain the Si on top of the IG will also be converted to metallic conductivity thereby serving as perfectly aligned gate contact.

Process steps.

high purity silicon slice (p type).

↓

deposit c. 1000 A° Si_3N_4 over whole slice

↓

deposit c. 10,000 A° SiO_2 - EOS decomposition

↓

Apply photoresist mask & etch through the SiO_2 but not the Si_3N_4

Etch rate of SiO_2 = 1000 A°/min, Si_3N_4 = 100 A°/min } Buffered HF

↓

Deposit c. 1000 A° Si over whole slice.

By evaporation technique

↓

29

Figure 8.1 My lab notebook entry describing the steps taken to produce the first transistor incorporating the self-aligned silicon gate process.

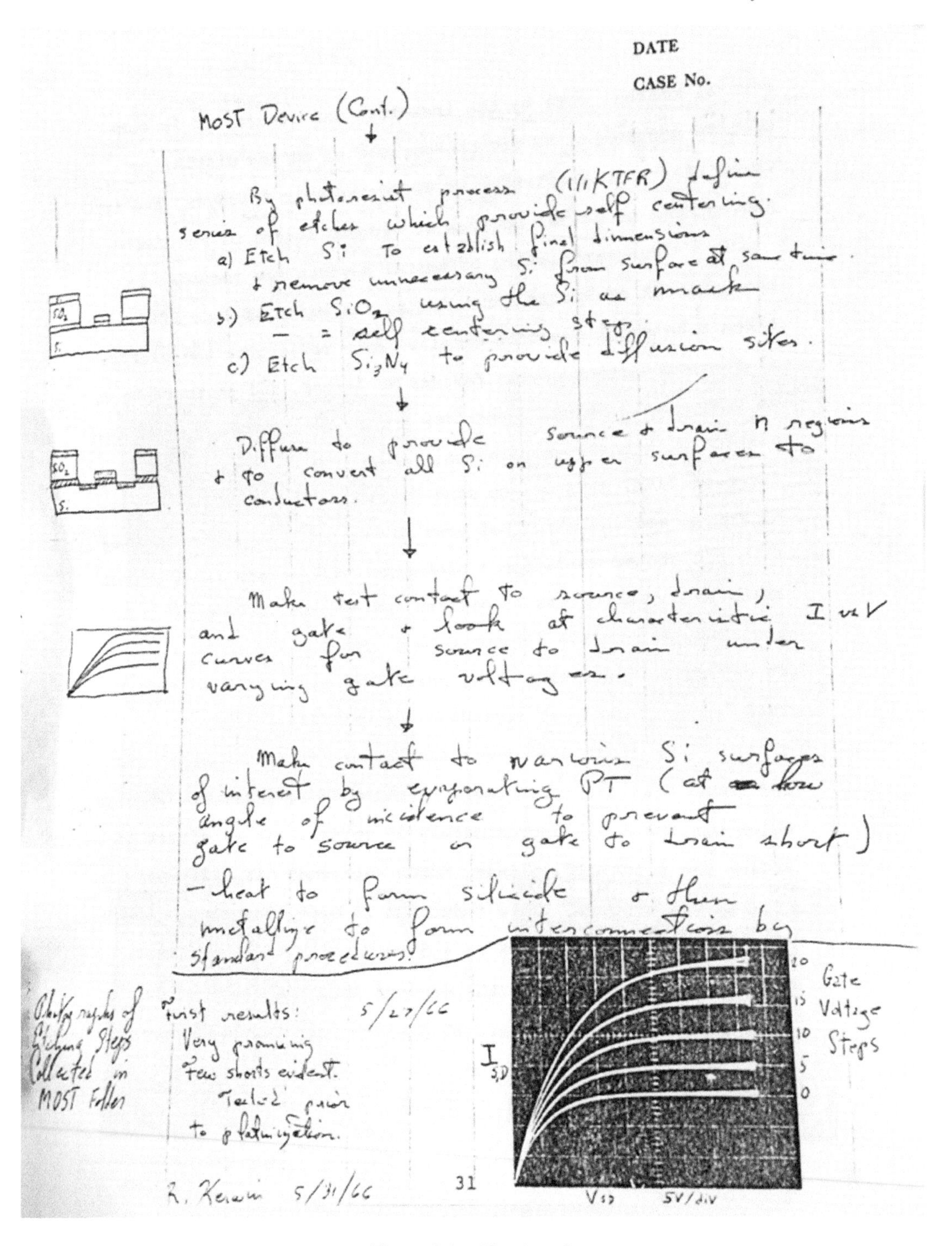

DATE

CASE No.

MOST Device (Cont.)

↓

By photoresist process (1/1 KTFR) define series of etches which provide self centering.

a) Etch Si to establish final dimensions & remove unnecessary Si from surface at same time.

b) Etch SiO_2 using the Si as mask
— = self centering step.

c) Etch Si_3N_4 to provide diffusion sites.

↓

Diffuse to provide source & drain n regions & to convert all Si on upper surfaces to conductors.

↓

Make test contact to source, drain, and gate & look at characteristic I vs V curves for source to drain under varying gate voltages.

↓

Make contact to various Si surfaces of interest by evaporating PT (at low angle of incidence to prevent gate to source or gate to drain short)
— heat to form silicide & then metallize to form interconnections by standard procedures.

Photographs of Etching Steps Collected in MOST Folder

First results: 5/27/66
Very promising
Few shorts evident.
Tested prior to platinization.

R. Kerwin 5/31/66

31

Figure 8.1 (Continued)

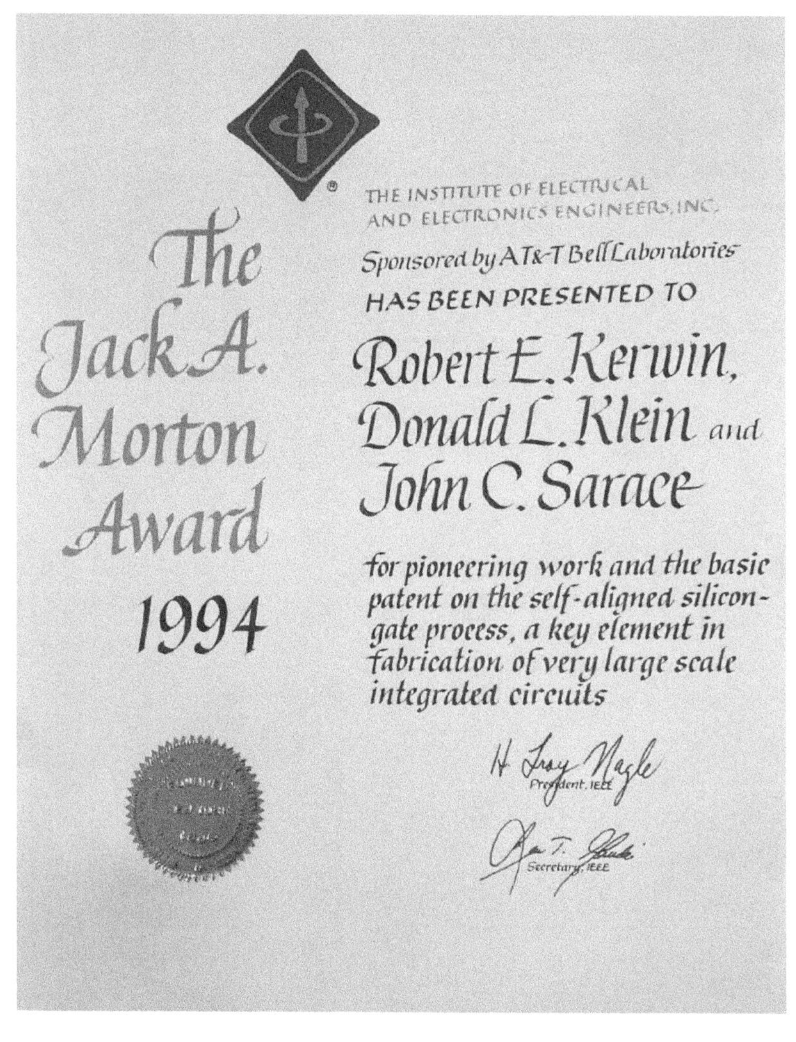

Figure 8.2 Photo of the three co-patentees (from left to right), Robert E. Kerwin, Donald L. Klein, and John C. Sarace, taken at the award ceremony where IEEE presented the Jack Morton award (top photo) at the International Electron Devices Meeting, San Francisco, 1994.

References

[1] Kerwin, R.E. (1969). Method for making MIS structures. US 3475234.
[2] Sarace, J.C., Kerwin, R.E., Klein, D.L., and Edwards, R. (1968). Metal – nitride – oxide – silicon field effect transistors with self aligned gates. *Solid State Electronics* II (7): 653–660.

Chapter 9

The Application of Ion Implantation to Device Fabrication

The Early Days

Alfred U. MacRae

MacTech Consulting, Seattle, WA, USA

9.1 Introduction

The technologies used to manufacture electron devices have changed considerably during the 75 years since the invention of the transistor. That is not surprising since the capability of the devices has evolved from the simple individual point-contact diodes and transistors to the present complex integrated circuits that contain tens of billion transistors. The major change has been in the printing of features that define the devices and their interconnects. The features in the masks and thus the dimensions of the features on the devices have gotten increasingly smaller in the pursuit to follow Moore's Law. Advances in photolithography have been impressive, evolving from the definition of the features using visible light to the presently used shorter wave length extreme ultraviolet light. It is not obvious how smaller features will be printed on future integrated circuits. Perhaps the drive to stack multiple layers of interconnects and transistors will be the driving force behind increases in silicon integrated circuit (SIC) complexity. If so, Moore's Law will continue to guide the increase in complexity of these circuits.

The techniques for the formation of the junctions and ease of making contacts to silicon have evolved also. In the 1950s and 1960s, gases or glasses containing the desired impurities, usually boron and phosphorous, were used as a pre-dep to dope the semiconductor substrate material, typically followed by diffusion into the bulk semiconductor. Then, William Shockley of Bell Labs filed a

75th Anniversary of the Transistor, First Edition. Edited by Arokia Nathan, Samar K. Saha, and Ravi M. Todi.

prophetic patent [1] in 1954, which is recognized as the initiation of a different doping process, namely the shooting of energetic ions of the desired impurities into the substrate, which we now call "ion implantation." Studies of this technique were initiated in the subsequent 10 years, with emphasis on the 600–1000 °C annealing to reduce the damage introduced by the slowing down of the energetic ions and the activation of the implanted impurities; channeling, the guiding of the ions in open channels of the crystal lattice; and profiling of the distribution of the implanted ions. Fortunately, theoretical studies [2] of the implanted ion depth and profiles guided much of these early studies. These studies included the fabrication of diodes and transistors, but reproducibility was limited. From my perspective, a brief review of this early work [3] was an important factor in highlighting the silicon device possibilities of ion implantation and may have stimulated further studies in Japan, Europe, and the United States.

9.2 Device Fabrication

In late 1967, Jack Morton, a demanding Vice President of the Bell Telephone Laboratories, Inc. (Bell Labs) Electronic Components Development Area, heard about the device possibilities of the Bell Labs invented ion implantation process during a trip to Europe, phoned to a direct report in Bell Labs, Murray Hill, New Jersey, and ordered that an ion implantation device effort be started immediately. For reasons that are still not clear to me, I was recruited from the Bell Labs Fundamental Physics Area, promoted to Supervisor, and told to initiate an ion implantation effort before Morton returned from Europe. Understandably, I was hesitant to leave the Fundamental Physics Area where I had an active, heavily supported program on studies of the fundamental properties of surfaces, but I said "yes, let's go." I was provided with an engineer, a technician, lab space, and an open order book. Since there were no commercial ion implantation systems available, it was necessary to design and build one. After discussing the anticipated equipment with friends, I made a rough design of an ion implantation apparatus and went ahead with the acquisition of the necessary equipment. We noticed an Austin, TX-based Accelerator, Inc. advertisement in a Physics Today issue for a 150 keV accelerator that was designed as a neutron generator using the $B_{11}(p, \alpha)$ reaction [4]. We visited Accelerators, Inc. in Austin and were impressed with the simplicity and flexibility of their equipment and their willingness to work closely with us. They used a radio-frequency (rf) cavity as an ion source, which we recognized as a simple source for many ions including our expected boron, phosphorous, and arsenic ions. In addition, the high voltage was generated by simple voltage doubling circuits. We ordered an off-the-shelf 150 keV accelerator from them immediately and it arrived quickly. We went ahead learning how to use the equipment and implanted such ions as oxygen into silicon wafers. We even created SiO_2 buried layers and considered how an oxide buried layer might be used in integrated circuit structural designs. I well remember a respected semiconductor scientist chastising me for leaving the Bell Labs fundamental physics research area to apply ion implantation to make devices – after all, the ions create damage in the silicon. This was not an uncommon view of ion implantation at the time and had to be overcome.

We worked closely with the Accelerator, Inc staff and designed a 300 keV accelerator, purchased a 300 kVolt power supply, obtained a big magnet for mass selection off the floor of an American Physical Society equipment show, purchased vacuum pumps and beam lines, made electrical scanning plates, and our shop made a target holder. We decided that a 20–300 keV accelerator would enable us to make conventional devices that required an implanted ion range of less than 0.5 microns. Accelerators, Inc. was anxious to collaborate with us, since we convinced them that our ion implantation equipment could easily serve as the precursor of future orders, and that happened. Accelerators, Inc became the prominent vendor of moderate energy, i.e. less than 300 keV, implantation equipment for several years. They shipped their first ion implant equipment to Hughes in Newport Beach, CA, in 1970. With arrival of the equipment at Bell Labs, we assembled the system by mid-1968 and were ready to start our device studies (Figures 9.1 and 9.2).

Figure 9.1 Bell Labs, 1968 ion implantation system high voltage end.

Figure 9.2 Bell Labs, 1968 ion implantation system target end.

While we tested out and calibrated the new ion implantation equipment, I set four challenging objectives for our group;

- Develop a process to make reproducible, high-quality silicon diodes.
- Develop a process to make high-frequency bipolar transistors.
- Develop a process to enable the growth of high-quality epitaxial layers over high dose, subsequently amorphous silicon used for buried collector layers.
- Introduce ion implantation technology into the (AT&T) Western Electric Co. semiconductor device factories in Reading and Allentown, PA.

Note that *research* was not included in this list. I had done research earlier and now my objectives were to make something practical and to introduce into a manufacturing environment.

After gaining confidence in implanting 30–300 keV boron and phosphorous into silicon and determining that the doping concentration was reproducible and was uniform over the area of 3″ OD wafers, we made it known in Bell Labs in mid-1968 that we were willing to make experimental devices. The first one was a self-aligned gate MOSFET with Schottky barriers [5], using 150 keV boron ions at a dose of 1.4×10^{14} ions cm^{-2}. This was followed by making a *Low Power Bipolar Transistor Memory Cell* [6] using a double implant of 50 and 150 keV boron ions with patterned photoresist serving as the implantation mask. This work was the subject of a talk given at the 1969 Solid State Circuits meeting and was the recipient of the 1969 conference Outstanding Paper Award [7]. This device work was greatly facilitated by the ready availability of Murray Hill clean room facilities and technicians who were skilled in fabricating devices and integrated circuits.

Despite the success of our early ion implant device work, there were still vocal skeptics. We listened to complaints that the equipment was too expensive and that its footprint was too large to fit in a factory clean room. I then went ahead and designed and had made a vertical 60 keV implanter with a spinning disc mechanical scan, with the objective that the apparatus could be rolled into a lab containing conventional door frames. We obtained considerable use of both the 60 and 300 keV implanters.

Fortunately, several Bell Labs engineers, who designed devices and integrated circuits for manufacture in the Reading, PA Western Electric, Co (WeCo) manufacturing plant had immediate applications for our Murray Hill ion implantation activity. They pulled the need for ion implantation and were enthusiastic supporters of the technology. The first, and by far the simplest application, was high resistance resistors for a linear integrated circuit. Resistors fabricated with conventional diffusion technology occupy considerable space on the chip and were a challenge to control. The WeCo Reading engineers did the pattern etching in the oxide, drove the wafers the 100 miles to us in Murray Hill, where we implanted boron with a dose of $1.5 \times 10^{12} cm^{-2}$ at 150 keV to produce the resistors. We then annealed the damage during the growth of a 0.1 micron oxide over the implanted area and drove the wafers back to the Reading, PA WeCo integrated circuit facility where the metallization of the circuits was completed and the chips packaged and tested and then inserted into communication's equipment that was shipped to a Bell System Operating Company. As you can well imagine, our Murray Hill group celebrated this 1968 achievement. While it is dangerous to make such claims, this application of ion implantation may be the first manufactured bipolar integrated circuit that ended up in shipped equipment. This early success [8] was followed by fabricating 200, 1500, and 3000 Ω resistors on several linear integrated circuits. Much to our delight, we were able to demonstrate a tight, reproducible ±3% spread in the resistor values and their temperature coefficients, something unheard of in a typical factory setting. Plus, we concluded that every implanted ion resulted in an electrically active boron atom. This simple application gained us considerable support in the Reading Bell Labs and WeCo organizations plus it generated credibility when I requested an increased budget and ability to hire traditional talented engineers/physicists/chemists.

This application was quickly followed by a request to produce a hyperabrupt junction diode – a voltage variable capacitor, which was a much greater challenge than making a simple resistor [9]. Again, the spread of the voltage necessary to achieve the desired capacitance was ±3%. These devices had a tenfold capacitance control over its voltage range and were used in a FM modulator that was part of a factory-shipped communications system. Similar uniformity and reproducibility were obtained in making Junction field-effect transistors for use as current limiters in several integrated circuits destined for shipment in communications equipment.

As we gained confidence in using ion implantation to make these unusual devices, we tackled the need to make an impact ionization avalanche transit time (IMPATT) avalanche double drift diode to produce a 50 GHz oscillator having one watt output. This device is essentially two complementary avalanche diodes in series with a tightly controlled doping profile requiring a double implant of 80 and 200 keV boron ions [10].

Even though we focused on making bipolar devices and integrated circuits, we also used ion implantation to make metal-oxide-semiconductor (MOS) devices, especially complementary MOS (CMOS) structures [8] that required precise doping to control the n-type transistors in the p-type tubs. Little did we appreciate at the time that we were almost a decade ahead of our time in pioneering the use of ion implantation to make easily manufactured CMOS integrated circuits.

We also used ion implantation to make the defect free [11], low leakage, challenging 10^6 diode array for the imaging device used in PICTUREPHONE®. Previous attempts to make this device by conventional techniques were difficult due to the undesirable and numerous defects in the resultant display. Subsequently, a team of Bell Labs and WeCo engineers then designed and had made several 30 keV ion implanters, the PR-30, to make these diode arrays, expecting that a new factory was needed to make these arrays. This service was heavily marketed and was viewed as a future major service offering of the Bell Operating companies. In the meantime, the CCD light imaging device was invented [12], obviating the need for the silicon diode array. PICTUREPHONE never did achieve the expected customer acceptance. Of course, now we use a similar service that utilizes the power of our personal computers.

During this time frame, we made double implant [13], arsenic emitter, 0.3 μm boron base, high-frequency transistors [14] with f_T of 8.7 GHz, hFE >1000, which were destined for long-distance transmission products, driven by PICTUREPHONE.

We also solved the problem of growing epitaxial films over heavily doped ion implanted buried collector layers for the common junction isolated bipolar integrated circuits [15]. While silicon crystalline defects are annealed out at ~850 °C for the typical dose of 10^{12}–$10^{15}\,cm^{-2}$, this was not so for the high doses of ~$10^{16}\,cm^{-2}$ required for the buried collector of ~$10\,\Omega\,square^{-1}$ sheet resistance. At these high doses, the silicon becomes amorphous and the crystalline perfection was not restored even after long-time, high temperature annealing and thus it was not possible to grow the necessary high-quality epitaxial films on top of the buried layer arsenic implants. This problem was solved by consuming the amorphous silicon in a thermally grown oxide prior to the epitaxial film growth. Interestingly, while the arsenic atoms diffused into the substrate silicon, the defects in the amorphous silicon were not plowed ahead of the growing oxide. It turned out that this process was widely used worldwide for many years [16].

As a result of these early device successes, we concluded that an important use of ion implantation was the use of low energy, 30 keV ions as a pre-dep followed by diffusion and that patterned photoresist was a usable mask for the energetic ions.

The references summarize our early contributions to this technology [15, 17].

9.3 Summary

The use of ion implantation has increased considerably since its invention in 1954. Our Bell Labs activity pioneered in the design of the implanter equipment and its broad use for the fabrication of discrete silicon devices and integrated circuits, but most importantly, we were able to transfer its acceptance and use into the factory environment. We were fortunate to have access to the integrated circuit fabrication facilities and technicians of Bell Labs and WeCo as well as the talented scientists in my Murray Hill, NJ Bell Labs organization and in the Bell Labs and WeCo of Reading, PA.

Acknowledgments

It would be remiss by not acknowledging the important contributions of the Hughes, Newport Beach, CA group on the widely used application of ion implantation to threshold voltage adjustment of MOS transistors in integrated circuits [18].

References

[1] Shockley, W. (1957). Forming semiconductive devices by ionic bombardment. US Patent 2,787,564, filed 28 October 1954, issued 2 April 1957).

[2] Lindhard, J., Scharff, M., and Schiott, H.E. (1963). *Matematisk-fysiske Meddelelser*, vol. 33, 1. Danske Videnskabernes Selskab.

[3] Gibbons, J.F. (1968). Ion implantation in semiconductors –Part I: Range distribution theory and experiments. In *Proceedings of the IEEE* 56 (3): 295–319 (March 1968). https://doi.org/10.1109/PROC.1968.6273.

[4] Rutherford, E. (1911). The scattering of α and β particles by matter and the structure of the atom. *Philosophical Magazine* 21: 669. N. Bohr Dr. phil. (1913). On the theory of the decrease of velocity of moving electrified particles on passing through matter. p.10.

[5] Lepselter, M.P., MacRae, A.U., and MacDonald, R.W. (1969). SB-IGFET, II: An ion implanted IGFET using Schottky barriers. In *Proceedings of the IEEE* 57(5): 812–813 (May 1969). https://doi.org/10.1109/PROC.1969.7089.

[6] Hodges, D.A., Lepselter, M.P., Lynes, D.J. et al. (1969). Low-power bipolar transistor memory cells. *IEEE Journal of Solid-State Circuits* 4 (5): 280–284. https://doi.org/10.1109/JSSC.1969.1050016.

[7] Hodges, D., Lepselter, M., MacDonald, R. et al. (1969). Low power bipolar transistor memory cells. *IEEE International Solid-State Circuits Conference*. Digest of Technical Papers, Philadelphia, PA, USA, (19–21 February 1969), pp. 194–194, https://doi.org/10.1109/ISSCC.1969.1154688.

[8] MacRae, A.U. (1971). Device fabrication by ion implantation. *Radiation Effects* 7 (1–2): 59–63.

[9] Foxall, G.F. and Moline, R.A. (1969). Advantages of ion implantation in fabricating hyperabrupt diodes. *International Electron Devices Meeting*, Washington, DC, USA (29–31 October 1969). https://doi.org/10.1109/IEDM.1969.188193.

[10] Seidel, T.E. and Scharfetter, D.L. (1970). High-power millimeter wave IMPATT oscillators with both hole and electron drift spaces made by ion implantation. In *Proceedings of the IEEE* 58 (7): 1135–1136 (July 1970). https://doi.org/10.1109/PROC.1970.7861.

[11] Pickar, K.A., Dalton, J.V., Seidel, H.D., and Mathews J.R. (1971). Electrical Properties of Silicon Diode Array Camera Targets Made by Boron Ion Implantation. *Applied Physics Letters* 19: 43–44. https://doi.org/10.1063/1.1653815.

[12] Boyle, W.S. and Smith, G.E. (1970). Charge Coupled Semiconductor Devices. *Bell System Journal* 49: 1587.

[13] Parrillo, L.C., Payne, R.S., Davis, R.E. et al. (1980). Twin-tub CMOS - A technology for VLSI circuits. *International Electron Devices Meeting*, Washington, DC, USA, (8–10 December 1980), pp. 752–755. https://doi.org/10.1109/IEDM.1980.189946.

[14] Fujinuma, K., Sakamoto, R., Abe, T. et al. (1970). *Japan Society of Applied Physics* 39 (Supplement 71): 801.

[15]MacRae, A.U. (1971). *Proceedings of the II International Conference on Ion Implantation in Semiconductors, Physics and Technology, Fundamental and Applied Aspects, Garmisch-Partenkirchen*, Bavaria, Germany (24–28 May 1971), pp. 329.

[16] MacRae, A.U. (1975). Method of forming buried layers by ion implantation. US Patent 3,895,965, (22 July 1975).

[17] MacRae, A.U. (1971). Device fabrication by ion implantation. *Radiation Effects* 7: 50.

[18] Dill, H.G., Toomba, T.N., and Bauer, L.O. (1971). Proceedings of the II International Conference on Ion Implantation in Semiconductors, Physics and Technology, Fundamental and Applied Aspects, Garmisch-Partenkirchen, Bavaria, Germany (24–28 May 1971), pp. 315.

Chapter 10

Evolution of the MOSFET

From Microns to Nanometers

Yuan Taur

Electrical and Computer Engineering Department, University of California, San Diego, CA, USA

10.1 Introduction

Since the invention of the bipolar transistor in 1947, there has been an unprecedented growth of the semiconductor industry, with an enormous impact on the way people work and live. In the last 40 years or so, by far the strongest growth area of the semiconductor industry has been in the silicon very-large-scale-integration (VLSI) technology in which the building block is the metal-oxide-semiconductor field-effect transistor (MOSFET). The idea of modulating the surface conductance of a semiconductor by the application of an electric field was first conceived in 1926 [1]. However, early attempts to fabricate a surface-field-controlled device were not successful because of the presence of large densities of surface states which effectively shielded the semiconductor surface from the influence of an external field. The first MOSFET on a silicon substrate using SiO_2 as the gate insulator was fabricated in 1960 [2].

Advances in lithography and etching technologies have enabled the industry to scale down transistors in physical dimensions, and to pack more transistors in the same chip area. Such progress, combined with a steady growth in chip size, resulted in an exponential growth in the number of transistors and memory bits per chip. The technology trends up to 2020 in these areas are illustrated in Figure 10.1. Dynamic random-access memory (DRAM) chips consisting of one-transistor memory cells [3] make the main memory of today's computers [3]. Circuit techniques are utilized to refresh and maintain the charge stored on a reverse-biased p–n junction. One remarkable feature of silicon devices that fueled the rapid growth of the information technology industry is that their speed increases and their cost decreases as their size is reduced. The transistors manufactured in 2020 were 10 times faster and occupy

75th Anniversary of the Transistor, First Edition. Edited by Arokia Nathan, Samar K. Saha, and Ravi M. Todi.

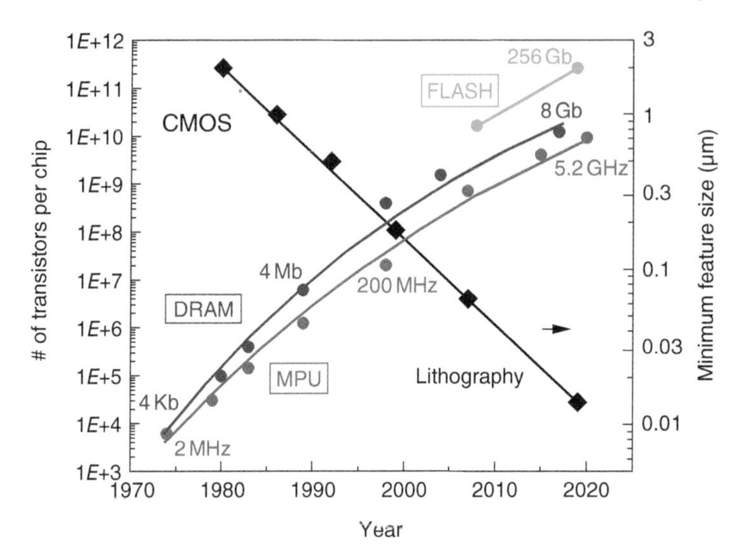

Figure 10.1 Trends in lithographic feature size, number of transistors per chip for DRAM and microprocessors (MPU), and number of memory bits per chip for Flash.

less than 1% of the area of those built 20 years earlier. This is illustrated in the trend of microprocessor units (MPUs) in Figure 10.1. The increase in the clock frequency of microprocessors is the result of a combination of improvements in microprocessor architecture and transistor speed.

With the MOSFET size shrinking by 100–1000 times from microns to nanometers over the past 50 years, the basic transistor structure and process also underwent a number of evolutionary changes over the same period. This article reviews those structural changes and the innovative processes that delivered them. It is organized into three time periods: Before 1980, From 1980 to 2000, and After 2000.

10.2 The Early Days: Before 1980

MOSFETs before 1980 are of the scales of a few microns. A key breakthrough was the invention of integrated circuit (IC) [4], a tiny silicon chip containing hundreds of transistors, diodes, and resistors. This microcircuit made possible the production of electronic systems with higher operating speeds, capacity, and reliability at a significantly lower cost.

10.2.1 From Aluminum Gate to Polysilicon Gate

In the beginning, MOSFETs were fabricated with aluminum as the gate material. Since aluminum melts at a temperature much lower than the 800–1000 °C required for the source–drain anneal, the source–drain must first be implanted with a mask and annealed, as shown in Figure 10.2a. After that, the gate oxide is grown and aluminum is deposited. A second mask is then used to pattern aluminum. The aluminum gate must fully cover the region between the source and drain. Any ungated region cannot be strongly inverted by the gate field, which adversely degrades the on current of the MOSFET. To ensure that there is no underlap between the gate and the source–drain regions, the gate mask is designed to account for the overlay error between the two masks. This results in excessive overlap capacitances in some of the device regions on the chip, and therefore a slower switching speed.

In the polysilicon gate process [5] outlined in Figure 10.2b, the gate is patterned first with a mask. Then, the source–drain regions are implanted, self-aligned to the gate as the polysilicon gate blocks the

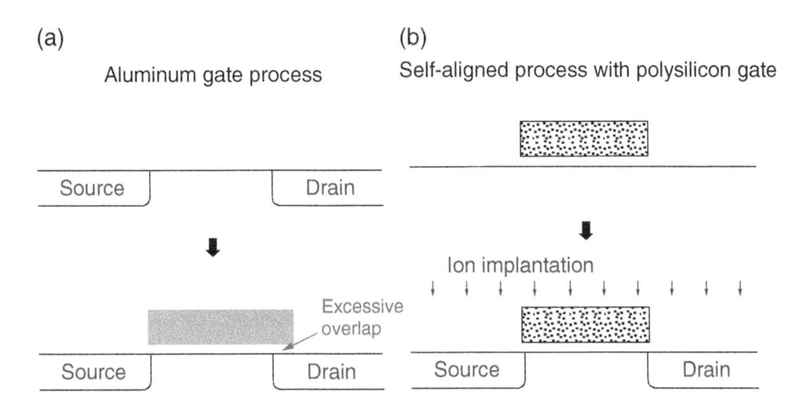

Figure 10.2 (a) Top: Source–drain region formation with a blocking mask. Bottom: Aluminum gate deposition and etching with another mask. (b) Top: Polysilicon gate deposition and etching with a gate mask. Bottom: Source–drain implantation with no mask.

implant from the silicon region under it. After that, source–drain anneal can be performed with the polysilicon gate in place. Self-aligned polysilicon gate process is a key breakthrough in the MOSFET technology that led to higher switching speeds because the parasitic gate-to-source/drain overlap capacitance is minimized.

10.2.2 From pMOS to nMOS to CMOS

Early on in the MOSFET technology, pMOS were used with an n-type substrate because they are easier to isolate as the field oxides are usually positively charged which turns the silicon region under the field oxide n-type. pMOS, however, has a lower performance than nMOS because of the lower hole mobility. It was soon figured out that nMOS can be isolated by adding a field implant to raise the p-type concentration under the field oxide so that it does not invert to n-type in spite of the positive oxide charge. This enabled the industry to switch to nMOS on a p-type substrate.

A major milestone in the MOSFET technology development is the invention of the CMOS (Complementary MOS) circuit [6]. It consists of n-channel and p-channel MOSFETs connected in parallel at the input and in series at the output such that there is no steady-state conducting path between the power supply terminals. Although CMOS occupies a larger area and has a longer delay time as the input signal has to turn on/off two MOSFETs, it becomes the key constituent of logic circuits because of its negligible standby power consumption. In a CMOS chip, nMOS are fabricated in p-type background, called p-well, whereas pMOS are fabricated in n-well, as shown in Figure 10.3. The substrate can be either n- or p-type, lightly doped well below the levels needed for the wells.

10.3 From 1980 to 2000

In the time period between 1980 and 2000, the size of MOSFET shrinked from ~2 micron to ~200 nm. Its structure underwent a number of evolutionary changes, as discussed below.

10.3.1 From n^+ Poly Gate to Dual Poly Gates

For MOSFETs of channel lengths 500 nm or longer, n^+-doped polysilicon gates are used for both nMOS and pMOS. For nMOS, the work function of n^+ poly gate yields a low (magnitude-wise) threshold voltage; hence, additional p-type channel implant is made to raise the threshold voltage to the

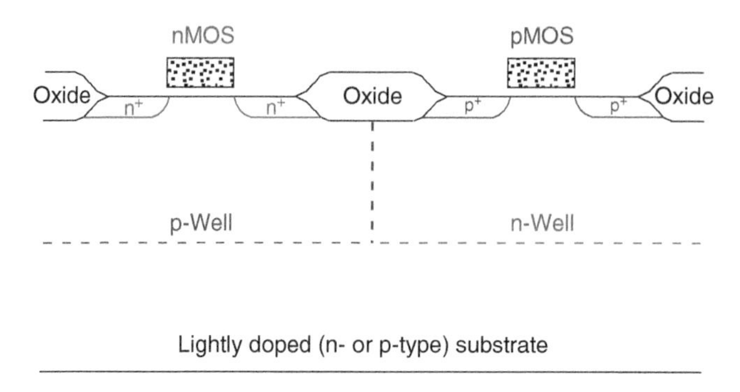

Figure 10.3 Schematic cross-section of CMOS, consisting of nMOS in p-well and pMOS in n-well in a silicon substrate.

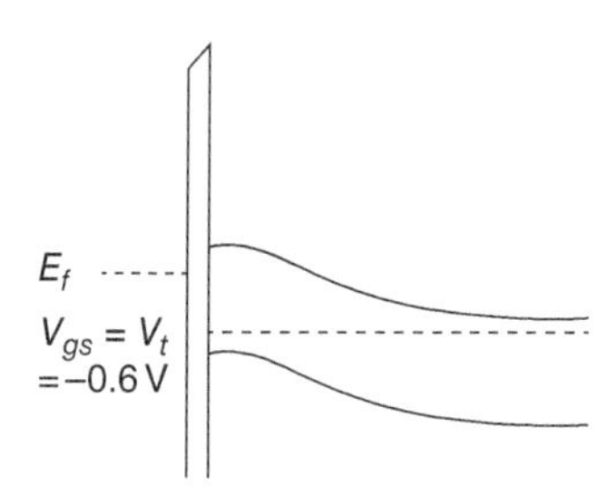

Figure 10.4 Band diagram at threshold of a buried channel pMOS with n^+ poly gate.

desired magnitude. For pMOS, on the other hand, the work function of n^+ poly gate yields too high (magnitude-wise) a threshold voltage. To bring it down, a counter-doped channel with p-type dopant is needed. This, however, reverses the surface field at the threshold condition and turns the pMOS into a buried channel device, with the band diagram shown in Figure 10.4. In general, it is more difficult to control the threshold voltage and the short-channel effect (SCE) of a buried channel MOSFET. It hinges on cancelation of the opposite types of dopants at certain precise depths. Below 500 nm channel length, even lower threshold voltages are required (see Figure 10.8) while higher n-well doping densities are needed to control the SCE; therefore, it is no longer practical to use n^+ poly gates on pMOS.

CMOS technologies below 500 nm use dual polysilicon gates, namely, n^+-doped polysilicon gates for nMOS and p^+-doped polysilicon gates for pMOS [7]. Ideally, after patterning of the undoped polysilicon gates, the n^+/p^+ gates can be doped with the same implants as the n^+/p^+ source–drain. One problem that caused the delay of p^+ poly adoption is boron penetration from the p^+ poly through the gate oxide when annealed in certain hydrogen containing ambient. Once in the substrate, boron can shift the pMOS threshold magnitude down to far below the designed value. One solution is to form a thin nitrided layer of the gate oxide which then acts as a barrier to boron diffusion.

10.3.2 Self-Aligned Silicide

The sheet resistance of heavily-doped source–drain regions or polysilicon gates is of the order of 50–500 Ω square^{-1}. This often results in excessive series resistance between the MOSFET channel and metal contacts. It leads to the degradation of switching currents and gate delays. A self-aligned silicide process has been implemented below a MOSFET channel length of ~1 μm to reduce the sheet resistance to 5–10 Ω square^{-1}. As depicted in Figure 10.5, conformal oxide or nitride films are deposited by CVD (Chemical Vapor Deposition) over the wafer following the source–drain formation and anneal

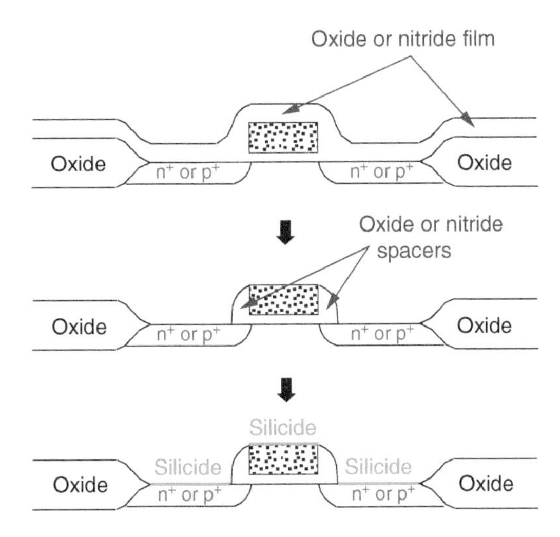

Figure 10.5 Self-aligned silicide process.

(Figure 10.2b). The film is then directionally etched in an RIE (Reactive Ion Etching) process. This leaves the oxide or nitride spacers along the gate edges. A metal film (Ti, Co, or Ni) is then blanketly deposited over the wafer. A subsequent anneal forms silicide, e.g. $TiSi_2$, where the metal film is in contact with silicon or polysilicon. The unreacted metal films over the spacers and the field oxide are then stripped in a selective wet etch afterward [8].

10.3.3 Sallow Trench Isolation

The field isolation shown in Figure 10.5 is called LOCOS (Local Oxidation) or Semi-ROX (Semi-Recessed Oxide) formed by selective oxidation using a patterned nitride mask. While it produced a tapered transition (called the Bird's Beak) between the field and the active regions with gentle topography for ease of subsequent processing (e.g. gate etching), the tapered regions took up too much space that impacted the circuit density. Below the 250 nm CMOS generation, a shallow trench isolation (STI) process is developed to replace the LOCOS process [9]. Figure 10.6 shows that isolation regions of variable widths are etched into silicon, followed by refilling with CVD oxide. A CMP (Chemical Mechanical Polish) process is then employed to planarize the wafer. To prevent "dishing" of a wide isolation region during the CMP operation, dummy patterns may be required to subdivide any wide isolation regions into a multitude of narrower isolation regions.

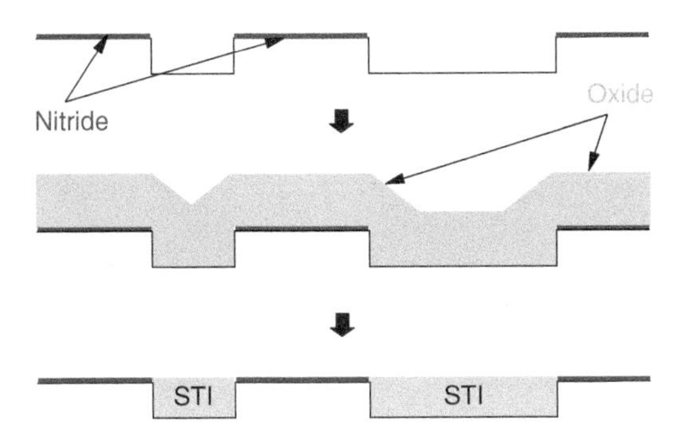

Figure 10.6 Shallow trench isolation process.

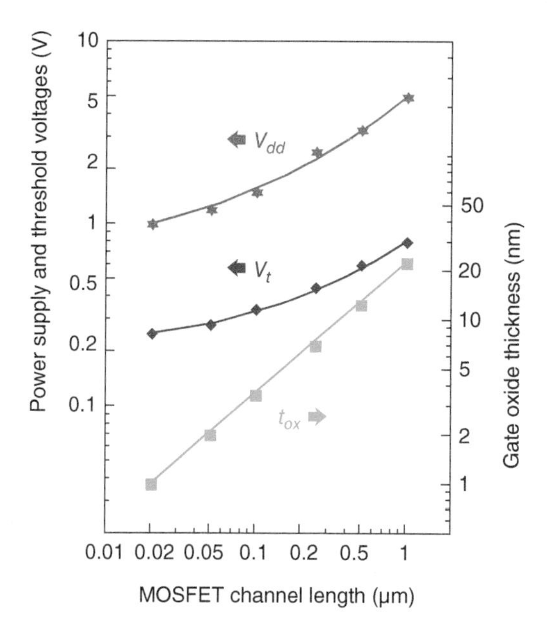

Figure 10.7 Trends of power-supply voltage, threshold voltage, and gate oxide thickness versus channel length for CMOS technologies from 1 to 0.02 μm.

10.3.4 Power Supply and Threshold Voltages

When the MOSFET channel length is scaled down through the VLSI technology generations, the power supply voltage and the threshold voltage, as well as the gate oxide thickness, are also reduced, as shown in Figure 10.7 [10]. The rate of the change in voltages, however, is much slower than the rate of change in channel length. This is because when turning off a MOSFET with the gate voltage (V_{gs}) below the threshold voltage (V_t), the source-to-drain current follows an exponential dependence, $I_{ds} \propto \exp[q(V_{gs} - V_t)/mkT]$. Here, $kT/q \approx 26\,\text{mV}$ is the electron thermal voltage at room temperature, and m is a dimensionless constant ≥ 1 that has to do with the ability of the gate voltage to incrementally modulate the potential at the silicon surface (ψ_s), namely, $m = \Delta V_{gs}/\Delta\psi_s$. A frequently used factor is the subthreshold swing defined by $SS = (\ln 10) \times mkT/q \approx 80\,\text{mV decade}^{-1}$. This figure is independent of the MOSFET dimension, and therefore independent of technology generations. To maintain a 100–1000 times reduction of the off current, I_{ds} at $V_{gs} = 0$, below the I_{ds} at $V_{gs} = V_t$, a minimum V_t of ~0.2V is required. The on current, on the other hand, is a linear function of V_{gs}, and is proportional to $V_{dd} - V_t$ at $V_{gs} = V_{dd}$. The circuit delay, CV_{dd}/I_{ds}, is then $\propto 1/(1 - V_t/V_{dd})$. For high-performance logic circuits, therefore, the power supply voltage, V_{dd}, cannot be reduced much below 1 V, as indicated in Figure 10.7.

10.3.5 Lightly Doped Drain

When the nMOS channel length is scaled below 1 μm while V_{dd} remains at 5V, the device often suffered from hot-electron degradation occurring at the high field region near the drain. A lightly doped drain (LDD) structure (Figure 10.8) was developed to alleviate that [11]. After gate patterning, an intermediate dose n-type implant is made to form lightly doped source–drain regions of ~$10^{18}\,\text{cm}^{-3}$ doping density. Nitride or oxide spacers are then formed similar to that described in Figure 10.5. A high-dose n-type implant is then made to form deeper source–drain regions of doping level ~$10^{20}\,\text{cm}^{-3}$, followed by the self-aligned silicide process. With the LDD process, the field near the drain is reduced, but the MOSFET current is traded off as a result.

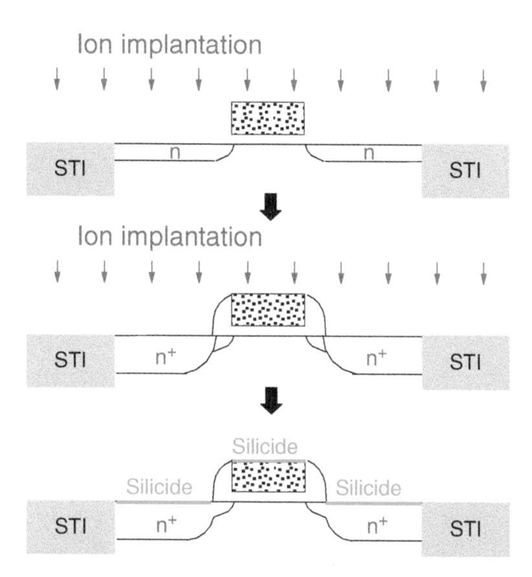

Figure 10.8 Lightly doped drain (LDD) process. Top: medium-dose source–drain implant, middle: spacer formation and then high-dose source–drain implant, bottom: self-aligned silicide formation like in Figure 10.5.

Later on, with the reduction of power supply voltage to 3.3V and then 2.5V, LDD is no longer needed. But the process to form shallow-deep source–drain junctions remained applicable in a "source–drain extension" structure. In that case, the shallow source–drain junctions are not lightly doped. There is no trading off of the MOSFET current. A shallow source–drain junction next to the gated channel is beneficial from the SCE point of view. But the deeper source–drain regions are required for silicide and contact formation. The source–drain extension process is practiced in pMOS as well.

10.3.6 Short-Channel Effect (SCE)

For a given process that builds the vertical structure of a MOSFET, there is a minimum channel length (or gate length) below which the device cannot be turned off properly. SCE stems from the effect of the source and drain on the channel potential. It is governed by the 2-D Poisson's equation subject to the boundary conditions of the device region. An analytical solution in the subthreshold region yielded a key parameter, called the scale length λ, that satisfies the following implicit equation [12]:

$$\frac{1}{\varepsilon_{ox}}\tan\left(\frac{\pi t_{ox}}{\lambda}\right)+\frac{1}{\varepsilon_{si}}\tan\left(\frac{\pi W_{dm}}{\lambda}\right)=0. \tag{10.1}$$

The significance of λ is that it gives the minimum channel length $L_{\min}\approx 2\lambda$, for the SCE to be tolerable. In the above equation, ε_{si} and ε_{ox} are the permittivities of silicon and the gate dielectric of thickness t_{ox}, respectively, and W_{dm} is the maximum depletion depth in silicon as depicted in Figure 10.9. For a uniformly doped silicon substrate of doping N_a,

$$W_{dm}\approx\sqrt{\frac{2\varepsilon_{si}\left(2\psi_B\right)}{qN_a}}, \tag{10.2}$$

where $\psi_B=(kT/q)\ln(N_a/n_i)$ with $n_i\approx 10^{10}\,\text{cm}^{-3}$, the intrinsic carrier density of silicon at room temperature.

The factor m associated with the subthreshold swing discussed earlier can be expressed in terms of W_{dm} [10],

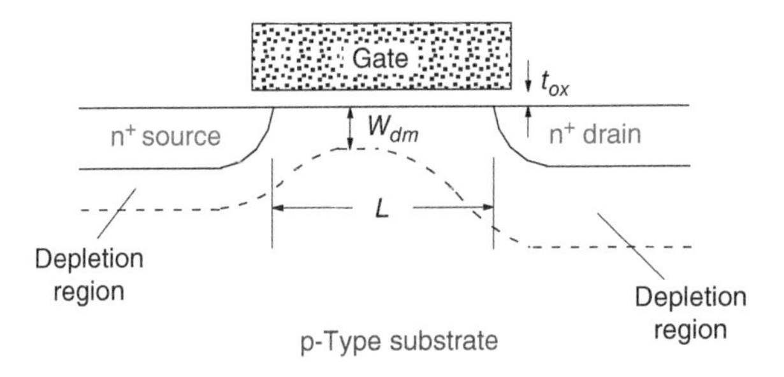

Figure 10.9 Schematic MOSFET cross section showing W_{dm}, the depletion depth under the gate at threshold condition.

$$m = 1 + \frac{\varepsilon_{si} t_{ox}}{\varepsilon_{ox} W_{dm}}. \tag{10.3}$$

Since $SS = (\ln 10) \times mkT/q$, m should not be much larger than one. Typically, $(\varepsilon_{si} t_{ox})/(\varepsilon_{ox} W_{dm}) < 0.5$. This implies that the angle of the first tangent in Eq. (10.1) is in the first quadrant while that of the second tangent is in the second quadrant. In the limit of $t_{ox} \ll W_{dm}$, Eq. (10.1) has an explicit solution,

$$\lambda \approx W_{dm} + \frac{\varepsilon_{si}}{\varepsilon_{ox}} t_{ox} = mW_{dm}. \tag{10.4}$$

To scale down λ hence $L_{\min}$, both t_{ox} and W_{dm} must be reduced. The reduction of t_{ox} with MOSFET scaling is plotted in Figure 10.7. To reduce W_{dm}, the body doping N_a needs to go up by the square of the scaling factor, as suggested by Eq. (10.2). An example is $W_{dm} = 0.1\,\mu\text{m}$ for $N_a = 10^{17}\,\text{cm}^{-3}$, a good choice for 0.3 μm CMOS technology.

10.3.7 Evolution of the Doping Profile

For a uniformly doped MOSFET, the threshold voltage is given by

$$V_t = V_{fb} + 2\psi_B + \frac{qN_a W_{dm}}{C_{ox}}, \tag{10.5}$$

where V_{fb} is the flatband voltage and $C_{ox} = \varepsilon_{ox}/t_{ox}$ [10]. To obtain the required values of V_t in Figure 10.7, n^+ polysilicon gate is used for nMOS (and vice versa for pMOS) such that

$$V_{fb} = -\frac{E_g}{2q} - \psi_B \tag{10.6}$$

and the first two terms on the RHS of Eq. (10.5) nearly cancel each other. The last term can be shown to be

$$V_t \approx \frac{qN_a W_{dm}}{C_{ox}} \approx 2(m-1)(2\psi_B) \tag{10.7}$$

using Eqs. (10.2) and (10.3). This implies that there lack degrees of freedom to adjust the V_t of a uniformly doped MOSFET, regardless of its channel length.

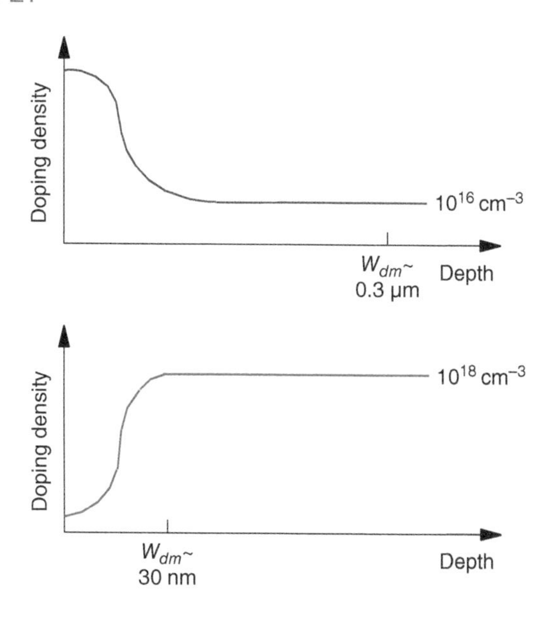

Figure 10.10 Vertically nonuniform doping profiles: high-low for MOSFETs longer than 0.5 μm, low-high for MOSFETs shorter than 0.5 μm.

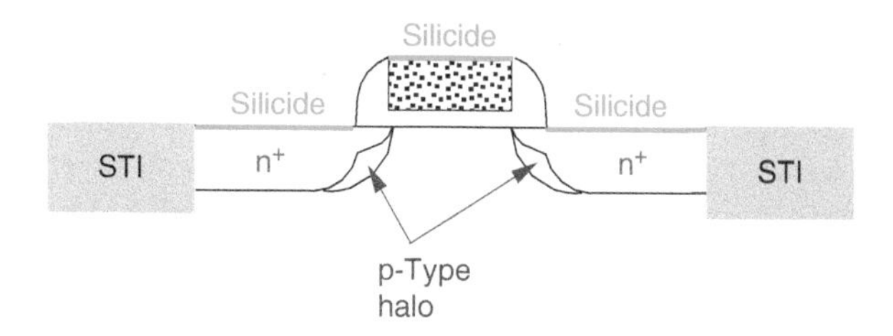

Figure 10.11 HALO: laterally nonuniform doping for mitigating SCE.

The V_t of Eq. (10.7) is approximately 0.5V as $m \sim 1.3$ and $\psi_B \sim 0.4$V. This value is suitable for the $L = 0.5\,\mu$m CMOS generation of V_{dd} = 3.3V in Figure 10.7. Above that, high-low doping profile (Figure 10.10 top) was employed to deliver a higher V_t to go with the higher V_{dd}. Below that, low-high doping profile (Figure 10.10 bottom) was used to obtain a lower V_t for the lower V_{dd}. The extreme limit of the low-high doping profile becomes a ground-plane MOSFET in which W_{dm} equals the depth of the undoped region near the surface [10]. Shown in Figure 10.10 are examples of the vertically nonuniform doping. For channel lengths of 100 nm and below, laterally nonuniform doping, called "halo" in Figure 10.11, has been practiced to help mitigating the SCE [11]. Halo regions can be formed by adding a medium dose p-type implant (sometimes angled implants) along with the source–drain extension implant described in Figure 10.8.

10.4 The Latest: After 2000

10.4.1 High-κ Gate Dielectrics

Below a thickness of ~4 nm, tunneling current through the gate oxide (SiO_2) becomes significant, as shown in Figure 10.12 [13]. While the gate current of an individual MOSFET of t_{ox} = 2 nm is still orders of magnitude below the level of the on-current, the total gate current of all the turned-on MOSFETs in

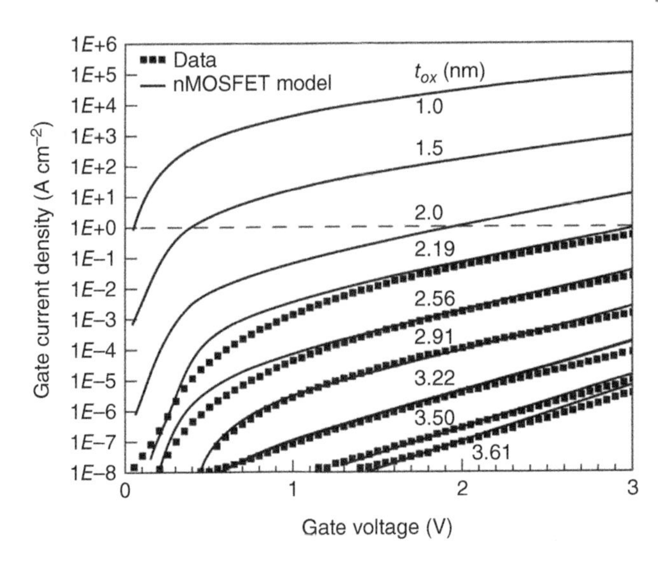

Figure 10.12 Measured (dots) and calculated (solid lines) tunneling currents in thin-oxide polysilicon-gate MOS devices. The dashed line indicates a tunneling-current level of 1 A cm^{-2}.

the standby state over the entire chip can be unacceptably high. To alleviate this problem, the industry developed high-κ gate dielectrics, e.g. HfO_2 which has a dielectric constant of $\approx$20 [14]. A physically thick high-κ gate dielectrics can deliver the same capacitance as that of a sub-1 nm SiO_2, while having much lower tunneling currents. The effect of high-κ gate dielectrics on the scale length λ is given by Eq. (10.1). In the limit of $t_{ox} \ll W_{dm}$, λ is approximated by Eq. (10.4) which shows that a higher ε_{ox} helps reduce the t_{ox} term. For the case of very high κ, however, the general Eq. (10.1) for λ must be used. It gives a limiting value of $\lambda = 2\ t_{ox}$, regardless of κ, when the angle of the first tangent in Eq. (10.1) approaches $\pi/2$. Physically, this happens when the tangential field dominates over the perpendicular field in the 2-D MOSFET geometry such that k no longer matters. The practical implication is that the physical thickness of high-κ gate dielectrics still needs to be $\leq L_{min}/4 = \lambda/2$, no matter how high κ is.

10.4.2 Replacement Metal Gate

While the dual (n^+/p^+) polysilicon gate technology led to self-aligned MOSFETs (Figure 10.2) with suitable work functions, there is a shortcoming from polysilicon depletion effects. The highest level polysilicon gates can be doped to is $N_p \sim 10^{20}$ cm^{-3}. The depletion width in the polysilicon gate is $W_p = Q_g/(qN_p)$ where Q_g is the sheet charge density in the polysilicon as well as in the silicon. Q_g is related to the oxide field E_{ox} through Gauss's law, $Q_g = \varepsilon_{ox} E_{ox}$. Modern CMOS devices often operate at a maximum oxide field of 5 MV cm^{-1}, which corresponds to a sheet electron density of $Q_g/q = 10^{13}$ cm^{-2}. For these values, the polysilicon depletion width is $W_p \sim 1$ nm. Its effect is equivalent to adding 0.3–0.4 nm to the gate oxide thickness. When the gate oxide is thick, e.g. for MOSFET channel length of 0.1 μm and above in Figure 10.7, the polysilicon gate depletion effect is tolerable. However, when the gate oxide thickness t_{ox} is scaled to 2 nm, or even 1 nm and below with high-κ dielectrics [for high-κ, the equivalent oxide thickness EOT is defined as $(\varepsilon_{SiO2}/\varepsilon_{ox})t_{ox}$], the performance loss due to polysilicon depletion is too much to bear. The industry then switched back to metal gates with proper work functions [14]. But, as illustrated in Figure 10.2, it is essential that the source–drain regions are self-aligned to the gate. The new metal gate technology is then implemented with the replacement gate process outlined in Figure 10.13 [15]. Following standard polysilicon gate etching, source–drain implant, anneal, spacer, and silicide formation, the device structure is covered with CVD oxide and then polished

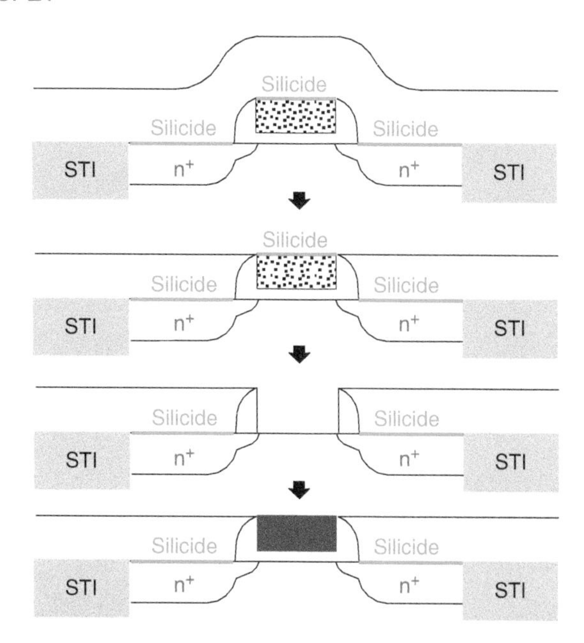

Figure 10.13 Replacement gate process.

down to expose the top of the polysilicon gates. After that, the dummy gates are etched out, and the high-κ gate insulator and the final metal gates are deposited and planarized with low-temperature processes. This retains the self-alignment between the source–drain regions and the gate. Different metals with work functions delivering the desired threshold voltages for nMOS and pMOS, respectively, may be needed.

10.4.3 FinFET

When the channel length is scaled to $L = 20\,\text{nm}$ or so, bulk MOSFETs faced a number of difficulties. First, coping with the SCE demands that W_{dm} be reduced by raising the body doping to $10^{19}\ \text{cm}^{-3}$ and above. This goes against the downward trend of V_t to accommodate the reduction of V_{dd} in Figure 10.7. Even the extreme low-high or the ground-plane doping profile cannot do the job. Second, with a fixed SS of $(\ln 10)\times mkT/q \approx 80\,\text{mV decade}^{-1}$, a minimum V_t of ~0.2V is required for the off current to be acceptably low. This in turn prevents the downscaling of V_{dd} with L, resulting in a rapid increase of the chip power. Third, in the minimum geometry MOSFETs in SRAM (Static Random Access Memory) cells, random fluctuations of the number of discrete dopant atoms can cause cell failure in a large array. When a MOSFET is scaled down by a factor of α (<1), the volume of its depletion region goes down by α^3 while the doping density goes up by $1/\alpha^2$ according to Eq. (10.2). The total number of dopant atoms in the depletion region then goes down by α. At $L \sim 20\,\text{nm}$, the number of dopant atoms in the depletion volume is down to a hundred or so. Random fluctuation of this number gives rise to a threshold voltage variation of 1σ ~70mV for a minimum width MOSFET [16]. In a large SRAM array, some cell may fall outside of the 6σ deviation from the mean, rendering a read or write failure of the cell. Facing the above difficulties in scaling bulk MOSFETs below 20nm, the industry migrated to double-gate MOSFETs.

The concept of double-gate MOSFETs was first conceived in a patent filing as early as 1935 [17]. Fundamentally, they differ from bulk MOSFETs in that the active transistor region is a thin semiconductor film of thickness t_{si} (Figure 10.14) sandwiched between two gates with the same voltage applied. Since the potential in the semiconductor film is controlled by the gate field from both sides, the

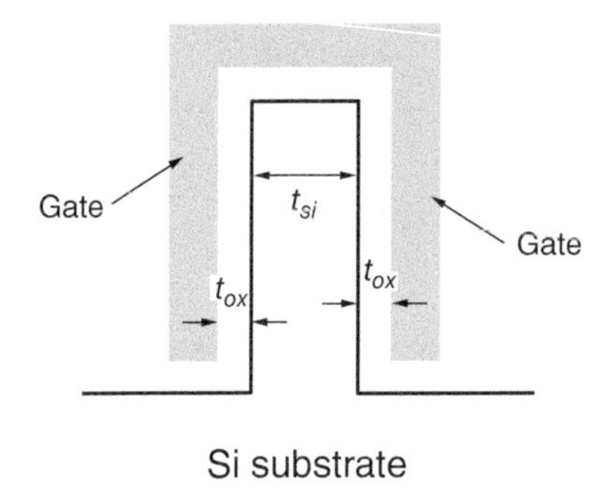

Figure 10.14 FinFET cross section along a cut perpendicular to the channel between the source and drain. The MOSFET width is the height of the silicon pillar.

subthreshold swing has the ideal value of $SS = 60\,\text{mV decade}^{-1}$, equivalent to $m = 1$ in the bulk case. Furthermore, dopant number fluctuation problem is eliminated as no doping is required in the semiconductor to deal with the SCE. Instead, the SCE is controlled by t_{si} as described by the following scale length equation [12]:

$$\tan\left(\frac{\pi t_{ox}}{\lambda}\right)\tan\left(\frac{\pi t_{si}}{2\lambda}\right) = \frac{\varepsilon_{ox}}{\varepsilon_{si}}. \tag{10.8}$$

In general, both angles of the tangents are in the first quadrant, hence $\lambda \geq 2t_{ox}$ as well as $\lambda \geq t_{si}$ for the lowest order (longest) solution.

Double-gate MOSFETs cannot be fabricated using the conventional planar IC technology, especially with the source–drain regions self-aligned to both gates. A version of the double gate MOSFET was fabricated in 1989 [18], which was later given the name FinFET (Figure 10.14). To go beyond the bulk MOSFET limit of 20 nm, however, the thickness of the silicon pillar t_{si} needs to be thinner than 10 nm because of the scale length requirement mentioned above. The manufacturing of FinFET into VLSI products was accomplished in 2012 [19] by applying a sublithographic patterning technique patented in 1980 [11]. It utilizes a spacer process similar to that depicted in Figure 10.5. A dummy "gate" is patterned and etched to create two sidewalls at the edges of the gate, then the dummy gate is stripped to leave only the sidewalls as the masks to further etch the underlying silicon into fins. Since the spacer width is controlled by the thickness of the film deposited over the dummy gates, it can be narrower than the lithographic resolution. Fins of widths 5 nm or below have been routinely produced in VLSI chips.

10.4.4 Planarized Interconnect Levels

As more and more transistors are packed into a single chip, and more and more functions the chip performs, an increasing number of interconnect levels are needed. The old versions of contact hole and metal wiring scheme for one or two levels of interconnect (Figure 10.15a) can no longer work. Too much topography makes the next level difficult to etch. The industry migrated to the fully planarized interconnect levels as shown in Figure 10.15b. CMP was employed extensively to achieve planarization at every level. Single-sized contact holes are enforced. Furthermore, to minimize the interconnect resistance, aluminum has been replaced by copper with the insertion of diffusion barriers. Today's microprocessors have over ten levels of interconnects of varying widths and thicknesses to accomplish local wires among transistors and global wires among functional blocks.

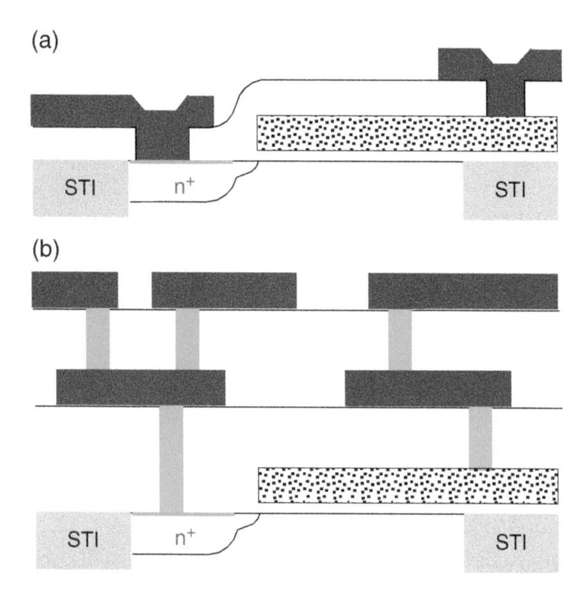

Figure 10.15 (a) Cross section of a single-level metal/contact with no planarization. (b) Cross section of a planarized multilevel interconnect scheme with contact studs.

10.5 Conclusion

MOSFET has gone through many structural changes from its inception to today's nanometer-scale transistors in VLSI chips containing over 10 billion elements. Those technological advances made the modern computers smaller, faster, and lower cost. They also enabled the Internet, smartphones, CMOS cameras, GPS systems, etc. The invention of the transistor 75 years ago has ushered in the information age with a tremendous impact on our everyday life into the twenty-first century.

References

[1] Lilienfeld, J.E. (1926). Method and apparatus for controlling electric currents. US Patent 1,745,175, filed 1926.
[2] Kahng, D. and Attala, M.M. (1960). Silicon–silicon dioxide field induced surface devices. *Presented at IEEE Device Research Conference*, Pittsburgh (26–29 June).
[3] Dennard, R.H. (1967). Field-effect transistor memory. US Patent 3,387,286, filed July 1967.
[4] Kilby, J.S. (1959). Miniaturized electronic circuits. US Patent 3,138,743, filed February 1959.
[5] Kerwin, R.E., Klein, D.L., and Sarace, J.C. (1967). Method for making MIS structures. US Patent 3,475,234, filed March 1967.
[6] Wanlass, F. and Sah, C.T. (1963). Nanowatt logic using field-effect metal-oxide-semiconductor triodes. *IEEE ISSCC Digest of Technical Papers*, pp. 32–33. Philadelphia, PA (20–22 February).
[7] Wong, C.Y., Sun, Y.C., Taur, Y. et al. (1988). Doping of n+ and p+ polysilicon in a dual-gate CMOS process. *IEEE IEDM Technical Digest*, pp. 238–241. San Fransisco, CA (11–14 December).
[8] Ting, C.Y., Iyer, S.S., Osburn, C.M., Hu, G.J., and Schweighart, A.M. (1982). The use of TiSi2 in a self-aligned silicide technology. *Presented at ECS Symposium on VLSI Science and Technology*, pp. 224–231. Detroit, MI (18–21 October).
[9] Davari, B., Koburger, C., Furukawa, T. et al. (1988). A variable-size shallow trench isolation technology with diffused sidewall doping for submicron CMOS. *IEEE IEDM Technical Digest*, pp. 92–95. San Fransisco, CA (11–14 December).
[10] Taur, Y. and Ning, T.H. (2022). *Fundamentals of Modern VLSI Devices*, 3e. United Kingdom: Cambridge University Press.

[11] Ogura, S., Codella, C.F., Rovedo, N., Shepard, J.F., and Riseman, J. (1982). A half-micron MOSFET using double-implanted LDD. *IEEE IEDM Technical Digest*, pp. 718–721. San Fransisco, CA (13–15 December).

[12] Frank, D.J., Taur, Y., and Wong, H.-S. (1988). Generalized scale length for two-dimensional effects in MOSFETs. *IEEE Electron Device Lett.* 19: 385–387.

[13] Lo, S.-H., Buchanan, D.A., Taur, Y., and Wang, W. (1997). Quantum-mechanical modeling of electron tunneling current from the inversion layer of ultra-thin-oxide nMOSFETs. *IEEE Electron Device Lett.* 18: 209–211.

[14] Robertson, J. and Wallace, R.M. (2015). High-*k* materials and metal gates for CMOS applications. *Mater. Sci. Eng. R* 88: 1–41.

[15] Chang, C.-P., Vuang, H.-H., Baker, M.R. et al. (2000). SALVO process for sub-50 nm low V_T replacement gate CMOS with KrF lithography. *IEEE IEDM Technical Digest*, pp. 53–56. San Fransisco, CA (10–13 December).

[16] Frank, D.J., Taur, Y., Ieong, M., and Wong, H.-S. (1999). Monte Carlo modeling of threshold variation due to dopant fluctuations. *Symp. VLSI Technology Digest of Tech. Papers*, IEEE, pp. 169–170. Kyoto, Japan (14–16 June).

[17] Heil, O. (1935). Improvements in or relating to electrical amplifiers and other control arrangements and devices. British Patent 439,457, filed 1935.

[18] Hisamoto, D., Kaga, T., Kawamoto, Y., and Takeda, E. (1989). A fully depleted lean-channel transistor (DELTA). *IEEE IEDM Technical Digest*, pp. 833–836. Washington, DC (3–6 December).

[19] Auth, C., Allen, C., Blattner, A. et al. (2012). A 22 nm high performance and low power CMOS technology featuring fully-depleted tri-gate transistors, self-aligned contacts, and high density MIM capacitors. *IEEE Symposium VLSI Technology*, pp. 131–132. Honolulu, HI (12–14 June).

Chapter 11

The SOI Transistor

Sorin Cristoloveanu

IMEP-LAHC, Grenoble Institute of Technology & CNRS, Grenoble, France

11.1 The Beginnings

Silicon-on-insulator (SOI) chips are already in all our smartphones and will soon invade our cars and homes. With Jules Verne being away, nobody else could predict, 60 years ago, the adventurous story of SOI technology.

But this is not the reason why SOI has been invented. Shortly after the fabrication of the first metal-oxide-semiconductor (MOS) circuits, it was realized they were badly hurt by the radiation effects. Energetic particles from cosmic rays or human-manufactured rays generate undesirable electron–hole pairs along tens of micrometers trail in the transistor thickness. Resulting photocurrent spikes act as leakage currents responsible for charge collection and logic upsets. The only way to alleviate the transient radiation effects was to reduce the thickness of the device.

Early 1960s was the wonderful time of The Beatles rise. But what was terrifying was the other music, that of the cold war orchestrated by Brejnev. SOS distress signals alerted on the threat of a nuclear war. Given the vulnerability of complementary metal-oxide-semiconductor (CMOS) circuits to radiations, the military ordered: Integrate them in a thin silicon layer formed on top of an insulator. It happened that the first member of the SOI club preserved the acronym SOS: Silicon On Sapphire [1].

Being an expensive technology, SOS was developed with the financial benediction of the defense and aerospace industry. The work by Cullen's group at RCA was most influential [2].

75th Anniversary of the Transistor, First Edition. Edited by Arokia Nathan, Samar K. Saha, and Ravi M. Todi.

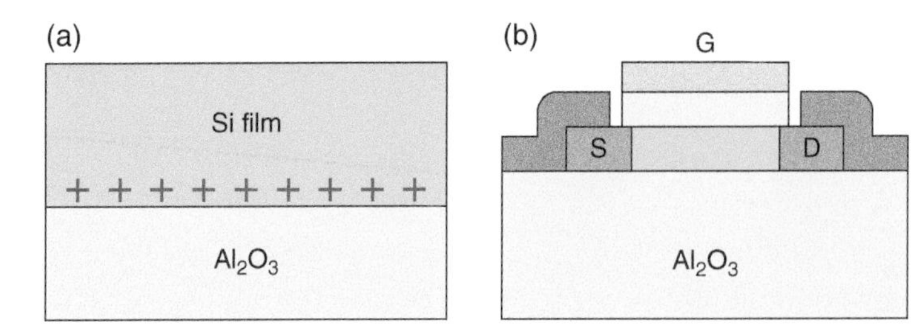

Figure 11.1 (a) SOS structure with the defective interfacial layer to be healed (Red Cross). (b) The venerable pioneer of SOS MOSFETs.

The heteroepitaxial growth of silicon layers on sapphire substrate is adversely affected by the mismatch of their thermal and crystallographic parameters. A 50-nm-thick interfacial layer with miserable quality was inevitable at the Si–Al_2O_3 interface. In order to screen the impact of the defective region, the silicon film had to be relatively thick, around half a micron (Figure 11.1a).

Over the years, considerable efforts were aimed at suppressing the transition layer and reducing the film thickness. A breakthrough came from the *Solid-phase epitaxial regrowth* (SPER) process. Silicon ions were implanted to render amorphous the Si film, except for a thin surface layer that acted as a seed during subsequent epitaxial regrowth. Additional innovations like the double-SPER process could achieve 100-nm-thick films with improved carrier mobility and lifetime [3]. These embellishments made the SOS wafers even more expensive.

In the meantime, the SOS transistor (Figure 11.1b) revealed promising assets: simple processing, perfect dielectric isolation for latch-up elimination, drastically reduced leakage and parasitic capacitance, and enhanced radiation hardness. Typical SOI mechanisms induced by the floating body have been quickly understood [4].

However, it was long admitted that SOS was not the best choice. In the late 1970s, the decline of the SOS reign was precipitated by the rise of Separation by Implantation of Oxygen (SIMOX) technology.

11.2 The Renaissance

In 1978, Katsu Izumi led his team at Nippon Telegraph and Telephone (NTT) Labs to fabricate an astonishing circuit on SIMOX wafer [5]. The genesis of this material is brutal: deep implantation of enormous amounts of oxygen ($\approx 2 \times 10^{18}\,\text{cm}^{-2}$) to eventually synthesize a continuous buried oxide (BOX). The horrible crystalline quality of the top Si film and Si–SiO_2 interface could be partially cured by long high-temperature anneals.

A revolutionary aspect was the presence of relatively thin Si film (200 nm) and BOX (400 nm) which opened the back gate (Figure 11.2a): biasing the substrate enabled the tuning of the threshold voltage, a major asset of fully depleted SOI (FD-SOI) transistors. With a back-gate action, SOI is nothing but an upside-down MOS structure where the BOX plays the role of gate dielectric. We simply applied two probes (source and drain) on the Si overlay and this is how the unbelievable pseudo-MOSFET (Ψ-MOSFET) was discovered [6]. It is still the undisputable method for the characterization of electrical properties in as-grown SOI materials.

The performance of test CMOS circuits on SIMOX was promising and incentive to pursue material developments. Dedicated ion implanters and furnaces became available. Peter Hemment organized a dynamic European Consortium in competition with American and Japanese organizations. IBIS Corporation brought SIMOX on the marketplace.

The illuminating idea came from Yannis Stoemenos during his sabbatical year at LETI in Grenoble. He demonstrated that annealing at extremely high temperature (>1300 °C) can clear the silicon film of

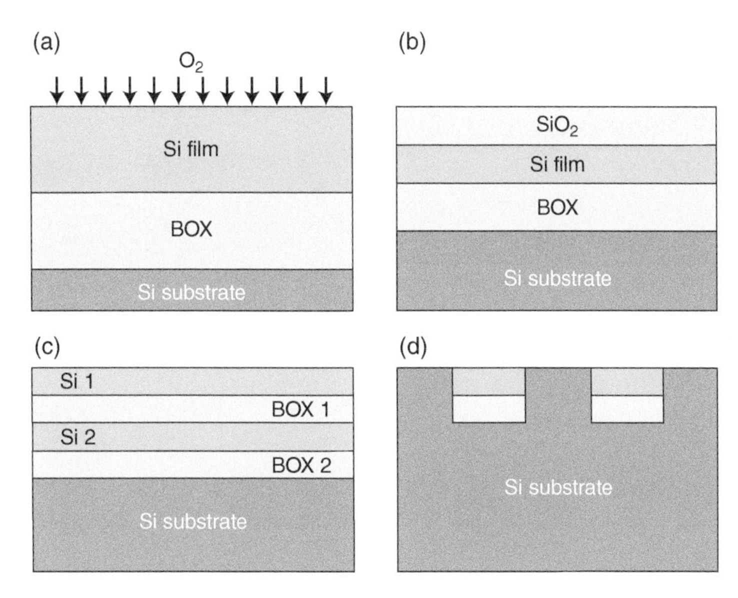

Figure 11.2 SIMOX variants: (a) regular structure, (b) ITOX with thin film and BOX, (c) double BOX, and (d) isolated SIMOX islands.

crystal defects leaving a sharp Si–SiO_2 interface and a leakage-free BOX [7]. SOITEC Company was founded by A-J. Auberton-Hervé and J-M. Lamure with the initial goal of commercializing top-quality SIMOX.

SIMOX was somehow handicapped by its long and costly processing, low throughput (<50,000 wafers per year), and rather rigid thickness of film and BOX. The ITOX variant was conceived to lower the implant dose and reduce the layer thickness to 100 nm range (Figure 11.2b) [8]. The double SOI structure in Figure 11.2c was formed by two implantation steps at different energies [9]. The Si layer sandwiched between the two oxides may serve for interconnects, extra gates, or wave guides. SIMOX islands isolated in an ocean of silicon were equally demonstrated (Figure 11.2d).

Material magicians uncovered astonishing SOI structures to compete with SIMOX.

Wafer Bonding (WB or BESOI): The process consists in mating two wafers, at least one of which is oxidized (Figure 11.3a). The bonded structure is thinned down by etching and grinding to reach the target thickness of the film. Etch-stop layers (SiGe, porous Si, by junctions) are helpful but insufficient for achieving ultrathin layers. WB technology is dedicated to thick power devices and sensors.

Zone Melting Recrystallization (ZMR): A polycrystalline or amorphous Si layer is deposited on top of an oxidized wafer. High-temperature sources (lamps, lasers, beams, or heat heaters) are scanned across the surface to erase the grain boundaries and associated defects (Figure 11.3b). The limited extension of the monocrystalline islands is a blocking issue.

Epitaxial Lateral Overgrowth (ELO): Single-crystal silicon is epitaxially grown through seed windows opened on the oxidized bulk-Si wafer (Figure 11.3c). Since the growth proceeds laterally and vertically, a thinning step is necessary unless the CELO variant of tunnel epitaxy through a confined cavity is implemented (Figure 11.3d) [10]. ELO is fit for 3D integration as it does not require post-growth recrystallization at high temperature.

Full Isolation by Porous Oxidized Silicon (FIPOS): Selected P-doped regions in a N-type wafer are converted into porous silicon by anodic reaction. Thanks to their huge surface-to-volume ratio, the porous regions are preferentially oxidized and transform in BOX and lateral isolation (Figure 11.3e) [11].

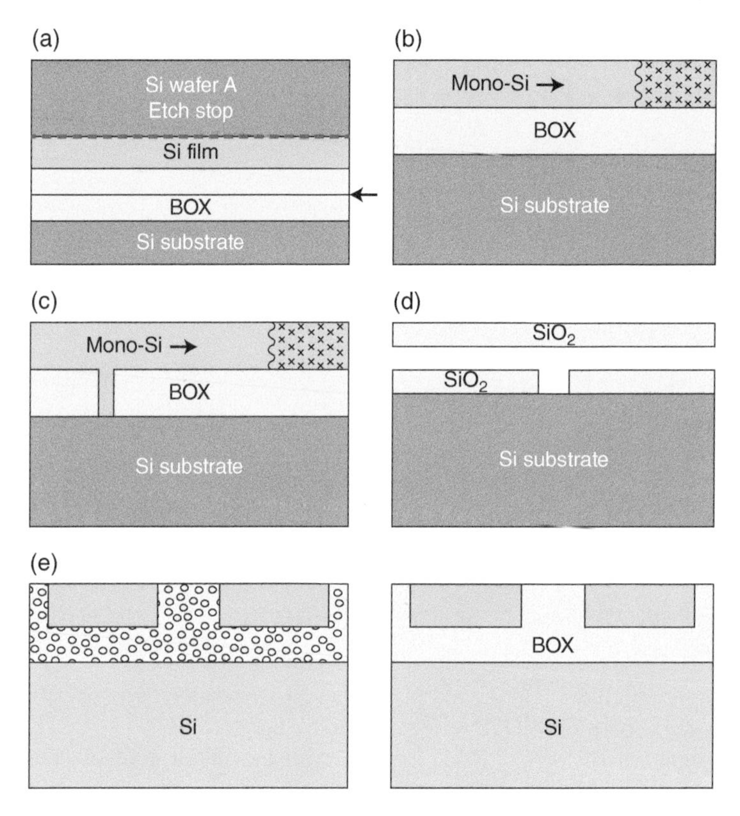

Figure 11.3 Other flavors of SOI wafers: (a) wafer bonding (arrow) and etch-back, (b) ZMR, (c) ELO, (d) CELO, and (e) FIPOS.

The 1980s was a glorious decade with unlimited creativity in materials science and device physics. The annual IEEE International SOI Conference was the grand scene for innovations, cheering and of course wild fights between the competing camps. Jerome Lasky, Witek Maszara, Ulrich Gösele, and their colleagues pushed the progress in WB technology [12–14]. Gerry Neudeck remained faithful to the ELO process and demonstrated double-layer SOI devices stacked on top of each other [15]. Atsushi Ogura invented the CELO variant used nowadays to combine Ge, III–V, and silicon heterointegration [16]. George Celler at Bell Labs and Jean-Pierre Colinge at CNET were fanatics about ZMR before thinking better.

Jerry Fossum and Hyung-Kyu Lim elaborated the theory of gate coupling and threshold voltage modulation 40 years ago [17]. After completing his PhD and returning home, Lim did not wait too long before becoming the boss of Samsung; SOI helps. An ultrathin SOI film behaves as a quantum well where subband splitting takes place [18]; Yasuhisa Omura-san showed the resulting increase in threshold voltage [19]. Dimitris Ioannou was tackling the transient floating-body effects (FBE) [20]. We discovered the principle of *volume inversion* in double-gate MOSFET which stipulates that in a thin body the mobile carriers are no longer confined at the interface [21]. Jean-Pierre Colinge invented the gate-all-around (GAA) MOSFET omnipresent today in nanowire and nanosheet devices [22]. A hard-to-believe three-layer image processor was fabricated at Mitsubishi by Akasaka's team via monolithic integration in 1987, too early for volume production [23]. All these guys are very old or worst.

Unperturbed, bulk-Si CMOS continued its triumphal march from node to node which was kind of boring. Fortunately, SOI was around to entertain imagination and boost the elaboration of novel concepts, smart characterization techniques, and sometimes weird devices.

11.3 The Smart-Cut Dynasty

In early 1990s, too many SOI materials kept competing, which means that none was really convincing. The dilemma was fixed by Michel Bruel from LETI. He took a few days off in the Alps to debate with himself about SOI structures and came back with the genius idea of Smart-Cut [24].

The processing sequence is illustrated in Figure 11.4 and consists in transferring a premium-quality silicon film from one wafer (A) to another (P). The magic of Smart-Cut is the ability to detach a nanometer-thick layer from wafer A using hydrogen implantation as an atomic scalpel (ion cut).

In a first step, the BOX is formed by oxidizing the donor wafer. Then, hydrogen ions are implanted through the oxide to produce fine microcavities in the Si lattice, which result in a mechanically fragile planar zone. After bonding, the handle wafer P acts as a stiffener and redirects the pressure from vertical to lateral direction, such that the microcavities coalesce. Annealing or a mechanical force is used to split the wafers along the hydrogen-weakened zone, not at the original bonding interface. The microcracks act as a zipper, leaving a well-controlled Si film on wafer P. After exfoliation, the SOI wafer undergoes surface polishing and annealing to strengthen the bonded interface.

The silicon layer retains the excellent quality of the high-grade donor wafer. This donor wafer is recycled repeatedly; according to the sausage theorem, a slice is cheaper than the whole sausage. The handle wafer only serves as mechanical support and is less costly.

Smart-Cut technology is extremely flexible, providing a full range of wafers with Si film thickness from 5 nm to 2 μm and BOX thickness from 5 nm to 5 μm. Only conventional equipment is needed for mass production. 5 nm means 200,000 thinner than the initial thickness of the wafer. The film uniformity is incredible: ±4 Å across a 30-cm wafer is like traveling the world without seeing mountains taller than 5 cm. Low defectivity, sharp interfaces, and high carrier mobility are other unrivaled assets.

The success of Smart-Cut can be inferred from collateral damage. First, the competition was killed, all other SOI variants left the arena. As of 2023, SOITEC and licensed companies provide the quasi-totality of FD-SOI wafers on the market. Second, in 30 years of reign, no new candidate could challenge seriously Smart-Cut:

- Also based on WB, the Epitaxial Layer Transfer (ELTRAN) used a thin porous-Si buffer [25], rather than hydrogen implants, to define the splitting region. The Si film was grown directly on the porous template. Uniformity and throughput were fatal.
- Silicon-On-Nothing is Thomas Skotnicki's SON [26]. Sacrificial SiGe regions are grown on silicon wafer and receive in turn a thin epitaxial Si layer. Removal of SiGe leaves cavities (the "nothing" part

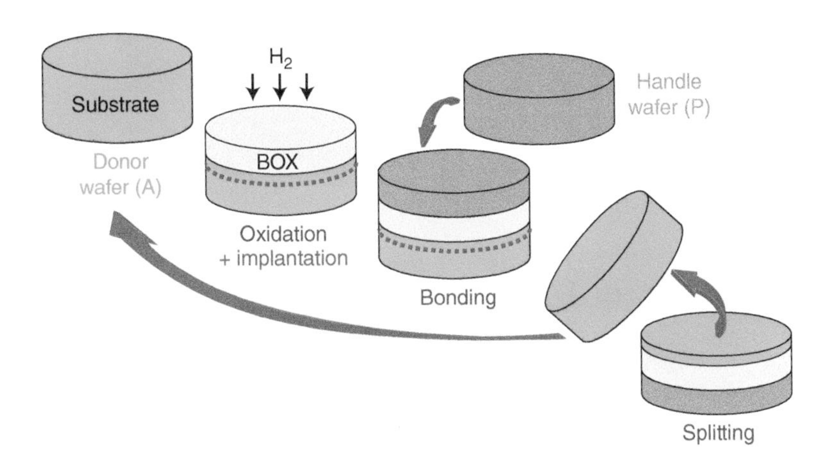

Figure 11.4 Principal steps in the fabrication of SOI wafer by Smart-Cut.

of the SON) which can be filled with oxide leading to localized SOI (as in Figure 11.2d). SON could not comply with the variability criteria of billion-transistor chips but serves as a building brick for GAA nanowire and nanosheet MOSFETs.

- Silicon-On-Diamond for high-temperature circuits had a brief and innocent existence [27].

Many are the offsprings of the original SOI structure. Smart-Cut can replace the silicon overlay with more talented semiconductors for speed (germanium, SiGe, InGaAs, strained layers, etc.), power (SiC by SmartSiC, GaN, …) or photonic devices. The faithful SiO_2 BOX is not immutable either: Al_2O_3, AlN, and diamond have superior thermal conductivity for alleviating the self-heating issues. On the other hand, ferroelectrics or silicon nitride promote the BOX from a passive to an active role, enabling nonvolatile buried memory and universal memory.

The engineering of the silicon substrate is a success story originated from a seemingly insane idea of destroying the crystalline quality of the BOX–substrate interface. The trap-rich SOI promoted by Jean-Pierre Raskin is actually a brilliant concept that conquered the RF market. Adding interface traps prevents the activation of a parasitic channel by BOX charges and preserves the high resistivity of the substrate [28].

Two more applications of Smart-Cut deserve attention: (i) Piezoelectric-On-Insulator (POI) comprising a $LiTaO_3$ film on high-resistivity substrate and (ii) transfer of layers containing already processed circuits from a silicon wafer to another substrate (3D integration, RF devices). Although Smart-Cut encouraged new materials to bond together, the divorce of the old couple, silicon and silicon-oxide, is seemingly impossible. Figure 11.5 summarizes the SOI story over the past 60 years.

The transistor architecture benefited from the progress in SOI technology and became thinner and shorter. Partially depleted (PD) MOSFETs won the first set against their fully depleted (FD) cousins. Ghavan Shahidi pushed at IBM the implementation of PD-SOI technology demonstrating performance/power/area advantages over bulk-Si CMOS. Lower power for equivalent performance or higher speed at same biasing were undisputable arguments. PD-SOI was adopted in mainstream CMOS and high-performance processors became available on the marketplace in 1995. AMD and Global Foundries continued in 2001 with a successful deployment of high-end power-PC microprocessors. These

Figure 11.5 60 years of SOI already.

accomplishments could not continue indefinitely for the limits of miniaturization are more or less the same in PD-SOI and bulk-Si MOSFETs.

It has been obvious for a long time that FD-SOI is best equipped for downscaling. Electrostatic considerations led Konrad Young to the definition of an *intrinsic* length, function of film thickness, that governs the ultimate scaling limit [29]. A common thumb-rule stipulates that a Si film of thickness t_{si} can accommodate FD-SOI MOSFETs with correct characteristics down to a minimum length of $L = 4t_{si}$. In multiple-gate transistors, the minimum achievable length is divided by the number of gates, or the film thickness can be relaxed correspondingly. Imagine a cargo traveling on an ocean just a few meters deep. Without waves or tsunamis, the cargo moves faster with reduced consumption. This is exactly what happens in ultrathin SOI transistors where the electrons are more mobile and most parasitic effects are eliminated.

A peculiar short-channel effect in FD-SOI transistor is the penetration of the electric field from source and drain into the body via the BOX [30]. The fringing field reduces the threshold voltage and contributes to DIBL (drain-induced barrier lowering). This is why a thin BOX (<20 nm) is unavoidable for further downscaling.

A state-of-the-art FD-SOI transistor is schematically pictured in Figure 11.6a. The film and BOX are 6 and 20 nm thick. The body is undoped and, in general, there are no halos, pockets, or LDD regions. A highly doped ground-plane underneath the BOX serves as back gate and is biased from the surface. The source and drain are raised by epitaxial regrowth with in situ doping to alleviate the series resistance. Each transistor is isolated from the next one by STI. 90% of the process modules, including the gate stack, are imported from bulk-Si CMOS. The transistor duplication by sort of copy-and-paste process is the foundation of the monolithic integration of 3D circuits (Figure 11.6b) [31, 32].

Before or just after the turn of the millennium, many research groups decided to switch from standby monitoring of SOI progress into active participation. Francisco Gamiz developed Monte-Carlo-in-Granada simulations, in competition with his friends from Udine (Luca Selmi and David Esseni). Ron Schrimpf and colleagues in Vanderbilt scrutinized the permanent effects of cumulated radiation dose. Gérard Ghibaudo (Grenoble), Babis Dimitriadis (Thessaloniki), and Cor Claeys (IMEC) enriched knowledge in noise and other characterization techniques. Joao-Antonio Martino put together an enthusiastic group at the University of Sao Paolo. Outstanding contributions came from the teams led by Shinichi Takagi and Toshiro Hiramoto at the University of Tokyo, by Jong-Ho Lee at Seoul National University, and by Tamara Rudenko at the Academy of Science in Ukraine. The problem is that I cannot enumerate all who would deserve it.

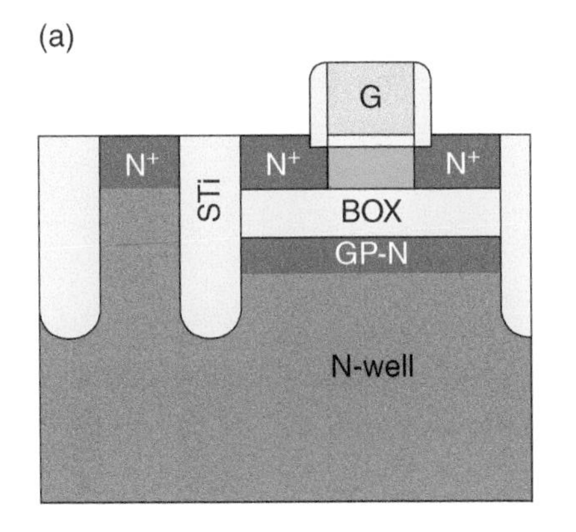

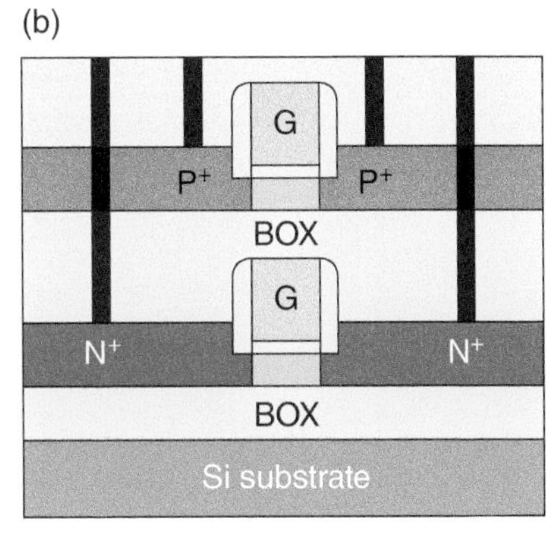

Figure 11.6 Schematics of (a) N-channel FD-SOI MOSFET with ground plane and (b) monolithic 3D circuit with two MOSFETs on top of each other.

In the last decade, stunning device concepts fueled by FD-SOI came out. Colinge invented the junctionless (JL) MOSFET with inherent ultrathin body [33]. We demonstrated the four-gate MOS transistor, where each gate can independently turn on and off the current [34]. A number of capacitorless dynamic random access memory (DRAM) variants use the MOSFET to store and read the information [35]. Band-modulation devices [36, 37], electrostatic doping [38], and tunneling FETs [37] all take advantage of ultrathin SOI films.

Inspired from several SOI workshops organized in Grenoble, the annual EuroSOI conference has formally been initiated 20 years ago in Granada. It is sponsored by IEEE and travels around Europe, attracting specialists from all continents. Alexei Nazarov organized sister SOI workshops in Ukraine. In parallel, Carlos Mazuré created the SOI Industry Consortium where designers and technologists started to talk to each other.

Along with Global Foundries, IBM, and Freescale in the United States, STMicroelectronics and some Japanese sumos developed ambitions in FD-SOI circuits. Technology modules, models, and design libraries were elaborated to conquer the market. Everyone expected FD-SOI will replace the dying bulk-Si CMOS before 2010. Conservative management delayed the switch, letting Intel to make a big surprise by introducing their fin field-effect transistors (FinFETs), soon adopted by major companies. This explains how FD-SOI missed the chance to be a premier technology.

However, FD-SOI did not give up and today the market is booming. All cellphones carry a big piece of SOI; soon all cars will rely on it. It is the technology for low-power, IOT circuits, reconfigurable devices, and so on.

11.4 Special Mechanisms in FD-SOI MOSFET

An FD-SOI transistor is a nano-box, perfectly isolated, where very interesting effects take place. Reminded hereafter are those our dear 75-year-old MOS patriarch never heard of.

11.4.1 Interface Coupling

The properties of one channel depend on the quality and biasing of the opposite interface. The primary coupling effect is the tuning of the front-channel threshold voltage V_T via the back gate (i.e. ground-plane). In ON state, a positive back-gate voltage V_{BG} is used to lower the threshold voltage such as the drive current is enhanced and/or the operating voltage is reduced (Figure 11.7a). Instead in OFF state, a negative bias V_{BG} increases V_T, enabling a remarkable drop in leakage current.

The coupling factor depends on the thickness of gate dielectric and BOX: $\Delta V_T/\Delta V_{BG} \approx -t_{ox}/t_{box}$ [17]. Figure 11.7b shows the linear variation $V_T(V_{BG})$; the thinner the film, the wider the coupling range. With a couple of volts change in V_{BG}, the threshold voltage is shifted by 200 mV. Since the subthreshold swing in FD-SOI MOSFET is close to the Boltzmann limit of 60 mV per decade at room temperature, the OFF current and the standby power are reduced by three orders of magnitude. Only FD-SOI transistors can enjoy this tuning capability which is instrumental in low-voltage, low-power circuits [39].

The drive current also benefits from a marked improvement in carrier mobility for $V_{BG} > 0$, due to the shift of carrier centroid from the interface toward the middle of the film where scattering is attenuated [40].

An intriguing consequence of inter-channel coupling is the violation of the so-called "universal mobility" law [41]. Magnetoresistance measurements showed that biasing the back channel in weak inversion and scanning the top gate leads to two distinct mobility values for the same electric field [27]. The reason is that two carrier profiles, centered near the front or near the back interface, can correspond to an identical average field.

The characterization techniques had to evolve in order to account for defect coupling. From an effective density of defects, measured by charge pumping or noise, it is not obvious to sort out the contributions of each interface [27, 31].

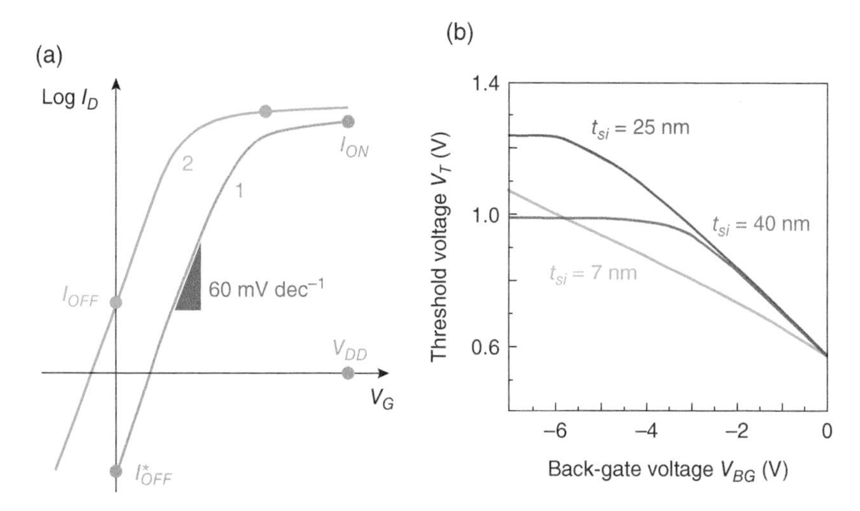

Figure 11.7 (a) Transfer characteristics illustrating the benefit of threshold voltage tuning on leakage current and drive current. (b) Front-channel threshold voltage versus back-gate voltage in SOI MOSFETs with thin and ultrathin body.

11.4.2 Supercoupling

We all learnt that a positive voltage on the top gate induces electrons in the front channel whereas a negative V_{BG} supplies the back channel with holes. This classical image is demolished in sub-10-nm-thick films by the supercoupling effect which denies the coexistence of electron and hole populations [42]. This counterintuitive mechanism makes one gate winning over the other. Despite opposite polarities on the two gates, only electrons or holes fill the body, not both. A positive gate voltage does not necessarily secure an electron channel! The argument is that the electric field needed to maintain electron and hole channels facing each other increases in thinner films up to unsustainable value.

The supercoupling concept was validated with the 4-gate MOSFET (inset in Figure 11.8) which features two longitudinal N^+ contacts and two lateral P^+ contacts [34]. The electron and hole currents

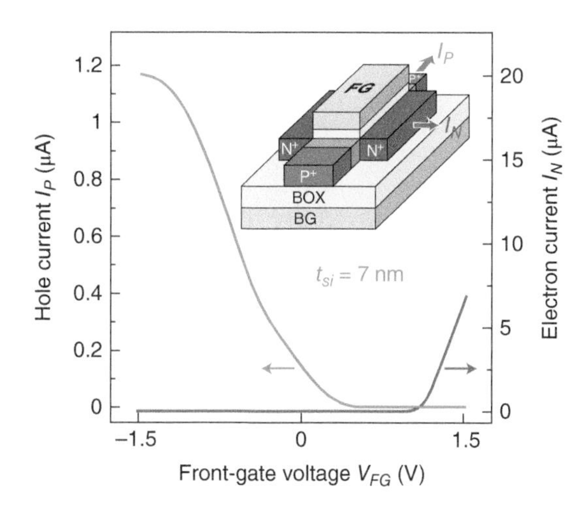

Figure 11.8 Electron and hole currents versus gate voltage monitored in an ultrathin 4-gate FET (inset). The hole current induced by the back-gate voltage (V_{BG} = −10V) disappears once the front channel is enriched with electrons.

could be probed simultaneously. In Figure 11.8, a back-gate voltage $V_{BG} = -10\,V$ was selected to supply holes which initially flow through the whole body (at $V_G = -1.5\,V$). Increasing V_G first removes the holes from the front channel (at $V_G = 0\,V$) and then activates an electron channel ($V_G = +1\,V$). As soon as the electron population starts prevailing, the holes are kicked out from the body and their current drops to zero [43].

Supercoupling has contrasted implications. It makes the threshold voltage tuning more efficient and inhibits the floating-body effects (FBE). The family of electrostatically doped devices takes advantage of supercoupling but, in turn, bilayer devices (single-transistor DRAMs and TFETs) cease to operate.

11.4.3 Volume Inversion and Electrostatic Doping

In a classical MOSFET, the inversion charge is confined at the Si–SiO_2 interface. This is no longer true in a thin double-gate transistor where the minority carriers spread from one interface to the other, occupying the whole volume [21]. The electrons flowing away from the interfaces have higher mobility. The sister effect is volume accumulation.

Volume inversion and accumulation also occur in single-gate ultrathin transistors. The carrier distribution is more homogeneous in thinner devices and looks like a virtual doping. This electrostatic doping relies on free carriers rather than dopants [38]. As compared to conventional chemical doping, the electrostatic doping is not rigid, its concentration and polarity being gate-controlled. The possibility to emulate P–N junctions in a FD body just by adding gates is a recent paradigm shift with fascinating applications (Section 11.5.3).

11.4.4 Floating-Body Effects

The FBE are historical properties of PD SOI MOSFETs. The inherent neutral region of the body accommodates the collection of majority carriers. Without a body contact, the adjustment of the majority carrier concentration to reach steady state is a long process. The excess or deficit of majority carriers gives rise to peculiar FBE that hurt or improve circuit operation. In FD-SOI, there is no neutral region but the back gate can in principle accumulate majority carriers and reactivate some FBE [44].

The holes generated by impact ionization increase the body potential, and hence lower the threshold voltage, giving rise to a kink in the output $I_D(V_D)$ characteristics. The kink effect is still visible in 25-nm-thick FD-SOI MOSFET provided that the back gate is biased in accumulation (Figure 11.9a). No kink is observed with the back gate grounded.

Impact ionization is equally responsible for hysteresis and latch in the subthreshold transfer $I_D(V_G)$ characteristics [45]. These features are eradicated in ultrathin transistors.

The tunneling current through the gate oxide is another source of majority carrier collection leading to GIFBE (gate-induced FBE) [46]. The lowering of the threshold voltage is here reflected by an anomalous second peak in the transconductance curve (for $V_G = 1.2\,V$ in Figure 11.9b).

Yet another FBE is triggered by band-to-band tunneling at the gate extremities (as in tunnel FETs). The transfer curve shows an ample hysteresis, attractive for memory applications (Figure 11.9c), which gives rise to a meta-stable dip (MSD) in transconductance [47]. The MSD effect is observed at low drain voltage which excludes impact ionization.

The bottom line is that the kink, hysteresis, GIFBE, and MSD effects are attenuated as the body is thinned down. There is no way to activate them in a 7-nm-thick MOSFET where the supercoupling effect forbids the accumulation of majority carriers.

However, a subsisting nuisance is the parasitic bipolar transistor in very short channels. It is visible in the OFF region where the band-to-band tunneling current at junctions forms the base current. The bipolar amplification results in a significant collector (drain) current that acts as a terrible leakage current (Figure 11.9d) [48].

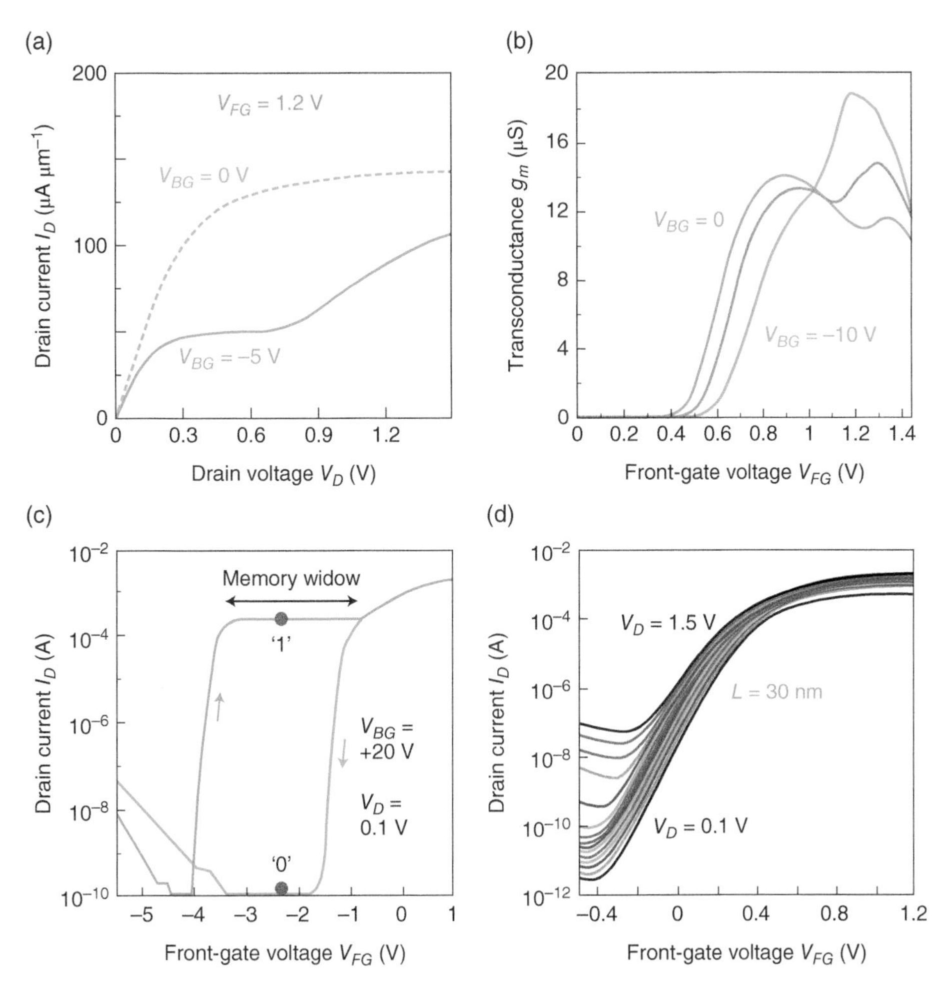

Figure 11.9 Persistent floating-body effects in thin FD-SOI MOSFETs. (a) Kink, (b) GIFBE, (c) MSD, and (d) parasitic bipolar transistor.

11.4.5 Thickness Effects

Careful measurements on FD-SOI MOSFETs with 4–15 nm thickness concluded that the electron and hole mobilities are rather constant and far superior to the typical values in bulk transistors [27]. Monte Carlo simulations backed by experiments show the existence of a little peak in electron mobility around 3 nm thickness [49, 50]. Only in thinner films does the mobility drop vertiginously due to potential fluctuations generated by surface roughness [51].

An FD-SOI MOSFET thinner than 10 nm is more than a transistor, it is a quantum well [52]. The energy quantization results in discreet subbands located above the conduction band. Interestingly, the few electrons that occupy the upper subband exhibit a mobility higher than the theoretical value in bulk silicon [27].

A stronger band bending is required for a given concentration of minority carriers. This explains the increase in threshold voltage $V_T \sim 1/t_{si}^2$, for example $\Delta V_T = 100\,\text{mV}$ in 4 nm thick body [19]. It is worth noting that the threshold voltage roll-off in short MOSFETs mitigates the quantum effect.

The above enumeration of FD-SOI mechanisms is just a homeopathic introduction. A detailed description is included in [27].

11.5 A Selection of Innovating Devices

FD-SOI is a perfect platform for the fabrication of nanowires, tunneling transistors, qubits, and other ultrathin devices. The selection below is totally subjective.

11.5.1 The Junctionless Transistor

The JL MOSFET is a thin semiconductor film with extremely high doping all the way from the source to the drain (Figure 11.10a). The role of the MOS gate is to switch the current off by creating full depletion underneath [33]. With a doping concentration beyond $10^{19}\,cm^{-3}$, the film thickness must necessarily be less than 10 nm. It is the progress of SOI that made the JL concept viable before the advent of nanowires. The thickness criterion is somehow relaxed in GAA devices for the depletion mechanism proceeds from all directions and is more efficient.

The JL attractiveness comes from their unrivaled simplicity that avoids the burden of processing source/drain terminals, in particular when ion implantation is ineffective, like in vertical nano-pillars and 2D materials.

A very high doping has natural drawbacks: poor carrier mobility, normally on operation with negative threshold voltage, random doping fluctuations, etc. Lowering the doping level enables normally off operation ($V_T > 0$), higher mobility, and thicker films. However, the series resistance becomes excessive imposing the overdoping of the terminals. The source and drain homojunctions demolish the beauty of the pure JL transistor which now operates as an accumulation-mode MOSFET.

A recent flavor of JL transistors features core-shell architecture (Figure 11.10b) [53]. Wrapping a heavily doped ($3 \times 10^{19}\,cm^{-3}$) and ultrathin (4 nm) core in an undoped shell (3 nm thick) eliminates the JL weaknesses. The combination of doping-induced electrons in the core with the very mobile ones attracted by the gate in the shell results in normally off operation, excellent effective mobility, and high drive current, comparable with that of undoped FD-NW MOSFETs and much higher than the core alone is capable of. The threshold voltage is still defined by the size and doping of the core, which also guarantees the absence of junctions. In turn, the shell supplies an unlimited number of high-mobility electrons and attenuates variability and noise issues.

11.5.2 Floating-Body 1T-DRAM Memory

The downscaling of classical DRAMs is beneficial to the transistor (higher read current) but devastating to the capacitor (loss of stored charge). The capacitor miniaturization being a handicap, a radical solution is to eliminate it. A capacitorless or floating-body DRAM aims at storing and reading the

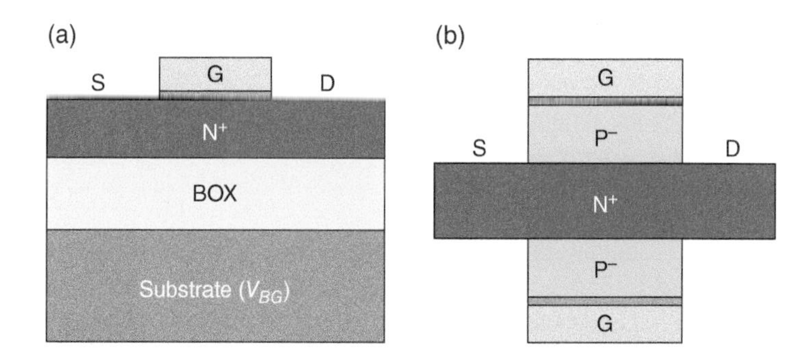

Figure 11.10 (a) Genuine junctionless transistor on FD-SOI. (b) Double-gated core-shell variant.

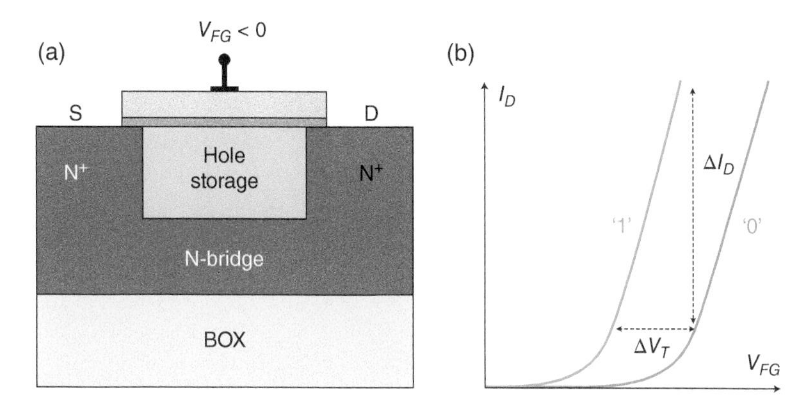

Figure 11.11 (a) Floating-body A2RAM memory cell and (b) corresponding transfer characteristics in "1" and "0" states.

charge in a single transistor (1T-DRAM, Figure 11.11a). There are numerous variants that take advantage of the body isolation in an FD-SOI MOSFET [35]. In memory state "1," majority carriers are generated by various mechanisms (BTB tunneling, impact ionization, or parasitic bipolar transistor) and stored within the body. In state "0," the majority carriers are swept away which increases the threshold voltage and lowers the read current (Figure 11.11b).

The enemy is the supercoupling effect which opposes the coexistence of electrons and holes in a thin body; 1T-DRAMs with good performance are 50 nm or thicker. On the other hand, gate-length scaling hurts the retention capability.

The A2RAM cell (Figure 11.11a) circumvents supercoupling by implementing a physically doped bridge at the bottom of an undoped body [54]. In state "0," the upper region of the body is empty (FD). A negative gate voltage, used to read the memory state, can deplete the bridge and cancel the drain current. Conversely, state "1" is programmed by generating holes via BTBT. Stored in the body, these holes absorb the gate field and preserve the bridge intact: the drain current is high.

The MSDRAM variant makes use of the MSD effect introduced in Figure 11.9c. The physical bridge is here replaced by an electron channel induced by the back gate. The merit is the tunability of the conductivity in the virtual bridge; the disadvantage again comes from supercoupling. Integrated on top of an engineered BOX with oxide–nitride–oxide composition, the MSDRAM pioneered the "universal" memory that combines volatile and nonvolatile retention. Electrons or holes can be stored permanently in the nitride layer, close to the drain or to the source. These four nonvolatile states together with the two volatile states of the 1T-DRAM led to eight distinct current levels [55].

11.5.3 Reconfigurable Devices with Electrostatic Doping

The electrostatic doping is the foundation of tunneling FET (TFET) and impact-ionization transistor (I-MOS). It is also efficient in tuning the resistance of the drift region in ultrathin LDMOS or converting Schottky contacts into N or P-type ohmic terminals [27].

An elementary device is the virtual diode (inset in Figure 11.12). It is actually an undoped SOI layer provided with end contacts (N^+ and P^+) and front and back gates. When the two gates receive voltages with opposite polarities, the regions underneath become electrostatically doped (N^* and P^*) and a mid-body junction turns out of nothing. Surprisingly, this electrostatic diode follows in many respects the classical diode theory. Refinements include the nonhomogeneous in-depth profiles of carriers across the film thickness. The novelty is that a junction made up of free electrons and holes behaves as a metallurgical junction. While the latter is rigid, the doping of the electrostatic diode can be continuously

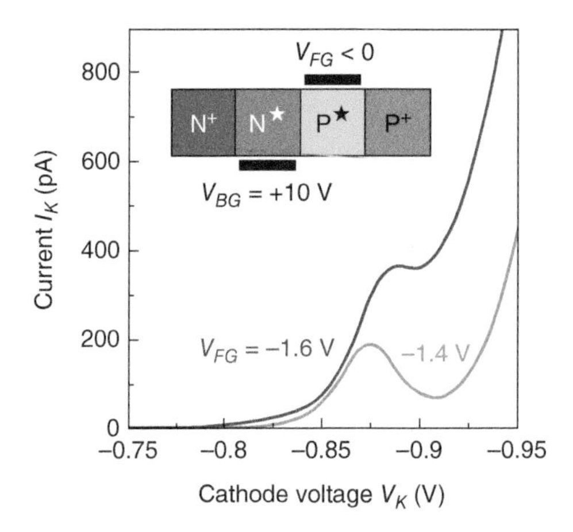

Figure 11.12 Experimental *I–V* characteristics of an electrostatic diode fabricated on 7 nm thick FD-SOI and operated in Esaki tunneling mode.

varied by adjusting the gate voltages. The experimental curves are well behaved: the forward current spans over 10 decades and the reverse current is extremely low, ideal for photodetectors [38].

A striking case is the highly doped Esaki diode emulated in undoped SOI film. Increasing the gate voltages fulfills the strict conditions for BTB tunneling to occur (very high impurity concentrations, favorable band alignment, and abrupt junction). Complicated hardware (junction engineering) is just replaced by simple software (gate programming). The *I–V* characteristics reproduced in Figure 11.12 do reveal a region with negative differential resistance where the peak-to-valley ratio is respectable [56].

The electrostatic diode is also named Hocus-Pocus to emphasize magician-made manipulations and metamorphoses. Since the two gates can induce P-type, N-type, or intrinsic regions underneath, there are nine possible embodiments, from PN and PIN diodes to thyristor, each with an infinite number of doping concentrations [38]. Shown in Figure 11.13 is a "chameleon" device which operates as a TFET in reverse mode whereas in forward mode, it is a sharp-switching band-modulation transistor.

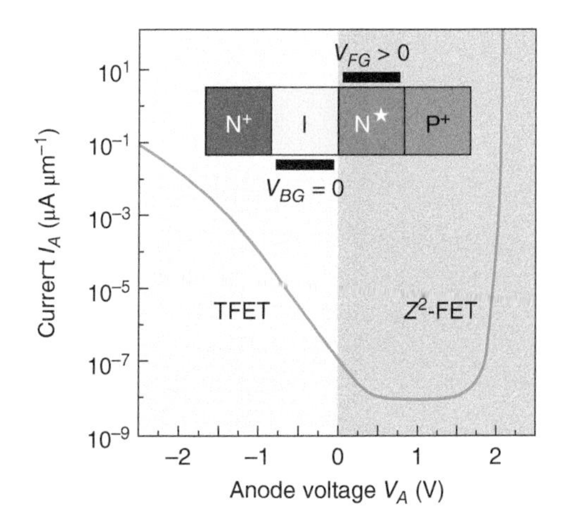

Figure 11.13 Experimental *I–V* characteristics of TFET and Z^2-FET emulated in a single FD-SOI device.

11.5.4 Band-Modulation Transistors

The device in Figure 11.14a is another embodiment of the Hocus-Pocus diode where the gates emulate a lateral NPNP structure. It reminds a thyristor although the basic mechanism is different. With forward bias, the holes available in the anode reservoir P^+ are ready for injection into the body but they face an energy barrier (Figure 11.14b). The electrons at the cathode N^+ have their own energy barrier. The device is blocked and the current remains negligible until the anode voltage reaches a critical value V_{ON}. A positive feedback loop leads to a sudden collapse of the energy barriers enabling the current to jump by 8–10 orders of magnitude [36, 37]. The waving of the energy bands from cathode to anode and their prompt collapse inspired the name "band-modulation."

Turning off the current requires the energy barriers to be rebuilt which results in a gate-controlled hysteresis (Figure 11.14c). The transfer characteristic $I_A(V_G)$ exhibits a similar sharp switch; the subthreshold swing is less than 1 mV per decade, unrivaled by any competing devices like TFET or ferroelectric MOSFET.

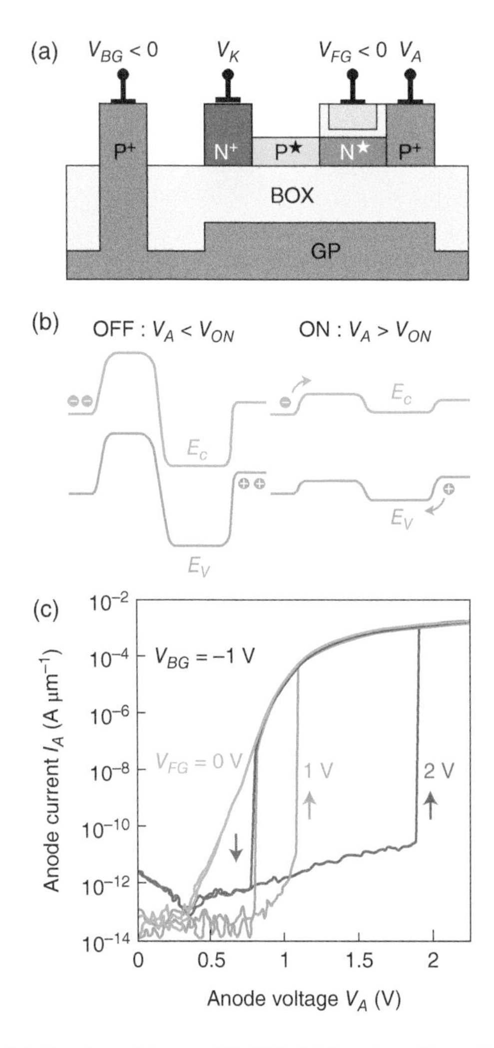

Figure 11.14 (a) Band-modulation transistor on FD-SOI, (b) band configuration in OFF and ON modes, and (c) experimental output characteristics.

The Z^2-FET version features front and back gates and operates at 1V. The following are applications already demonstrated [27].

Sharp Logic Switch: Fast pulses in the nanosecond range eliminate the detrimental hysteresis. A gate pulse of 0.5V amplitude is sufficient to abruptly switch the current on and off.

Memory: The Z^2-FET is intrinsically a DRAM cell. Memory state "1" is programmed by accumulating electrons underneath the gate. A positive read voltage applied to the anode kicks out the stored electrons; the discharge current triggers the device in ON mode with high current. In state "0," no electrons are stored, hence there is no discharge current and the cell remains blocked. The beauty of this Z^2-RAM is the higher performance and lower energy consumption achieved with faster access time [57]. Unaffected by super-coupling, a prototype 1MB matrix has been fabricated in sub-10-nm thick SOI. On the other hand, the ample hysteresis window in Figure 11.14c serves as a nonvolatile memory effect to design single-transistor static random access memory (SRAM).

ESD Protection: The Z^2-FET cumulates the attributes expected from an efficient protection against fast electrostatic discharges: quick response, reasonable drive current, negligible leakage, and tunability of the protection range through gate biasing. The device proved efficient against ESD pulses as short as 50ps.

Sensors: A novel scheme for photodetection uses incoming photons to produce electron–hole pairs in the ungated region that trigger the Z^2-FET. Light is sensed by the abrupt change in current from standby to ON mode. The response time indicates the light intensity. The principle of an ion-sensor is to replace the back-gate action with negative charges attached to the ungated surface. The turn-on voltage is extremely sensitive to the concentration of chemical species.

Other implementations of band-modulation feature two top gates or no gate at all. The latter structure (Z^3-FET) is commanded by twin ground-planes and makes available the free surface for sensing applications. Since the gate stack is omitted, the Z^3-FET is very robust and withstands higher voltages.

11.6 The Future

Nanoelectronic devices have more impact on our society than politics and religions. Whether this is admirable or devastating does not really matter; progress is ineluctable to satisfy the exponential demand. However, challenges are tough.

At *technology* level, the yield, variability, and reliability are probably more critical issues than the lithography resolution. The transistor *physics* points on intrinsic limitations: gate leakage, direct tunneling of electrons from source into drain, subthreshold leakage, threshold voltage roll-off, mobility collapse, and self-heating. The gate length of MOSFETs will shrink asymptotically to 5–10nm, a limit dictated by pragmatic considerations of cost and performance. Approaching this limit will not slow down the progress, neither the demand. The new rules of the chip game aim at implementing new functionalities and saving energy, exactly what the car and aircraft industry did after speed saturation.

Alternative materials will enter the production lines to assist silicon without threatening its supremacy. The Silicon Royalty will not abdicate.

The primary *economical* concern is the cost of fabs including production, design, test, and human resources. *Planet-conscious* actions focus on the saving of energy, water, and rare materials. In the last decade, the volume of mobile data increased 200 times whereas the production of electricity was limited to 20% growth. This disparity cannot continue. Missing electricity at home or in industry, in order to favor the data deluge, is not an option.

Our entire world is actually tending to full depletion. Oil, gas, and noble minerals will sometime soon be missing. Politics and finances are getting FD as the human relations became. We have more and more contacts in social media but never time to meet with and share a bottle of wine. Even our classrooms in semiconductors are deserted. Device physics and technology are too difficult and do not pay enough; bright students go to business school. Something has to be done, otherwise the stock of competent engineers will evaporate.

What is new is the *geopolitical* challenge. Any glitch in IC production has tragic impact on other industries. Overconcentration of fabs in Asia is considered to threaten sovereignty. The supply of advanced chips to a country may be conditioned by political interference. To counter the Chip War, ambitious Chips Acts are financed in the United States, Europe, and Japan. Europe will favor a long cherished technology: FD-SOI.

I believed in SOI for almost 50 years, a few more will not hurt. There is no doubt that SOI will be around for the celebration of the 150th anniversary of the transistor.

References

[1] Manasevit, H.M. and Simpson, W.I. (1964). Single-crystal silicon on a sapphire substrate. *J. Appl. Phys.* 35 (4): 1349–1351.

[2] Cullen, G.W. and Wang, C.C. (1978). *Heteroepitaxial Semiconductors for Electronic Devices*. Berlin: Springer.

[3] Cristoloveanu, S. (1987). Silicon films on sapphire. *Rep. Prog. Phys.* 50 (3): 327.

[4] Tihanyi, J. and Schlotterer, H. (1975). Influence of the floating substrate potential on the characteristics of ESFI MOS transistors. *Solid-State Electron.* 18 (4): 309.

[5] Izumi, K., Doken, M., and Ariyoshi, H. (1978). CMOS devices fabricated on buried SiO_2 layers formed by oxygen implantation into silicon. *Electron. Lett.* 14 (18): 593.

[6] Cristoloveanu, S. and Williams, S. (1992). Point contact pseudo-MOSFET for in-situ characterization of as-grown silicon on insulator wafers. *IEEE Electron Device Lett.* 13 (2): 102.

[7] Stoemenos, J., Jaussaud, C., Bruel, M., and Margail, J. (1985). New conditions for synthesizing SOI structures by high dose oxygen implantation. *J. Cryst. Growth* 73 (3): 546.

[8] Nakashima, S. and Izumi, K. (1990). Practical reduction of dislocation density in Simox wafers. *Electron. Lett.* 26 (20): 1647.

[9] Hemment, P.L.F., Reeson, K.J., Kilner, J.A. et al. (1987). Novel dielectric/silicon planar structures formed by ion beam synthesis. *Nucl. Inst. Methods Phys. Res.* B21 (1–4): 129.

[10] Ogura, A. and Fujimoto, Y. (1989). Novel technique for Si epitaxial lateral overgrowth: tunnel epitaxy. *Appl. Phys. Lett.* 55 (21): 2205.

[11] Tsao, S.S. (1987). Porous silicon techniques for SOI structures. *IEEE Circ. Devices Mag.* 3 (6): 3.

[12] Lasky, J.B. (1986). Wafer bonding for silicon-on-insulator technologies. *Appl. Phys. Lett.* 48 (1): 78.

[13] Maszara, P., Goetz, G., Caviglia, A., and McKitterick, J.B. (1988). Bonding of silicon wafers for silicon-on-insulator. *J. Appl. Phys.* 64 (10): 4943.

[14] Tong, Q.-Y. and Gösele, U. (1999). *Semiconductor Wafer Bonding: Science and Technology*. Wiley.

[15] Neudeck, G.W. (1990). Three-dimensional CMOS integration. *IEEE Circ. Devices* 6: 32.

[16] Czornomaz, L., Uccelli, E., Sousa, M. et al. (2015). Confined epitaxial lateral overgrowth (CELO): a novel concept for scalable integration of CMOS-compatible InGaAs-on-insulator MOSFETs on large-area Si substrates. *Tech. Dig. VLSI Technology Symp.*, Kyoto, Japan, p. T172.

[17] Lim, H.-K. and Fossum, J.G. (1983). Threshold voltage of thin-film silicon-on-insulator (SOI) MOSFETs. *IEEE Trans. Electron Devices* ED–30: 1244.

[18] Cristoloveanu, S. and Ioannou, D.E. (1990). Adjustable confinement of the electron gas in double-gate silicon-on-insulator MOSFET's. *Superlattice. Microst.* 8 (1): 131.

[19] Omura, Y., Horiguchi, S., Tabe, M., and Kishi, K. (1993). Quantum-mechanical effects on the threshold voltage of ultrathin-SOI nMOSFET's. *IEEE Electron Device Lett.* 14 (12): 569.

[20] Ioannou, D.E., Cristoloveanu, S., Mukkherjee, M., and Mazhari, B. (1990). Characterization of carrier generation in enhancement-mode SOI MOSFET's. *IEEE Electron Device Lett.* 11 (9): 409.

[21] Balestra, F., Cristoloveanu, S., Benachir, M. et al. (1987). Double-gate silicon on insulator transistor with volume inversion: a new device with greatly enhanced performance. *IEEE Electron Device Lett.* 8 (9): 410.

[22] Colinge, J.P., Gao, M.H., Romano, A. et al. (1990). Silicon-on-insulator gate-all-around device. In: *Technical Digest of IEDM*, 595–598. IEEE. http://doi.org/10.1109/IEDM.1990.237128.

[23] Nishimura, T., Ynoue, Y., Sugahara, K. et al. (1987). Three dimensional IC for high performance image signal processor. *IEDM'87 Conf. Proc.*, vol. 111.
[24] Bruel, M. (1995). Silicon on insulator material technology. *Electron. Lett.* 31 (14): 1201.
[25] Yonehara, T., Sakaguchi, K., and Sato, N. (1994). Epitaxial layer transfer by bond and etch back of porous Si. *Appl. Phys. Lett.* 64 (16): 2108.
[26] Jurczak, M., Skotnicki, T., Paoli, M. et al. (2000). Silicon-on-nothing (SON) – an innovative process for advanced CMOS. *IEEE Trans. Electron Devices* 47 (11): 2179.
[27] Cristoloveanu, S. (2021). *Fully Depleted Silicon–On–Insulator*. Amsterdam: Elsevier.
[28] Lederer, D. and Raskin, J.-P. (2005). New substrate passivation method dedicated to high resistivity SOI wafer fabrication with increased substrate resistivity. *IEEE Electron Device Lett.* 26 (11): 805.
[29] Young, K.K. (1989). Short-channel effect in fully depleted SOI MOSFET's. *IEEE Trans. Electron Devices* 36 (2): 399.
[30] Ernst, T., Tinella, C., Raynaud, C., and Cristoloveanu, S. (2002). Fringing fields in sub-0.1 μm fully depleted SOI MOSFETs: optimization of the device architecture. *Solid-State Electron.* 46: 373.
[31] Cristoloveanu, S. and Li, S.S. (1995). *Electrical Characterization of Silicon-on-Insulator Materials and Devices*. Boston: Kluwer Academic Publishers.
[32] Colinge, J.-P. (2004). *Silicon-on-Insulator Technology: Materials to VLSI*, 3e. Springer.
[33] Colinge, J.-P., Lee, C.-W., Afzalian, A. et al. (2010). Nanowire transistors without junctions. *Nat. Nanotechnol.* 5 (3): 225.
[34] Dufrene, B., Akarvardar, K., Cristoloveanu, S. et al. (2004). Investigation of the four-gate action in G^4-FETs. *IEEE Trans. Electron Devices* 51 (11): 1931.
[35] Bawedin, M., Cristoloveanu, S., Hubert, A. et al. (2011). Floating body SOI memory: the scaling tournament. In: *Semiconductor-on-Insulator Materials for Nanoelectronics Applications*, 393. Heidelberg: Springer.
[36] Salman, A.A., Beebe, S.G., Emam, M. et al. (2006). Field effect diode (FED): a novel device for ESD protection in deep sub-micron SOI technologies. *2006 International Electron Devices Meeting*, San Francisco, CA, USA (December 2006), pp. 1–4. http://doi.org/10.1109/IEDM.2006.346971.
[37] Cristoloveanu, S., Wan, J., and Zaslavsky, A. (2016). A review of sharp-switching devices for ultra-low power applications. *J. Electron Device Soc.* 4 (5): 215. https://doi.org/10.1109/JEDS.2016.2545978.
[38] Cristoloveanu, S., Lee, K.-H., Park, H.-J., and Parihar, M.S. (2019). The concept of electrostatic doping and related devices. *Solid-State Electron.* 155: 32.
[39] Clerc, S., Di Gilio, T., and Cathelin, A. (2020). *The Fourth Terminal*. Springer.
[40] Ohata, A., Bae, Y., Fenouillet-Beranger, C., and Cristoloveanu, S. (2012). Mobility enhancement by back-gate biasing in ultrathin SOI MOSFETs with thin BOX. *IEEE Electron Device Lett.* 33 (3): 348.
[41] Cristoloveanu, S., Rodriguez, N., and Gamiz, F. (2010). Why the universal mobility is not. *IEEE Trans. Electron Devices* 57 (6): 1327.
[42] Eminente, S., Cristoloveanu, S., Clerc, R. et al. (2007). Ultra-thin fully-depleted SOI MOSFETs: special charge properties and coupling effects. *Solid-State Electron.* 51: 239.
[43] Cristoloveanu, S., Athanasiou, S., Bawedin, M., and Galy, P. (2017). Evidence of supercoupling effect in ultrathin silicon layers using a four-gate MOSFET. *IEEE Electron Device Lett.* 38 (2): 107.
[44] Park, H., Colinge, J.-P., Cristoloveanu, S., and Bawedin, M. (2020). Persistent floating-body effects in fully depleted silicon-on-insulator transistors. *Phys. Status Solidi* A217 (9): 1900948.
[45] Ouisse, T., Ghibaudo, G., Brini, J. et al. (1991). Investigation of floating body effects in silicon-on-insulator metal-oxide-semiconductor field-effect transistors. *J. Appl. Phys.* 70 (7): 3912.
[46] Casse, M., Pretet, J., Cristoloveanu, S. et al. (2004). Gate-induced floating-body effect in fully-depleted SOI MOSFETs with tunneling oxide and back-gate biasing. *Solid-State Electron.* 48 (7): 1243.
[47] Bawedin, M., Cristoloveanu, S., Yun, J.G., and Flandre, D. (2005). A new memory effect (MSD) in fully depleted SOI MOSFETs. *Solid-State Electron.* 49 (9): 1547.
[48] Liu, F.Y., Ionica, I., Bawedin, M., and Cristoloveanu, S. (2015). Parasitic bipolar effect in ultra-thin FD SOI MOSFETs. *Solid-State Electron.* 112: 29.
[49] Gamiz, F., Roldan, J.B., Lopez-Villanueva, J.A. et al. (2002). Monte Carlo simulation of electron mobility in silicon-on-insulator structures. *Solid-State Electron.* 46 (11): 1715.
[50] Uchida, K., Koga, J., and Takagi, S. (2007). Experimental study on electron mobility in ultrathin-body silicon-on-insulator metal-oxide-semiconductor field-effect transistors. *J. Appl. Phys.* 102: 074510.

[51] Uchida, K. and Takagi, S. (2003). Carrier scattering induced by thickness fluctuation of silicon-on-insulator film in ultrathin-body metal–oxide–semiconductor field-effect transistors. *Appl. Phys. Lett.* 82 (17): 2916.
[52] Fossum, J.G. and Trivedi, V.P. (2013). *Fundamentals of Ultra-Thin-Body MOSFETs and FinFETs*. New York: Cambridge Univ. Press.
[53] Cristoloveanu, S. and Ghibaudo, G. (2023). The core-shell junctionless MOSFET. *Solid-State Electron.* 200: 108567.
[54] Rodriguez, N., Gamiz, F., Navarro, C. et al. (2012). Experimental demonstration of capacitorless A2RAM cells on silicon-on-insulator. *IEEE Electron Device Lett.* 33 (12): 1717.
[55] Chang, S.-J., Bawedin, M., Lee, J.-H. et al. (2014). Demonstration of unified memory in FinFETs. *Int. J. High Speed Electron. Syst.* 23 (3–4): 17.
[56] Lee, K.-H. and Cristoloveanu, S. (2019). Esaki diode in undoped silicon film. *IEEE Electron Device Lett.* 40 (9): 1346.
[57] Cristoloveanu, S., Lee, K.-H., Parihar, M.S. et al. (2018). A review of the Z^2-FET 1T-DRAM memory: operation mechanisms and key parameters. *Solid-State Electron.* 143: 10.

Chapter 12

FinFET

The 3D Thin-Body Transistor

Chenming Hu

Department of Electrical Engineering and Computer Sciences, University of California, Berkeley, CA, USA

12.1 The Show Stopper

The semiconductor industry seemed to be at a loss. In February 2021, the CFTO of Intel warned publicly that extrapolating the specifications of its latest central processing units (CPUs), the power consumption per chip area could soon exceed that of a nuclear reactor and then a rocket nozzle (Figure 12.1).

Clearly, the show stopper of Moore's Law was not going to be lithography. It was power dissipation. The exponential growth of power dissipation would not only be a nightmare for heat removal from the integrated circuit (IC) chips, it would also have been a disaster for Mother Earth considering how many chips would be used in smartphones and data centers. Architectural changes, such as multicore, could not reverse the projected orders-of-magnitude increase in power consumption. Both the cause and the solution of the power crises was the transistor.

12.2 The Cause of the Power Crises

Intel, being the undisputed leader of CMOS technology, was the company that had the confidence to sound the public warning. The expectation of runaway power dissipation was widely shared in the industry and the reason, rapidly growing transistor leakage current, was published knowledge. For those interested in the reason, it may be explained this way.

75th Anniversary of the Transistor, First Edition. Edited by Arokia Nathan, Samar K. Saha, and Ravi M. Todi.

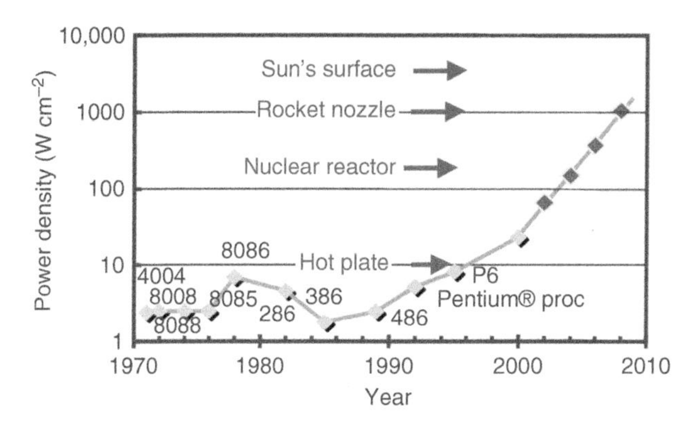

Figure 12.1 Dire projection of CPU chip power consumption.

1. The dynamic switching energy of a logic gate is $C \times V^2$, where C is the load capacitance (which will not be discussed in this article) and V is the IC operating voltage. To reduce the energy consumption, V must be reduced.
2. The speed of a logic gate depends on how fast the transistor can charge and discharge the load capacitance. It is proportional to the transistor on-state current (I_{on}), which is proportional to V-V_{th}. V_{th} is the transistor threshold voltage. To reduce energy consumption, V must be reduced. To increase speed, V_{th} must be reduced too.
3. Unfortunately, it was not possible to reduce V_{th} any more In fact, increasing V_{th} seemed unavoidable. The transistor standby power is proportional to the off-state or leakage current (I_{off}), which is proportional to $10^{-Vth/S}$. S, the "swing," was projected to rise fast with transistor shrinking, due to a phenomenon called the short-channel effect (SCE), if the gate oxide thickness, T_{ox}, cannot be reduced. Rising S and decreasing V_{th} would make I_{off} and the standby power rise exponentially. Hence, the comparison to nuclear reactor or rocket.

For decades, engineers knew that S can be held at a constant 60 mV by reducing T_{ox} in proportion to the transistor gate length, L_g. That straightforward rule had enabled transistor shrinks since 1970. By 1999, the semiconductor industry had concluded, correctly, that the oxide thickness (T_{ox}) reduction would slow down and stop around 1 nm (three oxide molecules thin).

If T_{ox} is not reduced in proportion to L_g, I_{off} (the current at zero V_g) will rise exponentially as illustrated by computer simulation in Figure 12.2a. This fact had been explained with the conceptual

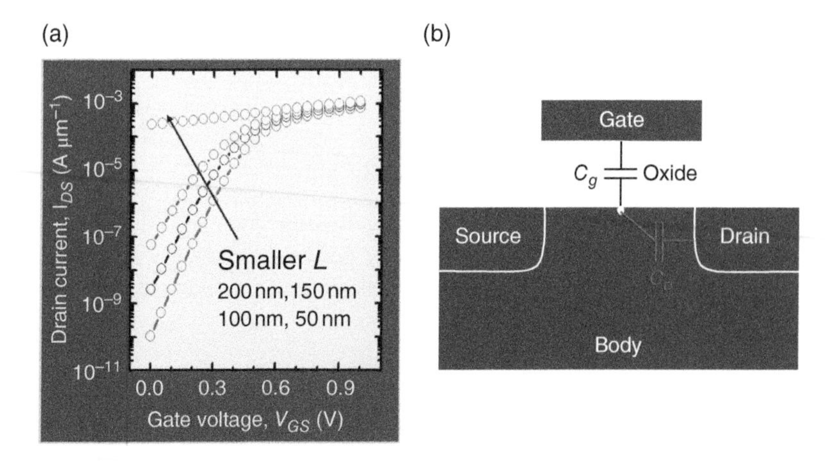

Figure 12.2 (a) Illustration of the exponential rise of I_{off} with decreasing L_g if T_{ox} is not reduced. (b) Capacitor-network explanation of the rise of the current at zero V_g.

capacitor-network model shown in Figure 12.2b. MOSFET current flows along the semiconductor-oxide interface. Through the capacitor C_g, the gate voltage controls the potential barrier at the interface and therefore controls the current. If T_{ox} can be reduced in proportion to gate length, C_g remains much larger than C_d and all is fine. If not, C_d becomes comparable to C_g, and the drain can turn on the current as if it were a second gate. Therefore, the end of T_{ox} reduction is the cause of the huge I_{off} in Figure 12.2a and the power crises in Figure 12.1. Or is it the *real* cause?

12.3 The Real Cause of the Power Crises

My own journey started with researching the tunneling current and reliability of thin gate oxide with my students since 1984 [1, 2]. We fabricated and studied very thin gate oxide a decade ahead of industry needs. Growing very thin thermal oxide on silicon could be done in our university laboratory very well. We correctly predicted that the oxide reliability (time-dependent dielectric breakdown lifetime) is many orders of magnitude better than the prevalent industry prediction. We also found that T_{ox} reduction would be limited by direct quantum tunneling, not reliability. That research gave me early and clear understanding of the limitations of T_{ox} scaling.

Simultaneously, my colleague, Prof. Ping Ko, and I and our students were developing more accurate analytical models of the three-dimensional electric field and potential profiles in the transistor body to model hot-carrier effect and SCE [3–5]. Those studies gave me this important insight. The really bad leakage current paths in a short-channel MOSFET do not hug the semiconductor-oxide interface, but run below the interface as shown in Figure 12.3. I modeled the submerged nature of the leakage current paths by changing Figures 12.2b and 12.3 [6].

Assuming that we found a magical dielectric and could reduce T_{ox} (the gap between gate and semiconductor in Figure 12.3) to zero, the leakage path is still far away from the gate electrode and the drain can still overpower the gate and induce large leakage current. Therefore, reducing T_{ox} was not sufficient. Furthermore, and fortunately, it was neither necessary. I realized that the ultimate solution to rising I_{off} was to remove the semiconductor that is more than ten nanometers away from the gate electrode. In other words, we needed to make transistors having very thin bodies such as the three structures in Figure 12.4.

Figure 12.4a could be built with silicon-on-insulator (SOI) substrates but the silicon film must be 10 nm or less while the available SOI substrates had silicon film thickness ranging from 100 nm to over several microns. I considered the multi-gate structures in Figure 12.4b, c more attractive because I_{on} would be two times larger and I_{off} would be much smaller for the same body thickness and T_{ox} than 12.4a. The structure in 12.4b seemed difficult to make to me. The structure in 12.4c could be fabricated with advanced lithography and etching tools, but not easily with the limited university laboratory resources.

Several industry laboratories had fabricated structures in the late 1980s at larger dimensions. Toshiba researchers reported ultrathin SOI transistor [7] and also a vertical surrounding gate transistor [8]. Hitachi researchers created an SOI fin structure on bulk substrate called DELTA [9]. These had all looked interesting to me but had not been adopted for production probably because the fabrication methods or benefits were not yet convincing enough.

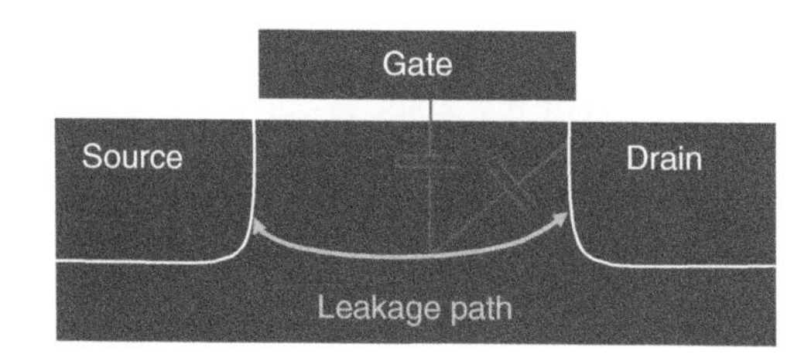

Figure 12.3 The correct capacitor network for understanding the short-channel transistor leakage current.

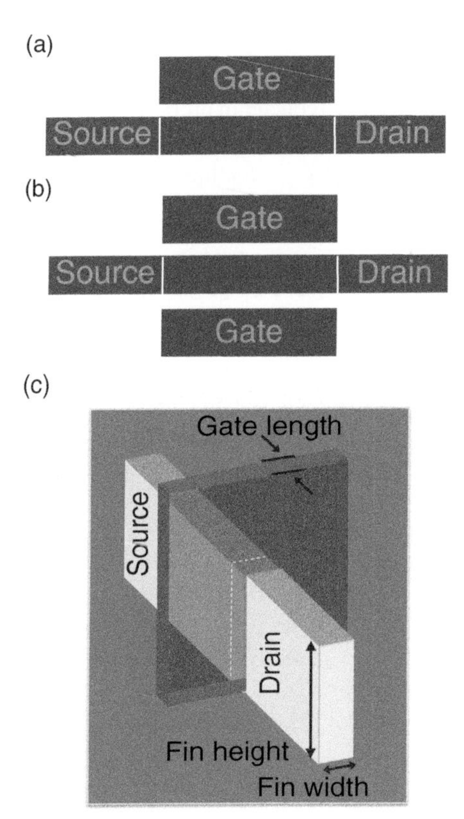

Figure 12.4 Three ways to build thin-body transistors. (a) Is ultrathin-body SOI (UTB-SOI) or fully depleted SOI (FDSOI). (b) Is double-gate and (c) Is FinFET.

12.4 A DARPA Request for Proposal

One day in 1996, my colleague Prof. Jeff Bokor told me that he had heard from a friend about a request for proposal from DARPA. I read the request and was amused by its title: "25 nm Switches." The text gave me the clear impression that an idea proposal would be a 25 nm MOSFET. Apparently, DARPA was expecting proposals of more exotic new devices because the consensus of the semiconductor industry was that MOSFET ICs were facing power crises even at 90 nm. I immediately thought that FinFET and UTB-SOI (Figure 12.5a, c) together would be a strong proposal because I could provide convincing explanation for why these structures would have excellent I_{off} and I_{on} at 25 nm and why they could be scaled much further beyond 25 nm (by scaling the body thickness with the gate length). I was confident that we could fabricate 25 nm UTB-SOI but wanted to include FinFET because it is the best solution.

There was a timing problem. The proposal deadline was days away and I would leave for a research conference in Japan the next day. Prof. Bokor, Prof. Tsu-Jae King Liu, and I quickly met. The meeting lasted about five minutes and we agreed that I would write the technical part of the proposal. I used the 10 hours on the plane the next day to write the technical part and sketch out the structures of FinFET and UTB-SOI with pen and paper. From my hotel in Japan, I faxed the pages to my colleagues at Berkeley.

Our proposal was selected for funding by DARPA. A team of a dozen students, professors, and industry visiting researchers started the research in 2017. We finally had the resources to demonstrate FinFET.

Figure 12.5 Photos of researchers. Digh Hisamoto is second from right in the top photo. In the middle photo front sitting row, Cathy Huang is first from left; front standing row, Choi is fourth from left. Bottom photo from left: Jeff Bokor, Tsu-Jae King Liu, Chenming Hu.

12.5 The Challenges and Team Work

To provide a solution for preventing the power crises, it was not enough to fabricate a fin-shaped transistor that is better than a 2D transistor. After all, similar structures had been published before. To achieve that goal, we had to convince at least one leading semiconductor company to spend a few hundred million US dollars to make what Intel would later call "the most radical shift in semiconductor technology in 50 years" [10]. From my experience of introducing the first industry standard transistor model for IC simulation and the hot-carrier reliability and thin-oxide reliability testing methodologies, I had learned that the industry needs to see crystal-clear data and overwhelming long term benefits to be convinced. Most importantly, FinFET must be seen as manufacturable and the experimental result must be clear and convincing with the relevant gate length.

We wanted to make the gate length much shorter than what the semiconductor industry considered possible. We must also make the transistor leakage current and on-state current clearly much better than the best production transistors of the day and estimated that the fin thickness must be less than 25 nm. The L_g and fin thickness dimensions were much smaller than what the best photolithography tools were capable of printing. We had to get the best out of electron-beam lithography. Prof. Bokor led that effort and contributed to others. We also wanted to use gate oxide thicker than the production T_{ox} (2 nm) to prove clearly that the end of T_{ox} reduction will not lead to power crises.

Furthermore, I wanted to demonstrate FinFET as a solution to another fundamental limit to transistor scaling – random dopant fluctuations (RDF). Dopants had to be added to the 2D transistor body to suppress SCE and I_{off}. As transistor's area shrank, the number of dopants in a transistor body drops. Suppose that there are only 10 atoms in the transistor body on average, some transistors may have 9 or 11 dopants, some may have 7 or 13 due to statistical variation. Such a large variation in body dopant concentration would lead to large V_{th} variation. This statistical phenomenon is called RDF. RDF had already been observed to degrade the yield of large SRAM circuits due to V_{th} variations. I was certain that the FinFET structure alone can suppress I_{off} so well that it offered a unique solution to break the RDF limitation: do not add significant doping to the FinFET body. No dopants, no random dopant fluctuation! Without heavy body doping, we needed another way to adjust V_{th} so that FinFET would have low leakage current at zero V_g and high I_{on} at low V_g. Prof. King Liu had pioneered the use of Si_xGe_{1-x} as a gate electrode with adjustable work-function and gate work-function adjustment was the solution for adjusting V_{th}. Although introducing this no-dopant innovation would add fabrication difficulty, I believed that FinFET must provide solutions to all known fundamental limitations of MOSFET scaling in order to be a convincing solution.

The source/drain junction depth was considered by many a third limit of CMOS scaling. The junction depth in Figure 12.3 must be kept small relative to the gate length. Otherwise, the fringing electric field between the vertical walls of the junctions and the channel exacerbates SCE. But, it could not be reduced much more without degrading contact resistance. This junction depth limit is overcome in all thin-body transistors because the doped source and drain thickness would be determined by the thin-body thickness, not junction depth, as shown in Figure 12.4. Prof. King-Liu's SiGe provided the material for the raised source and drain that reduced contact resistance.

Every team member contributed to the project by developing process, fabricating transistors, using TCAD simulation to guide the transistor design, or analyzing data. However, each wafer fabrication lot required a single person to carry out about a hundred fabrication steps through a period of several months. When the person succeeded in producing good transistors, they had earned the right to measure and publish the performance of the transistors as the lead author. Three researchers stood out for fabricating good transistors. They were Digh Hisamoto, Xuejue Huang, and Yang-Kyu Choi.

After DARPA had selected our 25 nm transistor proposal for funding, I received a letter from Digh Hisamoto, the Hitachi researcher who made the DELTA transistor [9]. Hitachi would support him to be a visiting researcher at a university for a year. He requested to spend the year in my research group. I gladly invited him to join the FinFET project. He became the researcher that carried the first lot of FinFET wafers through fabrication. He was an experienced and skillful researcher and produced FinFET with good I_{off} although I_{on} was lower than expectation [11]. The second researcher that carried wafer lot was Xuejue Huang. She was a talented and hardworking student. Her FinFET met all the targets described earlier. Good FinFET performance was convincingly demonstrated with several gate lengths between 18 and 45 nm and body thickness between 15 and 30 nm. The low S, low I_{off}, and high I_{on} were clear and compelling (see Figure 12.6). Xuejue's presentation at the 1999 International Electron

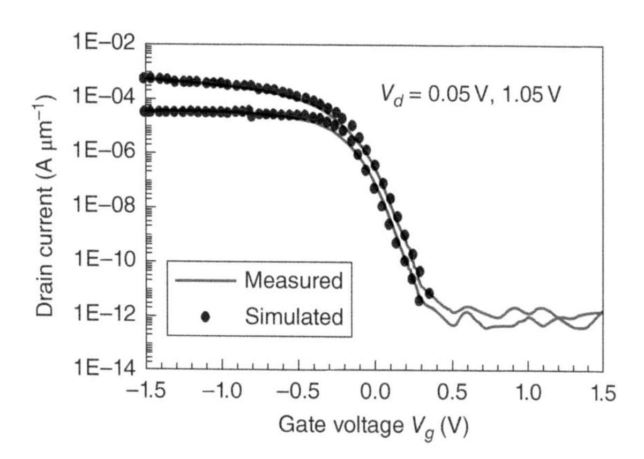

Figure 12.6 45 nm gate-length FinFET with 2.7 nm SiO_2 gate oxide, 66 mV/decade swing, and pA/μm leakage, and excellent I_{on}. Source: Huang [12]/IEEE.

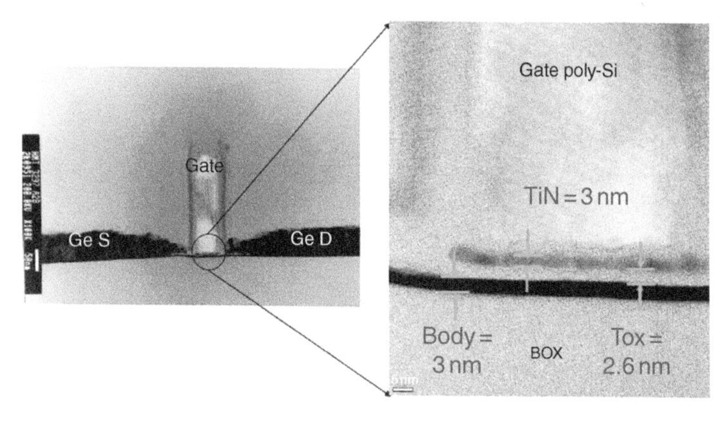

Figure 12.7 30 nm UTB-SOI (FDSOI) leakage was suppressed with 3 nm thin body [13] with raised source and drain. Source: Choi et al. [13]/IEEE.

Devices Meeting (IEDM) [12] awed the audience and the news of a breakthrough new transistor was reported by many trade and lay press.

After IEDM, Xuejue and I were invited to present and discuss the results at Intel's Silicon Valley site. In the following year, I was invited to do the same at its Oregon site twice. That was when I felt that FinFET could change the course of semiconductor technology.

In 2001, another student, Y-K Choi reported 30 nm gate-length low-leakage UTB-SOI shown in Figure 12.7 [13] and 20 nm gate-length FinFET with 10 nm fin thickness [14]. Figure 12.8 illustrates our message that MOSFET had arrived at new era of thin-body transistors; every 1 nm reduction of the thin-body thickness reduces the leakage current by nearly an order of magnitude. That is how the power crises suggested in Figure 12.1 was averted.

12.6 Further Advancements by Industry

The FinFET research convincingly demonstrated a new rule for continuing reduce future transistor gate-length: keep the channel or body thickness less than half of the gate length. In quick succession, IBM [15], AMD [16], TSMC [17], Samsung [18], and other companies used their fabrication resources

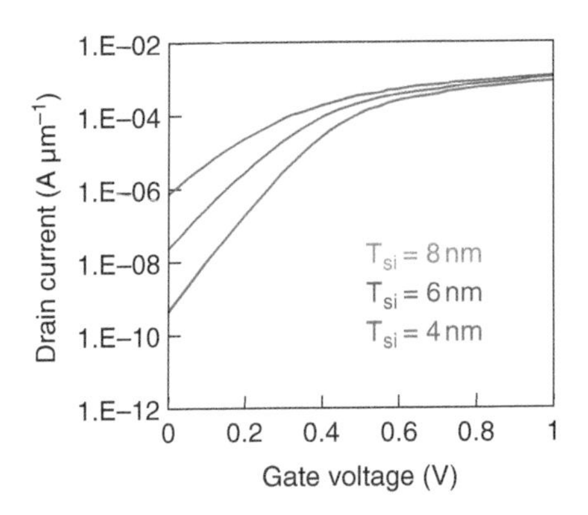

Figure 12.8 Each nm reduction of the body thickness reduces the leakage current of a 10 nm gate-length UTB-SOI by about an order of magnitude in the new era of thin-body transistors.

to improve on the Berkeley FinFET with excellent results at gate length as small as 10 nm. These industry reports completely proved the manufacturability of FinFET. Importantly, Samsung and university partners fabricated FinFET on bulk silicon substrate in 2003 [18]. In 2004, Samsung researchers demonstrated a way of making the thin body structure shown in Figure 12.4b by epitaxial growth [19]. It was called multi-bridge-channel MOSFET and would later be known also as GAA (gate-all-around) or nanosheet transistor. Meanwhile, 2D technology continued to suffice with the help of strained silicon, high-κ metal-gate, and multicore architecture.

A watershed event was Intel's adoption of FinFET for 22 nm production in 2011 [10]. TSMC basically ended its new 2D 20 nm technology because of power consumption problem in favor of 16 nm FinFET technology. Intel and the largest foundries (TSMC, Samsung, Globalfoundries, and UMC) all introduced 14 nm FinFET technologies. FinFET continued to enable the 12, 10, 7, 5, 4, and 3 nm technology nodes.

FinFET dramatically improved the performance and power consumption of electronics. Using it, smartphones can now be used for streaming videos and making free video chats for hours. In the past decade, US data center electricity consumption stayed basically constant while data generation rate increased 50 times. Artificial intelligence (AI) became widely accessible. FinFET increased NVIDIA Graphics Processing Unit (GPU) performance and power to the point that GPU can enable many deep learning (AI) applications. NVIDIA entered the top 10 list of most valuable companies in the world in 2021. Figure 12.9 is what the company website says about its AI solutions.

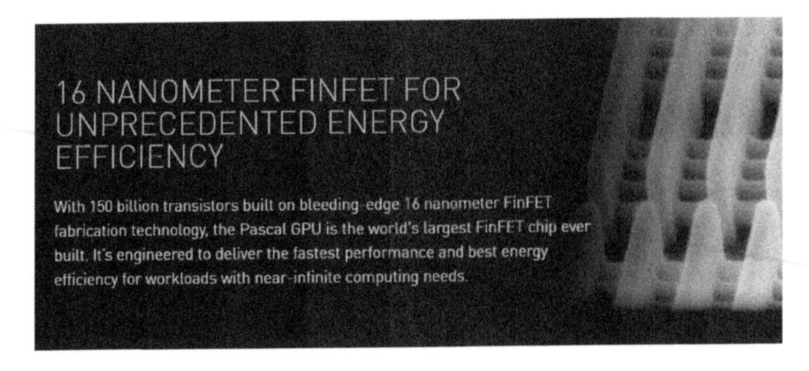

Figure 12.9 Snapshot of a 2017 NVDIA webpage titled AI Solutions. The latest GPU chip has 80 billion FinFETs.

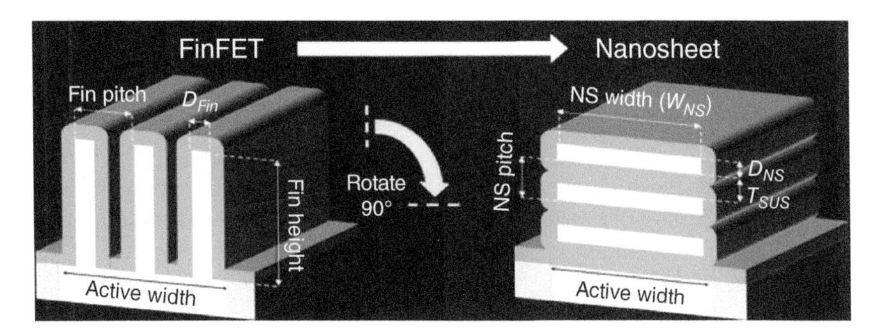

Figure 12.10 FinFET becomes GAA (gate-all-around or nano-sheet). Source: IBM/https://www.ibm.com/blogs/research/2019/12/nanosheet-technology-ai-5g// last accessed 28 March 2023.

With each new technology node, FinFET body was made taller and thinner. Taller makes footprint smaller, which is the reason of going 3D. Thinner suppresses I_{off} in the shrinking transistors. It has become hard to fabricate even taller and thinner FinFET body. At 3 nm node (for Samsung) or 2 nm node (for TSMC), thin body are turned 90° to become GAA, also known as nanosheet, as shown in Figure 12.10. GAA body is fabricated by epitaxial growth [19]. GAA body width is variable unlike the fixed height of the FinFET body. GAA is an excellent new 3D thin-body transistor. A very narrow GAA is called a nano-wire transistor.

Future thin-body thickness may be reduced to single-molecule thickness. Transition-metal-dichalcogenides (TMD) such as MoS_2 and WSe_2 are 2D semiconductors. Unlike 3D semiconductors such as Si and III-V compounds, a single layer of MoS_2 is a semiconductor crystal with attractive bandgap energy and carrier mobilities, and no surface dangling bounds. If desired, multiple layers can be stacked and held together by the weak van der Waals force between layers. Monolayer 2D semiconductor may be ideal for future thin-body transistors. Figure 12.11 illustrates a MoS_2 thin-body transistor with a 1 nm-diameter metallic carbon-nanotube gate [20]. The turn-off characteristics are very encouraging with on/off ratio of one million in 1 V of V_g span.

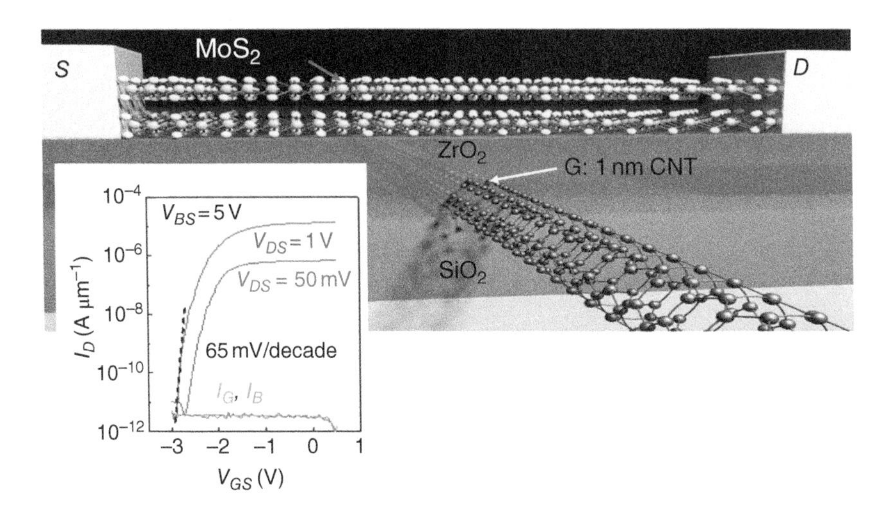

Figure 12.11 Nanometer thick 2D semiconductors may be excellent thin-body materials of future transistors. Source: Adapted from Desai et al. [20].

12.7 Conclusion

FinFET averted a projected power crises and perhaps end of the Moore's Law attributable to the end of oxide thickness reduction. FinFET proved that future transistor scaling could and should be driven by body thickness reduction. FinFET also demonstrated the manufacturability of 3D transistor, which adds a third dimension for density growth. The 3D and thin-body benefits are scalable; they can still provide many future generations of technologies by thinning the body and building upward. Higher transistor densities enable AI and future new technologies. This is the long-term legacy of FinFET.

References

[1] Chang, C., Brodersen, R.W., Liang, M.S., and Hu, C. (1983). Direct tunneling in thin gate-oxide MOS structures. *IEEE Transactions on Electron Devices* ED-30: 1571–1572.

[2] Chen, I.C., Holland, S., and Hu, C. (1985). Electrical breakdown in thin gate and tunneling oxides. *IEEE Transactions on Electron Devices* 20: 413–422. (1985). Also *IEEE Journal Solid-State Circuits*, pp. 333–342.

[3] Hu, C. (1983). Hot-electron effects in MOSFETs. *Tech. Digest of 1983 IEEE International Electron Devices Meeting (IEDM)*, Washington, DC (5–7 December 1983), pp. 176–181. IEEE.

[4] Chan, T.Y., Ko, P.K., and Hu, C. (1985). Dependence of channel electric field on device scaling. *IEEE Electron Device Letters* 6: 551–553.

[5] Jeng, M.C., Ko, P.K., and Hu, C. (1988). A deep submicron MOSFET model for analog/digital circuit simulations. *International Electron Devices Meeting*, San Francisco, CA (11–14 December 1988), pp. 114–117. IEEE.

[6] Hu, C. (2010). *Modern Semiconductor Devices for Integrated Circuits*. Pearson. Chapter 7.

[7] Yoshimi, M., Wada, T., Kato, K., and Tango, H. (1987). High performance SOIMOSFET using ultra-thin SOI film. *International Electron Devices Meeting*, Washington, DC (6–9 December 1987), pp. 640–643. IEEE.

[8] Takahashi, H., Sunouchi, K., Okabe¸N. et al. (1988). High performance CMOS surrounding gate transistor (SGT) for ultra high density LSIs. *International Electron Devices Meeting*, San Francisco, CA (11–14 December 19888), pp. 222–225. IEEE.

[9] Hisamoto, D., Kaga, T., Kawamoto, Y., and Takeda, E. (1989). A fully depleted lean-channel transistor (DELTA)-a novel vertical ultra thin SOI MOSFET. *International Electron Devices Meeting*, Washington, DC, (3–6 December 1989), pp. 833–836. IEEE.

[10] Clark, D (2011). Intel, seeking edge on rivals, rethinks its building blocks. *Wall Street Journal*. p. A1.

[11] Hisamoto, D., Lee, W-C.,Kedzierski, J. et al. (1998). A folded-channel MOSFET for deep-sub-tenth micron era. *International Electron Devices Meeting*, San Francisco, CA (6–9 December 1998), pp. 1032–1037. IEEE.

[12] Huang, X., Lee, W-C., Kuo, C. et al. (1999). Sub 50-nm FinFET: PMOS. *International Electron Devices Meeting*, Washington, DC (5–8 December 1999), pp. 67–70. IEEE.

[13] Choi, Y.-K., Ha, D., King, T.-J., and Hu, C. (2001). Nanoscale ultrathin body PMOSFETs with raised selective germanium source/drain. *IEEE Electron Device Letters* 22: 447–448.

[14] Choi, Y.-K., Lindert, N., Xuan, P. et al. (2001). Sub-20 nm CMOS FinFET technologies. *International Electron Device Meeting*, Washington, DC (2–5 December 2001), pp. 421–424. IEEE.

[15] Kedzierski, J., Fried, D.M., Nowak, E.J. et al. (2001). High-performance symmetric-gate and CMOS-compatible V/sub t/asymmetric-gate FinFET devices. *International Electron Devices Meeting*, Washington, DC (2–5 December 2001), pp. 19.5.1–19.5. IEEE.

[16] Yu, B., Chang, L., Ahmed, S. et al. (2002). FinFET scaling to 10 nm gate length. *International Electron Devices Meeting*, San Francisco, CA (8–11 December 2002), pp. 251–254. IEEE.

[17] Yang, F.-L., Chen, H-Y., Chen, F-C. et al. (2002). 25 nm CMOS Omega FETs. *International Electron Devices Meeting*, San Francisco, CA (8–11 December 2002), pp. 255–258. IEEE.

[18] Park, T., Choi, S., Lee, D.H. et al. (2003). Fabrication of body-tied FinFETs (Omega MOSFETs) using bulk Si wafers. *2003 Symposium on VLSI Technology*, Kyoto (10–12 June 2003), pp. 135–136. IEEE.

[19] Lee, S.-Y., Yoon, E-J., Kim, S-M. et al. (2004). A novel sub-50 nm multi-bridge-channel MOSFET (MBCFET) with extremely high performance. *Symposium on VLSI Technology*, Honolulu, HI (15–17 June 2004), pp. 200–201. IEEE.

[20] Desai, S.B., Madhvapathy, S.R., Sachid, A.B. et al. (2016). MoS2 transistors with 1-nanometer gate lengths. *Science* 354 (6308): 99–102.

Chapter 13

Historical Perspective of the Development of the FinFET and Process Architecture

Digh Hisamoto

Research & Development Group, Hitachi Ltd., Kokubunji, Tokyo, Japan

13.1 Introduction

Since the presentation of the first MOSFET by D. Kahng and M. M. Atalla in 1960 [1], the evolution of devices has been carried out under the guiding principle of the scaling law proposed by R. H. Dennard [2]. The scaling law originally demanded uniform shrinkage for all device parameters, but various external factors lead to nonuniform shrinkage, which has had a strong impact on the evolution of device structures. The scaling of metal-oxide-semiconductor (MOS) field-effect transistors (MOSFETs) since around 1970 has reduced device size and dramatically increased the number of integrated devices. Based on the scaling law, the main interest in device research and development has changed from reliability to high-speed operation, and to power consumption reduction. Until around the 1980s, power supply voltage scaling was not performed for each generation due to compatibility with peripheral chips, and the same power supply voltage (e.g. 12 or 5V) was used for several generations. Ensuring reliability by improving the lifetime of devices became a top priority. From the 1990s to around 2000, complementary MOS (CMOS) integration was clearly superior to other devices, and the operating frequency was rapidly increased in pursuit of higher performance. Therefore, device scaling, especially miniaturization of channel length, has been aggressively pursued to drive this trend over other device parameters. As a result, after the year 2000, the power consumption of chips became critical even with CMOS due to the superimposed effect of increasing the number of integrated devices. This is because the off current increased due to the increase in the number of devices while the switching characteristics deteriorated. Furthermore, it is considered that variations in threshold voltage due to

75th Anniversary of the Transistor, First Edition. Edited by Arokia Nathan, Samar K. Saha, and Ravi M. Todi.

statistical fluctuations of channel impurities prevented the reduction of power supply voltage. Since then, power consumption has become the main problem with scaling and has been the driving force for the introduction of fin field-effect transistor (FinFET), which is mandatory MOSFET device technology today.

In terms of manufacturing process technology, the FinFET structure appears to be quite different from that of conventional planar MOSFETs, giving the impression that planar technology is difficult to adapt. However, when we verified each process step, we found that there were almost no obstacles to introducing the structure because each step was modified conventional process technology. Rather, FinFET has promoted the evolution of manufacturing process technology because it derives excellent device characteristics by actively using the advantages of planar technology.

Here, we go back to the late 1980s when fin-channel structure was first invented and review the device requirements we envisioned as MOSFET scaling at that time. In addition, we will explain how the fin structure was derived as an answer to the aforementioned requirements by giving insight into the characteristics of the planar manufacturing process. Finally, we will summarize the technical concept inspired by the development of FinFET, which opens up the future of LSIs.

13.2 Requirements for the End of CMOS Scaling

Few electronic devices have battled reliability more than the MOSFET. MOSFETs can be obtained with extremely simple processes and structures in exchange for using the interface between ultrathin insulating film and a semiconductor as the channel. This is why reliability considerations have become inevitable for MOSFETs. Also, because of this, the limitations of the MOSFETs have been discussed since the invention.

In the late 1980s, discussion of scaling limits mainly increased due to the emergence of LSIs that used the GaAs compound semiconductor as a switching device. MOSFETs were said to be replaced by compound semiconductor devices due to their inferior basic performance (such as carrier mobility). Therefore, we decided to explore how far the performance could be improved by scaling the Si MOSFET in device physics. Papers on device scaling limits were very helpful in this study, especially one by R. E. Keyes [3]. That paper listed the following summarized items as limiting factors for MOSFETs:

- Dielectric breakdown
- Minimum channel length
- Minimum oxide thickness
- Random fluctuation of impurity number
- Saturation velocity limit

Dielectric breakdown, minimum channel length, the thinning limit of the gate insulating film, or saturation velocity limit were widely known even at that time, but little was known about impurity fluctuations. Keyes drew images like the checkerboard shown in Figure 13.1, and predicted that device characteristics would become unstable due to impurity fluctuations at the end of miniaturization. He expected this to be the limiting factor for scaling. Looking at the subsequent history, the effect of impurity fluctuations on LSIs began to be intently analyzed in the late 1990s when the impurity concentration increased, and it got serious in the 2000s when device technology reached the 45 nm generation [4, 5]. In contrast to all other device parameters, the scaling law suggested that only the impurity concentration should be increased, which would lead to future bankruptcy. Considering that the channel concentration exceeds $10^{18}\,cm^{-3}$ (far in the future at that time), in addition to unstable device characteristic, significant deterioration could be predicted in junction breakdown voltage and inversion layer mobility. These would bring scaling to an end. Therefore, we decided to work on reducing the impurity concentration while scaling. If the impurity concentration can be lowered, the electric field

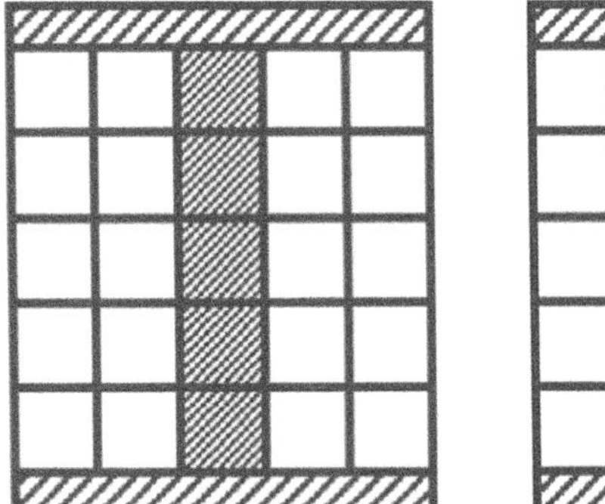
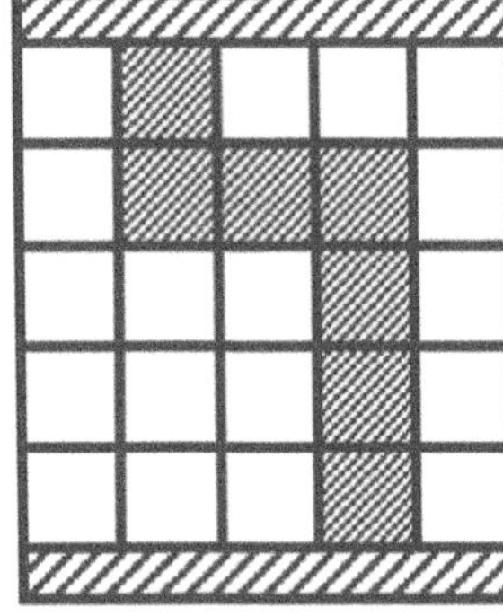

Figure 13.1 Impurity fluctuation images in scaled MOSFET according to Keyes [3]. Two example cases are picked up for reference. The channel is divided into a mesh and modeled to express that the current path changes due to the difference in impurity concentration.

generated in the semiconductor should tend to be lower; so many items related to reliability should work in the direction of making it easier to extend the life of the device. However, countermeasures against the short-channel effect must be considered separately because it is in principle contrary to the scaling law. Therefore, we thought that lowering the concentration of the channel would be synonymous with suppressing the short-channel effect.

The double-gate structure we focused on as a solution to this problem was not a new idea. An old example can be found in Kahng's paper on the history of MOSFETs [6]. He listed the patents of J. E. Lilienfeld and O. Heil as the first FETs along with the first transistor and the first MOSFET. In this 1935 O. Heil patent, a double-gate FET was already described [7]. Unlike integrated devices, discrete devices must deal with the back side of the device, so it may have been natural to consider a double-gate structure. In the 1980s, silicon on insulator (SOI) wafers based on separation by implanted oxygen (SIMOX) technology were at the research level but were put into practical use, and many reports were made in this field. Simulations and experiments had begun to report that the short-channel effect could be suppressed by the field effect of the gate instead of forming a potential barrier by channel impurity doping and excellent switching characteristics could be obtained by thinning the SOI [8–10]. Double-gate device was the most promising solution because the effects could work effectively. Based on this consideration, Figure 13.2 shows the roadmap we devised at the end of the 1980s [11]. The gate electrode should be effectively utilized to lower the channel impurity concentration to obtain a stable threshold voltage while advancing scaling. If we proceed further, we will end up with a double-gate structure.

Thus, we thought about the operation of a double-gate device. In the case of a double-gate device, since the channel is thin, the potential difference between the two gate electrodes sandwiching the channel is directly applied in the vertical direction of the channel. As a result, the thinner the channel, the stronger the surface electric field. However, by synchronously operating the double gates, the vertical electric field can be weakened even if the channel is thin and intrinsic. Therefore, ideal switching characteristics can be achieved at the same time as the highest efficiency in both the on-state and off-state. From these considerations, the required conditions that should be met for future scaled MOSFET structures are:

- Lightly doped or intrinsic semiconductor channel for stable device operation.
- Double-gate structure for suppression of the short-channel effects.
- Short-circuit between double gates for good operating characteristics.

We decided to search for a device structure that satisfies these conditions.

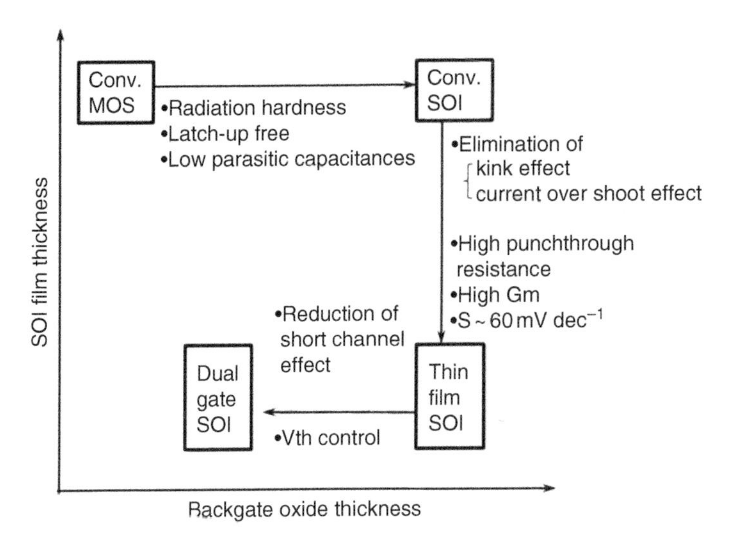

Figure 13.2 Around 1990, we established a roadmap for CMOS device technology based on requirements for the end of CMOS scaling. The path to double-gate MOSFETs (dual-gate at the time) was predicted. Source: Takeda et al. [11]/IEICE.

13.3 Restrictions of Planar Process Technology

First, we considered a structure in which a buried gate is placed under a conventional SOI MOSFET. Because the device was too large compared to conventional planar MOSFETs, it was thought that sufficient performance could not be obtained for the area. Therefore, we decided to review the planar process, considering why conventional planar MOSFETs are small and have such a high affinity with manufacturing processes.

We found that unlike bipolar devices, planar MOSFETs do not have detailed structure inside the substrate other than a large well substrate electrode (well structure). Most of the device elements, including the gate electrode, are only formed as wiring structures stacked on the substrate surface. The planar process forms elements by repeating deposition, patterning, and processing perpendicular to the wafer surface. Therefore, it is the perfect manufacturing technology for structures that are stacked on a substrate, such as MOSFETs.

To see the actual evolution of MOSFET structures, an overview of the device structure transition from the first MOSFET to FinFET is shown in Figure 13.3. It could be said that the introduction of LOCOS and Si-gate technology in the 1970s established MOSFETs by using the planar process. That is, the device active region and the device isolation region can be simultaneously formed in one process using the pattern (L) in Figure 13.3 placed on the boundary between them. The gate electrode can be formed by the pattern (FG) in Figure 13.3. Ion implantation is performed using the delineated gate pattern, thereby forming diffusion layer electrodes in a self-aligned manner without new patterning. All elements required for integrated devices such as isolations, channels, and gates can be formed by using two patterns, L and FG, once each. The contact pattern is also an important element, but since self-aligned contact (SAC) technology was already in practical use at that time, it was omitted here from the geometric considerations. This simple structure helps reduce the gate length, which improves device performance and helps significantly reduce unnecessary parasitic capacitance and resistance compared with other devices. Also, it has had a great effect on device miniaturization and processing reproducibility in manufacturing. Since then, scaling has improved with the same basic structure, such as the lightly doped drain (LDD) structure using spacer technology and the self-aligned silicide (SALICIDE)

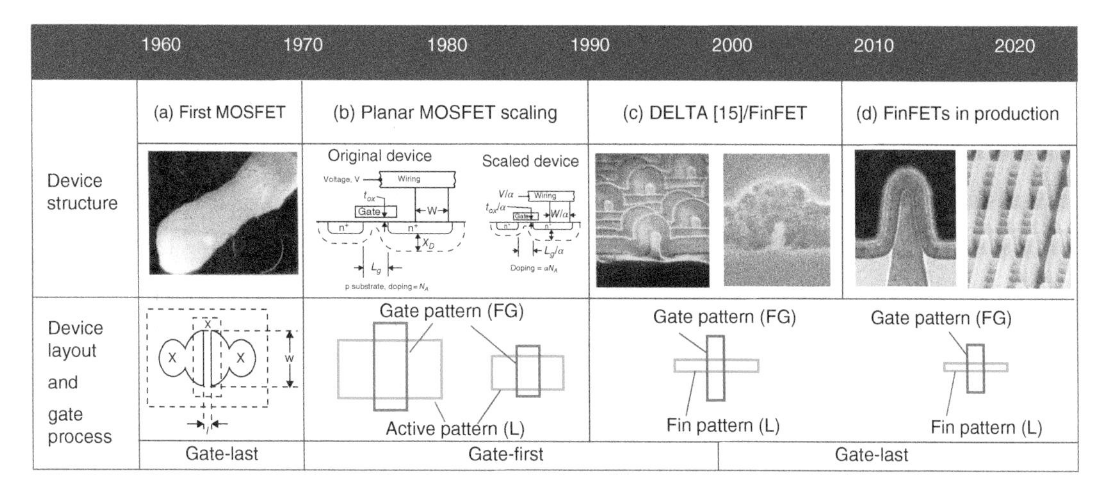

Figure 13.3 MOSFET evolution trends and their mask layout design and gate fabrication process. (a) SEM picture of the first MOSFET by Kahng and Atalla [1], Sze and Lee [12], and Kahng [13], (b) the images of device scaling according to Dennard's scaling law. Source: Adapted from Dennard et al. [2] and Frank et al. [14], and (c) SEM picture of the first Fin-channel device (DELTA) [15] and FinFET [16]. (d) SEM images of volume production FinFETs [17, 18]. In principle, there are two manufacturing methods for MOSFETs with two-layer patterns: gate-first and gate-last. To demonstrate the scalability, we have shown that the fin-channel structure is applicable to both methods.

process for reducing resistance. Therefore, we were convinced that this consideration should be applied to the realization of the double-gate structure.

Assuming planar or linear channels, there are three possible double-gate MOSFET structures that can be integrated using semiconductor planar processes: planar-channel, vertical–channel, and fin-channel structure. This was later reviewed by H. S. P. Wong and became widely known [19].

Since these structures assume double-gate configuration, the first two requirements are satisfied, so it is the third term that really matters. Considering the third term, we can see that the planar-channel type and the vertical-channel type have stacked electrodes, so a third wiring pattern other than L and FG is required to connect these electrodes. This is because the planar process technology does not have the technology to pattern the stacked structure at once. Even worse, the third pattern becomes a contact-hole pattern for connecting to the buried layer. The size of such a contact pattern can be several times the channel thickness. In practice, for example, the current channel thickness (fin width) is around 10 nm, while the contact diameter is as large as 20–30 nm. This is the reason that makes the planar or vertical-channel double-gate MOSFET overwhelmingly larger than the planar MOSFET, and we noticed its incompatibility with scaling.

However, in the fin-channel structure, the gates can be connected on the fins by simply patterning the gate beyond the fin. Therefore, the third pattern is not required for electrode connection. In other words, only a fin structure with a channel parallel to the substrate satisfies the above three propositions without being subject to these restrictions.

Concerns arising from the use of the fin-channel structure were closely related to the three-dimensional device configuration. When applying this structure for the first time, we were particularly concerned about how the top and bottom of the fin would affect the device properties. These were structures not found in planar MOSFETs. We verified the characteristics using a 3D device simulator that had just been put into practical use [20]. Unlike today, the available memory was limited, and it was difficult to secure a sufficient number of nodes even if mainframe computers were used. In addition, SOI simulation had just started, and the method of processing floating nodes in the simulation was

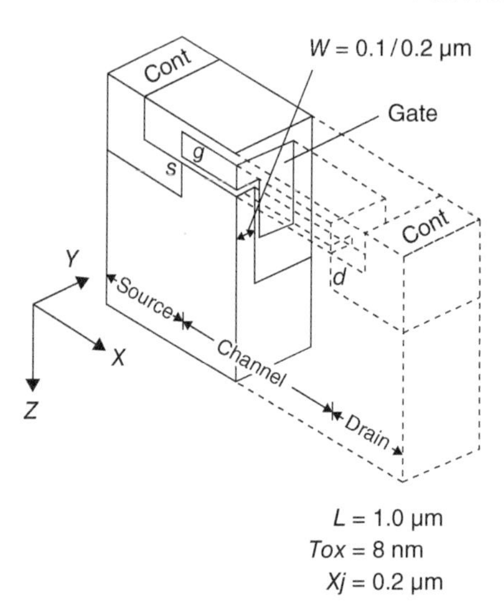

Figure 13.4 Simulated 3D device structure. In order to reduce the number of memories, all structures were composed of cuboids and a boundary condition of mirror symmetry was set at the center of the fin. The bottom of the fin was connected to the substrate.

not mature. Therefore, we created a device model system with a simple structure (Figure 13.4) and performed simulations. To avoid creating floating nodes in this structure, we tied the bottom of the fin to the substrate to provide a potential path. As a result, looking back now, what we first simulated was the bulk FinFET structure itself, later made possible by the maturation of shallow trench isolation (STI) technology.

Thanks to the simulation with this structure, we were able to obtain useful knowledge for designing an evolved device structure. We found that the channel potential is controlled by the gate electrodes from both sides if the fin width is narrowed and the device operation is not affected by the substrate. This means that the fin structure works well in both active and isolation regions and can be applied for device integration in either field isolation or SOI structures.

For the actual device prototyping, we originally intended to use SOI wafers. However, SOI wafers made by SIMOX in the 1980s had metal contamination, so we stopped using them in prototypes. Instead, we devised an isolation method that improved on LOCOS and developed the first fin structure double-gate device, DELTA (fully depleted lean channel transistor) [15]. In the latter half of the 1990s, wafer bonding and smart cut technology were established, and high-quality SOI wafers became available. Therefore, in prototyping folded-channel MOSFET [16], later called FinFET [21], SOI wafers have been used as a simple isolation method. For fin-channel devices, the choice between a bulk or SOI wafer is primarily considered a process option rather than a device characteristic of the active area, so there were little restrictions on wafer choice in device design.

Regarding the upper part of the fin, the quantitative effect is sufficiently small when the channel impurity concentration is low and the fin width is narrower than the depletion layer width. Therefore, the structural design decision whether to form the gate or put the insulator on top of the fin can be mainly based on the requirements of the manufacturing process. In developing DELTA, the gate insulating film was thick enough for processing compared to the height of the fin, so the gate was also placed on the top surface and used as a channel to widen the channel width. In developing FinFET, we left the thick insulating film and used only the two sides as the channel. This is because we wanted to use a new material for the gate-stack to give priority to the controllability of the threshold voltages by the work function (more on this later). Therefore, the question of using a double gate or triple gate is not essential in terms of device operation but merely a processing option again. Various discussions have been held on this point, but since FinFET began to be mass-produced, they have hardly been discussed.

13.4 Prompted Device/Process Technology Evolution by FinFET

FinFET as a single device, characterized by substrate electrode isolation, is one of the known double-gate structures and does not require new mechanisms for device operation. However, its excellent switching characteristics and low power consumption performance spurred the multi/many-core MPUs and GPUs when viewed as LSI systems. This is also believed to have been the driving force behind domain-specific hardware such as mobile APUs.

On the other hand, process technology has been driving progress in more practical ways. Patterning technology is a prime example. In fin-channel structures, the patterning of the fin must be less than the gate length (i.e. the minimum feature size in previous LSIs). Therefore, we had to solve this problem when developing the fin-channel structure. We focused on the arrangement of the active layer and the gate electrode layer. These layers were arranged orthogonally in the majority of the chip in order to increase the density of the elements and the density of the wiring layer. Therefore, the active layer was laid out so that each pattern was almost parallel. Given the design constraints of parallel placement (especially pitch uniformity), there were ways to obtain patterns with linewidths below the patterning limit of lithography. For example, phase shift and sidewall spacer technologies were convenient. In fact, fin patterning was performed using spacer processing in the first prototype until lithography technology was established. Later, FinFET was commercialized using self-aligned double patterning (SADP) with spacer technology. It takes advantage of the parallel arrangement of the fin-channel structure, which is suitable for manufacturing. FinFET supported miniaturization by evolving from SADP to self-aligned quadruple patterning (SAQP) during the development delay of extreme ultraviolet (EUV) technology. In addition, we believe that it will lead lithography technology as multiple patterning technology when exposing with EUV.

Another feature is the gate fabrication process. In demonstrating the fin-channel structure, the development of DELTA in the late 1980s first aimed to demonstrate that the fin could be formed by a planar process and that the structure would function as a double-gate MOSFET. The point of the device design was to create a structure as common as possible with existing planar MOSFETs. Therefore, as shown in Figure 13.3, we adopted the usual gate-first process, which was a mature process.

Later, in the late 1990s, when FinFET was developed with the aim of demonstrating scalability beyond the 20 nm generation, device designs were carried out to demonstrate the feasibility of the technology required for advanced fin-channel structures with advanced scaling. For gate electrodes, the work function of the gate material should be used to control the threshold voltage. This meant the introduction of high-κ/metal gates and the gate-last process required to manufacture them. Thus, we have demonstrated the gate-last process for fin-channel structures, paving the way for the introduction of new gate materials. As shown in Figure 13.3, all FinFETs in volume production have a high-κ/metal gate realized using the gate-last process.

A further feature of FinFETs in manufacturing technology is the height difference between the fin and the gate electrode due to the upright fin shape. To alleviate the height difference, we developed a fin lifting pad when we demonstrated the first Dynamic Random Access Memory (DRAM) application [22]. This can be thought of as the beginning of Middle of the Line (MOL), an intermediate layer connecting metal wiring layers and fins, in addition to the widely used Front End of Line (FEOL) and Back End of Line (BEOL) processes. With the introduction of MOL, the degree of freedom in LSI design has improved, and, for example, the area of basic logic gates has been reduced. This led to the development of DTCO (Design Technology CO-optimization), which is now established as an important process and design technology. We believe that as 3D device technology (such as 3D stack structures) evolves, its importance will increase.

13.5 Conclusion

Since the product installation began in 2011 and major foundries launched around 2015, most high-speed CPUs, MPUs, GPUs, and mobile devices have been made with FinFET. In this paper, we showed that FinFET is a device structure that has arrived as an inevitable result of the evolution of MOSFET scaling, both in terms of MOSFET device performance and planar process fabrication technology.

We devised the first fin-channel device, DELTA, in the late 1980s and demonstrated its scalability with FinFET in the late 1990s. Many people doubted its productivity because of its appearance. Since around 2000, many papers on FinFET have been published by many institutions. However, in many cases, those groups had no experience with DELTA, so instead of evolving FinFET, they firstly confirmed the effects of fin-channel structure as we did in DELTA. Then, they have moved on to fin-channel device evolution.

Already four or five device generations have passed since the launch of FinFET products. Advances in process technologies such as lithography, etching, CVD, and MOL have made it widely recognized that complex 3D structures can be realized in planar processes. As shown in this paper, device evolution has been driven primarily by process technology requirements. Therefore, we believe that these technologies developed to realize the 3D structure of fins will lead to technologies that support the direction of future integration, such as the evolution from miniaturization to 3D stacked structures.

References

[1] Kahng, D. and Atalla, M.M. (1960). Silicon-silicon dioxide field induced surface devices. *IRE-AIEE Solid-State Device Res. Conf.*, Pittsburgh, PA. Carnegie Inst. of Technol.

[2] Dennard, R.H., Gaensslen, F.H., Yu, H.N. et al. (1974). Design of ion-implanted MOSFET's with very small physical dimensions. *IEEE J. Solid State Circuits* SC-9: 256.

[3] Keyes, R.W. (1975). Physical limits in digital electronics. *Proc. IEEE* 63 (5): 740–767.

[4] Pelgrom, M.J.M., Duinmaijer, A.C., and Welbers, A.P.G. (1989). Matching properties of MOS transistor. *IEEE J. Solid State Circuits* 24 (5): 1433–1440.

[5] Mizuno, T., Okumtura, J., and Toriumi, A. (1994). Experimental study of threshold voltage fluctuation due to statistical variation of channel dopant number in MOSFET's. *IEEE Trans. Electron Devices* 41 (11): 2216–2221.

[6] Kahng, D. (1976). A historical perspective on the development of MOS transistors and related devices. *Trans. Electron Devices* 23 (7): 655–657.

[7] Heil, O. (1935). Improvements in or relating to electrical amplifiers and other control arrangements and devices. British Patent 439457, filed 1935 issued 1935.

[8] Malhi, S.D.S., Lam, H.W., Pinizzotto, R.F. et al. (1982). Novel SOI CMOS design using ultra thin near intrinsic substrate. *Tech. Dig. IEDM*, San Francisco, CA, pp. 107–110. IEEE.

[9] Sekigawa, T., Hayashi, Y., Ishii, K., and Fujita, S. (1985). XMOS Transistor for a 3D-IC. Abstract of SSDM, C-3-9 LN, Tokyo, Japan, p. 14.

[10] Colinge, J.P. (1986). Subthreshold slope of thin-film SOI MOSFET's. *Electron Device Lett.* 7 (4): 244–246.

[11] Takeda, E., Hisamoto, D., and Nakamura, K. (1991). A new SOI device – DELTA–, structure and characteristics. *IEICE Trans.* E74 (2): 360–368.

[12] Sze, S. and Lee, M.-K. (2013). *Semiconductor Devices, Physics and Technology*, 3e. Wiley.

[13] Kahng, D. (1961). *Silicon-Silicon Dioxide Surface Devices*, 583–596. Technical Memorandum of Bell Laboratories.

[14] Frank, D.J., Dennard, R.H., Nowak, E. et al. (2001). Device scaling limits of Si MOSFETs and their application dependencies. *Proc. IEEE* 89 (3): 259–288.

[15] Hisamoto, D., Kaga, D., Kawamoto, Y. et al. (1989). A fully depleted lean-channel transistor (DETA) –a novel vertical thin SOI MOSFET. *Tech. Dig. IEDM*, Washington, DC, p.833. IEEE.

[16] Hisamoto, D., Lee, W.-C., Kedzierski, J.et al. (1998). A folded-channel MOSFET for deep-sub-tenth micro era. *Tech. Dig. IEDM*, San Francisco, CA, pp.1032–1034.

[17] Auth, C., Allen, C., Blattner, A. et al. (2012). A 22nm high performance and low-power CMOS technology featuring fully-depleted tri-gate transistors, self-aligned contacts and high density MIM capacitors. *Symposium on VLSI Technology Digest of Technology Papers*, Honolulu, HI, pp.131–132.

[18] https://www.tsmc.com/japanese/dedicatedFoundry/technology/logic/l_16_12nm.

[19] Wong, H.-S.P. (2002). Beyond the conventional transistor. *IBM J. Res. Dev.* 46 (2/3): 133.

[20] Masuda, H., Toyabe, T., Hagiwara, T., and Ushiro, Y. (1984). High speed three dimensional device simulator on a super computer: CADDETH. *IEEE International Symposium on Circuits and Systems, Proceeding*, Montreal, Canada, pp. 1163–1166.

[21] Hisamoto, D., Lee, W.C., Kedzierski, J. et al. (2000). FinFET –a self-aligned duble-gate MOSFET scalable to 20nm. *Trans. Electron Devices* 47: 2320.

[22] Hisamoto, D., Kimura, S.I., Kaga, T. et al. (1991). A new stacked cell structure for giga-bit DRAMs using vertical ultra-thin SOI (DELTA) MOSFETs. *Tech. Dig. IEDM*, Washington, DC, pp. 959–961.

Chapter 14

The Origin of the Tunnel FET

Gehan A. J. Amaratunga[1,2]

[1]Electrical Engineering Division, Department of Engineering, University of Cambridge, Cambridge, UK
[2]Zhejiang University – University of Illinois Urbana Champagne Institute (ZJUI) and School of Information and Electronic Engineering, Zhejiang University, Haning, China

14.1 Background

This is an account of how the Tunnel FET (TFET) in silicon came to be proposed and experimentally verified. It will also give some idea of the related device concepts based on tunneling which were being explored at the time (early to mid-1990s). It is not my intention to chart "invention" of the device but rather how the concept evolved. I will also introduce the current interest in the TFET for low-power electronics.

The origin of the concept of tunneling between heavily doped p–n regions is from Esaki [1]. In 1990, I took on the lecturing of the final year Quantum Electronics Course (Paper 8) in the Electrical and Information Sciences Tripos (EIST) at Cambridge. At the time, EIST was the option for specialization in Electrical Engineering after two years of general engineering at Cambridge. Admission to the general (combined) engineering course still remains the only way to study engineering at Cambridge. The course included quantum mechanics based on the Schrodinger equation and the solutions to it, from which Bloch waves, crystals, and semiconductors. The first electronic device which was covered was the Esaki diode [1]. The students were given as a handout the original Esaki paper, perhaps one of the most succinct papers written on electronic devices. From this they progressed to resonant tunneling devices which was a subject of active research at the time.

The Esaki diode had as its main innovation the negative differential resistance which could be obtained in the forward mode of the diode. However, it also has a tunnel current in the reverse mode

75th Anniversary of the Transistor, First Edition. Edited by Arokia Nathan, Samar K. Saha, and Ravi M. Todi.

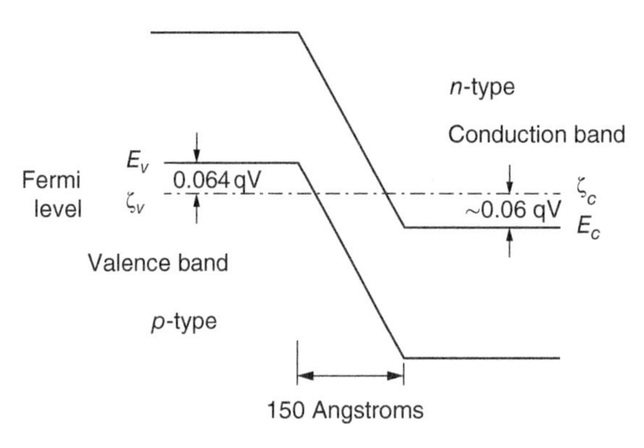

Figure 14.1 The original band diagram from the Esaki tunnel diode paper. Source: Esak [1]/American Physical Society. The width depletion region in the p+–n+ Ge junction is narrow enough (~15 nm estimated) to allow tunneling of electron through it. (i) With a positive bias (n – volts with respect to p) electrons can tunnel from the occupied conduction band states in the n to empty states in the valence band of the p. But with increasing bias occupied, conduction band states align with the bandgap and current drops. This is the origin of the Negative Differential Resistance (NDR) feature in the Esaki diode. With further increase of bias, there will be standard p–n junction forward operation and the current will rise again. (ii) With negative bias (n + volts with respect to p) electrons in the valence band of the p region can tunnel into unoccupied states in the conduction band of the n region. This can commence at $V > 0$ and keep increasing. There is no NDR region. Unlike in a conventional diode, in the NDR region and reverse conduction region, current is unipolar and carried by electrons.

(Figure 14.1) where electrons in the valence band of the p-region tunnel into the empty states of the conduction band of the n-region. In tunnel mode, both the forward and reverse currents are unipolar. The same concept of tunneling through a very narrow depletion region formed by a very heavily doped semiconductor region and a metal is termed Zener tunneling. It is the key enabler of linear metal semiconductor contact resistance in all semiconductor devices and integrated circuits to this day. It is fair to say that teaching of the Esaki diode inspired me to think about tunneling between heavily doped p and n regions as a device concept.

At the time, I was also reading current research in devices which exploited tunneling and would introduce significant papers to the students. One of these was the single electron memory developed in my senior colleague Haroon Ahmed's laboratory in Cambridge [2]. An elegant realization of a single electron memory was reported later by Stephen Chou [3]. It relied on the tunneling of a single electron from the channel into a nanosized poly-Si island to shift the potential significantly above the thermal threshold (26 meV at 300 k) to prevent further tunneling and to alter the threshold of the MOS channel. These all drew on the great excitement around the concept of the single electron transistor at the time. Here, a gate voltage is applied to Si islands (which would now be termed quantum dots) in which the band states are discretized and tunneling from a filled state to an empty state of a single electron is controlled [4].

14.2 Conception

In 1993, I took on a new research student Michael Reddick to explore device concepts based on tunneling. Michael had graduated in EE from Auburn University in the United States and arrived in Cambridge with a white Ford Mustang. Given nearly all students and faculty got around the town on bicycles, Michael's arrival made an impact! In the previous year I had taken on a PhD student, Florin Udrea, to work on power devices, initially sponsored by Philips Research Labs in the United Kingdom. David Coe at Philips encouraged us to explore the emerging trench gate IGBT structure.

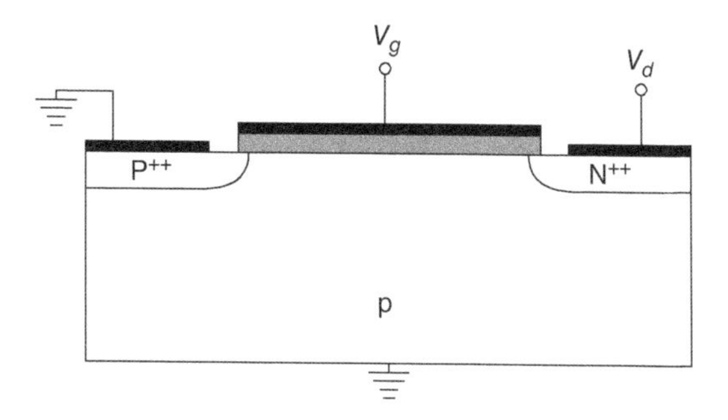

Figure 14.2 The Tunnel FET structure from [6]. The heavily doped p and n regions, as per source drain regions is PMOS/NMOS FETs, separated by an MOS gate on a p Si substrate. Source: From Reddick and Amaratunga [6].

As part of that research, we realized that the MOS inversion layer can act as a gate-controlled emitter in a bipolar device in 1993. This led to a new power device structure, the Inversion Layer Emitter Thyristor (ILET) [5]. Following on from this, it suddenly struck me one morning at breakfast that if an inversion layer can act as a gate-controlled bipolar emitter, then in principle a gate-controlled MOS channel should also be able to act as a heavily doped region in a tunnel diode. I discussed this further with Michael and he proceeded to explore the concept through simulation. The test structure we arrived at for simulation was an NMOS transistor where the n^+ source was replaced by a p^+ region, Figure 14.2.

Simulations did indeed show that an inversion layer could act as an n^+ region in an n^+–p^+ junction in reverse bias mode. With the p+ region negatively biased with respect to the n+ source, the current–voltage characteristic showed an exponential dependency, as per a tunnel diode, but as a function of gate voltage, Figures 14.3 and 14.4. Importantly, no current was seen if the band to band tunneling model was not included in the simulation, confirming the current was not due to avalanche breakdown. The three-terminal behavior in essence was transistor action based on tunnel current, the TFET. However, in Si with a bandgap of 1.1 eV, the currents available at 300 K with typical MOSFET voltages were low.

14.3 Realization

As experimental validation, a number of TFET structures were fabricated in a 2 μm CMOS process with the source and drain implant masks suitably altered so that the MOS gate was flanked by the doping region of opposite polarity with concentrations of the standard source/drain regions (in both NMOS and PMOS). The structures were realized as part of a test chip fabricated at Fuji Electric in Japan, with whom I had a collaborative research program on novel MOS gated power devices. Luckily, the Fuji process engineers accommodated our strange variations to the source/drain masks. The experimental TFET structures confirmed the simulation results. A gate-controllable Zener tunnel current could be achieved, Figure 14.5. But, as in the simulations, the current levels were low and similar to those in the subthreshold region in a conventional MOSFET of the same dimensions.

We published the results in 1995 [6]. Given the low current levels achieved, the MOS TFET was received as an interesting concept, but of little practical impact on Si integrated circuits at the time. In the publication, we used the title Surface Tunnel Transistor in Si. This was in recognition of the contemporaneous term used by the group from NEC in Japan who were working independently of us, and we became aware of prior to publication, on gate-controlled (non-inversion layer) tunnel diodes in GaAs and had demonstrated negative differential resistance in forward mode [7]. Our interest was in reverse mode operation with a MOS gate-controlled channel.

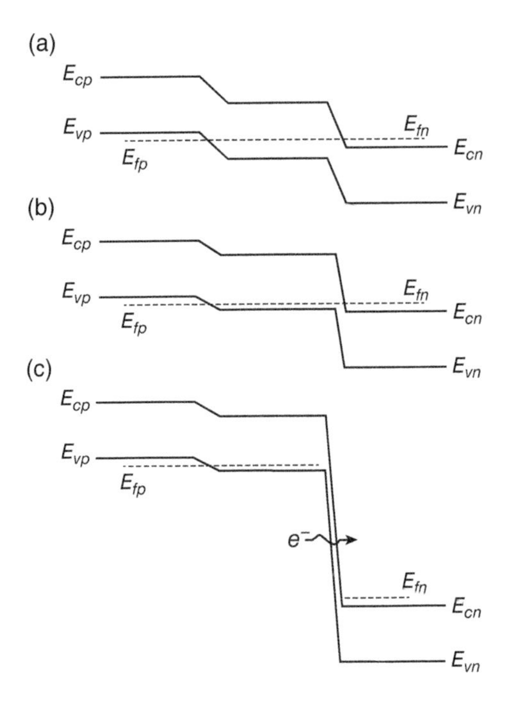

Figure 14.3 Band diagram for the p+–p–n+ Tunnel FET structure in Figure 14.2. (a) Equilibrium energy levels with $V_d = V_s = V_g = 0V$. (b) $V_g < 0$ bringing channel into accumulation mode. (c) Conduction mode with $V_d > 0$ and large enough to reduce the depletion width between the drain and the accumulation channel to enable tunneling of valence band electrons in the p region into the empty states of the conduction band in the n+ drain region. This is the same mode of operation as the Esaki diode in reverse bias. Source: From Reddick and Amaratunga [6]/American Institute of Physics. Note: In this Tunnel FET structure, there is no NDR region in forward bias operation as in the Esaki diode. The accumulation channel cannot be made heavily n+ for the rated voltage of the MOS gate oxide.

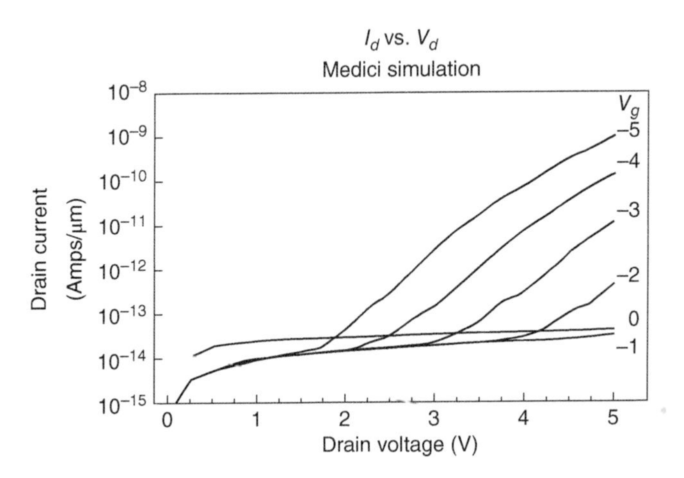

Figure 14.4 Simulation of the output characteristics of the Tunnel FET structure in Figure 14.2, Source: From Reddick and Amaratunga [6]/American Institute of Physics. Output current is only seen when the band to band tunneling model for Si is included in the simulation.

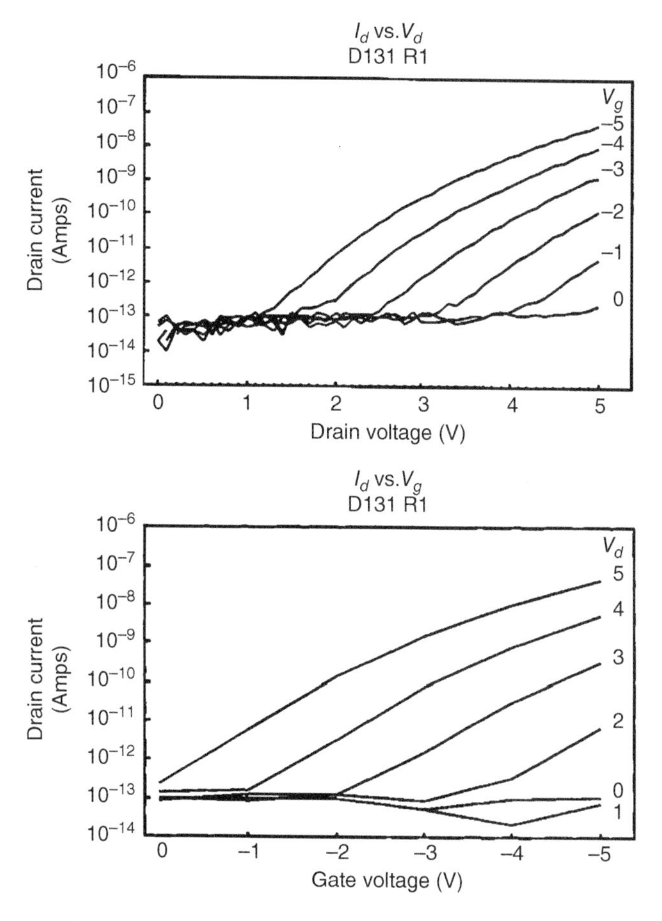

Figure 14.5 The measured output and gate transfer characteristics from experimental Si Tunnel FET devices made in a 2 μm CMOS process in 1994, Source: From Reddick and Amaratunga [6]/American Institute of Physics. The experimental device width was 84 μm. At rated 4–5 V operation, the measured current levels agree well with those predicted by simulation.

14.4 Relevance

In time, low power circuit innovations which utilize response in the subthreshold region [8] and the limitations posed by the thermal response of the subthreshold current in scaled CMOS awakened interest in the TFET. This has been extensively reviewed by Ionescu and Riel [9] and Seabaugh and Zhang [10]. Additionally, transistors in lower bandgap materials such as carbon nanotubes and graphene [11] and the advent of 2D-metal dichalcogenide semiconductors [12] have made the TFET an attractive option. In most of the current TFET structures, the tunneling takes place at the source end of the channel, Figure 14.6, with a depletion mode channel which is taken into accumulation mode to turn on the device, similar to the original concept.

14.5 Prospects

In Si, the fact that the TFET current is typically in the subthreshold region makes it unviable for high-frequency operation in standard digital integrated circuits. However, it comes into focus when

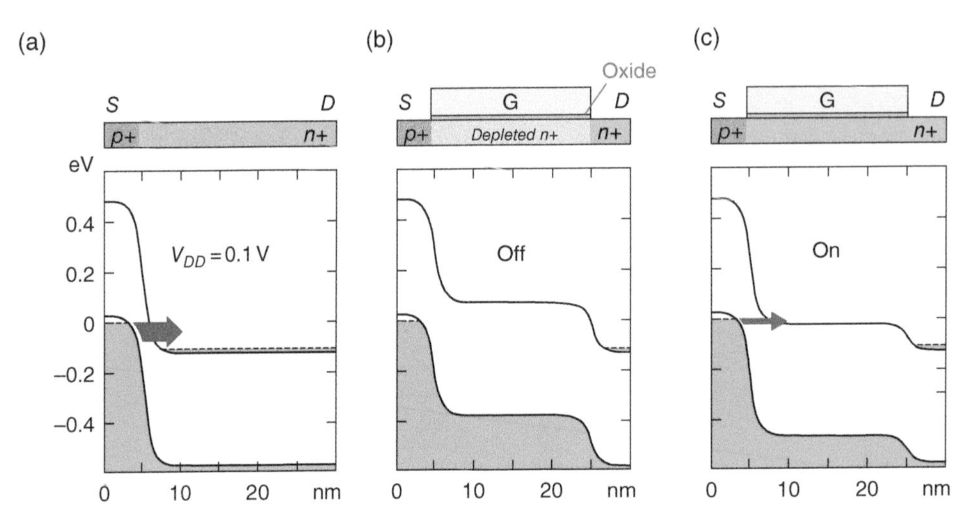

Figure 14.6 The Tunnel FET structure commonly used at present, Source: From Seabaugh and Zhang [10]/IEEE. The tunnel junction is formed at the source. (a) Only junction. (b) With an MOS gate. The substrate is n and tied to the n+ drain (instead of being p and tied to the p+ source in Figure 14.2). (c) A positive gate voltage on V_g takes the channel into accumulation and electrons from the p+ source valence band tunnel into empty states in the conduction band of the n-channel, Esaki diode reverse mode. This configuration has the advantage of all-positive unipolar voltage operation with the source grounded. The substrate is at high voltage and has to be isolated as in an n-well CMOS process. This can be readily achieved in an SOI process.

considering ultralow power electronics (analog and digital) in applications such as wireless sensing and energy harvesting [13, 14]. A device which was serendipitously arrived at may still have some practical future.

References

[1] Esaki, L. (1958). New phenomenon in narrow Germanium p-n junctions. *Phys. Rev.* 109: 603–604.

[2] Nakazato, K., Blikie, R.J., Cleaver, J.R.A., and Ahmed, H. (1993). Single electron memory. *Electron. Lett.* 29 (4): 385–385.

[3] Guo, L., Leobandung, E., and Chou, S.Y. (1997). Single electron memory operating at room temperature. *Science* 275 (5300): 649–651.

[4] Kastner, M.A. (1992). The single electron transistor. *Rev. Mod. Phys.* 64 (3): 849–858.

[5] Udrea, F. and Amaratunga, G. (1994). The inversion layer emitter thyristor – a novel power device concept. *Proc. 6th Int. Symp. Power Semiconductor Devices and ICs*, Davos (31 May 1994 – 2 June 1994). IEEE.

[6] Reddick, W.M. and Amaratunga, G.A.J. (1995). Silicon surface tunnel transistor. *App. Phys. Lett.* 67: 494–496.

[7] Umera, T. and Baba, T. (1994). First observation of negative differential resistance in surface tunnel transistors. *Jap. J. App. Phys.* 33: L207–L2010.

[8] Vittoz, E. and Fellrath, J. (1977). CMOS analog integrated circuits based on weak inversion operations. *IEEE J. Solid State Circuits* 12 (3): 224–231.

[9] Ionescu, A. and Riel, H. (2011). Tunnel field-effect transistors as energy-efficient electronic switches. *Nature* 479: 329–337.

[10] Seabaugh, A.C. and Zhang, Q. (2010). Low voltage tunnel transistors for beyond CMOS logic. *Proc. IEEE* 98 (12): 2095–2110.

[11] Zhang, Q., Fang, T., Xing, H. et al. (2008). Graphene nanoribbon tunnel transistors. *IEEE Electron Device Lett.* 29 (12): 1344–1346.

[12] N. Oliva, J. Backman, jL. Capua, M. Cavalieri, M. Luisier, A. M. Ionescu; $WSe_2/SnSe_2$ vdW heterojunction Tunnel FET with subthermionic characteristic and MOSFET co-integrated on same WSe_2 flake, npj 2D. *Mater. Appl.* 4 (5), 2020.

[13] Cavalheiro, D., Moll, F., and Valtchev, S. (2018). *Ultra-Low Input Power Conversion Circuits based on Tunnel-FETs*, 1e. River Publishers.

[14] Trivedi, A.R., Carlo, S., Mukhophadyay, S. (2013). Exploring Tunnel-FET for ultra-low power analog applications: a case study on operational transconductance amplifier. *Proc. DAC'13, ACM*, Austin, TX (29 May 2013 – 7 June 2013). IEEE.

Chapter 15

Floating-Gate Memory

A Prime Technology Driver of the Digital Age

Simon M. Sze

National Yang Ming Chiao Tung University, Hsinchu, Taiwan, ROC

15.1 Introduction

The floating-gate memory (FGM) effect was discovered [1] by Dawon Kahng[1] and Simon Sze in 1967. At that time, we were members of the Technical Staff in the Semiconductor Device Laboratory of Bell Laboratories (now Nokia Bell Labs) in Murray Hill, New Jersey, USA. Our mission was to study high-speed transistors and to develop new device concepts. We knew that the magnetic core memory (MCM) was used in mainframe computers and communication equipment. Although MCM was nonvolatile [2], it suffered from many drawbacks such as a large form factor, high power consumption, and long access time. In addition, MCM was not compatible with the existing semiconductor technology.

Because of these drawbacks, we were interested in exploring the possibility of developing a nonvolatile memory device using semiconductor technology. We tried to combine various semiconductor device building blocks (e.g. Schottky contact, metal-oxide-semiconductor [MOS] capacitor, heterojunction) to form the memory device. However, theoretical analysis showed that none of these combinations could function properly as a nonvolatile memory. During one of our luncheons at the Murray Hill Cafeteria, Kahng ordered a *four-layered* cheese cake for dessert. We looked at the cake and came up with the idea of employing a metal-layer (the "floating gate") embedded in the oxide between the top metal gate and the channel of a conventional MOS field-effect transistor (MOSFET) [3].

[1] I am saddened to report that Dr. Kahng passed away in 1992.

75th Anniversary of the Transistor, First Edition. Edited by Arokia Nathan, Samar K. Saha, and Ravi M. Todi.

The gate stack was a *four-layered* structure of metal-upper insulator-floating gate-lower insulator. The "floating gate" (FG) would serve as the charge storage layer and it would be surrounded by insulators to minimize charge leakage. The name "floating gate" referred to the absence of a direct electrical contact to the metal layer, its potential is floating.

We did a theoretical analysis of the device characteristics, and an experimental device structure was designed by us and fabricated by my two technical assistants, George Carey and Andy Loya. For the floating gate material, Marty Lepselter, a Group Supervisor in our Laboratory, suggested zirconium (Zr), because the surface of Zr could be oxidized easily to form ZrO_2 as the upper insulator layer. The measured results of the structure were consistent with the theoretical analysis and one of the first FGM structures had a storage time of longer than one hour. We submitted the paper describing the *discovery* of the FGM effect to the *Bell System Technical Journal* on *16 May 1967* and the paper was published on 1 July 1967 [(1967). *BSTJ*, 46: 1288–1295].[2]

Originally, we thought of the FGM structure only as a replacement for MCM. The applications of the device have gone far beyond what we or anyone else thought it would be [4, 5]. In 1983, FGM was adopted by Nintendo in their game console to facilitate the restart of a game. In 1984, FGM was used as BIOS (Basic Input and Output System) in the personal computer to activate it when the system was first turned on. Subsequently, FGM has enabled the invention or development of *all* advanced digital systems: from digital cellular phone to cloud computing to internet of things. These digital systems, in turn, have significantly improved the quality of life for billions of people around the world.

15.2 The Charge-Storage Concept

Since 1990, the FGM has served as the most important *nonvolatile memory* for the global electronics industry. The cross-sectional view of the first FGM structure is shown in Figure 15.1. The floating gate M (1) is sandwiched between a tunnel oxide I (1) and a blocking oxide I (2). The operation of FGM is based on the amount of charge stored in the floating gate to determine the on-state or off-state of the MOSFET.

An important *limiting case* of FGM is the pseudo FGM or the *charge-trapping memory* (CTM). When the thickness of the floating gate M (1) is reduced to zero and charges are stored in the upper insulator I (2), FGM becomes CTM. Thus, CTM is based on the same MOS structure and the same charge-storage concept as the FGM. The main difference is the *change* of the charge-storage materials.

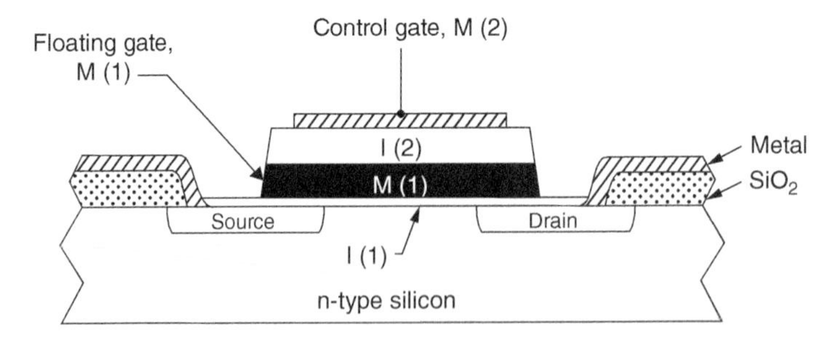

Figure 15.1 Cross-sectional view of the first floating-gate memory structure [1] in 1967.

[2] On *5 June 1967*, Bell Laboratories also filed a patent entitled "Field Effect Semiconductor Apparatus with Memory Involving Entrapment of Charge Carriers." The patent was granted on 10 March 1970, patent number 3,500,142.

15.2.1 The Floating-Gate Memory

The band diagrams for the basic FGM operations of programming, storage, and erase are shown in Figure 15.2. When a positive voltage is applied to the control gate M (2), electron transport across the first insulator I (1) is possible via Fowler–Nordheim tunneling, Figure 15.2a. If the insulators I (1) and I (2) are sufficiently thick, the charge in the floating gate can be stored for a long time, Figure 15.2b. When a negative voltage is applied to M (2), electrons can transport out of the floating gate via Fowler–Nordheim tunneling, Figure 15.2c.

As mentioned in the Introduction, an experimental structure was made using 5 nm SiO_2 as I (1), 100 nm Zr as M (1), and 100 nm ZrO_2 as I (2). When a positive voltage pulse of 50 V with a pulse duration of 0.5 μs was applied to the control gate M (2), about 10^{12} electrons cm^{-2} were transported to and stored in the floating gate. The stored charge caused a large threshold voltage shift and the device was "on" with a channel current of 0.25 mA. When a large negative voltage pulse was applied to the control gate, the stored charge was depleted and the device was "off." The applied gate voltage pulse and the drain current are shown in Figure 15.3. From the slope of the "on" state, we can estimate the retention time (i.e. the time when the stored charge decreases to 50% of its initial value) of a few hundred milliseconds. This result can be considered as the *first demonstration of the EEPROM* (electrically erasable programmable read-only memory) operation.

The threshold voltage shift ΔV_T is shown in Figure 15.4. ΔV_T is given by $|Q|/C_{FC}$, where $|Q|$ is the magnitude of the stored charge and C_{FC} is the capacitance between the control gate and the floating gate.

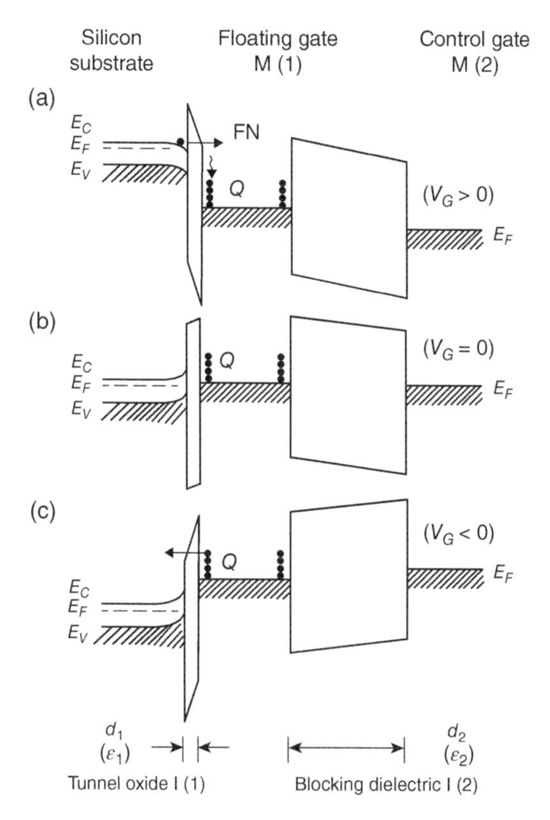

Figure 15.2 Energy band diagrams of a floating-gate memory with a semiconductor-insulator-metal-insulator-metal sandwich [1]. (a) Programming mode, positive voltage is applied to the control gate. (b) Storage mode, voltage is removed. (c) Erase mode, negative voltage is applied to the control gate.

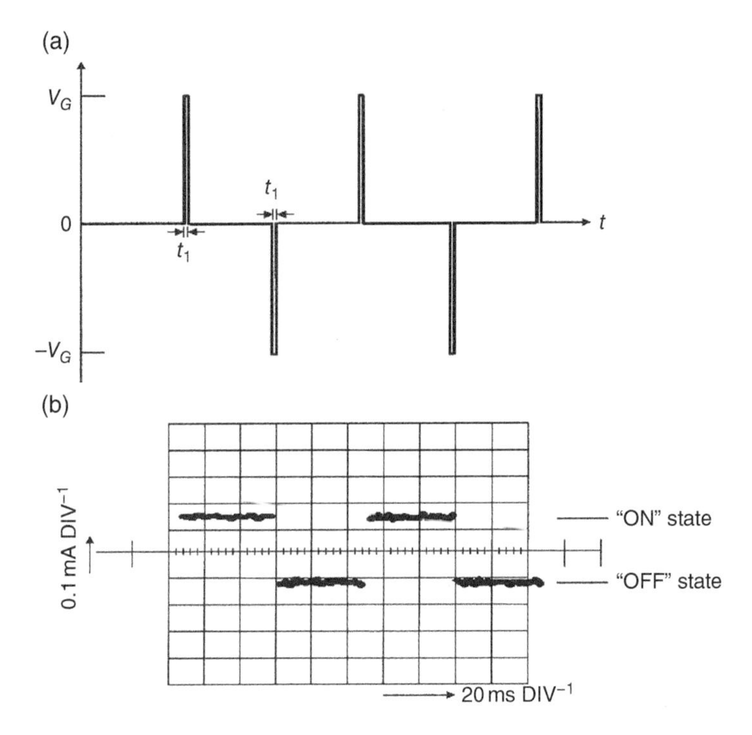

Figure 15.3 First demonstration of an electrically erasable programmable read-only memory (EEPROM) [1]. (a) Applied gate voltage pulse. (b) Source–drain current for $V_G = \pm 50$ V, $t_1 = 0.5$ μs.

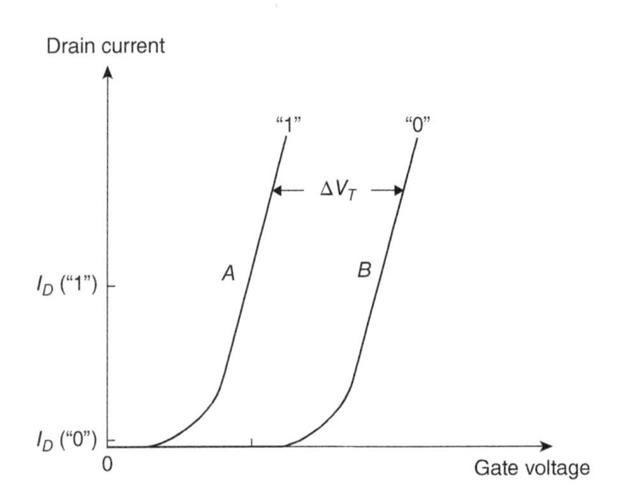

Figure 15.4 Current–voltage curves of a floating-gate memory structure when a negative charge Q is stored in the floating gate (curve A), and when there is no charge stored in the floating gate (curve B).

15.2.2 Pseudo FGM or the Charge-Trapping Memory

The band diagrams for the programming and erasing operations of CTM are shown in Figure 15.5. This is an MNOS (metal-nitride-oxide-silicon) memory [6, 7]. We note the band diagrams are similar to those shown in Figure 15.2. In the programming operation, Figure 15.5a, a nitride layer is used as an efficient material to trap electrons as current passes through it via modified Fowler–Nordheim

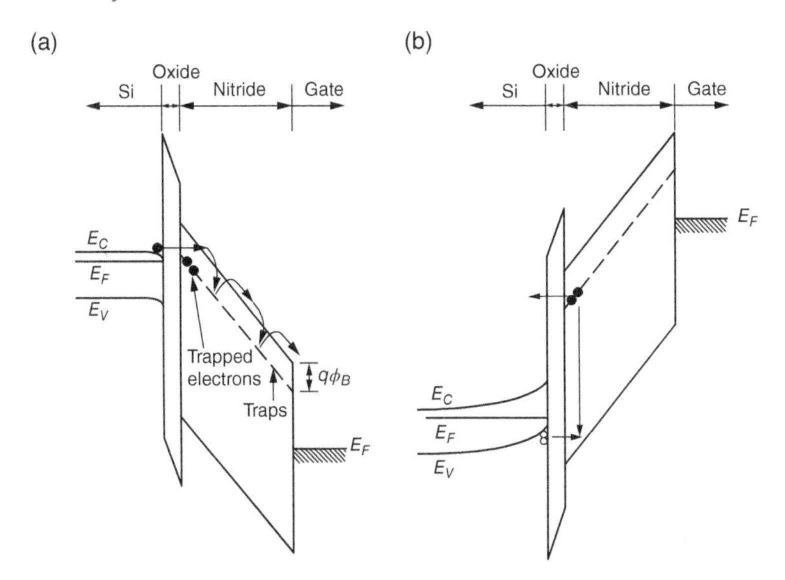

Figure 15.5 Energy band diagrams of a charge-trapping memory which is a limiting case of the floating-gate memory. (a) Programming mode. (b) Erase mode.

tunneling and Frenkel–Poole transport. In the erasing operation, Figure 15.5b, holes are tunneling from the silicon substrate to neutralize the trapped electrons in the nitride layer.

An improved version of MNOS is the MONOS (metal-oxide-nitride-oxide-silicon) [8] or SONOS (silicon-oxide-nitride-oxide-silicon) in which an additional blocking oxide layer is placed between the control gate and the nitride layer. This oxide layer can prevent electron injection from the gate to the nitride layer during erasing operation and improve device performance. The SONOS has replaced the MNOS but the operation principle remains the same.

15.3 Early Device Structures

In 1971, Frohman–Bentchkowsky developed the FAMOS (floating-gate avalanche-injection MOS) [9] shown in Figure 15.6a. This structure is an EPROM (erasable programmable read-only memory), it has a floating gate but no control gate. The stored charge in the floating gate came from hot electrons generated at the avalanche region near the drain that were injected over the semiconductor–insulator barrier (3.1 eV). Since FAMOS has no control gate, the stored charge cannot be erased electrically and requires the use of ultraviolet light to do the erasing.

In 1976, Iizuka and his coworkers studied the SAMOS (stacked gate avalanche injection MOS) [10], shown in Figure 15.6b. This structure is an EEPROM, and it looks essentially the same as Figure 15.1. However, the injection mechanism is due to avalanche instead of Fowler–Nordheim tunneling. Because of the relatively thick oxide between the floating gate and the channel, a substantial improvement of the retention time was obtained. However, for a typical EEPROM operation, we need two devices per cell, i.e. an EEPROM and a selection MOSFET. Therefore, the cell size is relatively large.

In 1984, Masuoka and his coworkers developed the Flash memory [11], shown in Figure 15.7. Along the *A*–*A*′ section, it is the basic floating gate configuration. However, along the *B*–*B*′ section, an erase gate is added, and this gate is connected to many cells in series. When a voltage is applied to the erase gate, a whole block of memory cells is erased simultaneously, thus the name "Flash." Since there is only one device per cell, Flash memory has the advantage of higher density, lower cost, and higher scalability compared to EEPROM.

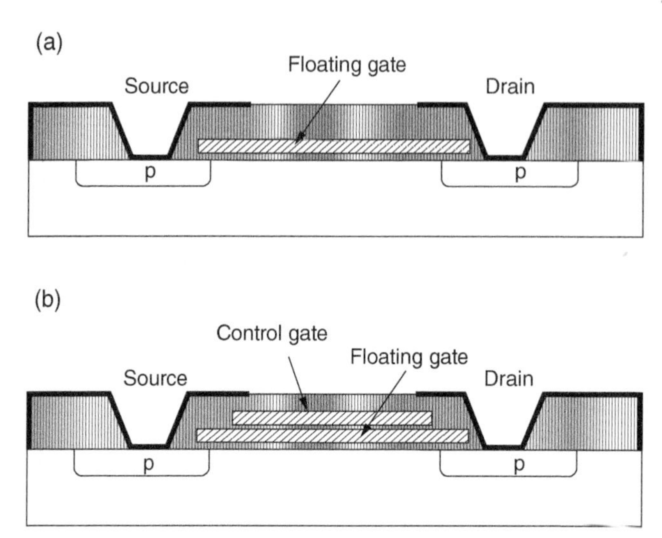

Figure 15.6 Two early floating-gate memories: (a) FAMOS [9] and (b) SAMOS. Source: Adapted from Iizuka et al. [10].

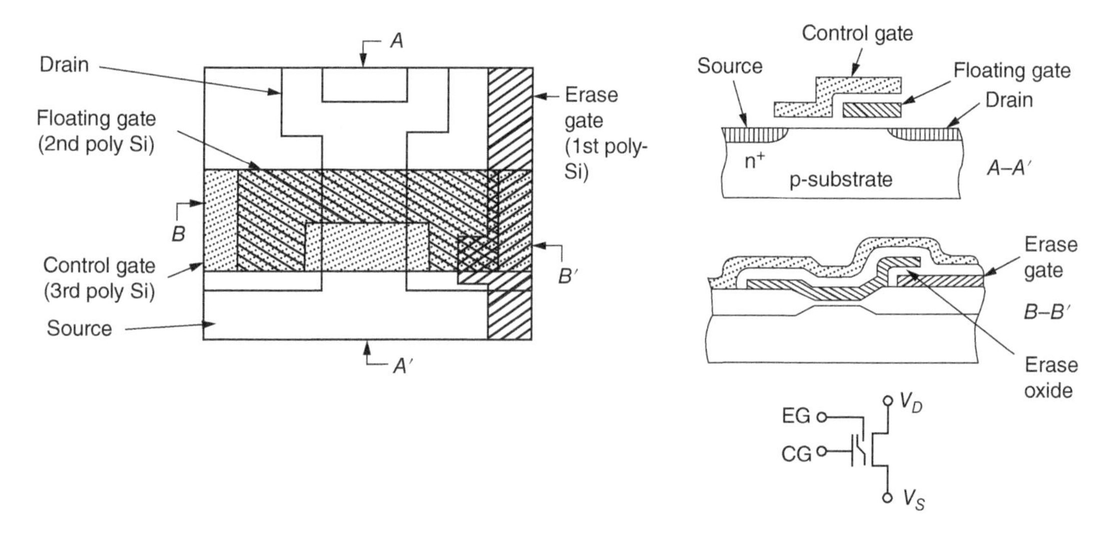

Figure 15.7 Flash memory proposed in 1984. In the erase operation, a whole block is erased, thus the name Flash. Source: Adapted from Masuoka et al. [11].

In 1985 and 1987, Masuoka and his coworkers proposed the NOR-Flash architecture and the NAND-Flash architecture, respectively [12, 13]. They are shown in Figure 15.9. In the NOR Flash, each cell is directly connected to the word line and the bit line of the memory array, while NAND Flash cells are arranged in series within small blocks. Thus, while NOR Flash offers faster random access, NAND Flash can have significantly higher packing density than NOR Flash.

Today, EEPROM, NOR Flash, and NAND Flash are the three principal nonvolatile-semiconductor-memory products. EEPROM is used where bit alterability is required, NOR Flash is mainly for code storage, and NAND Flash for high-volume data storage. Figure 15.8 shows the market share of the

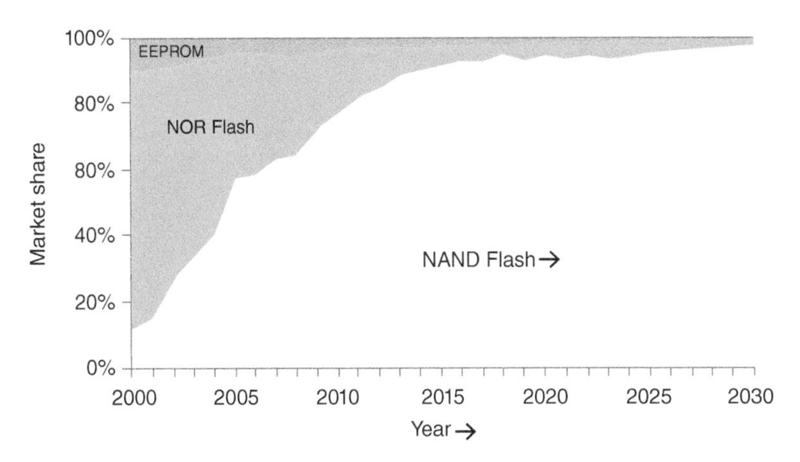

Figure 15.8 Market share of three nonvolatile-semiconductor-memory products from year 2000 to 2022 and projected to 2030. Source: Adapted from Gartner [15].

aforementioned products [15] from the year 2000 to 2022 and projected to 2030. Due to the demand of high-volume storages for solid-state drives, cloud computing, internet of things, and big data, NAND Flash has had the largest market share since 2005. We expect that NAND Flash will reach 98% market share in 2030.

15.4 Multi-Level Cells and 3D Structures

In 1995, Bauer and his coworkers proposed the multi-level cell (MLC) [14]. MLC is in contrast to the single-level cell (SLC) technology in that MLC allows multiple bits to be stored per cell. The advantage of MLC is higher density than the standard SLC. A comparison of the voltage distributions between a SLC (1 bit cell^{-1}) and an MLC (2 bits cell^{-1}) is shown in Figure 15.9. To read the bit from the SLC, only a single comparison with a reference voltage is required, whereas in an MLC (2 bits cell^{-1}), three threshold voltages are required. Because of the reduced voltage margins, MLC is more likely to have errors. For a 3 bits MLC and a 4 bits MLC, we need $2^3 = 8$ and $2^4 = 16$ voltage distributions, respectively, or 7 and 15 threshold voltages, respectively, further increasing the readout logic and the error probability.

To increase the number of memory cells and to reduce the cost of memory cells per package, 3D (3-dimensional) structures are proposed. There are two approaches to the 3D structures: multi-chip stacking and multilayer integration. Figure 15.10 shows an example of a 16-chip stacked NAND Flash using through-silicon-via (TSV) technology [16]. The main advantages of multi-chip stacking are the increase of the memory density for a given plane area, and a more relaxed design rule can be used for a given technology generation.

For the multilayer integration, a cross-sectional view of a 3D floating-gate NAND memory block is shown [17] in Figure 15.11. The microphotograph of a memory chip shown in Figure 15.12 is for a 1.33 Tb 4 bit cell^{-1} 3D 96-layered Flash memory [18]. The design rule is 30 nm and the chip footprint in 158 mm^2. Therefore, the memory density in a plane is 8.4 Gb mm^{-2}, which is much higher than a 2D (2-dimensional) 128 Gb floating-gate 2 bits cell^{-1} NAND Flash memory with a design rule of 16 nm and a similar footprint (memory density is only 0.72 Gb mm^{-2}). The multilayer integration has the same advantages as that for the multi-chip stacking. The multilayer integration has an additional advantage in its automatic layer-to-layer alignment, while for the multi-chip stacking, one needs to do extremely precise chip-to-chip physical alignment [17–19].

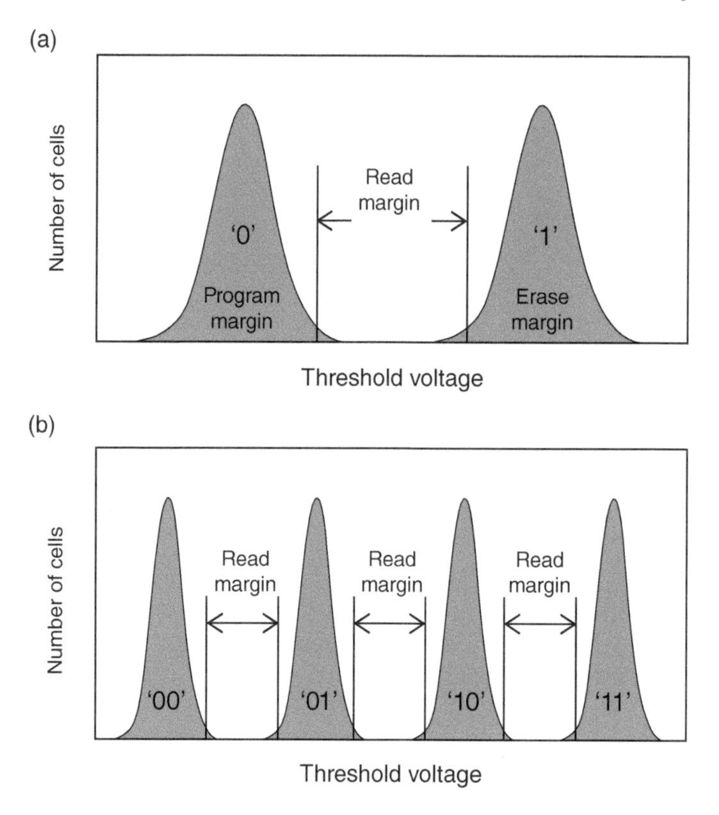

Figure 15.9 Voltage distributions for (a) single-level cell with 1 bit cell^{-1} and (b) multi-level cell with 2 bits cell^{-1}. Source: Adapted from Bauer et al. [14].

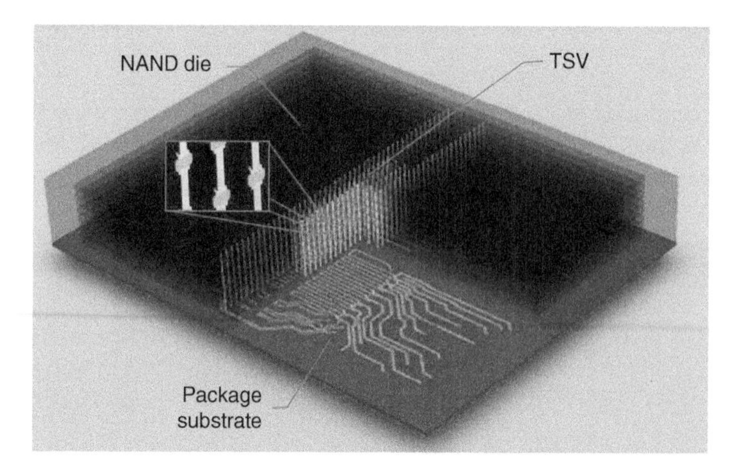

Figure 15.10 An example of a 16-chip 3-dimensional structure: a multi-chip stacking with through-silicon-via (TSV). Source: Beica [16]/IEEE.

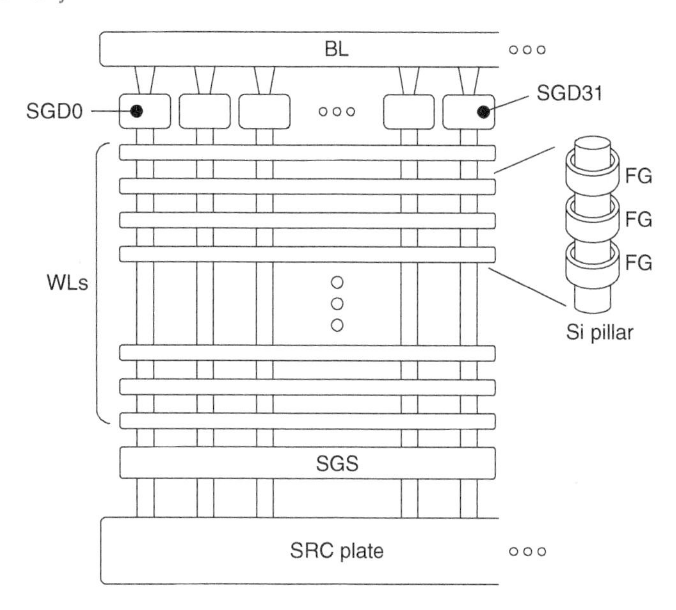

Figure 15.11 Cross-sectional view of 3D floating-gate NAND Flash memory block in multilayer integration. Source: Tanaka et al. [17]/IEEE.

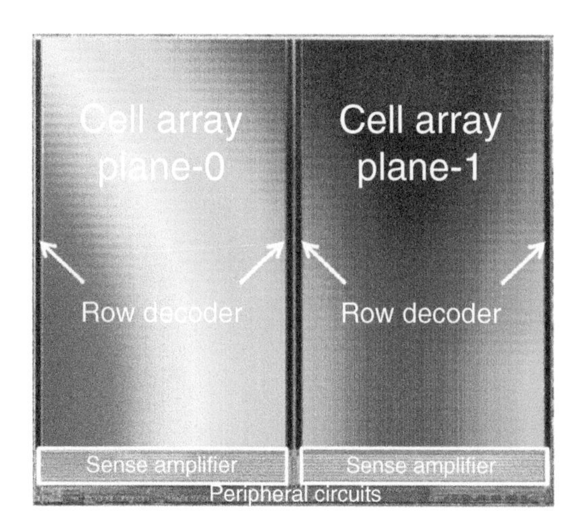

Figure 15.12 Microphotograph of a 1.33 Tb 4 bit cell^{-1} 3D floating-gate NAND Flash memory in multilayer integration. Source: Tanaka et al. [17] and Shibata et al. [18]/IEEE.

15.5 Applications

FGM has the unique combination of nonvolatility, in-system rewritability, ruggedness, high density, low power consumption, small form factor, and compatibility with CMOS processes. Since 1990, FGM has enabled the invention or development of *all advanced digital systems* by providing long-term information storage. Figure 15.13 shows the cost per gigabyte [15] of hard disk drive (HDD) and NAND Flash from 1985 to 2022 and projected in 2030. We note that the cost of NAND Flash has dropped

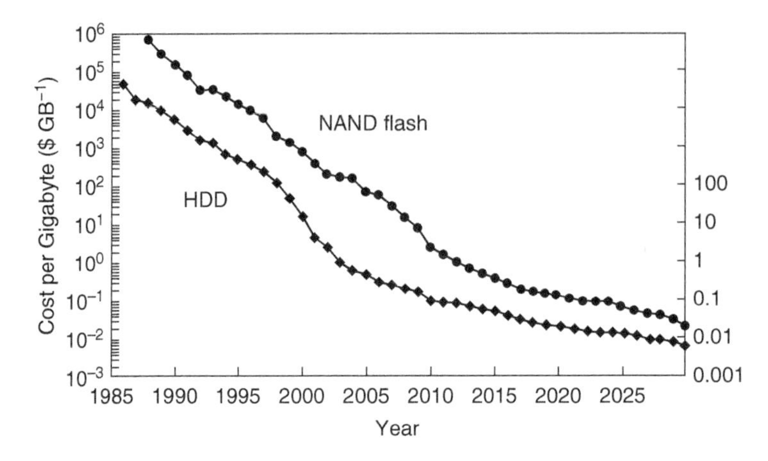

Figure 15.13 Cost per gigabyte (GB) of hard disk drive (HDD) and NAND Flash from 1985 to 2022 and projected to 2030. The cost/GB for NAND Flash will drop to $0.02 in 2030. Source: Data taken from Gartner [15].

rapidly from $600,000 GB^{-1} in 1987 (for 256 kb SLC units) to around $0.1 GB^{-1} in 2022 (for 512 Gb MLC units), and is expected to drop to $0.02 GB^{-1} at 2030. Although the cost of HDD is still lower than the NAND Flash-based solid-state drive (SSD), but at such a cost level, SSD becomes very attractive to replace HDD for ultrabook computer, big data, cloud computing, digital archiving, and numerous other applications.

Currently, there are billions of FGM-based-digital systems being used around the world. The following is a list of some important applications [20]:

- Communications
  - Over 5 billion FGM-based cellular phone subscribers in the world
  - Bluetooth, pagers, network systems, etc.
- Computing
  - Billions of FGM-based tablet computers, USB memory sticks
  - Artificial Intelligence
  - Big Data
  - Cloud Computing
  - Internet of Things
  - Robotics
- Consumer Electronics
  - Billions of digital TVs, DVD players, MP3 music players
  - Digital cameras, digital camcorders, electronic books
  - Smart IC cards
  - Bar-code readers
- Energy Conservation
  - FGM-based microcontroller units (MCUs) for buildings will save 50% electricity in 2030
  - Smart sensors, smart grids, and smart systems to minimize energy consumption
  - FGM-based brushless motors can increase energy efficiency of household appliances by 30% or more
- Health Care
  - Wearable medical devices
  - Implantable systems – pacemakers, defibrillators
  - Biomedical systems-on-a-chip
  - Micro-surgery (daVinci robot)

- Transportation
  - 50–100 FGM-based MCUs are installed in every modern car
  - Next-generation automobiles – autonomous self-driving cars
  - FGM-based systems in every modern aircraft and marine vessel to improve performance and travel safety

Without FGM, none of the above systems could have been invented or developed. Because of the growing demand, more FGMs are produced than any other semiconductor devices. In 2021 alone, the world produced more FGM (5×10^{21}) than the number of transistors (bipolar and MOS types) ever produced over all time [21]. In 2022, the number of FGM cells will be the equivalent of 4 trillion cells for every man, woman, and child in the world, making FGM a truly indispensable electronic device for the Digital Age.

15.6 Scaling Challenges

The lower curve in Figure 15.14 shows the reduction of the half pitch [22] of 2D FGM cells from 1995 to 2022 and projected to 2030. We note that the half pitch decreases rapidly from 360 nm in 1995 to 12 nm in 2022. It is expected that the half pitch will remain at 12 nm due to scaling challenges.

At the decananometer regime, the scaling challenges include [23]:

- Retention (i.e. the ability to retain its information for 10–20 years) will require a tunnel oxide thickness of 6–7 nm. This requirement makes it difficult to improve the programming/erase speed or to reduce the applied voltages.
- Endurance (i.e. the number of program/erase cycles a memory device is able to perform) is limited by the generation of traps and charge trapping in the tunnel oxide during the high-field programming and erase operations. Typically, the endurance is 10^5–10^6 cycles for an SLC Flash memory, and 10^3–10^4 cycles for an MLC (2 bits cell^{-1}) Flash memory.
- Interference of neighboring cells which will cause cell-to-cell cross talk, V_T shift, and reduction of memory window margins.
- Reduction of coupling ratio (i.e. the ratio of the capacitance between the control gate and the floating gate to the total capacitance) due to parasitic capacitances as the cell-to-cell spacing is reduced.
- Reduction of number of electrons in the floating gate. At 20 nm gate length, there are only about 50 electrons in the floating gate. The charge loss tolerance is around 5 electrons! At 12 nm, there

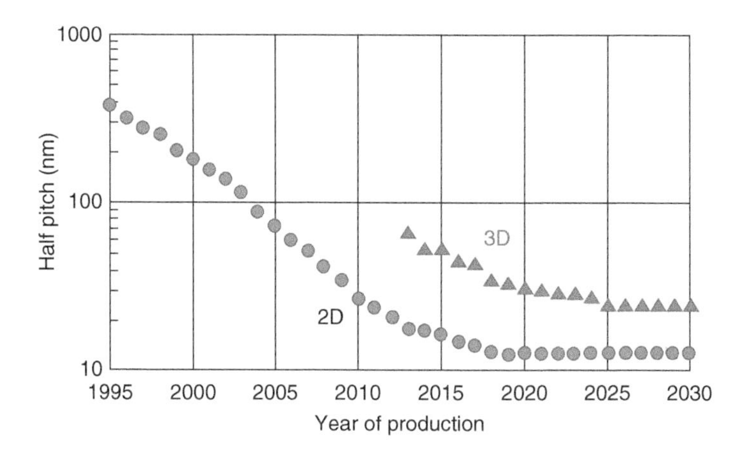

Figure 15.14 Reduction of half pitch for 2D and 3D NVSM from 1995 to 2022 and projected to 2030. By 2030, the half pitch will be 12 nm for 2D and 20 nm for 3D. Source: IEEE [22]/American Scientific Publishers.

are only about 15 electrons; so a reduction of one electron will cause a significant change in threshold voltage.
- Dielectric leakage which will cause retention problem.
- Variability of substrate doping, line edge roughness, and gate granularity, which will affect device performance and reliability.
- Random telegraph noise (RTN) due to the capture and emission of a channel electron by a single tunnel oxide trap near the substrate surface.

RTN will cause drain current fluctuation and threshold voltage shift, $\Delta V_T = q\,(1/C_{FS} + 1/C_{FC})$, where q is the unit charge, C_{FS} is the floating gate/substrate capacitance, and C_{FC} is the floating gate/control gate capacitance. As the device area decreases, C_{FS} and C_{FC} are reduced causing a large V_T shift which will affect the readout operation.

To solve or minimize these scaling challenges, many circuit innovations are proposed including the error-correction code (ECC), parallel/shadow programming, wear-level management, data compression schemes, and incremental step pulse programming (ISPP) to tighten V_T distribution with an intelligent algorithm. However, even with these circuit innovations, the smallest half pitch for 2D designs is estimated to be 12 nm. Below 12 nm, all 2D designs will suffer from severe reliability problems [22, 23].

As mentioned in Section 15.4 that by adopting 3D multi-chip stacking or multilayer integration, we can significantly increase the memory density and relax the design rule. As shown in the upper curve in Figure 15.14 for the 3D designs, the volume production of memory chips started in 2013 with a half pitch of 64 nm; it will be reduced to 30 nm in 2022 and 24 nm in 2030. On the other hand, the number of stacked chips or layers will be increased from 16–32 in 2013 to 96–144 in 2022 and to 192–320 in 2030. We expect to have 4 Tb (4×10^{12} bits) 3D memory chips in 2030.

For the three principal NVSM products shown in Figure 15.8, we expect that the dominant cell type for EEPROM, NOR Flash, and 2D NAND Flash will remain to be *FGM* in the foreseeable future. However, after 2022, the dominant cell type for super-high-density 3D NAND Flash (≥516 Gb) will be the pseudo FGM or *CTM* (Section 15.2.2). The main reason is that CTM does not have a larger 3D pillar due to the presence of the floating gate as shown in Figure 15.11; therefore, CTM has a higher scalability in 3D design.

15.7 Alternative Structures

In the past 55 years, many non-charge-storage nonvolatile-memory structures have been proposed to hopefully circumvent the scaling challenges. Furthermore, the electronics industry is in constant search of an ideal memory or the so-called "Unified Memory" which can satisfy simultaneously three requirements: high speed, high density, and nonvolatility (Figure 15.15). At the present time, such a memory has not been developed. For the FGM or the CTM, it has high density and nonvolatility, but the program/erase speed is relatively low (of the order of µs). DRAM has high speed (~10 ns) and relatively high density, but it is volatile. SRAM has very high speed (~5 ns), but suffers from very low density and volatility.

Many nonvolatile memory devices have been proposed for the "Unified Memory." The followings are a few potential candidates.

- FeRAM (ferroelectric random access memory) is based on the remnant polarization in perovskite materials [24].
- PCRAM (phase change RAM) is based on reversible phase conversion between the amorphous and the crystalline state of a chalcogenide glass, which is accomplished by heating and cooling of the glass [25].
- RRAM (resistance RAM) is based on change in resistance with applied electric field. Resistance switching from low to high resistance or the other way around has been found in materials such as lead zirconium titanate (PZT) and tantalum pentoxide (Ta_2O_5) [26].

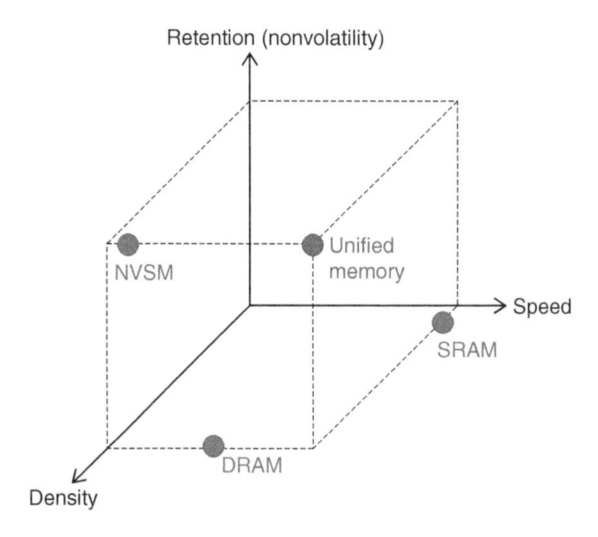

Figure 15.15 The Unified Memory having high speed, high density, and nonvolatility (i.e. high retention time). Such a memory has not been developed, and the three existing memory devices are shown with their key attributes.

- STT-MRAM (spin-torque-transfer magnetic RAM) is based on spin polarization that can be used to control the magnetic orientation of layers in an MRAM cell [27].

All the above candidates have very simple structures, and most are two-terminal devices. The difficult part is to find the right materials or material combinations that could fulfill the requirements of high speed, high density, and nonvolatility. In addition, the materials must be compatible with integrated-circuit technology, since the logic devices will be CMOS. We hope the FGM-inspired search of novel nonvolatile memories [28, 29] will successfully lead to the realization and commercialization of the "Unified Memory."

15.8 Conclusion

We have presented a review of the historical development of the FGM and projected its future trends to the year 2030. In the past 55 years (1967–2022), FGM has evolved from a charge-storage concept to FAMOS, SAMOS, Flash memory, MLCs, and 3D structures. Since 1990, FGM has become a *prime technology driver* of the *Digital Age*, enabled the invention or development of *all advanced digital systems*, and brought *unprecedented benefit* to humankind.

As the device dimension is reduced to the decananometer regime, FGM faces many serious scaling challenges such as the interference of neighboring cells, reduction of stored charges, and random telegraph noise. We are confident that our device scientists will find ways to meet these challenges and may even develop a "Unified Memory" with low cost, high performance, and high reliability for future digital systems. These systems, in turn, will continue to *enrich* and *improve* the *quality of life* for all the people around the world.

Acknowledgments

I wish to thank Prof. H. S. Koo of the Taipei University of Marine Technology for his critical reading of the manuscript. I have benefited from suggestions made by Prof. T. C. Chang of the National Sun Yet-sen University, Dr. H. Kuan and Dr. C.Y. Lu of Macronix International Co., Dr. Nicky Lu of Etron

Technology Inc., and Profs. Y.M. Li, P.T. Liu, and T.Y. Tseng of the National Yang Ming Chiao Tung University. I am further thankful to Mr. Norman Erdos for technical editing of the manuscript, and Ms. J.M. Hsu for preparation of the illustrations and various revisions of the manuscript. Finally, I am grateful to my mentors at Bell Laboratories, Dr. R.M. Ryder and Dr. G.E. Smith, for providing a stimulating and challenging environment in which I, a young device scientist in the 1960s, was welcomed, inspired, and abundantly assisted. Without their help, the floating-gate memory effect could not have been discovered in the spring of 1967.

References

[1] Kahng, D. and Sze, S.M. (1967). A floating gate and its application to memory devices. *Bell Syst. Tech. J.* 46: 1288.

[2] Wang, A. and Woo, W.D. (1950). Static magnetic storage and delay line. *J. Appl. Phys.* 21: 49.

[3] For a discussion of MOSFET and nonvolatile memory devices, see for example:Sze, S.M. and Lee, M.K. (2013). *Semiconductor Devices: Physics and Technology*, 3e. Hoboken, NJ: Wiley.

[4] Lu, C.Y. and Kuan, H. (2009). Nonvolatile semiconductor memory revolutionizing information storage. *IEEE Nanotechnol. Mag.* 3: 4.

[5] Coughlin, T. (2016). Storage in media and entertainment: the flash advantage. *Flash Memory Summit,* Session IT-8, Santa Clara, CA (8–11 August 2016).

[6] Szedon, J.R. and Chu, T.L. (1967). Tunnel injection and trapping of electrons in Al-SiN-SiO2-Si (MNOS) capacitor. *Abs. Solid-State Device Res. Conf.*, p. 631.

[7] Wegener, H.A.R., Lincoln, A.J., Pao, H.C. et al. (1967). The variable-threshold transistor, a new electrically-alterable, nondestructive READ-only storage device. *International Electron Devices Meeting*, Washington, DC, USA (18–20 October 1967), p. 70. https://doi.org/10.1109/IEDM.1967.187833.

[8] Kashavan, B.V. and Lin, H.C. (1968). MONOS memory element. *International Electron Devices Meeting*, Washington, DC, USA (23–25 October 1968), pp. 140–142. https://doi.org/10.1109/IEDM.1968.188066.

[9] Frohman-Bentchkowsky, D. (1971). Memory behavior in a floating-gate avalanche-injection MOS (FAMOS) structure. *Appl. Phys. Lett.* 18: 332.

[10] Iizuka, H., Masuoka, F., Sato, T., and Ishikawa, M. (1976). Electrically alterable avalanche injection type MOS read-only memory with stacked gate structure. *IEEE Trans. Electron Devices* ED-23: 379.

[11] Masuoka, F., Asano, M., Iwahashi, M., Komuro, T., and Tanaka, S. (1984). A new flash E^2PROM cell using triple polysilicon technology. *International Electron Devices Meeting*, San Francisco, CA, USA (9–12 December 1984), pp. 464–467. https://doi.org/10.1109/IEDM.1984.190752.

[12] Masuoka, F., Asano, M., Iwahashi, H., Komuro, T., and Tanaka, S. (1985). A 256K flash EEPROM using triple polysilicon technology. *IEEE International Solid-State Circuits Conference. Digest of Technical Papers*, New York, NY, USA (13–15 February 1985), pp. 168–169. https://doi.org/10.1109/ISSCC.1985.1156798.

[13] Masuoka, F., Momodomi, M., Iwata, Y., and Shirora, R. (1987). New ultra high density EPROM and flash EEPROM with NAND structured cell. *International Electron Devices Meeting*, Washington, DC, USA (6–9 December 1987), pp. 552–555. https://doi.org/10.1109/IEDM.1987.191485.

[14] Bauer, M., Alexis, R., Atwood, G. et al. (1995). A multilevel-cell 32 Mb flash memory. *International Solid-State Circuits Conference*, San Francisco, CA, USA (15–17 February 1995), pp. 132–133. https://doi.org/10.1109/ISSCC.1995.535462.

[15] Gartner. (2022). Gartner, adjusted by Macronix International Co.

[16] Beica, R. (2015). 3D integration: applications and market trends. *IEEE 2015 Int. 3D System Integration Conf.*, Sendai, Japan, p. 77.

[17] Tanaka, T., Helm, M., Vali, T. et al. (2016). 7.7 A 768Gb 3b/cell 3D floating-gate NAND flash memory. *International Solid-State Circuits Conference*, San Francisco, CA, USA (31 January–4 February 2016), pp. 142–144. https://doi.org/10.1109/ISSCC.2016.7417947.

[18] Shibata, N., Kanda, K., Shimizu, T. et al. (2020). 1.33 Tb 4b/cell 3D NAND flash memory on a 96-word-line-layer technology. *IEEE J. Solid State Circuits* 55 (1).

[19] Sivaran, S. (2016). Storage class memory: learning from 3D NAND. *Flash Memory Summit,* Keynote 4, Santa Clara, CA (8–11 August 2016).

[20] The Information Network (2022). Gartner, adjusted by Macronix International Co.
[21] World Semiconductor Trade Statistics (WSTS) (2022).
[22] IEEE (2020). *The International Roadmap for Devices and Systems*. IEEE.
[23] Compagnoni, C.M., Spinelli, A., Lacaita, A.L. et al. (2012). Emerging constraints to NAND flash memory reliability. In: *Nonvolatile Memories: Materials, Devices, and Applications* (ed. T.Y. Tseng and S.M. Sze). Stevenson Ranch, CA: American Scientific Publishers.
[24] Wu, S.Y. (1974). A new ferroelectric device, metal-ferroelectric-semiconductor transistor. *IEEE Trans. Electron Devices* 21: 499.
[25] Ovshinsky, S.R. (1968). Reversible electrical switching phenomena in disordered structures. *Phys. Rev. Lett.* 21: 1450.
[26] Hickmott, T.W. (1962). Low-frequency negative resistance in thin anodic oxide film. *J. Appl. Phys.* 33: 2669.
[27] Berger, L. (1996). Emission of spin waves by a magnetic multilayer traversed by a current. *Phys. Rev. B* 54: 9353.
[28] Cappelletti, P. (2015). Non volatile memory evolution and revolution. *International Electron Devices Meeting*, Washington, DC, USA (7–9 December 2015), pp. 10.1.1–10.1.4. https://doi.org/10.1109/IEDM.2015.7409666.
[29] Lee, S.-H. (2016). Technology scaling challenges and opportunities of memory devices. *International Electron Devices Meeting*, San Francisco, CA, USA (3–7 December, 2016), pp. 1.1.1–1.1.8. https://doi.org/10.1109/IEDM.2016.7838026.

Chapter 16

Development of ETOX NOR Flash Memory

Stefan K. Lai

Woodside, California, USA

16.1 Introduction

ETOX (Erasable Programmable Read Only Memory Tunnel Oxide) NOR flash memory was not invented in a single moment, but was developed through evolution of device concepts over many years in search of the ideal semiconductor memory. ETOX NOR flash was not ideal but was first to provide key capabilities of nonvolatility, in system alterability, and low cost in one product. It played a key role in enabling cellular devices, which went from simple cell phones in the early 1990s to now ubiquitous smartphones in hands of billions of people. Its success also served as a stepping stone to the development of lower cost but slower NAND flash memories [1]. After years of development challenges, NAND flash emerged as the mass market solution for first music and then picture and video storage in mobile devices to now high-performance solid-state drives in computing applications. NOR flash is still used in most electronic devices but at much lower bit density.

16.2 Background

Both EPROM (Erasable Programmable Read Only Memory) and EEPROM (Electrically Erasable Programmable Read Only Memory) can trace their origin to a 1967 Bell System Technical Journal article by D. Kahng and S. Sze entitled "A Floating Gate and Its Application to Memory Devices" [2]. EPROM was first developed at Intel by Dov Frohman in 1970 [3]. The EPROM operating principle is

75th Anniversary of the Transistor, First Edition. Edited by Arokia Nathan, Samar K. Saha, and Ravi M. Todi.

simple: when high electric field is applied across a MOS source/drain and its channel, hot electrons are generated. These hot electrons are directed by the vertical field to overcame the silicon dioxide energy barrier and be transferred onto the floating gate that is insulated by a high-quality silicon dioxide. Stored charge on the floating gate is sensed by measuring the conductance of the transistor channel under floating gate. To erase an EPROM, ultraviolet light of high enough energy can excite trapped electrons on the floating gate to overcome the silicon dioxide barrier. Before EPROM, ROM (*R*ead *O*nly *M*emory) and PROM (*P*rogrammable *R*ead *O*nly *M*emory) were available but they did not provide flexibility when program codes needed to be changed. EPROM provided this flexibility and became the perfect companion product to the newly introduced microprocessor, allowing developers to build and update electronic systems. EPROM became a big and highly profitable business for Intel.

As EPROM business grew, the very process of UV erase and reprogramming became a burden especially in high-volume system products. Customers wanted the capability to change information in system, instead of using the cumbersome EPROM UV erase/reprogram process. Technologically, the challenge was to replace the UV erase with an electrical mechanism to reliably remove electrons from the floating gate. FN (*F*owler *N*ordheim) tunneling [4] was determined to be the electrical erase solution. When a high electric field is applied across silicon dioxide, the energy barrier goes from trapezoidal to triangular shape, and this energy barrier can become so thin that electrons can tunnel through it. EEPROM based on a tunnel oxide was first reported by Eli Harari from Hughes Microelectronics in 1976 [5] and Intel developed its own version called FLOTOX (*Flo*ating Gate *T*unnel *Ox*ide) [6]. The concern at that time was whether silicon dioxide could survive the high tunneling electric fields. Reliability studies showed silicon dioxide would eventually breakdown after repeated stress, but this limitation was deemed acceptable if reprogramming was limited to few number of "cycles."

The biggest drawback of EEPROM was large cell area. EPROM programming is a selective process: programming occurs when voltage is applied to gate and drain terminals. EPROM lends itself to a simple array architecture where only selected cells are programmed in X–Y matrix. On the other hand, EEPROM tunneling is a bidirectional process. In a simple X–Y array, EEPROM programming of one cell can disturb other cells. To address this problem, EEPROM was designed with a two-transistor cell: a select transistor plus a memory transistor. The select transistor is required to isolate the memory transistor from other cells during programming/erase and read operations. Furthermore, in a typical EEPROM cell layout, the tunnel oxide area was separate from the floating gate transistor that had a thicker gate oxide. With two transistors and a tunneling area, the layout of an EEPROM cell becomes large compared to that of a simple single transistor EPROM cell. This cell size disadvantage led to high EEPROM unit cost. EEPROM market never grew to be large, but it does live on. Since the tunneling process requires very low power, EEPROM is used extensively in applications requiring low-density memories like contactless cards. NAND flash [1] is actually a variation of EEPROM. Instead of one select and one memory transistor, NAND flash unit cell has 2 select transistors and serial memory transistors in the range of 32 in early days to now hundreds, so that select transistor is not a cost limitation.

Customers wanted the new electrical erase capability of the high-price EEPROM at EPROM prices. In early 1980s, the race was on across semiconductor industry to develop a new low-cost electrically erasable memory. I joined Intel in 1982 during which time experienced technologists were leaving Intel to form their own companies. I was chartered to find a solution to EEPROM scaling problem. My background was in silicon dioxide research. Memory devices and manufacturing were new to me. My inexperience might allow me to take a fresh look at what could or could not be done with no preconceived ideas. After a year of learning and a failed EEPROM scaling project, I asked a very simple question: we already had a device that programed well that was in high volume production, why didn't we start from there? The challenge was to enable electrical erase. FN tunneling was already proven in EEPROM, how do we make it work for EPROM?

The solution was simple. The gate oxide in EPROM was too thick for tunneling: what if we thinned down the gate oxide to be same as EEPROM to allow electrical erase? ETOX (*E*PROM *T*unnel *Ox*ide) was born. With a thin floating gate oxide, we had two choices for erase electrode: the control gate or the source/drain. Erase from the gate would require a negative voltage which was impossible to implement with circuit technology available at that time. The other option was erase from the source with drain reserved for programming. There was a well-known problem of gate induced drain leakage (GIDL) [7]: when there was a high vertical field across a junction, there would be a high current flowing from the junction into substrate, and hot holes could be generated and trapped. The proposed solution was to introduce a phosphorous implant to grade the junction to produce a lower junction field. Furthermore, since the floating gate was only required to erase to neutral state, a lower vertical field would be needed. The big concern was the distribution of the erase cell threshold voltage (*Vt*). There were many factors contributing to *Vt* variability, with tunneling area being an important one. The thought at that time was since tunnel area would be defined by lateral diffusion of dopants from poly silicon edge to underneath gate polysilicon, it would be a well-controlled process not defined by lithography. All these came together in the form of an ETOX invention disclosure 16 January 1984 [8].

There was only one way to prove it: build something. In those days, it was easier to do experimental work. There was an existing EPROM development program and ETOX could ride on it. Small ETOX arrays were fabricated by middle of 1984. To the delight of the ETOX team, we observed a tight erase *Vt* distribution in small arrays. Erase function was demonstrated with 12V on source, along with an acceptable substrate current and an erase time in the 100 milli-second range.

16.3 Not the Perfect Solution

We were able to demonstrate functional cells with a small team and relatively low R&D effort. The next step was to demonstrate product viability. By 1985, the ETOX team had a working 64Kb flash test memory. But to everyone's surprise, the Intel EPROM business unit was not interested in commercializing ETOX flash fearing flash would cannibalize its existing EPROM business. In addition, they claimed that customers needed the EEPROM byte alterability that ETOX lacked. But the seed for flash memory was already sown: Fujio Masuoka-san of Toshiba introduced flash memory in 1984 IEDM [9]. The entire memory chip could be erased at once, like the "flash" of a camera. However, Toshiba was never able to make their flash product a commercial success.

In a last-ditch effort to get Intel to commercialize ETOX flash, Dr. Pashley, who headed Intel's Non-Volatile Memory Technology Development, had lunch with Dr. Gordon Moore. Dr. Moore instantly recognized the potential of the ETOX flash technology and shortly thereafter Dr. Pashley was offered the opportunity to lead an Intel internal startup business based on ETOX flash. However, at the last second, Intel management decided to go in another direction in the pursuit of a full function EEPROM by betting on a new EEPROM technology developed by Xicor Inc., whose founders came from Intel. Key to the Xicor technology was erasing from one poly silicon layer to another [10], instead of erasing through a uniform tunnel oxide between a silicon substrate and a floating gate. Rough grains of poly silicon enhanced the electrical field locally, enabling enhanced electron tunneling even in thicker and presumably more reliable silicon dioxide layer. It was a simple decision. ETOX was in its infancy: new concept developed by new inexperienced team. Xicor was led by veteran engineers well known to Intel management. Just overnight, the team had a new direction.

Now the question was raised: what to do with ETOX program? Dr. Pashley agreed to take his new EEPROM business assignment as long as the ETOX effort was kept alive and folded into this new business unit. He made a consequential decision: if the ETOX program could be kept small, he would protect it through technology demonstration. We did exactly that, putting majority of our resources on the joint development program with Xicor while keeping ETOX development going. This was not too difficult as while Xicor technology required significant new process development, ETOX rode easily on the EPROM baseline with tunnel oxide being the only new module required.

We put our best effort into the Xicor EEPROM joint development program. We studied poly-to-poly tunnel process in detail [10] and Xicor introduced a triple poly unit memory cell [11] that could be programmed and erased by poly-to-poly tunneling. In effect, it was three transistors in series with floating gate transistor in the middle. Programming was from drain poly to floating gate and erasing was from floating gate to source poly. This cell had select transistor built in, compared to a separate select transistor in FLOTOX, making it more compact than FLOTOX EEPROM. However, it was large compared to ETOX which had only one transistor. After two years, we successfully developed a manufacturing process ready for product demonstration of the Xicor cell. At the same time, the "skunkworks" ETOX program also made significant progress. The simpler ETOX cell and well proven EPROM manufacturing process was easy to study, optimize, and execute even with a small team. When it came time to develop long-term business plan, products manufactured using the triple poly Xicor cell were shown to be more expensive compared to its ETOX counterparts. For a memory business, cost is everything. Dr. Pashley received the approval from Intel Management to terminate the Xicor joint development and focus entirely on ETOX flash memory. As a sidenote, poly-to-poly erase lived on. It became the basis of embedded flash memory developed by SST (Silicon Storage Technology), which has become the industry standard embedded flash memory fabricated in multiple foundries [11].

16.4 ETOX Development Challenges

The next step in ETOX development was demonstrating large memory array functionality and reliability. It was pretty straightforward to convert EPROM test chip by connecting common source to external contacts. As expected, programming was straightforward like an EPROM. The big question was the erase *Vt* distribution. UV erase is self-limiting: electrons are removed from the floating gate to produce a neutral state. Electrical erase is not self-limiting: prolonged erase can remove sufficient electrons from floating gate to create a positive charge on the floating gate which makes the transistor conducting with zero gate voltage. This would short out a simple X–Y memory array. The erase *Vt* distribution has to be in a narrow range. It is limited on high end by gate drive for sufficient read current. On the low end, it must have a high enough threshold voltage to minimize leakage current with no cross talk in a memory array.

From the beginning, narrow erase *Vt* distributions were observed in small test arrays in hundreds of bits and good results were carried over to high-density arrays of tens of thousands of bits. In order to control the final distribution, an erase algorithm was developed. A lower erase voltage or shorter time electric pulse was first applied to common source and the erase *Vt* distribution of cells were measured. If any or all of the cells were not erased, more pulses would be applied with increasing voltage or time until all the cells showed minimum read current. Leakage current was measured with gate grounded. If there was a defect in a memory cell, an "over erase" would be observed and the cell would be deemed defective. Eventually, this algorithm was integrated into ETOX memories and transparent to users. From early days, there were a low density of defective cells and mostly good memory arrays were observed. In manufacturing, defective cells could be replaced by redundant cells and if not, the device would be a reject. Defective cells were seldom a yield limiter.

In the early days, ETOX development was based on advanced EPROM process development, reducing process R&D for ETOX. When good results on ETOX were demonstrated, the factory requested us to move the ETOX development to the existing EPROM manufacturing line. Disaster struck when first lot came out from the EPROM manufacturing line. The ETOX Erase *Vt* distribution, instead of in the 1–2 V range, widened to 4–6 V, with most cells in an over-erased condition. Comparing the ETOX development flow with combined ETOX/EPROM manufacturing flow quickly identified the culprit. For EPROM scaling, a new poly to poly dielectric was developed: a composite of *O*xide-*N*itride-*O*xide (ONO). Physical thickness of inter-poly dielectric was reduced from hundreds of nanometers to tens of nanometers while maintaining high capacitance and good electrical insulation. With the nitride layer, process temperature for the stack was lowered to about 900 °C. For comparison, in the standard EPROM production manufacturing process, a temperature above 1000 °C was used to produce a thick poly oxide on floating gate. High temperature oxidation of poly silicon produced large silicon grain size which was in fact the basis of enhanced tunneling in the Xicor EEPROM process mentioned above. With an ONO dielectric and a lower temperature process, poly silicon grains were smaller. This grain size difference was easily observed in high-resolution TEM images. We regained control of erase distribution by reverting back to the ONO process. We had, with little luck, found the "secret sauce" for ETOX technology. When results were first reported publicly [12, 13], industrial experts did not believe we could control the erase *Vt* distribution: since they themselves were not successful. We were the only company able to produce high volume flash memory products for first number of years.

One big challenge to nonvolatile memory technology is reliability. EPROM had been proven to be reliable with the floating gate showing little charge loss over a product life cycle of 10 years. Intrinsically, the silicon dioxide barrier should hold charge for billions of years. However, the issue was that there were always defects. FN tunneling erase introduced a new variable: high electric field stress. Data on FLOTOX memories showed after repeated high field stress on the tunneling silicon dioxide, defects were generated in the form of electron traps which would give rise to a phenomenon called window closing [14]. Furthermore, an increasing number of cells would suffer oxide breakdown with increasing program/erase cycles. ETOX had two advantages compared to FLOTOX. First, its use of EPROM programming required a low electric field across gate oxide. Second, with erase only required to create a neutral cell *Vt*, the applied electric field for ETOX erase was lower than FLOTOX. Another nonvolatile technology requirement was for cells to retain charge over 10 years even after repeated high field stress of program/erase cycling. Our reliability engineering team went to work to study the physics of stressed ETOX cells and developed a temperature screening method that proved to be capable of identifying weak cells that would have failed over 10 years. An additional problem identified was random over-erased bits: erratic bits [15]. Under rare conditions, hot holes were created and trapped in silicon dioxide, which could enhance tunneling to give lower erase threshold voltage. Hot holes might be neutralized in the next cycle and normal erase was observed. An algorithm was developed to weakly program erratic bits to move them into normal distribution. ETOX flash proved to be one of the most reliable nonvolatile memories ever produced in high volume.

16.5 Building a Business

With good results from the ETOX development and the termination of the Xicor program, the team finally had a single focus: build products and get customers excited about the new ETOX NOR flash memory. Dr. Pashley put together a well-coordinated team in multiple disciplines: process, manufacturing, reliability, design, product, marketing, etc. Emphasis was on fast parallel development taking reasonable risk, instead of serial development with rigid handoff. In a few years, we progressed from simple 64 Kbit demonstration vehicle to 256 Kbit product [13], both on a conservative 1.5 micron process; then on to a 1 Mbit device based on 1 micron technology. Every design worked first time, always fabricated on a process with proven reliability and manufacturability. On the other hand, market development faced one big hurdle: cost. Even though in principle, ETOX products could approach cost parity with EPROM, low volume of ETOX made products expensive with fully loaded cost. There were many applications in embedded systems which were willing to pay higher price for in system change capability, but volume was low in general. The marketing team did focus on two high-volume applications by approaching the leading companies in those fields. One was solid-state storage in portable computers. There were a lot of interest and discussions but ultimately, the complexity of change in operating system and relatively high cost of ETOX memory discouraged any bold change. The second was digital photography, which was demonstrated to one of the biggest photography companies in those days. Pictures were taken in CCD cameras, stored in memory cards which were then moved from camera to computer for digital processing. In this case, the company was very successful in their existing business and did not see the new market opportunity; especially if it cannibalized its highly profitable film business. Both of these applications were realized years later with NAND flash memory, which was cheaper than ETOX with smaller effective cell structures and the ability to scale first in 2D and eventually to 3D [16]. Reliability problems with NAND flash were addressed by system-level error detection and correction. NAND flash was too slow to be used as an in-system program code store; but it worked well as a data storage memory where delay due to error correction was not a problem.

The "killer" application for ETOX came with emergence of cellular phones. The story went as follows: a major cellular phone manufacturer had to junk high volume of cell phones based on ROM code store when a software bug made the units faulty. EPROMs were not practical due to the UV erase and labor-intensive process required for code change. All of a sudden, cell phone manufacturers recognized the higher product cost of ETOX flash compared to EPROM was insignificant compared to the code update labor cost. This one application made ETOX flash into a billion-dollar business.

With the success of its first products, the Intel ETOX flash business grew rapidly, quadrupling in revenue every year over the first four years. Intel faced a dilemma: it had an existing EPROM business that was under competitive cost pressure and an emerging flash business with almost no competition but small. Intel did not have enough manufacturing capacity to support both product lines. Intel management, working with Dr. Pashley, made the decision to phase out of the EPROM business and focus entirely on ETOX flash. This was one of the rare occasions in semiconductor history where a company killed a successful product line for a new and yet to be proven product. The decision was validated in a few years with rapid growth of ETOX flash business. Another factor was at that time, Intel was transitioning from a memory company to a microprocessor company. Even with the rapid growth of the ETOX flash memory business, it would always be a small part of a very successful microprocessor company.

The ETOX development team went through many changes over the years, riding up and down with semiconductor business cycles, but always with a single focus: follow Moore's Law and develop their next generation technology every two years. We tried to run as fast as we could. The team developed from 1986 through 2006 10 generations of fabrication technologies, achieving over 1000 times cell size reduction (Figure 16.1). Furthermore in 1995, the team introduced an MLC (*m*ulti-*l*evel-*c*ell) technology [17] which allowed storage of two bits of information in a single cell, effectively reducing product cost by half. This Intel MLC product proved the manufacturability of MLC in flash, and paved way for its adaption later in NAND products.

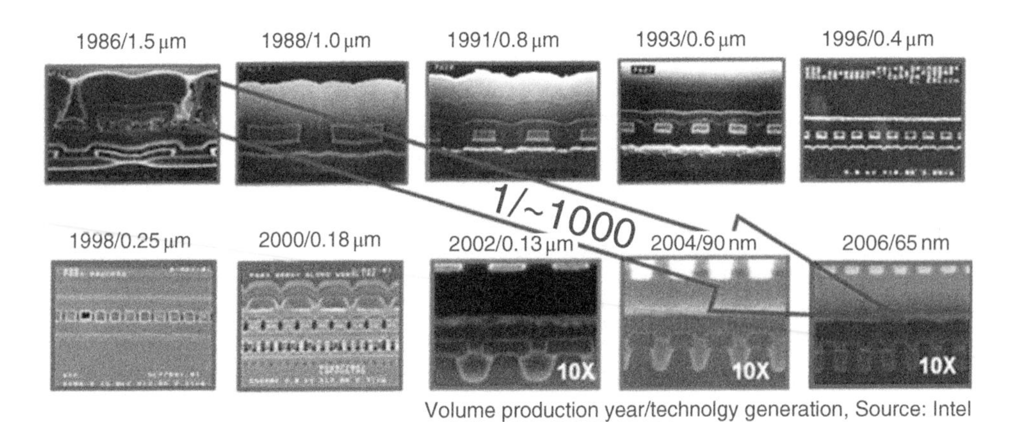

Figure 16.1 SEM micrographs of 10 generations of ETOX NOR flash memory. Cell size was 36 μm^2 in 1986 shrinking to 0.457 μm^2 in 2006 (0.0228 μm^2 if MLC is included).

16.6 Closing Words

The development of ETOX NOR flash memories and its subsequent growth into a billion-dollar business for Intel was the result of many talented people working together as a cohesive team with a common purpose: building a new kind of products that could change the world. The ETOX cell feasibility demonstration in 1984 was a small first step. The difficult work was to weave together many different and difficult steps in making from one then to million, then billion units. Every unit is required to work under all specified operating conditions, after program and erase ten thousand times and after 10 years of operation. These represent some of the most stringent requirements of any semiconductor memories. NAND flash memory had an easier path in reliability with system-level error correction and detection. On top of ETOX process technology development, there were continuous innovations on design, product features, testing methodology, reliability screening and last but not least, the marketing teams' continuous effort to find new applications and new customers. This process went on for about 20 years during which Intel EOTX NOR maintained leadership in the flash memory technology and market, serving as the first memory to wake up most electronic devices and systems. ETOX NOR flash was displaced by NAND flash in revenue in 2005 because as a storage memory, the need for more NAND bits is insatiable. I am most fortunate to have had a front row seat in this journey, going from a concept on paper to knowing that we, as a team, have had something to do with the little device everybody is holding in their hands. All these years, I was humbled by the thought that if we had started our first demonstration on a standard manufacturing EPROM process without the ONO module, the ETOX flash project would have been likely killed and the history of flash memory would be very different.

Acknowledgments

It is impossible to name all the people who had played important roles in this journey, and I want to thank them all. I do want to mention two people. The first is Dr. Pashley who was really the one person who put everything together, from convincing Intel management first and then the world later, and assembling a team to make it happen from the beginning, with my work being one part of it. And he has been a big help in this paper. The second person is Fujio Masuoka-san of Toshiba, who presented the first NOR flash paper in the 1984 IEDM [9], first 2D NAND paper in the 1987 IEDM [1], and first 3D NAND paper in the 2001 IEDM [16]. His competitive spirit made the whole journey so much more exciting.

References

[1] Masuoka, F., Momodomi, M., Iwata, Y., and Shirota, R. (1987). New ultra high density EPROM and flash EEPRON with NAND structure cell. *IEDM Technical Digest*, Washington, DC (6–9 December 1987), pp. 552–555. IEEE.

[2] Kahng, D. and Sze, S.M. (1967). A floating gate and its application to memory devices. *The Bell System Technical Journal* 46 (6): 1288–1295.

[3] Frohman-Bentchkowsky, D. (1971). A fully decoded 2048-bit electrically-programmable MOS ROM. *IEEE ISSCC Digest of Technical Papers*, Philadelphia, PA (17–19 February 1971), pp. 80–81. IEEE.

[4] Lenzlinger, M. and Snow, E.H. (1967). Fowler-Nordheim tunnelling into thermally grown SiO2. *Journal of Applied Physics* 40: 278–283.

[5] Harari, E. (1978). Electrically erasable non-volatile semiconductor memory. US Patent 4,115,914, division of Ser. No. 671,183.

[6] Johnson, W., Perlegos, G., Renninger, A. et al. (1980). A 16Kb electrically erasable nonvolatile memory. *ISSCC Digest of Technical Papers*, San Francisco, CA (13–15 February 1980), pp. 152–153. IEEE.

[7] Chan, T.Y., Chen, J., Ko, P.K., and Hu, C. (1987). The impact of gate-induced drain leakage current on MOSFET scaling. *IEDM Technical Digest*, Washington, DC (6–9 December 1987), pp.718–721. IEEE.

[8] Lai, S., Tam, S., and Dham, V. (1984). An electrically erasable EPROM cell. *Intel Invention Disclosure*.

[9] Masuoka, F., Assano, M., Iwahashi, H. et al. (1984). A new flash EEPROM cell using triple polysilicon technology. *IEDM Technical Digest*, San Francisco, CA (9–12 December 1984), pp. 464–461. IEEE.

[10] Lai, S., Dham, V., and Guterman, D. (1986). Comparison and trends in today's dominant E2technologies. *IEDM Technical Digest,* Los Angeles, CA (7–10 December 1986), pp. 580–583. IEEE.

[11] Guterman, D., Houch, B., Starnes, L., and Yeh, B. (1986). New ultra-high density textured poly-Si floating gate EEPROM cell. *IEDM Technical Digest,* Los Angeles, CA (7–10 December 1986), pp. 826–828. IEEE.

[12] Tam, S., Sachdev, S., Chi, M. et al. (1988). A high density CMOS 1-T electrically erasable non-volatile (flash) memory technology. *Symposium on VLSI Technology Digest of Technical Papers*, San Diego, California (10–13 May 1988), pp. 31–32.

[13]Kynett, V.N., Baker, A., Fandrich, M. et al. (1988). An in-system reprogrammable 256k CMOS flash memory. *ISSCC Digest of Technical Papers*, San Francisco, CA (17–19 February 1988), pp. 132–133. IEEE.

[14] Yeargain, J. and Kuo, C. (1981). A high density floating-gate EEPROM cell. *IEDM Technical Digest*, Washington, DC (7–9 December 1981), pp. 24–27. IEEE.

[15] Ong, T.C., Fazio, A., Mielke, N. et al. (1993). Erratic erase in ETOX flash memory array. *Symposium on VLSI Technology Digest of Technical Papers*, Kyoto (17–19 May 1993), pp. 83–84. IEEE.

[16] Endoh, T., Kinoshita, K., Tanigami, T. et al. (2001). Novel ultra high density flash memory with a stacked-surrounding gate transistor (S-SGT) structured cell. *IEDM Technical Digest*, Washington, DC (2–5 December 2001), pp. 33–36. IEEE.

[17] Bauer, M., Alexis, R., Atwood, G. (1995). A multilevel-cell 32Mb flash memory. *ISSCC Digest of Technical Papers*, San Francisco, CA (15–17 February 1995), pp. 132–133. IEEE.

Chapter 17

History of MOS Memory Evolution on DRAM and SRAM

Mitsumasa Koyanagi

New Industry Creation Hatchery Center (NICHe), Tohoku University, Sendai, Japan

17.1 Introduction

The LSI era using MOSFETs started in 1970 with the 1-kbit DRAM launched by Intel. In the past, magnetic core memories were used in the early stage of computer. However, magnetic core memories were replaced by DRAM early in the 1970s. Since then, the packing density and capacity of DRAM have continued to significantly increase until today, contributing to the evolution of IT society despite many attempts to replace it. Both data and instructions are stored in a main memory in a stored-program-type computer. DRAM has been used as a main memory for a long time. Therefore, DRAM is the key component in a computer system. A cache memory is also important in a computer system to increase the memory read/write operation speed between a processing unit (PU) and a main memory. SRAM has been used as an embedded cache memory in a processor chip. Thus, DRAM and SRAM have been important memories indispensable for a computer system. Revolutionary technologies of DRAM and SRAM in the past are described in this article.

17.2 Revolutionary Technologies in DRAM History

DRAM revolution has been achieved by the revolution of memory cell and innovative circuit technologies. Then, the memory cell revolution and innovative circuit technologies are discussed in this section.

75th Anniversary of the Transistor, First Edition. Edited by Arokia Nathan, Samar K. Saha, and Ravi M. Todi.

17.2.1 Evolution of DRAM Memory Cell

The packing density and capacity of DRAM have increased more than sixfold toward today's 16-Gbit DRAM over a half century since the advent of 1-kbit DRAM. Such a dramatic increase in the packing density and capacity of DRAM has been achieved by the evolution of memory cell. It was well known that a flip-flop circuit can be used as a basic memory cell circuit. However, it was difficult to reduce a memory cell size using a flip-flop circuit as a memory cell. Then, many people challenged to reduce the memory cell size and power consumption by decreasing the number of transistors composing of a flip-flop. Thus, four-transistor-type cell and three-transistor-type cell were proposed in 1-kbit DRAM. Intel used a memory cell with three p-channel MOS transistors since the n-channel MOS transistor was unstable in those days although four n-channel MOS transistors were used in four-transistor-type cell. These memory cells were eventually replaced by one-transistor-type cell (1T-cell) invented by Robert Dennard in 4-kbit DRAM [1]. The 1T-cell consisted of one MOS transistor and one MOS capacitor as shown in Figure 17.1. The invention of 1T-cell by Dennard significantly accelerated the evolution of DRAM in conjunction with his device scaling theory [2]. However, many DRAM people were skeptical at first about the 1T-cell since the signal read out of memory cell was significantly small (100–200 mV) as a result of charge sharing of the cell node and the bit-line (BL). This is due to much smaller cell capacitance (C_S = 20–50 fF) compared to the bit line (BL) capacitance (C_B = 50–600 fF). Hence, the 1T-cell was more susceptible to noise. Therefore, the evolution in DRAM memory cell structure was indispensable for the 1T-cell to be eventually employed in production.

17.2.2 Evolution in Planar-Type Memory Cell

A planar-type memory cell was used in the first 4-kbit DRAM using 1T-cell in the mid-1970s. However, it was a big issue in a simple planar cell that a storage capacitance and hence the signal charge stored in memory cell significantly decreased as scaling-down the memory cell size. Then, a planar-type memory cell had evolved in its structure as shown in Figure 17.2. A memory cell in Figure 17.2b was

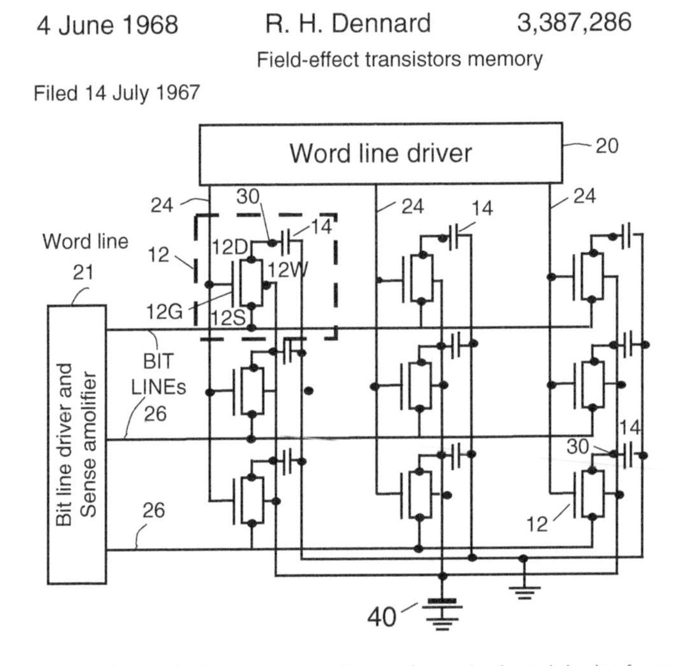

Figure 17.1 Array of one-device memory cells as shown in the original referenced patent.

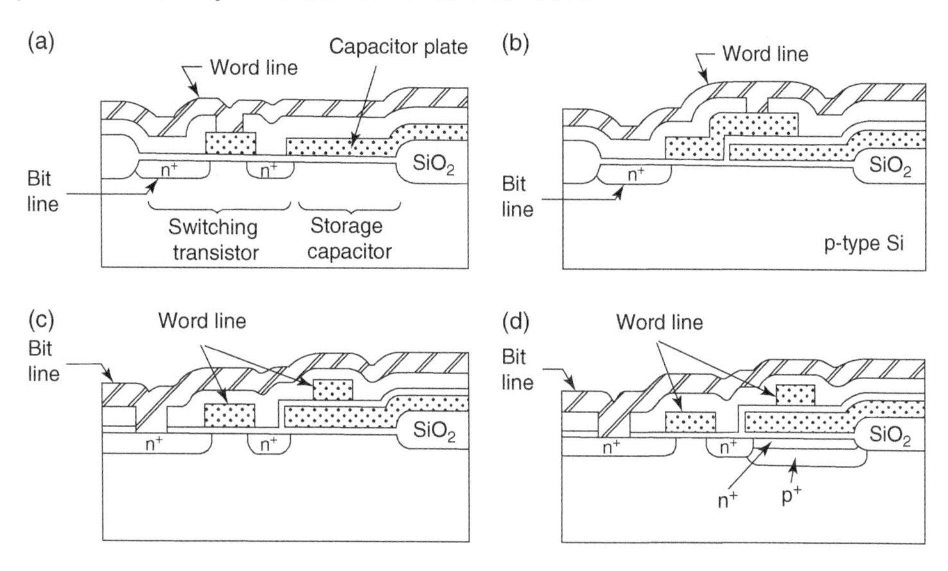

Figure 17.2 Various kinds of planar cell structures. (a) Simple planar cell. (b) Charge coupled cell. (c) Folded BL cell. (d) Folded BL cell with Hi-C.

called a charge coupled cell where the n^+ diffusion region was removed and the signal charge transfer between the switching transistor and storage capacitor was performed like a charge coupled device (CCD). This memory cell was firstly employed in a 16-kbit DRAM. A memory cell in Figure 17.2c was called a folded BL cell used in a 64-kbit DRAM where a new folded BL circuit architecture was firstly employed to cancel the fixed pattern noises in a memory cell array. A polycrystalline silicon word line crossed-over a storage capacitor in this memory cell so as to match the folded-bit-line architecture. This memory cell evolved to a folded BL cell with Hi-C (High Capacitance) as shown in Figure 17.2d where an MOS capacitor and a p–n junction capacitor were connected in parallel and in addition a silicon nitride was used as a capacitor insulator to increase the storage capacitance. However, it was difficult to further scale-down the memory cell size using planar-type memory cells although the memory cell structure had evolved as shown in Figure 17.2. Then a planar-type memory cell was eventually replaced by three-dimensional (3D) memory cells from 4-Mbit DRAM [3] although its life span was extended to 1-Mbit DRAM by changing the DRAM design from NMOS-based design to CMOS-based design.

17.2.3 Invention of Three-Dimensional Memory Cell

17.2.3.1 Stacked Capacitor Cell

The first 1-T cell was realized using one switching MOS transistor and one MOS capacitor as mentioned above. The amount of signal charges stored in the storage capacitor has to be maintained almost constant or can be only slightly reduced as the memory cell size is scaled down. However, MOS capacitor value and hence the amount of signal charges is significantly reduced as the memory cell size is reduced even if the capacitor oxide thickness is scaled-down. Then, Koyanagi prospected in 1975 that the 1-T cell with two-dimensional (2D) structure using a planar MOS capacitor eventually encounters the scaling-down limitation because we cannot reduce the MOS capacitor area according to the scaling theory. In addition, he pointed out the problem in 1-T cell using an MOS capacitor that the signal charges are seriously reduced due to the influences by the minority carriers generated in a silicon substrate. An inversion layer capacitance and a depletion layer capacitance are connected to the gate oxide capacitance in series in the MOS capacitor. The charges in the inversion layer and the depletion layer

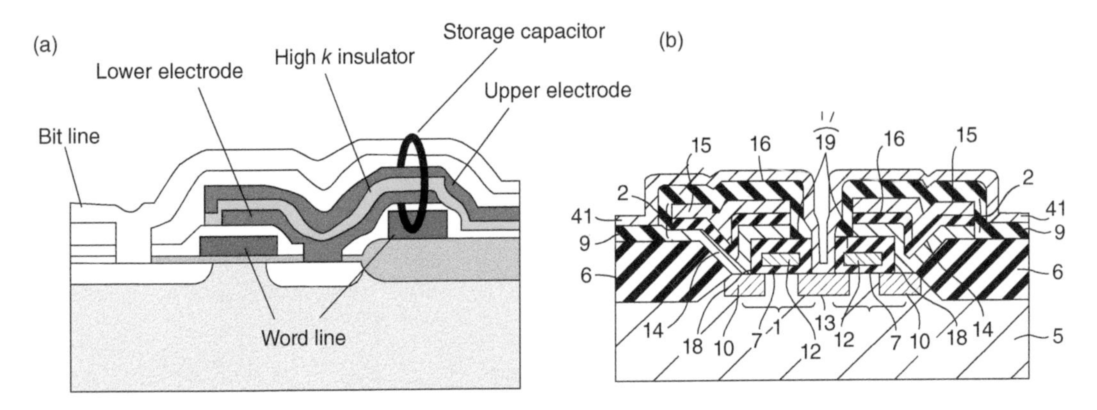

Figure 17.3 Cross-sectional structure of stacked capacitor DRAM cell and US patent. (a) Cross-sectional structure of stacked capacitor DRAM cell. (b) US Patent 4,151,607, filed 5 July 1977. Patented 24 April 1979.

are easily affected by the minority carriers which are thermally or optically generated or generated by the irradiation of energetic particles in a silicon substrate. Therefore, he prospected that the 1-T cell using an MOS capacitor encounters the scaling-down limitation due to the influences by the minority carriers although the size of switching MOS transistor can be reduced according to the scaling theory. Then, Koyanagi persisted in his confidence that a passive capacitor with the structure of electrode–insulator–electrode should be used as a storage capacitor of DRAM memory cell and such a capacitor should be three-dimensionally stacked on a transistor to reduce the memory cell size. Thus, Koyanagi invented a 3D stacked capacitor cell (STC) in 1976 to solve these problems in planar-type memory cells [4–8]. The basic structure of STC is shown in Figure 17.3 where a storage capacitor is three-dimensionally stacked on a switching transistor.

17.2.3.2 Background to Invention of Stacked Capacitor Cell

Mitsumasa Koyanagi had studied on the silicon surface and the inversion layer in MOS structure in his PhD research at Prof. Junichi Nishizawa's laboratory in Tohoku University during 1971–1974 [9]. The inventor of Flash memory, Fujio Masuoka, was a senior PhD student in Prof. Junichi Nishizawa's laboratory at that time. He often persuaded Koyanagi about the importance of semiconductor memory even after he joined Toshiba. Then, Koyanagi was strongly impressed by semiconductor memory. In his PhD research, he fabricated the impedance analyzer with the frequency range of 0.01 Hz–100 MHz to evaluate the electrical properties of the interface states and the inversion layer since such an impedance analyzer with the ultralow frequency capability was not available at that time. Masayoshi Esashi collaborated with him to fabricate the impedance analyzer who was also PhD student at other laboratory in Tohoku University at that time where he was involved in the research of sensor and MEMs technology. Koyanagi evaluated various kinds of capacitor elements including high-k (high dielectric constant) capacitors as a reference capacitor of this impedance analyzer. Eventually, he made a vacuum capacitor for a reference capacitor in which fin-type capacitor electrodes are encapsulated in a vacuum container. From these studies, he learned that an ideal capacitor with low loss should consist of metal electrodes and low loss insulator (MIM structure), 3D structure of capacitor electrodes is effective to increase the capacitance value and there is the trade-off between high-k and loss in the capacitor insulator. In addition, he knew through his PhD research that the charges in the inversion layer and the depletion layer are easily influenced by the minority carriers. Therefore, he had a question why the MOS capacitor with inversion capacitance and the depletion capacitance is used as the storage capacitor in 1-T cell when he firstly knew it in 1975 after joining Hitachi Central Research Laboratory (CRL). Then, he tried to eliminate the inversion capacitance and the depletion capacitance by employing the passive capacitor such as MIM capacitor for a storage capacitor and thus invented a 3D cell in 1976. He called this new 3D memory cell a STC. He was in the same group with Hideo Sunami in Hitachi CRL who invented a trench capacitor cell.

17.2.3.3 Fabrication of Three-Dimensional Stacked Capacitor Cell

A storage capacitor is three-dimensionally stacked on a switching transistor in the STC as shown in Figure 17.3. A passive capacitor with the structure of electrode–insulator–electrode is used as a storage capacitor. The bottom electrode of storage capacitor is connected to the source/drain region of switching transistor. Koyanagi proposed to use self-aligned contacts to connect the bottom electrode of storage capacitor and a BL to the source/drain of switching transistor. The self-aligned technique was also used for the formation of capacitor electrodes. We can dramatically reduce the memory cell area by three-dimensionally stacking the storage capacitor on the switching transistor. In addition, we can use a high-*k* (high dielectric constant) material as a capacitor insulator to increase the storage capacitance since a passive capacitor is used as a storage capacitor. This is also useful to reduce the memory cell size. Furthermore, we can solve the problem that the signal charges in the inversion layer and depletion layer are influenced by the minority carriers since an inversion capacitance is not used in a STC. He fabricated the first DRAM test chip with a STC using 3 μm NMOS technology in 1977 and presented a paper of STC in 1978 IEDM (IEEE International Electron Devices Meeting) [10]. Figure 17.4 shows the SEM cross section of a STC fabricated using 3 μm NMOS technology. It is clearly shown in the figure that a storage capacitor is stacked on a switching transistor and a self-aligned contact is successfully formed although plasma etching and RIE (reactive ion etching) were not available at that time. The self-aligned contact has been widely used in today's memory LSI's. In this STC, he employed polycrystalline silicon (poly-Si)-Si_3N_4-polycrystalline silicon (poly-Si) capacitor as a storage capacitor. So far, thermal SiO_2 was used as a capacitor insulator in a conventional 1-T cell with an MOS storage capacitor. In the STC, he used Si_3N_4 for the first time in DRAM instead of SiO_2 as a capacitor insulator to increase the storage capacitance. The dielectric constant of Si_3N_4 is approximately two times larger than that of SiO_2. He found that the leakage current of Si_3N_4 is significantly reduced by oxidizing its surface as shown in Figure 17.5. He also used a Ta_2O_5 film as a capacitor insulator for the first time. Ta_2O_5 has a dielectric constant five or six times larger than that of SiO_2. Therefore, we can greatly increase the storage capacitance although the leakage current is larger compared to those of SiO_2 and

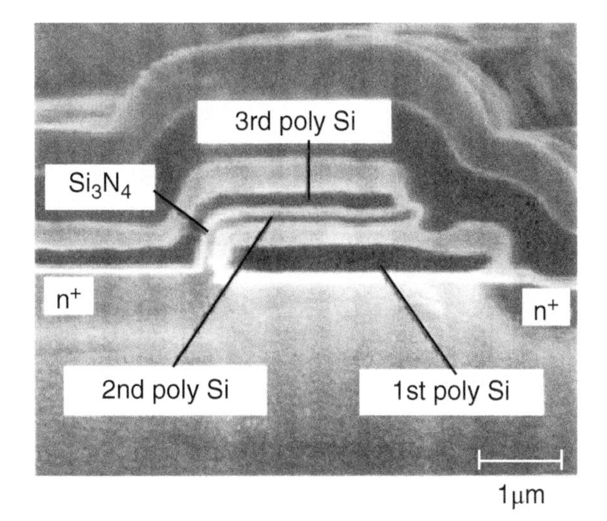

Figure 17.4 SEM cross section of stacked capacitor cell fabricated using 3 μm technology.

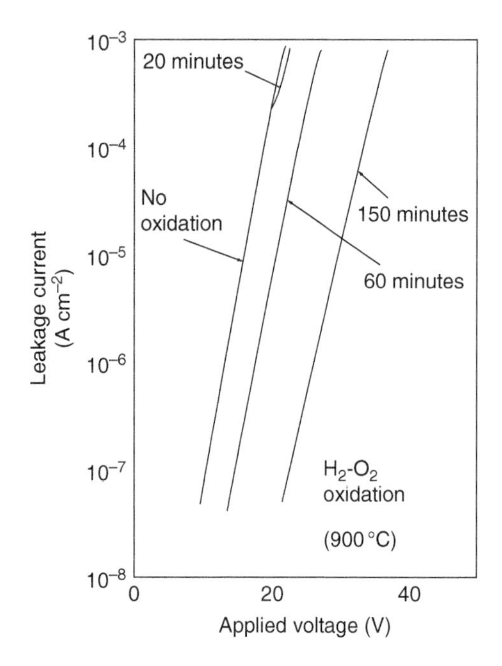

Figure 17.5 Oxidation time dependence of leakage current flowing through ON (oxidized nitride) capacitor insulator.

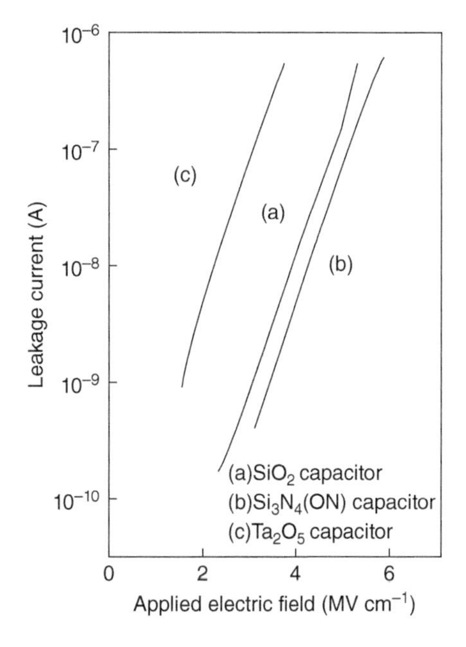

Figure 17.6 Leakage current-applied electric field characteristics of poly-Si–insulator–poly-Si capacitor structure.

Si_3N_4 as shown in Figure 17.6. He presented the first STC paper in IEDM in 1978 prospecting that the data retention characteristics of DRAM with the conventional planar (2D) cell are seriously degraded due to the soft-error. At that time, he believed that a STC was tolerant of soft-error since the signal charges stored in the passive capacitor are not influenced by the minority carriers generated in the

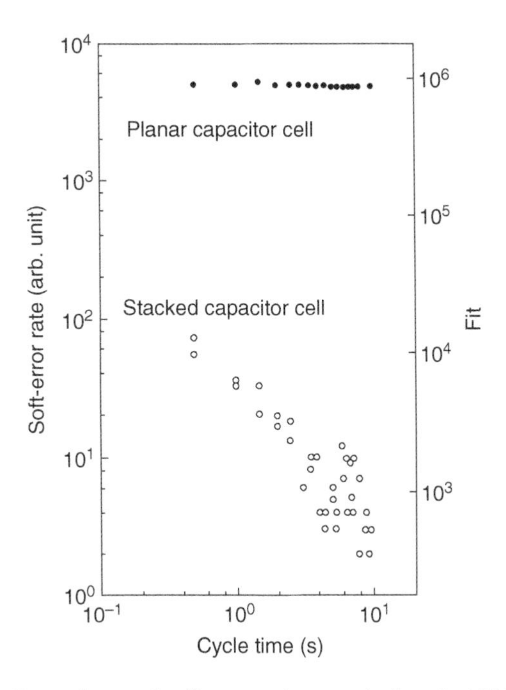

Figure 17.7 Dependence of soft-error rate on cycle time in 16 kbit DRAM chip.

substrate. Figure 17.7 shows the dependence of soft-error rate on cycle time in DRAM test chip. As he prospected, the soft-error rate was dramatically reduced by employing a STC while the data retention characteristics of DRAM with the conventional planar (2D) cell are seriously degraded due to the soft-error which is caused by the carriers generated in the silicon substrate by alpha-particle irradiation.

Eventually, he proposed three types of STCs as shown in Figure 17.8 [11]. A storage capacitor is stacked on the switching transistor and the BL in a top-capacitor-type cell, on the switching transistor in the intermediate-capacitor-type cell (original STC) and on the isolation oxide (LOCOS) in the bottom-capacitor-type cell. The top-capacitor-type STC cell is also known as capacitor-over-bit line (COB) STC and widely used in current DRAM [12]. We can use various kinds of materials for the capacitor insulator and electrodes, and can employ low-temperature processes in the formation of storage capacitor in COB-type STC since the storage capacitor is formed on the top of the memory cell. He fabricated 16-kbit DRAM using a STC with the oxidized Si_3N_4 (O/N) capacitor insulator in 1979 after the fabrication of the first stacked capacitor DRAM test chip [11]. Then, he tried to introduce a STC in 64-kbit DRAM production. However, it was too early to introduce a STC in DRAM production due to the cost issue. As a result, only the oxidized Si_3N_4 (O/N) capacitor insulator was employed in production.in 64-kbit DRAM production. Since then, the oxidized Si_3N_4 (O/N) capacitor insulator was widely used in DRAM production. A STC was employed in 1-Mbit DRAM production for the first time by Fujitsu [13]. Hitachi also employed a STC in 4-Mbit DRAM production. Figure 17.9 shows an SEM cross-sectional view of COB-STC cell for 4-Mbit DRAM fabricated by Hitachi [14]. Many other DRAM companies used a trench capacitor cell in the early stage of 4-Mbit DRAM production. However, a STC eventually came to occupy a major position in 4-Mbit to today's 8-Gbit DRAM. A STC has evolved by introducing the 3D capacitor structures with fin-type electrode [15] and cylindrical electrode [16] in conjunction with a capacitor electrode surface morphology of hemi-spherical grain (HSG) [17]. In addition to the introduction of 3D capacitor structure, a storage capacitor insulator with high dielectric constant (high-*k*) was employed in high-density DRAM's. In general, a leakage current of high-*k* material increases by high temperature processing. Therefore, a COB-type STC is suitable for introducing a high-*k* capacitor insulator since the storage capacitor can be formed by lower temperature process. Thus, Ta_2O_5 capacitor insulator was employed in 64-Mbit and 256-Mbit DRAM's with

(a)

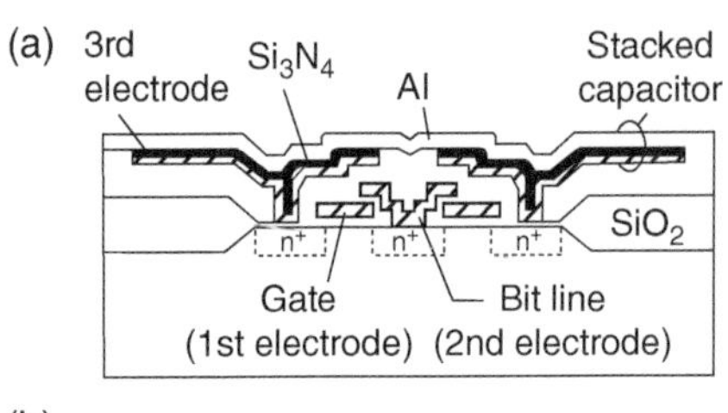

(b)

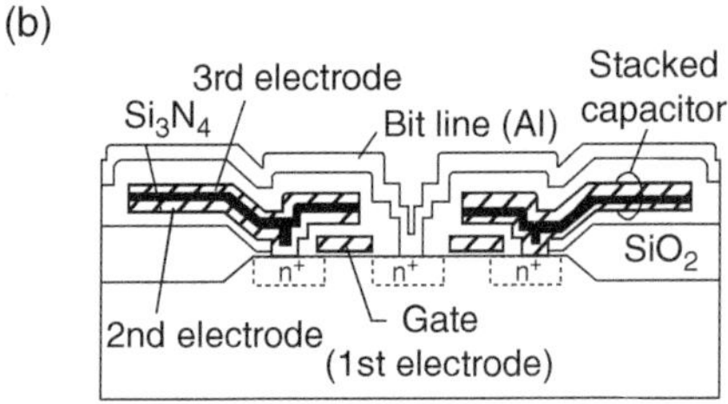

(c)

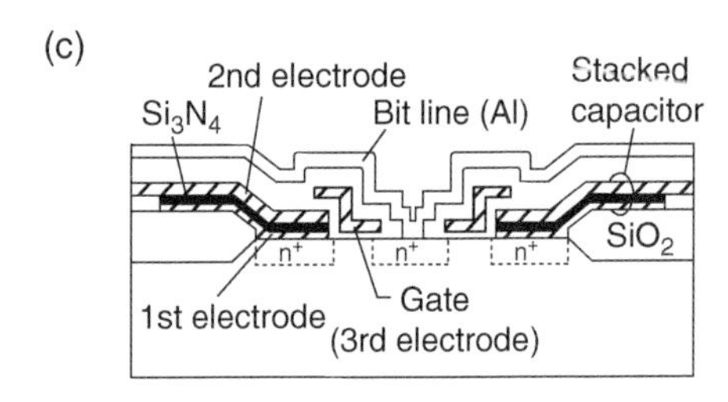

(d)

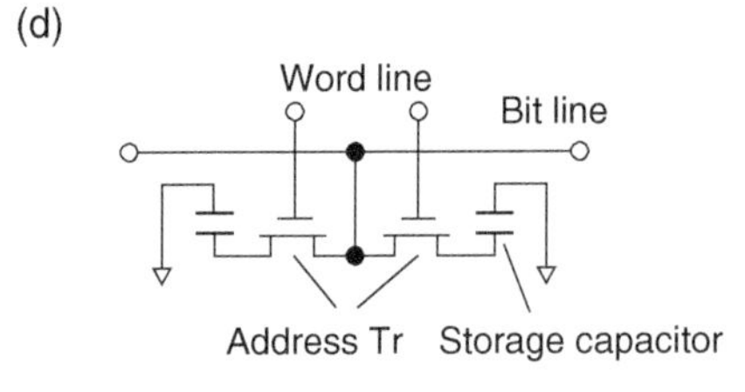

Figure 17.8 Cross-sectional structure for three kinds of stacked capacitor (STC) cell. (a) Capacitor-over-bit-line (COB) STC cell. (b) Original STC cell. (c) Capacitor-over-isolation (COI) STC cell. (d) Memory cell circuit (two bits).

COB-type STC. Since then, various kinds of high-*k* materials have been studied as storage capacitor insulators in high-density DRAM's beyond 1-Gbit. The concept of COB-type STC that various kinds of materials can be stacked on the switching transistor using lower temperature process has been carried on new memory devices with 3D structure such as Fe-RAM (Ferroelectric RAM), P-RAM (Phase Change RAM), R-RAM (Resistive RAM), and M-RAM (Magnetic RAM).

17.2.3.4 Trench Capacitor Cell

Trench capacitor DRAM cell was invented in 1975 by Hideo Sunami [18]. The basic structure of trench capacitor cell is shown in Figure 17.10 where a storage capacitor is formed into a deep trench in a silicon substrate. He had felt the dilemma of size vs. capacitance in one-transistor-type DRAM memory cell (1T-cell). In response to die size reduction to cope with a fourfold increase in memory capacity, memory cell size has been reduced by almost one-third for each generation. The cell size reduction through scaling alone lead to area reduction and a subsequent decrease in capacitance value. To cope with such a dilemma of size vs. capacitance, insulator thickness should be reduced by a factor of 10

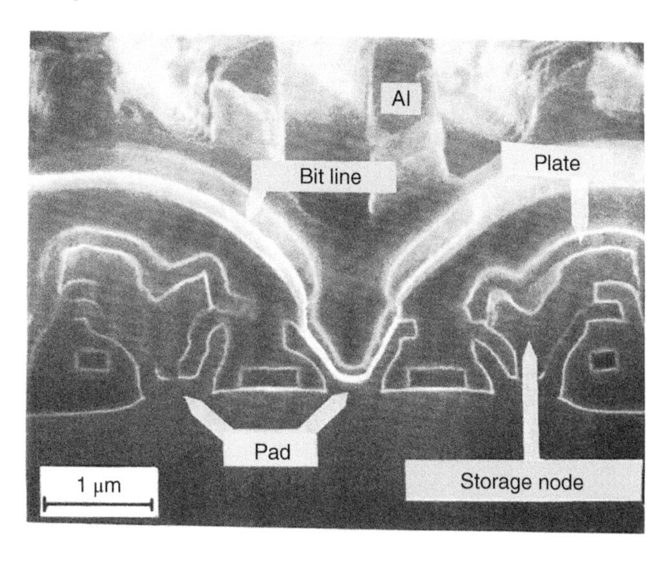

Figure 17.9 SEM cross section of COB-STC cell for 4 Mbit DRAM.

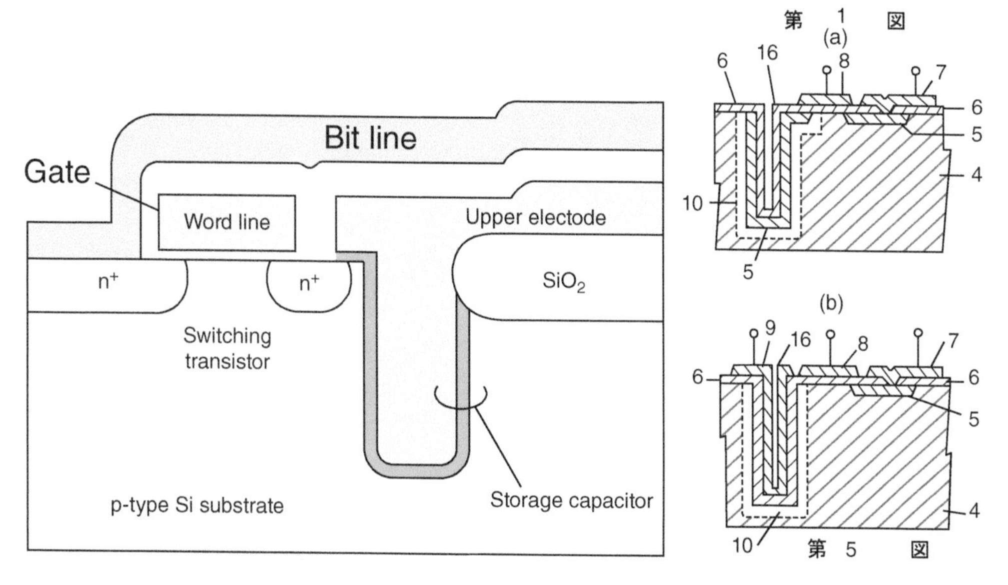

Figure 17.10 Cross-sectional structure of trench capacitor DRAM cell. (a) Cross-sectional structure of trench capacitor DRAM cell. (b) Japanese patent application (Tokugansho 50-53883, 1975) (H. Sunami and S. Nishimatsu).

from 100 nm in 1-kbit chips to 10 nm in 1-Mbit chips, getting dangerously close to dielectric field breakdown. Then, he got an idea of a trench capacitor DRAM cell when he took a glimpse at some conference presentations in 1974 from Texas Instruments introducing a highly efficient silicon solar cell with a steep trench and forecasting the upcoming issue of cell size vs. capacitance. He proposed to use a capacitor formed in a deep trench in a silicon substrate as a storage capacitor of 1T-cell [19]. He thought that by adjusting the trench depth, the capacitance value could be increased without increasing the capacitor footprint.

17.2.3.5 Background to Invention of Trench Capacitor Cell

At an early stage of Sunami' junior-high school age, he had hobbies of collecting butterflies and hand-making of short-wave radio receivers and transmitters. Then, he became to be familiar to electronic components such as vacuum tubes, transistors, resistors, capacitors, etc. He entered Tohoku University in 1963 and joined Hitachi CRL in 1969 after his graduation from Prof. Junichi Nishizawa's laboratory in 1969. He was offered a leave opportunity at Stanford University when he engaged in CCD research at Hitachi CRL. In the conference he attended during the stay in Stanford University, he was impressed by some presentation where truly vertical trench of aspect ratio of almost 5 was formed on (110)-oriented silicon surface since it had been difficult to form truly vertical trench on silicon surface in those days. He knew from his experience on CCD that surface characterization was a major tool to master the integrity and reliability of silicon devices. At that time, he heard that one-device and one-capacitor DRAM cell had been invented. Then, he got an inspiration of a first trench capacitor cell concept based on a combination of that DRAM cell and a cylindrical trimmer condenser used in radiofrequency transmitter together with his experiences on CCD research.

17.2.3.6 Fabrication of Trench Capacitor Cell

Sunami fabricated the first DRAM test chip with a trench capacitor cell in 1982 and presented a paper of trench cell named a corrugated capacitor cell (CCC) in 1982 IEDM [20, 21]. Figure 17.11 shows the SEM cross section of a trench capacitor cell fabricated using a plasma etching. He tried to apply his trench capacitor cell to 1-Mbit DRAM. Then, Hitachi developed a 1-Mbit DRAM using a trench capacitor cell in 1984 and presented a paper in 1984 ISSCC (International Solid-State Circuits Conference) [22]. An SEM cross section of a trench capacitor cell used in a 1-Mbit DRAM is shown in Figure 17.12. As be clear in the figure, the shape of trench capacitor was drastically improved as a result of the progress of plasma etching technology. The soft-error immunity was also significantly improved owing to the increased storage capacitance as shown in Figure 17.13. However, an inversion capacitance and depletion capacitance in the silicon substrate were still included in a simple trench capacitor cell and hence there was the possibility that the signal charges were seriously reduced due to the influences by the minority carriers generated in a silicon substrate. To avoid this, the inversion capacitance and the depletion capacitance should be excluded from the storage capacitance. A method to achieve this is to replace the inversion capacitance and the depletion capacitance by the accumulation capacitance or to form the p^{++} region or n^{++} region on the silicon

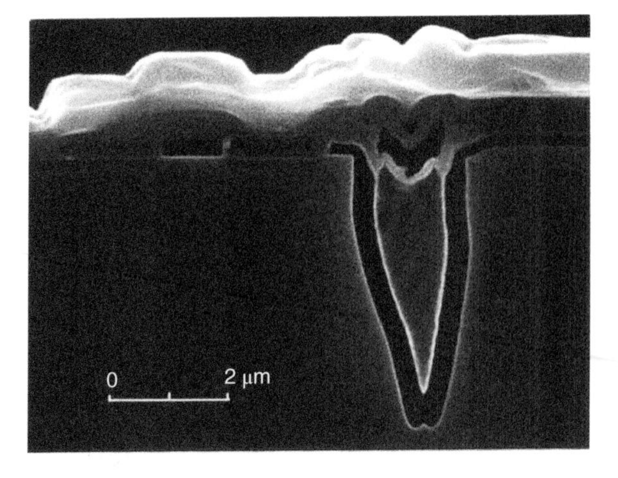

Figure 17.11 Cross-sectional structure of the first trench capacitor DRAM cell.

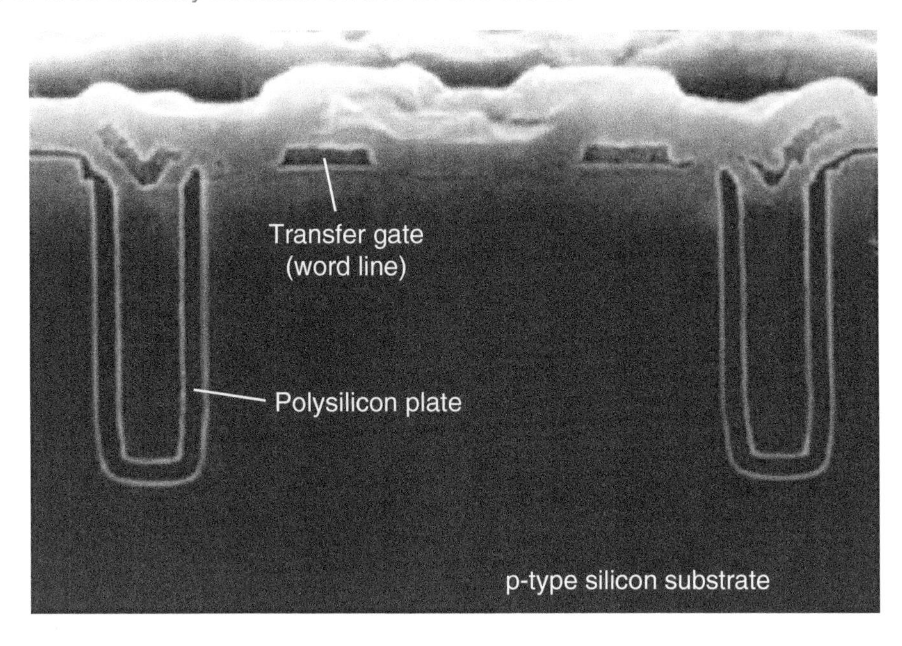

Figure 17.12 Cross-sectional structure of trench capacitor DRAM cell used in 1 Mbit DRAM.

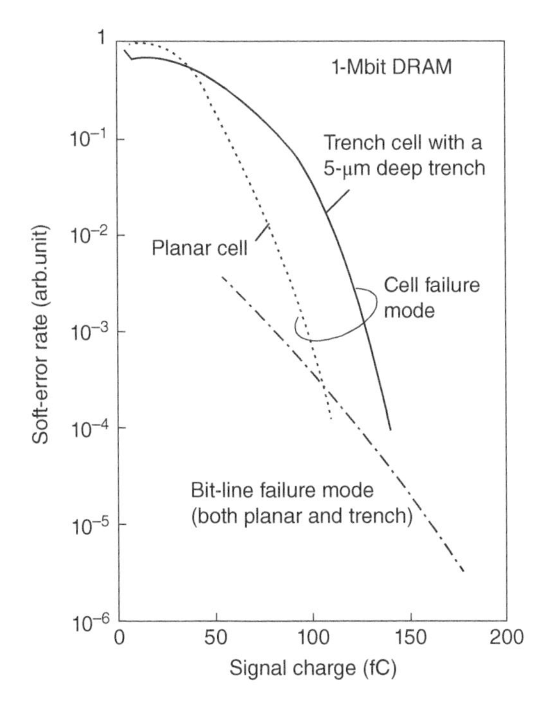

Figure 17.13 Soft-error rates of planar and trench cells.

surface of trench capacitor. Thus, a substrate-plate trench capacitor cell with the p^{++} region around a trench was employed in production. IBM, Siemens, and Toshiba employed a trench capacitor cell in 4-Mbit DRAM [23]. Currently, a trench capacitor cell has been used as a memory cell of embedded DRAM integrated in a processor chip [24].

17.2.4 Innovative Circuit Technologies for DRAM

An open-BL architecture as shown in Figure 17.14 was employed for a memory cell array in NMOS 4-kbit DRAM using 1T-cell. A signal voltage read out to the BL was very small (usually 100–200 mV) in such DRAM with an open-BL architecture because the cell capacitance (C_S = 20–50 fF) was much smaller than the BL capacitance (C_B = 50–600 fF). A small C_S and a large C_B result from a small cell size and a large number (128–512) of cells connected to a BL. Consequently, the read–write operation in the open-BL architecture was susceptible to noise. The innovative circuit technologies were necessary for dealing with such a small signal resulting from the destructive readout. Then, it was proposed to use a cross-coupled differential sense amplifier for the signal sensing/restoring [25]. A small signal read out of a memory cell to a BL was successively amplified and stored back to the memory cell by a cross-coupled differential sense amplifier. The refresh operation indispensable for DRAM was easily performed by activating this cross-coupled differential sense amplifier. However, it was difficult to completely solve the noise issues even though the cross-coupled differential sense amplifier was employed. Therefore, a more sophisticated circuit technology was necessary for stable operations of DRAM using 1-T cell. Then, the folded-BL architecture was invented by Kiyoo Itoh [26, 27] in 1974 in order to cancel array noise. In the conventional open-BL architecture, two BLs in a pair connected to a differential sense amplifier are separately arranged on two array conductors which causes electrical imbalances between the BLs. Moreover, if voltage bounces at the conductors are different, different voltages are coupled to the BLs via conductor-BL parasitic capacitances. Consequently, a differential noise that cannot be cancelled by the amplifier is generated between the pair of BLs. In the folded-BL architecture shown in Figure 17.15, the two BLs run close to one another and are parallel on the same conductor enabling the same noise voltage to be coupled to the BLs which can be cancelled by the amplifier. The folded-BL architecture was employed in 16-kbit DRAM. Beside the folded-BL architecture, several new circuit technologies were needed to increase the packing density, to decrease the chip size, to improve the performance, and to reduce the power consumption which are represented by the address-multiplexing scheme to reduce the package pin-count [28], a BL precharging scheme [29], the word "bootstrapping" and the redundancy technique [29], and so on. After the memory capacity at the research and development level reached 1–4 Gbit level in the mid-1990s, the focus has been on high-throughput DRAMs to bridge the more prominent DRAM-processor performance gap as represented by EDO (Extended Data Out) DRAM, pipe-lined DRAM, block-transfer-oriented protocol DRAM (Rambus DRAM), synchronous DRAM (SDRAM), and DDR (Double Data Rate) DRAM [30–38].

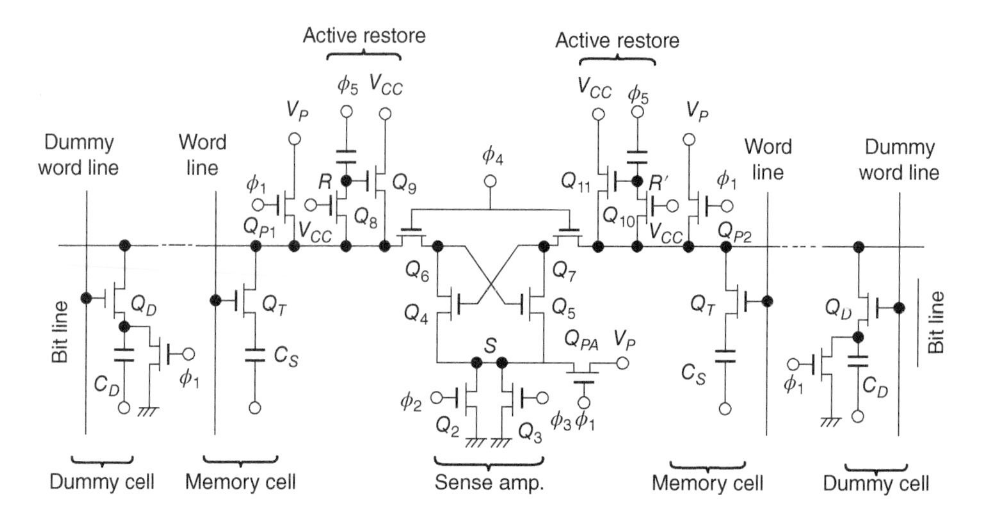

Figure 17.14 Open-BL architecture.

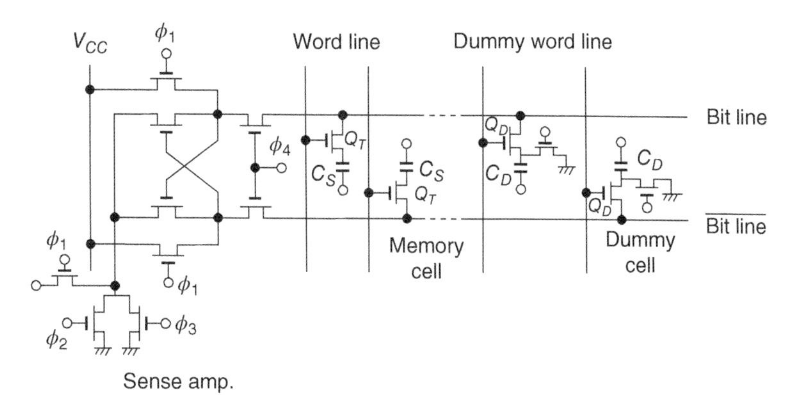

Figure 17.15 Folded-BL architecture.

The data are continuously read out by continuously changing the CAS (Column Address Strobe) signal after applying the RAS (Row Address Strobe) signal and then multibit data are simultaneously output in EDO DRAM as shown in Figure 17.16a. An EDO mode is often called as a hyper-page mode. A high-speed read out operation is achieved by pipeline processing of the operation to read out the data after applying the address signals in SDRAM as shown in Figure 17.16b. The read operation is performed at both edges of the external clock to increase the data rate in DDR DRAM. A prefetch method is employed in SDRAM and DDR DRAM as shown in Figure 17.16c. The data with the size of several bits are prefetched and then required data are output by multiplexing the prefetched data in a prefetch method. The name of DDR DRAM changes like DDR2, DDR3, DDR4, and so on, depending on the data size prefetched.

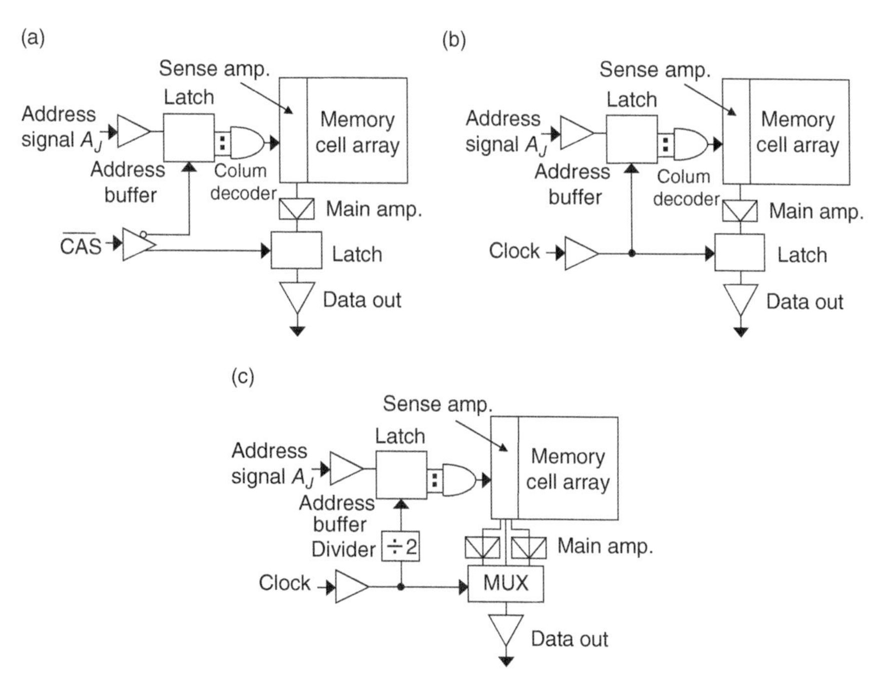

Figure 17.16 Read out methods in high-throughput DRAM. (a) EDO (asynchronous). (b) Two-stage pipeline (SDRAM). (c) Two bit prefetch (DDR/SDRAM).

17.2.5 Future of DRAM

The memory capacity and the data bandwidth of DRAM will continue to increase toward the future. The DRAM memory capacity will increase more than several tens of Gbits and the data bandwidth will exceed several tens of TBytes/s in a near future. The key to achieve such a tremendous improvement in DRAM memory capacity and performance is the evolution of memory cell after all. The evolution of DRAM memory cell so far is summarized in Figure 17.17 [19]. Mainly a STC has been employed in almost all high-density DRAM with a large memory capacity from 4-Mbit DRAM to 16-Gbit DRAM. A trench capacitor cell has been used in embedded DRAMs. This tendency will not change even in the future. The evolution of STC employed in DRAM production so far is shown in Figure 17.18 [39]. A cylinder-type STC has been used in DRAMs with the memory capacity of more than 1-Gbit. The cell size is reduced to less than 0.01 μm^2 by using a cylinder-type storage capacitor with extremely large aspect ratio of more than 30 [40]. In addition, the open-BL architecture has come back to achieve such a small cell size. The memory cell size in the open-BL architecture is described as $6F^2$ while that in the folded-BL architecture is $8F^2$ where F means the feature size. However, the memory cell with the cell size of $4F^2$ is necessary to further reduce the cell size down to 100 nm^2. A pillar-type STC with a vertical switching transistor as shown in Figure 17.19 is indispensable to achieve such a small cell size although a crucial issue of floating body in a vertical transistor has to be solved [40]. Another big issue is that the storage capacitance of memory cell has continuously decreased in each generation. Eventually, the storage capacitance value may be reduced down to 5 fF which gives rise to an extremely small signal read out on the BL. We will need an extra transistor to read the signal voltage stored in such a small storage capacitance. This implies that we have to change the memory cell from the one transistor and one capacitor cell to the two transistor and one capacitor.

Another approach to achieve a future high-density and high-performance DRAM is to stack many DRAM chips three-dimensionally, so-called 3D-DRAM. Koyanagi proposed a new 3D integration technology using a wafer-on-wafer bonding with through-silicon-vias (TSV's) in 1989 [41–44]. Then, he pointed out that the DRAM data bandwidth can be dramatically increased by simultaneously transferring a block of data with large size in parallel using many TSVs in 3D stacked DRAM. He succeeded

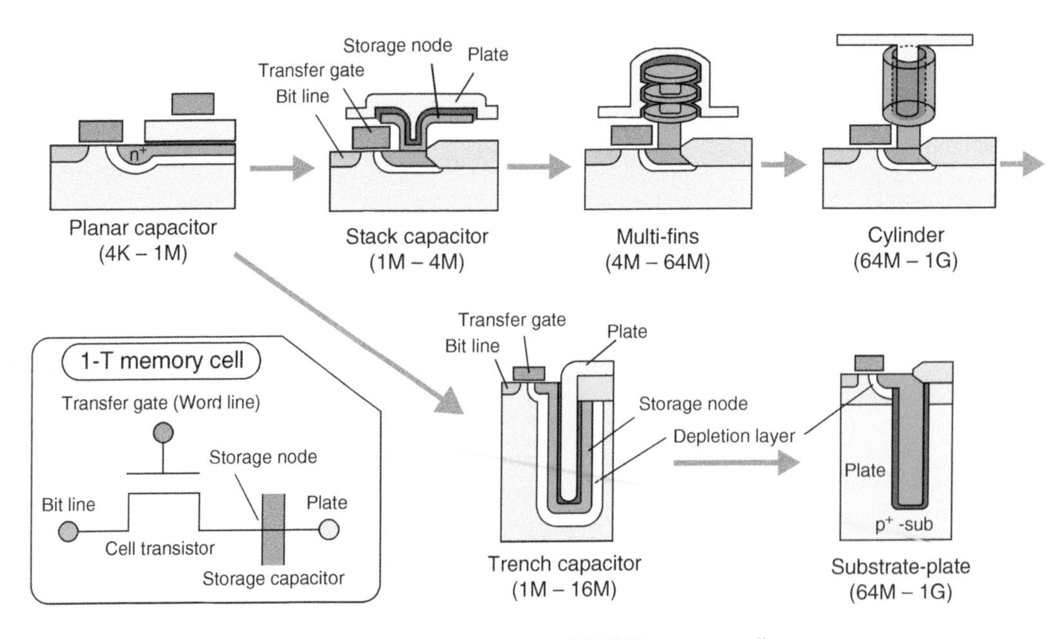

Figure 17.17 Revolution of DRAM memory cell.

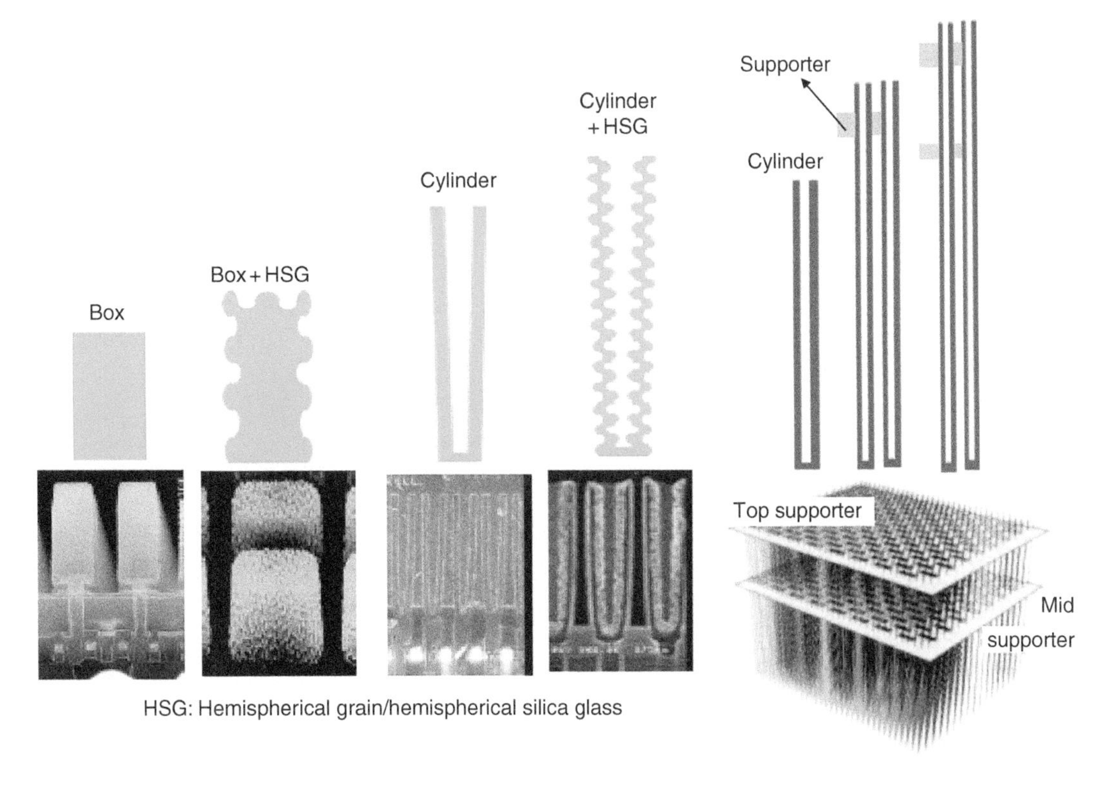

Figure 17.18 Revolution of stacked capacitor cell.

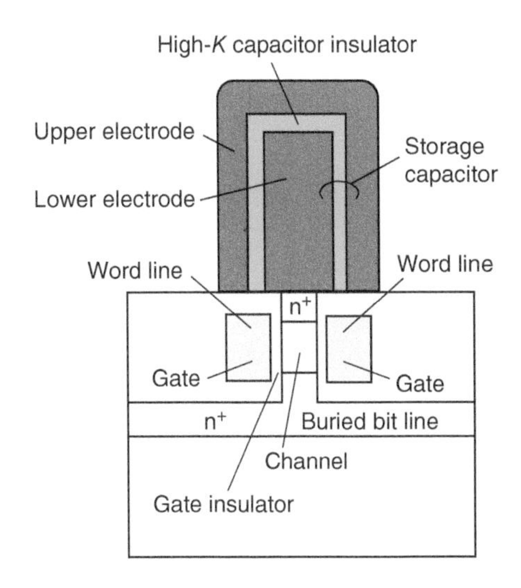

Figure 17.19 Pillar-type stacked capacitor cell with vertical switching transistor.

in fabricating a 3D-DRAM test chip with 10 memory layers as shown in Figure 17.20 in 2000 and confirmed a parallel data transfer through many memory layers [45]. Such a 3D-DRAM called HBM (High Bandwidth Memory) has been commercialized around 2015 by SK-Hynix, Samsung, and Micron [46–48]. HBMs with eight memory layers and one logic layer have been already in mass

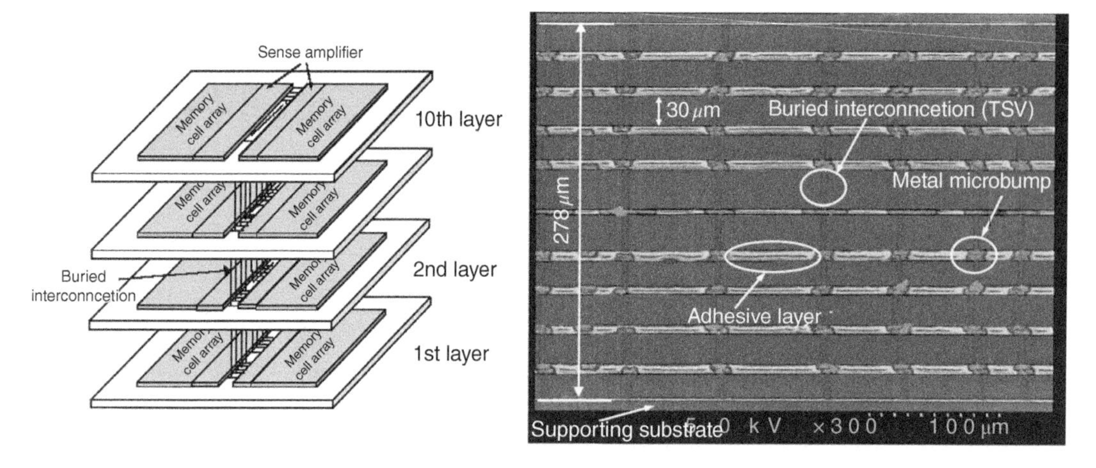

Figure 17.20 SEM cross section of 3D-DRAM test chip with 10 memory layers fabricated by wafer-on-wafer bonding technology with through-silicon-vias (TSV's).

production. The data bandwidth of HBM has already exceeded more than 1-TBytes/s [49]. The number of stacked layers in HBM will exceed several 10 layers in the future.

17.3 Revolutionary Technologies in SRAM History

SRAM has been utilized as a cache memory for a long time due to its high-speed capability although the memory density is lower than that of DRAM. It has been always required for SRAM as a cache memory to reduce the power consumption with increasing the memory capacity since it has occupied a considerable area in processor chip. Consequently, the SRAM also has evolved with the evolution of memory cell and innovative circuit technologies.

17.3.1 Evolution of SRAM Memory Cell

An SRAM memory cell consists of a bi-stable flip-flop and two access transistors. The SRAM memory cell has evolved by changing the load element of flip flop as shown in Figure 17.21. An n-channel MOS transistor has been used as a driver transistor of flip-flop. A depletion-type n-channel MOS transistor was used as a load transistor in the initial SRAM memory cell as shown in Figure 17.21a. This memory cell was not employed after 16-kbit SRAM due to its larger power consumption. The memory cell with resistor loads was introduced in a 16-kbit SRAM to reduce the power consumption by employing a polycrystalline (poly-Si) silicon resistor with a high resistance value as a load element as shown in Figure 17.21b. The poly-Si resistor was three-dimensionally stacked on the access transistor to reduce the memory cell size in this memory cell. However, this memory cell had the issues that it was sensitive to the noise and soft-error and not fast in speed since the load resistance was so high. Then, the poly-Si resistor with a high resistance was configured as a p-channel thin film transistor (TFT) in the memory cell shown in Figure 17.21c. The gate of this TFT was also formed by poly-Si which was tied to the gate of the opposite inverter. This p-channel TFT was also three-dimensionally stacked on the access transistor to reduce the memory cell size. A CMOS flip-flop with six transistors (6T) is used in the memory cell shown in Figure 17.21d where a p-channel MOS transistor is used as a load transistor. The CMOS 6T memory cell had not been employed in a SRAM for a long time due to its larger memory cell size and a latch-up issue although it provided a high speed and low power consumption. A CMOS flip-flop

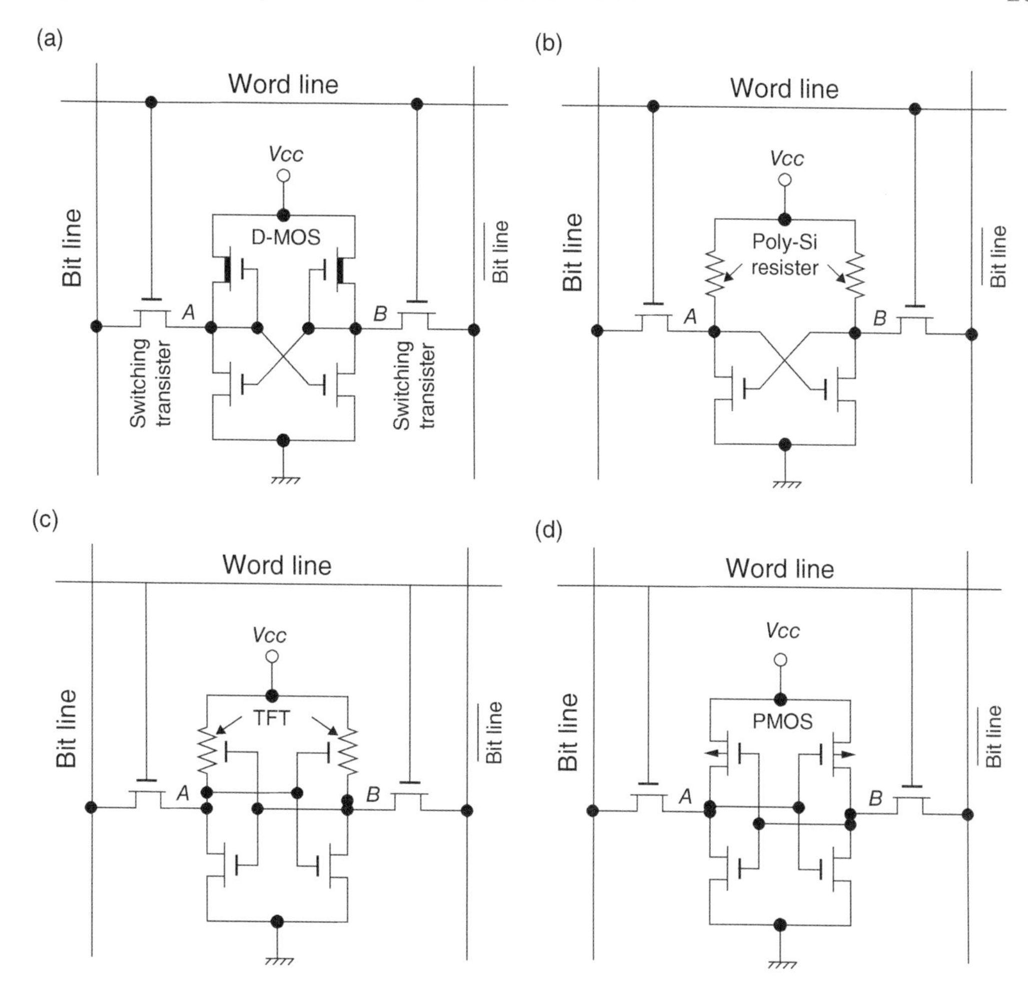

Figure 17.21 Various kinds of SRAM memory cells. (a) Depletion-MOS load type. (b) Poly-Si resistor load type. (c) TFT load type. (d) CMOS type.

was employed as a memory cell for the first time in 4-kbit SRAM by T. Masuhara and Y. Sakai et al. in 1980 [50, 51]. Since then, a CMOS flip-flop has continued to be used as a memory cell even in a current advanced SRAM. The improvements of packing density and power consumption in SRAM with CMOS memory cell has been implemented by the device scaling. Fin-FET is used as transistors of CMOS memory cell in current advanced SRAM to increase the read/write speed and to decrease the stand-by leakage current.

17.3.2 Innovative Technologies to Expand SRAM Capabilities

It has been continuously required for SRAM to increase the operation speed and decrease the power consumption. The address transition detector (ATD) and the address pre-decoder were employed to increase the operation speed and decrease the power consumption in early stage SRAM. The ATD generated control pulses by detecting the transition of address signals. The internal circuits were controlled by pulses generated by the ATD. The address signals initially decoded by the pre-decoder were input to the main decoder to generate control pulses for the word lines and BLs as shown in Figure 17.22.

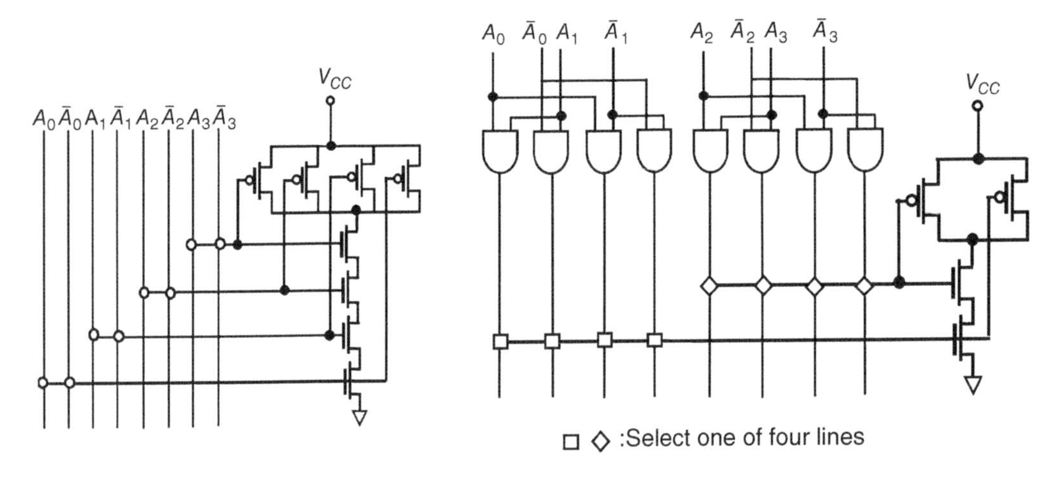

Figure 17.22 Address decoder circuit with pre-decoding method. (a) Basic decoder circuit. (b) Decoder circuit wit pre-decoding method.

The signal delay between the address buffer to the decoder were minimized by using this pre-decoding method. In addition, the BL pre-charge and equalizing method and the word line pulsing method were employed to reduce the power consumption for the read operation and the variable impedance method for the BL load was introduced to reduce the power consumption for the write operation. The BL pre-charge and equalizing circuit and the BL variable impedance circuit are illustrated in Figures 17.23 and 17.24, respectively.

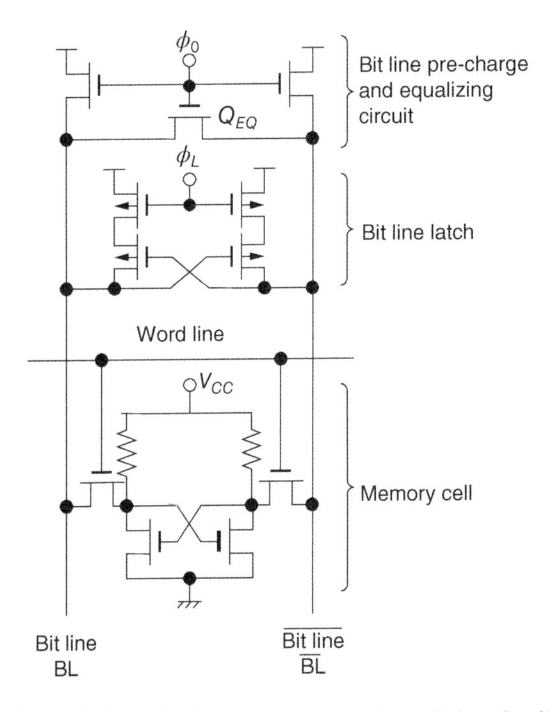

Figure 17.23 Bit-line pre-charge and equalizing circuit.

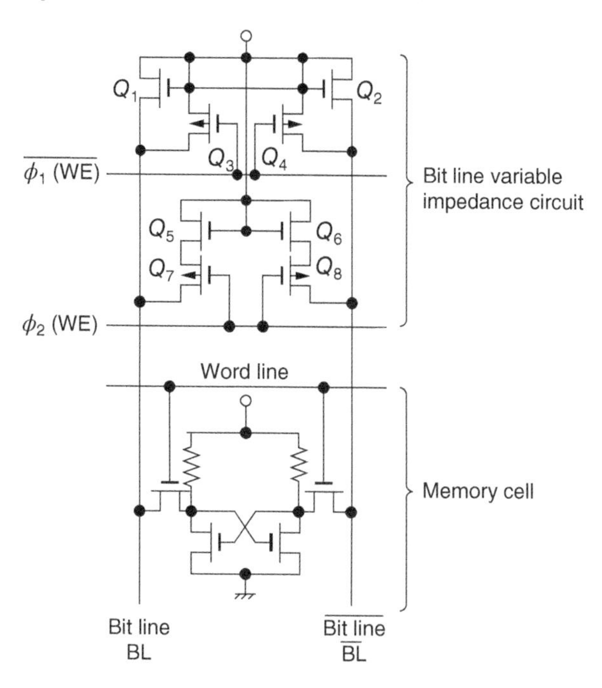

Figure 17.24 Bit-line variable impedance circuit.

It is the most important items to guarantee the stable read/write operation. However, it has become more and more difficult to guarantee the stable read/write operation as the memory cell size has been scaled down and the supply voltage has been lowered to increase the memory capacity and to decrease the power consumption. Especially, the increased threshold voltage variability resulting from the device scaling seriously degrades the 6T CMOS cell stability. To solve these issues, several innovative technologies have been introduced in recent advanced SRAM's. Different methods have been employed to solve the issues in read disturb and write disturb [52–56]. Read stability failures are caused for a bit cell to lose state due to charge injection from BL to internal nodes during a read operation. A static noise margin (SNM) is used to evaluate a worst-case read stability although a dynamic stability margin analysis is more accurate for modeling to absolute lowest voltages possible. We should also take read performance failures into account which are influenced by the variation in the bit cell read current, the BL RC, the word line RC, and sense amplifier mismatch. Then, a read assist circuit (R-AC) with the word line underdrive (WLUD) or suppressed word line (SWL) has been employed to improve the read stability which has been implemented with simple voltage divider in SRAM row decoder driver using tunable pull-down NMOS or PMOS transistors to lower the word line voltage during reading operation as shown in Figures 17.25 and 17.26 [52, 57]. We can equivalently increase β ratio of cell by lowering the word line voltage which improve the SNM. The β ratio is defined as the ratio of the pull-down NMOS transistor size to the access transistor size. The WL voltage should be carefully adjusted to tradeoff the read stability vs. write margin/performance in introducing the WLUD scheme because lowering the word line voltage results in a degraded write margin. Write margin failures occur when NMOS access transistors (pass gate transistors) are incapable of overwriting state held by the cross-coupled inverter pair due to the contention of NMOS access transistor with pull-up PMOS transistor. A static write voltage margin (WVM) is defined as the BL voltage at which the bit flips when the bit-line bar (BLB) is ramped down holding the BL at the supply voltage V_{cc}. It is usually optimistic compared with a dynamic write margin. A voltage collapse write assist (VC-WA) circuit and a negative BL write assist (NBL-WA) circuit have been employed to improve the write stability. A memory cell

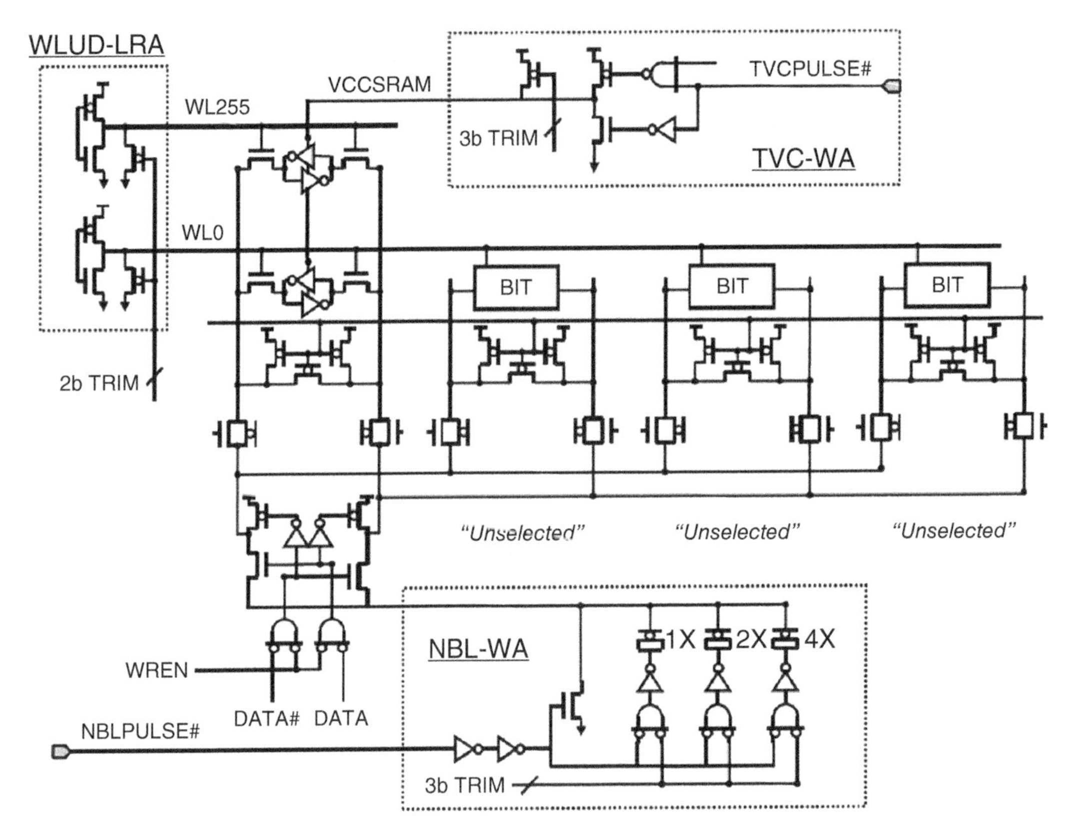

Figure 17.25 WLUD, NBL, and TVC assist circuit implementation.

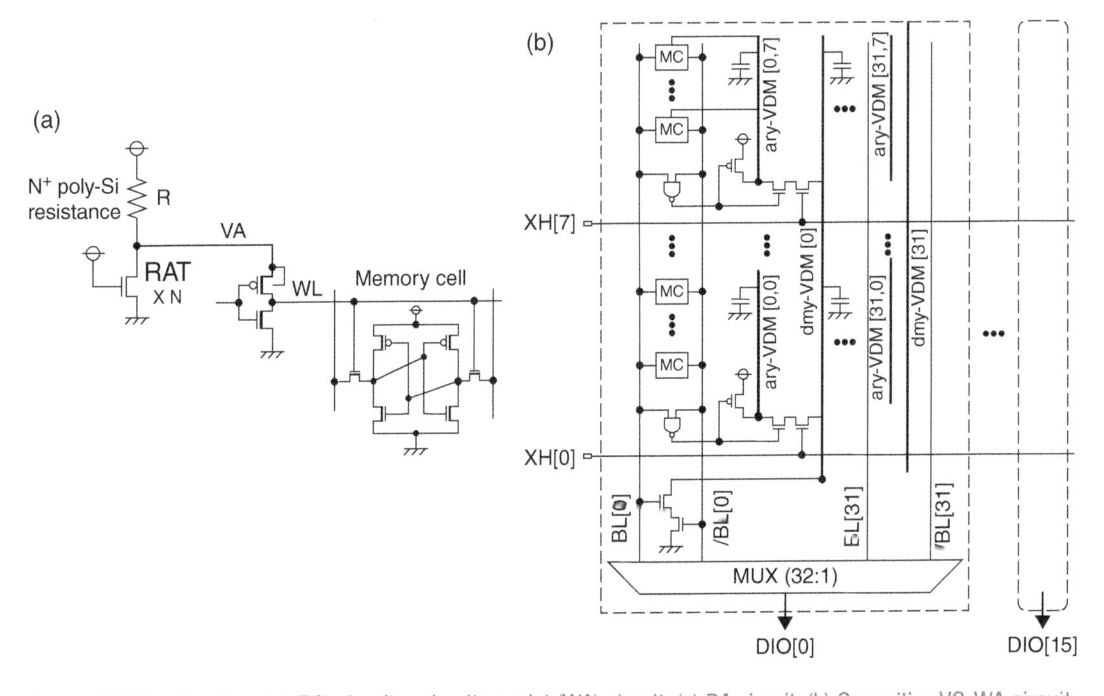

Figure 17.26 Read assist (RA) circuit and write assist (WA) circuit. (a) RA circuit. (b) Capacitive VC-WA circuit.

supply voltage (V_{cc}SRAM) is collapsed by a transient voltage collapse write assist circuit (TVC-WA) circuit in Figure 17.25 [52]. A capacitive VC-WA circuit is shown in Figure 17.26 where a memory cell supply voltage of ary-VDM (V_{AV}) is lowered by the making use of the capacitance ratio between the ary-VDM (Cav) and the additional dmy-VDM (Cdv) which requires a small Cav/Cdv. A divided ary-VDM scheme is used to decrease Cav. A γ ratio is equivalently reduced by collapsing the memory cell supply voltage and as a result we can avoid the write contention since the current drivability of pull-up PMOS transistor can be reduced. The γ ratio is defined as the ratio of the pull-up PMOS transistor size to the access transistor size. Unselected bit cells along the column have dynamic retention risks (lose state) in the TVC-WA circuit. The NBL-WA is a column-based write assist that temporarily increases gate overdrive on the SRAM access transistor by pulling the low BL to a negative voltage during write operation. A MOS transistor connected to the write driver/BL node is used as the coupling capacitor. The BL with a negative voltage breaks the write contention by increasing the access transistor current drivability as shown in Figures 17.25 and 17.26. However, it should be noted that a large negative voltage on the BL reduces the signal charge on the node capacitance.

It has been proposed to use memory cells with 8 transistors (8T cell) and 10 transistors (10T cell) to solve various conflicts in 6T cell. The memory cell circuits of 8T cell and 10T cell are illustrated in Figures 17.27 and 17.28. A dedicated read port with two NMOS transistors are added for a read operation in the 8T cell. The gate of read NMOS transistor is connected to one of storage nodes of flip-flop. The signal stored in a flip-flop can be directly read through the read NMOS transistor without disturbing node potentials of flip-flop. Therefore, the 8T cell ensures a sufficient read SNM. Consequently, we can

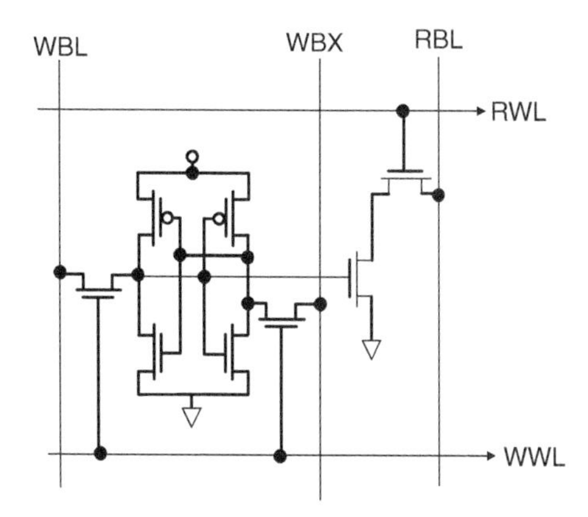

Figure 17.27 Memory cell circuit of 8T cell.

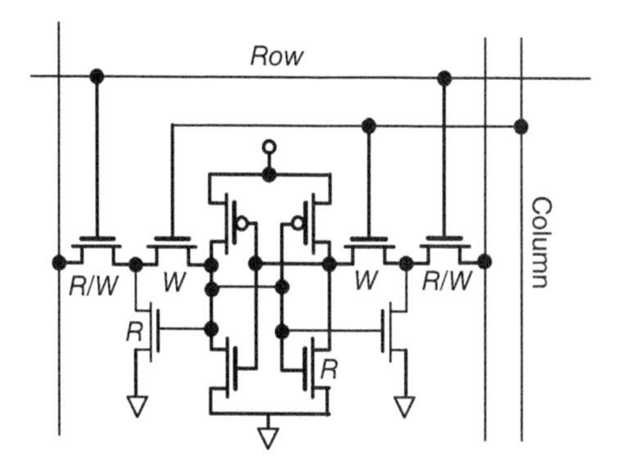

Figure 17.28 Memory cell circuit of 10T cell.

mitigate the conflicts among the cell current, the dynamic and leakage power, and the read stability in the 6T cell. In addition, the 8T cell enables a concurrent 1 read/1 write operation. However, we cannot use the differential sense amplifier in the 8T cell because the read operation is implemented in a single-ended read port. Therefore, we have to use a single-ended sense amplifier in the signal sensing instead of a differential sense amplifier in the 8T cell. This is the different tradeoffs for read performance in the 8T cell. In addition, the 8T cell cannot address the half-select issue in the write operation when requiring the column interleaving operation. Therefore, the tradeoff between WRM and SNM should be taken into account for the memory cell design. The half-select access issue can be solved by adding two more NMOS transistors to the 8T cell to configure the 10T cell. The pass gate for the write operation consists of two access transistors connected in series between the BL and the storage node, which are selected by row-and-column-decoding, respectively, while the pass gate for the read operation bypasses the column-decoded NMOS access transistor connected to the storage node. Therefore, the 10T cell can eliminate the read disturbance and the tradeoff between SNM, WRM, and cell current. In addition, the 10T cell can solve the column interleaving issue since it enables a cross-point access by selected row-and-column decoding [58]. The column interleaved architecture enables sense-amplifier sharing across multiple column since only one of many columns are active. This means that a sufficient area can be allocated for the sense amplifier. Therefore, it is not necessary in the 10T to scale down the sense amplifier area as rapidly as the bit cell area which enables the optimum design for the sense amplifier. Furthermore, increasing column-interleaving provides better array for the same offset voltage by amortizing the sense amplifier area over multiple columns. Column-interleaving also enables to use a simple single single-bit error correction coding (ECC) to address soft-errors [59]. It is only a problem in the 10T cell that we have to pay for the expense of cell area due to the increased number of transistors.

The address access restriction when attempting to increase read/write bandwidth per cycle is another issue in single-port SRAM. Multi-port SRAM's are used to avoid such access restrictions. Multi-port SRAM's such as 2-read/write (2RW) dual-port (DP) SRAM and 1-read/1-write (1R1W) two-port (2P) SRAM are widely used for recent SoCs, especially for image processing and high-speed communication devices because of their high bandwidth capabilities. A memory cell circuit is illustrated in Figure 17.29. This memory cell supports concurrent two reads, concurrent two write, and concurrent one read/one write. A true 2RW DP-SRAM provides the most flexible access characteristics for a 2-port memory, but introduces unique margin and timing challenges. Four kinds of memory accesses such as different row/different column, different row/same column, same row/different column, and same row/same column can be implemented in 2RW DP-SRAM.

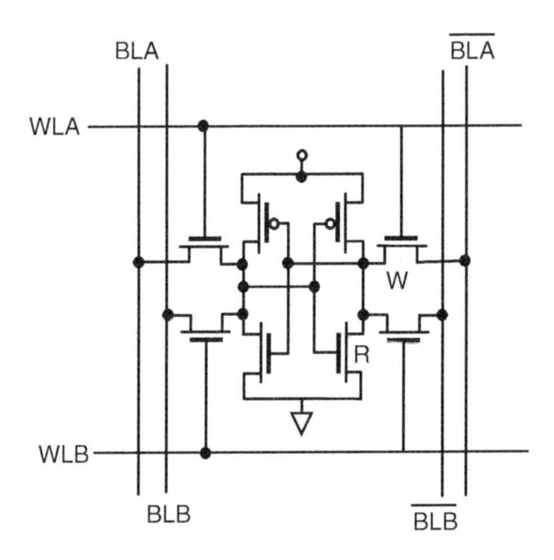

Figure 17.29 Memory cell circuit of 2-read/write (2RW) DP-SRAM.

The different row/different column access is identical to that of single port (SP) SRAM while different row/same column, same row/different column, and same row/same column accesses are unique to DP-SRAM. The same row access introduces the concurrent read or dummy read operations or the double dummy read on two active word lines in unaccessed cells while the same row/different column access introduces the single dummy read on one active word line interacting with read/write operations in the selected cells. A write access on both ports to the same cell is not commonly supported. The concurrent read or dummy read operations in the same row access reduces the noise margin (cell read stability) and requires a larger bit cell design (larger NMOS pass gate) or external circuit assists to stabilize the cell. The dummy read creates the additional contention with the write operation (read disturb write). The same row access also sets the worst-case read current since the dummy read colliding with the read operation in the same row leads to the lowest read current (read disturb write). In addition, we have to be careful for the word line skew sensitivity since DP-SRAM ports are often clocked independently (asynchronous access). Timing alignment of the disturb word line to the access word line in the same row access is critical to identify the worst-case timing and cell margins. The negative write and read boosting and the word line lowering have been employed to enhance the write and read capability in DP-SRAM [60]. A negative write bias on the BL is effective to improve the write capability as well as in a 6T cell. A negative voltage booster is combined with the local write buffer in the negative write boosting. When either of ports in DP-SRAM cell is activated, the write buffer pulls down the corresponding BL to 0V in the first half period of a write pulse then starts the booster to generate a negative pulse into the BL. The SNM in DP-SRAM is degraded when both ports turn on. However, the SNM can be improved by lowering the word line voltage as well as in 6T cell. In addition, pulling-down the cell ground level voltage V_{ss} to negative can enhance the cell current. One of the most significant issues in true multi-port SRAM is the larger memory cell size. Then pseudo multi-port SRAMs which realize multi-port access of 1R1W, 1RW1R, and 2RW using 6T SP SRAM bit cell was proposed to achieve the smaller area and the lower standby power than typical multi-port SRAM's using 8T DP SRAM bit cell. An example of pseudo 2RW DP SRAM is described in Figure 17.30 [61]. Newly introduced circuits of the double pumping clock generator (DPCG), the data sequencer (DS), and the dual output sense amplifier (DOSA) enable 2RW operation within one clock cycle. DPCG generates two TDEC pulses to access two addresses

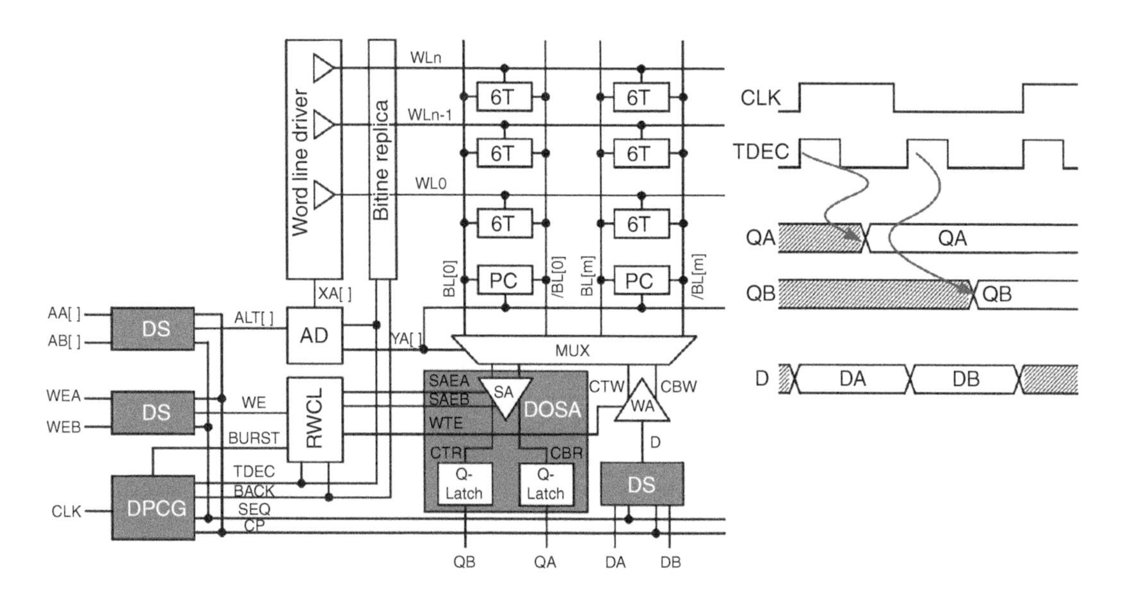

Figure 17.30 Block diagram and waveform of pseudo 2RW DP-SRAM.

in a single clock cycle. Then, DS transfers input data DA and DB to the write driver in order. DOSA has two independent Q-latches connected to the common read BLs which enable to read out data one after the other.

17.3.3 Future of SRAM

The importance of cache memory and register file will more and more increase toward the future because the gap of bandwidth between processor and main memory continues to increase. Therefore, much larger memory capacity will be required to cache memory and register file. The memory cell size of SRAM as a cache memory should be continuously scaled down toward the future to increase the memory capacity of cache memory. The transistor threshold variation in SRAM cell will become more serious in the future as a result of scaling down of the cell size. Therefore, more innovative technologies to reduce the cell size and power consumption and to enhance the performance and the operation stability should be proposed taking into account the tradeoffs among the 6T cell, 8T cell, 10T cell, and new memory cells with more cell transistors to break the scaling limit of memory cell. The device scaling is indispensable to reduce the cell size and power consumption. The transistor structure has been changed from the traditional 2D transistor to the 3D transistor Fin-FET and further changing to new 3D transistors such as Nanosheet FET, Forksheet FET, and eventually CMOS transistor CFET. These 3D transistors may help to solve several issues concerned with the scaling of SRAM memory cell [62]. Especially, it can be expected that SRAM cell size will be significantly reduced by employing CFET [63].

17.4 Summary

Memory capacity and performance of DRAM and SRAM have dramatically increased for half a century by developing many innovative technologies. The evolution of memory cells has greatly contributed to such spectacular advances in DRAM and SRAM. Then, innovative DRAM and SRAM technologies developed in the past have been reviewed referring to the background in which new memory cells had been invented. DRAM and SRAM will still continue to evolve toward the future as important memories indispensable for a computer system. To continue such evolution in DRAM and SRAM, innovative technologies should be continuously developed for the future.

References

[1] Dennard, R.H. (1968). USP-3387286 (4 June 1968).
[2] Dennard, R.H., Gaensslen, F.H., Yu, H.N. et al. (1974). Design of ion implanted MOSFET's with very small physical dimensions. *IEEE J. Solid State Circuits* SC-9: 256–268.
[3] Deleonibus, S. (2019). Marvels of microelectronic technology: the 1T-1C dynamic random access memory, from a groundbreaking idea to a business benchmark. *IEEE Electron Devices Soc. Newsl.* 26: 3–9.
[4] Koyanagi, M. and Sato, K. (1976). Japanese patent application. *Tokuganshо* 51: 78967.
[5] Koyanagi, M. and Sato, K. (1977). USP-4151607 (5 January 1977).
[6] Koyanagi, M., Sunami, H., Hashimoto, N. and Ashikawa, M. (1978). Novel high density, staked capacitor MOS RAM. *Extended Abstracts of the 10th Conference on Solid State Devices*, Tokyo (29–30 August 1978), pp. 348–351.
[7] Koyanagi, M., Sakai, Y., Ishihara, M. et al. (1980). 16-kbit Stacked Capacitor MOS RAM. IEICE Technical Report (Japanese), SSD80-30.
[8] Koyanagi, M. (2008). The stacked capacitor DRAM cell and three-dimensional memory. *IEEE SSCS News* 13: 37–41.
[9] Nishizawa, J., Koyanagi, M., and Kimura, M. (1974). Impedance measurements to study semiconductor surface and thin insulating films on its surface. *Japanese J. Appl. Phys.* Suppl. 2: 773–779.

[10] Koyanagi, M., Sunami, H., Hashimoto, N., and Ashikawa, M. (1978). Novel high density, staked capacitor MOS RAM. *IEDM Tech. Dig.*, Washington D. C. (4–6 December 1978), pp. 348–351.

[11] Koyanagi, M., Sakai, Y., Ishihara, M. et al. (1980). A 5-V only 16-kbit stacked capacitor MOS RAM. *IEEE Trans. Electron Devices* ED-8: 1596–1601. *IEEE J. Solid-State Circ.* SC-15, 661–666, 1980.

[12] Kimura, S., Kawamoto, Y., Kure, T. et al. (1988). A new stacked capacitor DRAM cell characterized by a storage capacitor on a bit-line structure. *IEDM Tech. Dig.*, San Francisco (11–14 December 1988), pp. 596–599.

[13] Takemae, Y., Ema, T., Nakano, M. et al. (1985). A 1Mb DRAM with 3-dimensional stacked capacitor cells. *ISSCC Dig. Tech. Papers*, New York (13–15 February 1985), pp. 250–251.

[14] Kimura, K., Shimohigashi, K., Etoh, J. et al. (1987). A 65-ns 4-Mbit CMOS DRAM with a twisted driveline sense amplifier. *IEEE J. Solid State Circuits* SC-22: 651–656.

[15] Ema, T., Kawanago, S., Nishi, T. et al. (1988). *3-dimensional stacked capacitor cell for 16M and 64M DRAMs. IEDM Tech. Dig.*, San Francisco (11–14 December 1988), pp. 592–595.

[16] Wakamiya, W., Tanaka, Y., Kimura, H. et al. (1989). Novel stacked capacitor cell for 64Mb DRAM. *Symp. VLSI Technol. Dig. Tech. Paper*, Kyoto (22–25 May 1989), pp. 69–70.

[17] Watanabe, H., Tatsumi, T., Ohnishi, S. et al. (1992). A new cylindrical capacitor using hemispherical grained Si (HSG-Si) for 256Mb DRAMs. *IEDM Tech. Dig.*, San Francisco (13–16 December 1992), pp. 259–262.

[18] Sunami, H. and Nishimatsu, S. (1975). Japanese patent application. *Tokugansho* 50: 53883.

[19] Sunami, H. (2008). The role of the trench capacitor in DRAM innovation. *IEEE SSCS News* 13: 42–44.

[20] Sunami, H., Kure, T., Hashimoto, N. et al. (1982). A corrugated for capacitor megabit dynamic mos memories cell (CCC). *Dig. Tech. Abstracts IEDM*, San Francisco (13–15 December 1982), pp. 806–808.

[21] Sunami, H., Kure, T., Hashimoto, N. et al. (1984). A corrugated capacitor cell (CCC). *IEEE Trans. Electron Devices* ED-31 (6): 746–753.

[22] Itoh, K., Hori, R., Etoh, J. et al. (1984). An experimental 1 Mbit DRAM with on-chip voltage limiter. *Technical Digest of IEEE Intern. Solid-State Circuits Conf.*, San Francisco (22–24 February 1984), pp. 282–283.

[23] Gruening, U., Radens, C.J., Mandelman, J.A. et al. (1999). A novel trench DRAM cell with a VERtIcal access transistor and BuriEd strap (VERI BEST) for 4Gb/l6Gb. *Dig. Tech. Abstr. IEDM*, Washington DC (5–8 December 1999), pp. 25–28.

[24] Fredeman, G., Plass, D., Mathews, A. et al. (2015). A 14nm 1.1Mb embedded DRAM macro with 1ns access. *ISSCC Dig. Tech. Pap.*, San Francisco (22–25 February 2015), pp. 316–317.

[25] Stein, K.U., Sihling, A., and Doering, E. (1972). Storage array and sense/refresh circuit for single-transistor memory cells. *ISSCC Dig. Tech. Pap.*, Philadelphia (16–18 February 1972), pp. 56–57.

[26] Itoh, K. (1975). USP-4044340 (29 December 1975).

[27] Ito, K. (2008). The history of DRAM circuit designs-at the forefront of DRAM development. *IEEE SSCS News* 13: 27–31.

[28] Ahlquist, C.N., Breivogel, J.R., Koo, J.T. et al. (1976). A 16K dynamic RAM. *ISSCC Dig. Tech. Pap.*, Philadelphia (18–20 February 1976), pp. 128–129.

[29] Itoh, K. (2001). *VLSI Memory Chip Design*. NY: Springer-Verlag.

[30] Jones, F., Prince, B., Norwood, R. et al. (1992). A new era of fast dynamic RAMs. *IEEE Spectr.* 29: 43–49.

[31] Kumanoya, M., Ogawa, T., and Inoue, K. (1995). Advances in DRAM interfaces. *IEEE Micro* 15: 30–36.

[32] Oshima, Y., Sheu, B.J., and Jen, S.H. (1997). High-speed memory architecture for multimedia applications. *IEEE Circ. Devices* 13 (1): 8–13.

[33] Watanabe, Y., Wong, H., Kirihata, T. et al. (1996). A 286mm2 256Mb DRAM with x32 both-ends DQ. *IEEE J. Solid State Circ.* SC-31 (4): 567–574.

[34] Takai, Y., Nagase, M., Kitamura, M. et al. (1994). 250 Mbyte/s synchronous DRAM using a 3-stage-pipelined architecture. *IEEE J. Solid State Circ.* SC-29 (4): 426–431.

[35] Yoon, H., Cha, G.-W., Yoo, C. et al. (1999). A 2.5-V, 333-Mb/s/pin, 1-Gbit, double-data-rate synchronous DRAM. *IEEE J. Solid State Circ.* SC-34 (11): 1589–1599.

[36] Fujisawa, H., Nakamura, M., Takai, Y. et al. (2005). 1.8-V 800-Mb/s/pin DDR2 and 2.5-V 400-Mb/s/pin DDR1 compatibly designed 1-Gb SDRAM with dual-clock input-latch scheme and hybrid multi-oxide output buffer. *IEEE J. Solid State Circ.* SC-40 (4): 862–869.

[37] Park, C., Chung, H., Lee, Y.-S. et al. (2006). A 512-Mb DDR3 SDRAM prototype with CIO minimization and self-calibration techniques. *IEEE J. Solid State Circ.* SC-41 (4): 831–838.

[38] Lee, K.-W., Cho, J.-H., Choi, B.-J. et al. (2007). A 1.5-V 3.2 Gb/s/pin graphic DDR4 SDRAM with dual-clock system, four-phase input strobing, and low-jitter fully analog DLL. *IEEE J. Solid State Circ.* SC-42 (11): 2369–2377.
[39] Woo, D. (2018). DRAM challenging history and future. *IEDM Short Course* 2: 1–55.
[40] Spessot, A. and Oh, H. (2020). 1T-1C dynamic random access memory status, challenges, and prospects. *IEEE Trans. Electron Devices* 67 (4): 1382–1393.
[41] Koyanagi, M. (1989). Roadblocks in achieving three-dimensional LSI. *8th Symposium on Future Electron Devices Tech. Dig.*, Tokyo (30–31 October 1989), pp. 55–60.
[42] Koyanagi, M., Nakamura, T., Yamada, Y. et al. (2006). Three-dimensional integration technology based on wafer bonding with vertical buried interconnections. *IEEE Trans. Electron Devices* 53: 2799–2808.
[43] Fukushima, T., Yamada, Y., Kikuchi, H., and Koyanagi, M. (2005). New three-dimensional integration technology using self-assembly technique. *IEDM Tech. Dig.*, Washington, DC (5–7 December 2005), pp. 359–362.
[44] Fukushima, T., Yamada, Y., Kikuchi, H. et al. (2007). New three-dimensional integration technology based on reconfigured wafer-on-wafer bonding technique. *IEDM Tech. Dig.*, Washington, DC (10–12 December 2007), pp. 985–988.
[45] Lee, K.W., Nakamura, T., Ono, T. et al. (2000). Three-dimensional shared memory fabricated using wafer stacking technology. *IEDM Tech. Dig.*, San Francisco (11–13 December 2000), pp. 165–168.
[46] Lee, D.U., Kim, K.W., Kim, K.W. et al. (2014). An exact measurement and repair circuit of TSV connections for 128GB/s high-bandwidth memory (HBM) stacked DRAM. *Symp. on VLSI Circuits Digest of Technical Papers*. http//doi/org/10.1109/VLSIC.2014.6858368.
[47] Sohn, K., Yun, W.-J., Oh, R. et al. (2017). A 1.2 V 20 nm 307 GB/s HBM DRAM with at-speed wafer-level IO test scheme and adaptive refresh considering temperature distribution. *IEEE J. Solid State Circuits* 52 (1): 250–260.
[48] Chun, K.C., Kim, Y.K., Ryu, Y. et al. (2021). A 16-GB 640-GB/s HBM2E DRAM with a data-bus window extension technique and a synergetic on-die ECC scheme. *IEEE J. Solid-State Circ.* 56 (1): 199–211.
[49] Chen, M.F., Tsai, C.H., Ku, T. et al. (2020). Low temperature SoIC bonding and stacking technology for 12-/16-Hi high bandwidth memory (HBM). *IEEE Trans. Electron Devices* 67 (12): 5343–5348.
[50] Masuhara, T., Minato, O., Sakai, Y. et al. (1980). 2Kx8b HCMOS static RAMs. *ISSCC Dig. Tech. Pap.*, San Francisco (13–15 February 1980), pp. 224–225.
[51] Sakai, Y., Hayashida, T., Hashimoto, N. et al. (1981). Advanced Hi-CMOS device technology. *IEDM Tech. Dig.*, Washington, DC (7–9 December 1981), pp. 165–168.
[52] Karl, E., Guo, Z., Ng, Y.-G. et al. (2012). The impact of assist-circuit design for 22nm SRAM and beyond. *IEDM Tech. Dig.*, San Francisco (10–13 December 2012), pp. 561–564.
[53] Ohbayashi, S., Yabuuchi, M., Nii, K. et al. (2006). A 65 nm SoC embedded 6T-SRAM design for manufacturing with read and write cell stabilizing circuits. *Symposium on VLSI Technology Dig. Tech. Papers,* Honolulu (13–15 June 2006).
[54] Yabuuchi, M., Nii, K., Tsukamoto, Y. et al. (2007). A 45 nm low-standby-power embedded SRAM with improved immunity against process and temperature variations. *ISSCC Dig. Tech. Pap.*, San Francisco (13–15 February 2007), pp. 326–327.
[55] Pilo, H., Barwin, J., Braceras, G. et al. (2006). An SRAM design in 65nm and 45nm technology nodes featuring read and write-assist circuits to expand operating voltage. *Symp. on VLSI Circuits Digest of Technical Papers,* Kyoto (15–17 June 2006).
[56] Yamauchi, H. (2010). A discussion on SRAM circuit design trend in deeper nanometer-scale technologies. *IEEE Trans. Very Large Scale Integr. (VLSI) Syst.* 18 (5): 763–774.
[57] Yabuuchi, M., Nii, K., Tsukamoto, Y. et al. (2007). A 45nm low-standby-power embedded SRAM with improved immunity against process and temperature variations, *ISSCC Dig. Tech. Pap.*, San Francisco (13–15 February 2007), pp. 326–327.
[58] Chang, I.J., Kim, J.-J., Park, S.P. et al. (2008). A 32kb 10T subthreshold SRAM array with bit-interleaving and differential read scheme in 90nm CMOS. *ISSCC Dig. Tech. Pap.*, San Francisco (3–7 February 2008), pp. 388–389.
[59] Sinangil, M.E., Verma, N., Chandrakasan, A.P. et al. (2009). A 45nm 0.5V 8T column-interleaved SRAM with on-chip reference selection loop for sense-amplifier. *IEEE Asian Solid-State Circ. Conf.*, Taipei (16–18 November 2009), pp. 225–228.

[60] Wang, D.P., Liao, H., Yamaguchi, H. et al. (2007). A 45nm dual-port SRAM with write and read capability enhancement at low voltage. *IEEE Intern. SOC Conf.*, Hsinchu, Taiwan (26–29 September 2007). http://doi/org/10.1109/SOCC.2007.4545460.

[61] Ishii, Y., Yabuuchi, M., Sawada, Y. et al. (2016). A 5.92-Mb/mm2 28-nm pseudo 2-read/write dual-port SRAM using double pumping circuitry. *IEEE Asian Solid-State Circuits Conf.*, Toyama, Japan (7–9 November 2016), pp. 17–20.

[62] Gupta, M.K., Weckx, P., Schuddinck, P. et al. (2021). A comprehensive study of nanosheet and forksheet SRAM for beyond N5 node. *IEEE Trans. Electron Devices* 68 (8): 3819–3825.

[63] Gupta, M.K., Chehab, B., Cosemans, S. et al. (2021). The complementary FET (CFET) 6T-SRAM. *IEEE Trans. Electron Devices* 68 (12): 6106–6111.

Chapter 18

Silicon-Germanium Heterojunction Bipolar Transistors

A Retrospective

Subramanian S. Iyer[1] and John D. Cressler[2]

[1] Department of Electrical and Computer Engineering, UCLA, Los Angeles, CA, USA
[2] School of Electrical and Computer Engineering, Georgia Tech, Atlanta, GA, USA

"If it can be done in silicon, it will be done in silicon."
—Woodall's rule

"If it can't be done in silicon, do it in silicon-germanium."
—Iyer's first corollary

"Why bother with silicon? Silicon + germanium is twice the fun!"
—Cressler's second corollary

18.1 Introduction (JDC)[1]

Happy, happy 75th birthday to the transistor! This remarkable little piece of quantum physics in action irrevocably altered the technological trajectory of planet Earth. In terms of its impact on civilization writ large, one could fairly argue that the invention of the transistor is the most important discovery in human history. In only 75 years, we have built over 10^{24} transistors, within 10 times of the total number of stars in the universe, all at a cost of nano-$ per device, despite being the most formidable manufacturing problem ever successfully solved. An exercise in the remarkable properties of exponential growth that silicon affords (thank you, Moore's Law)!

[1] Subramanian S. Iyer and John D. Cressler are co-lead authors for this article. Authorship of the various sections are as follows: Sections 18.1 (JDC), 18.5 (JDC), 18.7 (JDC),18.2 (SSI), 18.3 (SSI), 18.4 (SSI), 18.9 (SSI) 18.6 (JDC + SSI), and 18.8 (JDC +SSI).

75th Anniversary of the Transistor, First Edition. Edited by Arokia Nathan, Samar K. Saha, and Ravi M. Todi.

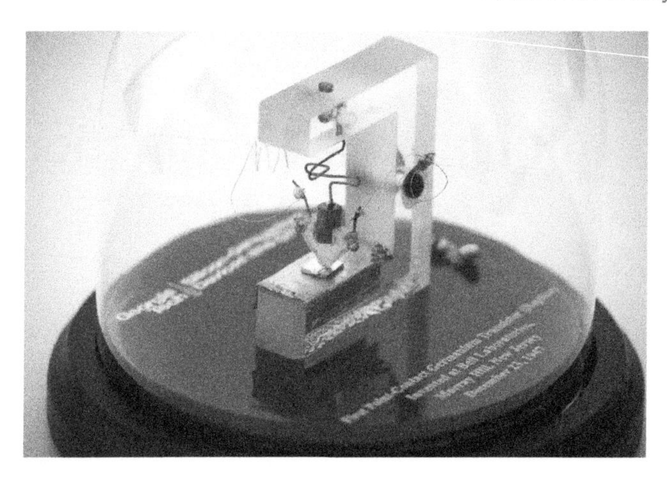

Figure 18.1 Scaled replica of the world's first transistor (a point-contact device). This model was handcrafted by Cressler's former student, now Dr. Milad Frounchi.

Ironically, both Bardeen and Brattain's first Point-Contact Transistor [1] (born on 16 December 1947, but not officially presented to Bell Labs management until 23 December 1947 – the now official date – see Figure 18.1) and Shockley's first Bipolar Junction Transistor [2] (BJT) (conceived on 23 January 1948, patent applied on 26 June 1948, patent granted on 25 September 1951) were made from germanium (Ge), not silicon (Si). Close cousins, even back then! The concept of the heterojunction bipolar transistor (HBT) dates to the original BJT patents filed by Shockley in 1948 [3]. Given that III–V semiconductors were not yet on the scene, it seems clear that Shockley envisioned the combination of Si (wider bandgap emitter) and Ge (narrower bandgap base) to form a SiGe HBT. Who knew? The basic formulation and operational theory of the HBT, for both the traditional wide bandgap emitter plus narrow bandgap base approach found in most modern III–V HBTs, as well as the drift-base (graded-base) approach used in SiGe HBTs today, was pioneered by Herb Kroemer, and was largely in place by 1957 [4–6]. It is noteworthy that Kroemer worked hard early on to realize a SiGe HBT, without success, ultimately pushing him toward the III–V material system for his heterostructure studies, a path that proved in the end to be quite fruitful, since he shared the Nobel Prize in physics in 2000 for his work in (III–V) bandgap engineering for electronic and photonic applications. While III–V HBTs (e.g. AlGaAs/GaAs) began appearing in the 1970s, driven largely by the needs for active microwave components in the defense industry, reducing the SiGe HBT to practical reality took 30 years after the basic theory was in place, due primarily to material growth limitations. More on this pesky problem in a moment.

The first functional SiGe HBT, grown by MBE, was demonstrated in December of 1987 by Subu Iyer and colleagues at IBM Thomas J. Watson Research Center (IBM Research) [7], 40 years, nearly to the day, after the first transistor. So happy, happy 35th birthday, SiGe HBT! Over the past 35 years, SiGe HBT technology has gone from lab curiosity to commercial reality (Figures 18.2–18.7). SiGe even has its own pet pronunciation (/sig-ee/), which can now even be found in the dictionary (joking!). From its introduction into commercial manufacturing in 1994[a] until today, 28 years later, frequency response has risen by over 10 times, a remarkable feat! SiGe HBTs can be found globally in performance-constrained analog, digital, RF, microwave, mm-wave, and near-THz circuits and systems, supporting a diverse set of applications. SiGe HBTs also possess a natural affinity for robust operation in so-called "extreme environments," which include exposure to intense space radiation and deep cryogenic temperatures, and they are now being used in integrated Si photonics (light + electrons on Si substrates). More on all these topics shortly.

[a] This is an example of something called "Chatterjee's 7-year rule" (after Pallab Chatterjee, then at TI) – it takes about 7 years after a material is introduced at IEDM for it to make it to manufacturing.

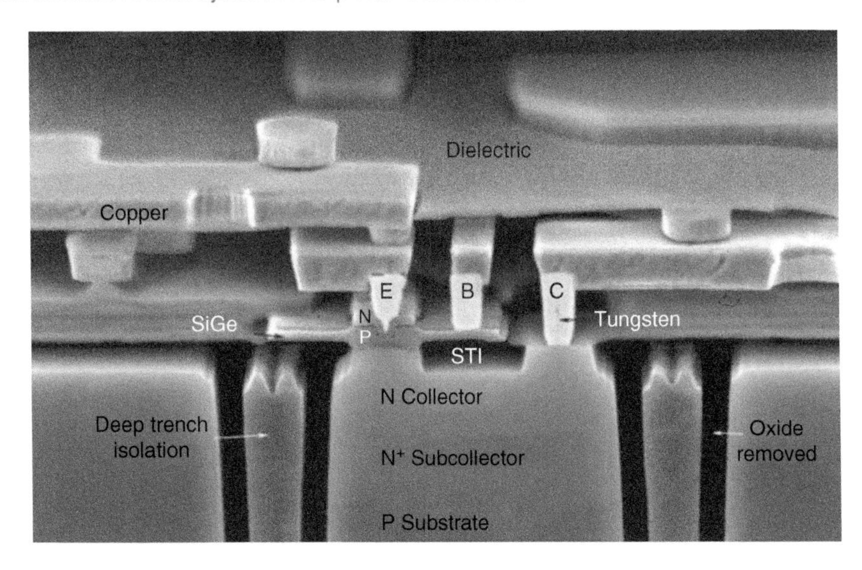

Figure 18.2 Decorated cross-sectional scanning electron micrograph of a SiGe HBT. The epitaxial SiGe alloy is shown in orange. Source: Courtesy of GlobalFoundries.

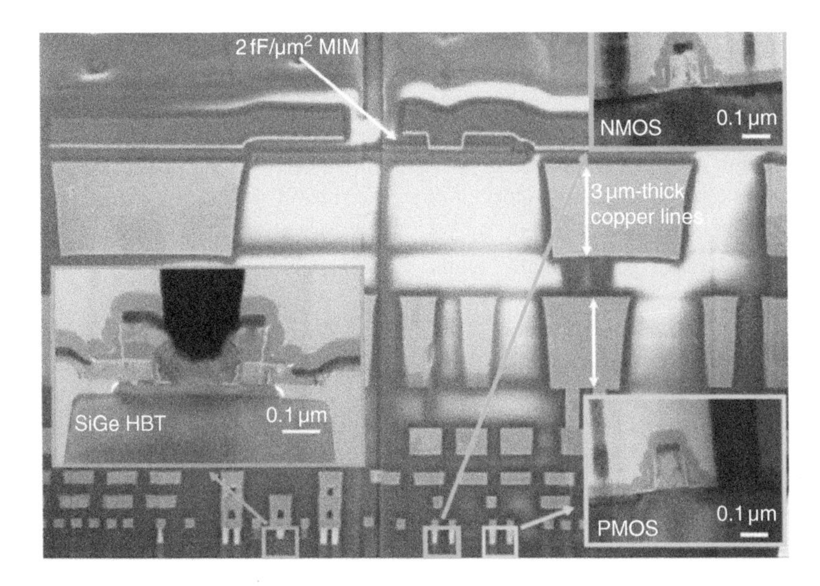

Figure 18.3 Cross-sectional scanning electron micrograph of a modern SiGe BiCMOS platform, showing SiGe HBT, CMOS devices (n-channel and p-channel), and the back-end-of-the-line metallization layers and passive elements. Source: Courtesy of P. Chevalier and ST Microelectronics.

In the present paper, we are going to do something a tad different than you might be expecting. We will spare readers the gory details of both theory and practice of the SiGe game ("look mom, no equations!") and instead focus on intuitive insights, notable anecdotes, and personal stories from these past 40 years. This history, though rich, is often little appreciated and sometimes even unknown except to a choice few. Stories matter. People matter. As Thomas Carlyle famously said, "History is the essence of innumerable biographies." We will keep our narrative style informal and fun (and use first person!), so hopefully it will be an enjoyable read! Each section will delineate which of us wrote it, rather than

Figure 18.4 The 125 mm Si MBE system that was used to build the first SiGe HBT and many other Si-based band engineered devices. This picture was taken on the manufacturing floor at VG. After the system was shipped, we did not have room in the lab to get a view like this.

try and force a consistent narrative voice upon the two of us. Needless to say, these stories we tell are from our own perspectives, which are inevitably centric to IBM Research (where we both worked), so for anyone with a alternate take, or who may recall things differently, well, forgive us, we mean well! For those readers that may yearn for a deep dive into the nuts and bolts of SiGe technology (trust us, it is an interesting subject!), there are plenty of places to turn [8–10]. A full suite of references can be found in these sources and hence for brevity we will only include key citations that align with our stories and themes. So please do not feel bad if we leave out your world-beating paper! Buckle up, here we go!

18.2 Some History from Early Days at IBM Research (SSI)

It was 1981, and I was a fresh PhD who had just completed my dissertation on how to dope Si using the then novel method of molecular beam epitaxy (MBE). I joined the IBM T.J. Watson Research Center right after my PhD and the very first thing my manager told me was that we would never do Si MBE at IBM! I did not say a word, but it haunted me. A lot. That was the time computer mainframes ruled the roost and IBM ruled the mainframe market. The IBM 360 Mainframes ran on bipolar ICs, ECL logic, and bipolar SRAMs. For their day, they ran blazingly fast, and a very large part of IBM focused on how to make them faster and faster. My senior colleague, Denny Tang (John's first-line manager at the time), explained why we used bipolars: "These things are like trucks – they can drive any load – nothing can come close." My colleague Hans Stork (John's second first-line manager) would often show his famous "bathtub" power delay curve (Figure 18.6), where the lowest power was limited by the base region delay – a point I still make to my students – and so clearly the easiest way to make the BJT more efficient and faster was to make the base thinner. But that was difficult to control as the base thickness was determined by the difference of two diffusions. How easy it would be to do this with MBE I thought – an epitaxially grown base with ultimate control.

Another interesting feature of the predominantly *npn* Si BJT world of those days was the use of a polysilicon emitter. The n+ doped polysilicon emitter was deposited on the ion implanted base and the interfacial oxide played a role in controlling the reverse injection of holes from the base into the emitter,

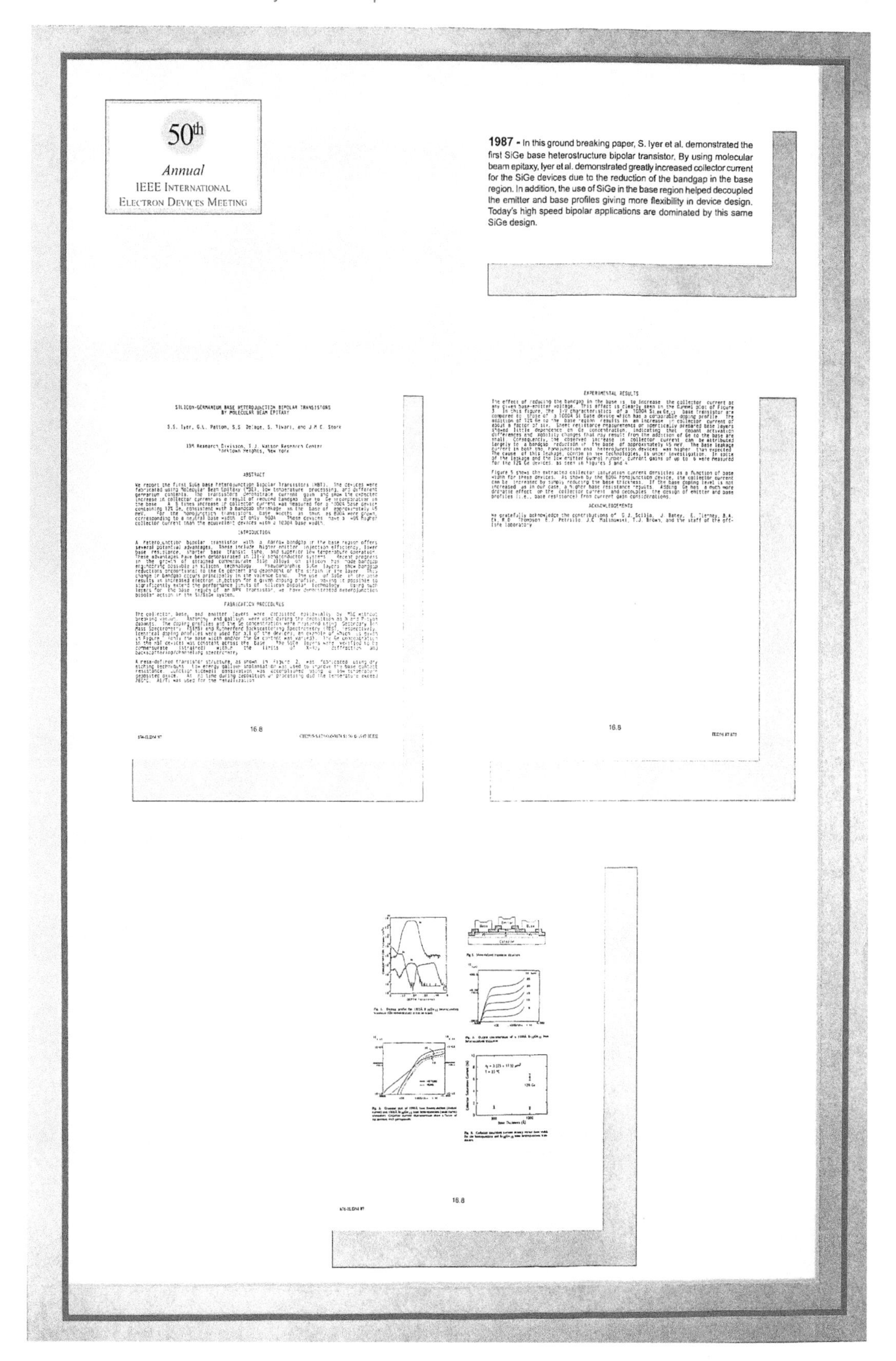

Figure 18.5 The IEDM poster on the first SiGe HBT at the 50th anniversary of IEDM (2004).

Figure 18.6 Patton and Iyer posing in front of the 50th anniversary IEDM display (2004).

Figure 18.7 Some of the key players of the early SiGe team at IBM: (seated are Hans Stork and David Harame; second row, standing: Gary Patton and Jack Sun; third row, standing: Joachim Burghartz, Jim Comfort, Eduoard de Fresart, and Emmanuel Crabbé), ca. 1992.

thereby increasing the emitter injection efficiency. However, this also led to a poor junction quality and presumably lower injection of electrons, a problem addressed by diffusing the dopant from the highly doped n+ polysilicon into the base, thus both providing for a diffused high-quality junction and improved injection of electrons into the base. For this process to work properly, the interface between the polysilicon and implanted base had to be just right, something our senior manager and IBM Fellow, Tak Ning (John's second-line manager), technically referred to as "the right stuff," the "rightness" of which was a matter of great debate. This was another thing I thought epitaxy could address.

I was not working directly on Si BJTs at that time (BUT, John was!). I had decided to learn about things like electromigration at submicron dimensions (BJT circuits required high currents, so electromigration was always an issue) and develop a self-aligned silicide (salicide) process for CMOS, and

I worked in a team that focused on metallurgy for ICs – something I knew little about and where I thought I would learn something new. The atmosphere at IBM Research in those days collegial and almost academic in feel. Our meetings were free form and full of debates and everyone was enthusiastic about sharing what they were doing and teaching others about what they had learned – a far cry from today's ultra-secretive atmosphere that pervades some companies. I learned all these nuances about Si BJTs from these hallway and lunch time discussions.

One such lunch discussion was a turning point for me. I was wandering in the cafeteria with my lunch tray, and I saw the Director of Semiconductor Technology, John Armstrong, sitting by himself at a table. I asked him if I could join him, and he graciously welcomed me. I introduced myself and I knew he had recently returned from an assignment at East Fishkill, where he oversaw Si BJT technology development. I gave him my simplistic (and perhaps a tad naïve) view of how I could build a better bipolar transistor with an epitaxial process, and how MBE would be the ideal method to test such ideas. He was fully engaged and asked what I would need to try this out. "An MBE system," I replied (duh!). "How much would that cost?" I had no idea but told him it would be about two million dollars (always be bold!). He was quiet for a while, and I thought I had blown it! Perhaps I should have said one million?! But then he surprised me. "Look Subu," he said, "you have been out of school for just a couple of years, and you want two million bucks – that's a lot of money and I think you will need to write me a memo of what you plan to do and why we should start this project." Music to my ears! I walked straight to the office of my second line manager, Nunzio Lipari (a theoretical physicist by training), and gave him the good news. He was flabbergasted. After he recovered, he said that this was a great opportunity and I needed to really work on this memo. More importantly, he said he would help me! To cut a long and extremely interesting story short, I was given every resource I asked for that was needed to build the world's first SiGe HBT.

18.3 SiGe Epitaxy and Making the First SiGe Transistor (SSI)

Si and Ge are unquestionably the primordial semiconductors. Moreover, they like each other, forming an intimate substitutional alloy at any composition. As you mix more Ge into Si, the average lattice constant increases and the band structure changes, going from Si-like to Ge-like, and band structure theorists have had a field day predicting the nuances of this new material system. To make useful devices, one has to grow these alloys on (large) Si wafers and therein, unfortunately, lies the rub. Ge atoms are a lot larger than Si atoms and the Ge lattice constant is about 4.2% larger than that of Si. While 4.2% may not sound like a lot, if you think about it, after about 25 lattice spacings, the lattice has shifted by about one full lattice constant. Yikes! This happens within about 10 nm. Hence, if you try to grow Ge on Si, there would need to be an extra column of atoms (called a misfit dislocation) every so often which is needed to accommodate the strain of the mismatched lattice constant. This sounds bad, and it is, but it causes havoc to the crystalline structure (misfits lead to threading dislocations which in turn cause leakage in minority carrier devices – read: bad news), the electronic properties at the SiGe/Si interface. The alloy itself would be rendered unstrained and bulk-like. In short, no good.

It turns out, however, that if you can somehow prevent the misfit dislocations from being generated in the first place, the interface would remain "clean" and the SiGe alloy itself would become compressively strained. There is another important effect that also takes place. When we compress the lattice constant in the transverse plane, the lattice elongates in the vertical plane (think of this simplistically as squeezing a balloon in one direction – the balloon elongates the other direction, which is known as Vegard's law in materials science). The result of this elongation is something we call tetragonal distortion, and the film is referred to as "coherently strained."

The folks at Bell Labs, notably Roosevelt People with the help of John Bean and others, had studied this phenomenon in some detail and explained these rather difficult concepts quite lucidly. The most relevant result of that study was that if the film is coherently strained, the indirect bandgap is reduced significantly compared to an unstrained alloy (tetragonal distortion is a good thing!) and in fact, a tetragonally strained SiGe alloy with 60% Ge could potentially (there were some pesky error bars in the calculations!) even have a bandgap smaller than Ge! This was important because of something called the Matthews-Blakeslee (M&B) thermodynamic stability limit (applied first to III–V semiconductors). Matthews and Blakeslee were legendary IBMers whom I had never met, but I knew people at IBM, who knew other people at IBM, who had met these gentlemen. Small world. M&B had calculated the thermodynamical thickness limit where a film could be coherently strained and not form misfit dislocations. Unfortunately, this limit was quite low. Fortunately for us, however, thermodynamics in this context is not as important as kinetics, and even though a film beyond this stability limit is thermodynamically unstable, it remains free of misfit dislocations because these misfit dislocations were incredibly lazy and take a long while to form and if you grow things at low enough temperatures.

This was a pivotal observation. You could, in principle, build a strained SiGe epi layer on Si and make it thick enough to be useful (5–10 times the Matthews-Blakeslee limit). The "in principle" caveat is important. You would need to grow the film at low enough temperature so that misfits do not nucleate. But at low temperatures, atomic mobility is also low (which is why misfits do not occur) and so it is difficult to place atoms at their assigned lattice locations, leading to poor material quality.

This is where MBE comes in! Surprise! Elemental Si and Ge MBE delivers a pure atomic flux of these elements in an ultrahigh vacuum environment. It is important that the Si and Ge flux hit the surface at a much higher rate than any impurity gases (oxygen, organics, etc.). This is easy to do in an MBE chamber (as long as you do not have a leak!). At high temperatures (>1000 °C), impurities do not stick, atomic surface mobility is high, and films of excellent material quality can be grown. As you lower the temperature, surface mobility reduces and impurity sticking increases. Hence our conundrum: How do you find the right temperature at which you have adequate mobility to grow a high-quality film, but low enough impurity sticking to avoid material defects. All while at the same time making it kinetically unfavorable to generate misfit dislocations. The beauty of MBE is that you can control the fluxes of the Si and Ge independently, as well as choose the substrate temperature independent of those fluxes. The growth rate, alloy composition, and substrate temperature are independently controlled, making MBE an ideal technique to explore the growth conditions that would give us the optimal material quality. I was pretty convinced that I could zero in on these requisite conditions within a few days after I got my MBE system working.

Big problem. Building a custom MBE system for SiGe epi is not cake! In those days, you could not just go to the shop and buy a turn-key MBE and press "go." While there were commercial III–V MBE systems available, Si MBE was a completely new animal and required a from-the-ground-up new design. We got the okay from IBM management to build this system in late 1982 and I immediately set to work. I selected a company in England called Vacuum Generators (VG for short) and began designing the system with them. I had not realized that when you order something at IBM that costs $2M, you cannot just go to any company and order it, you need to convince the purchasing department that this is a good investment. Imagine! Even after 40 years, I still have not figured this out why Purchasing departments have so much sway.

I had to invent specs for the system, the most important being that it was a 125 mm diameter wafer. Seriously? Seriously. In those days, IBM would march to their own tune as far as wafer sizes were concerned. Predictably, IBM Purchasing sent the specs out for competitive bid to three or four places. Only VG was bold enough to bid it! The other MBE houses most likely did not have a clue as to what I was asking them to build and how, and later on I found out that VG too was clueless! But they were game, and so was I.

The construction process required several trips to a place called East Grinstead, England, and I quickly got used to having a pint with my lunch. Hint – management never knew! The system itself was a dream come true. It had two growth chambers and each growth chamber had multiple electron beam sources, multiple sources for dopants, and additional chambers for surface analysis where we could look at the surface with electrons and X-rays *in-situ* to make sure we were growing high-quality material. I got my money's worth for the two million bucks! I told the folks at VG what I trying to do with this system and how it was going to change the way we made transistors. They were all in and went out of the way to do the best they could. At every visit, I could see the machine slowly take form. In late 1985, I was allocated space for the system in the lab, complete with new HEPA filters, clean hoods, and wet benches. I also was assigned a technician, Bruce Ek, easily the best technician I have ever had.

Finally, it was time to do the final factory inspection and accept the tool so that it could be shipped to IBM. I took Bruce along with me. I warned him on the way that this was the mother of all systems and he assured me that he had worked on big systems. As we walked into the factory floor, the machine was ready all pumped down and ready for the factory inspection. I turned to Bruce. His eyes were wide and his mouth gaping. Now he knew what big system was! "Subu," he said, "this thing is never going to fit in the lab. Me: "Don't worry, the system is only 21 ft. wide and will have at least 6 in. on either side. Plenty of room!" (See Figure 18.4.)

We began to put the system through a slew of mechanical tests to make sure things were working well. They were! That evening, while returning from dinner, we learned that there had been a terrible accident – the Space Shuttle Challenger had exploded in flight, killing the astronauts on board. Neither Bruce nor I said a word after we got that news. I recall that moment vividly even today.

Soon after the tool acceptance, two additional people joined our team. Sylvain Delage joined me as a post-doc to work on MBE. He had done his PhD in France at CNET building a version of a JFET that used an epitaxial cobalt silicide grid embedded in Si, and which was grown by MBE. Epitaxial silicides were a hot topic in those days. Sylvain played a key role in bringing up our system and developing processes for doping the semiconductor films with n-type (Sb) and p-type (Ga) dopants. I had studied doping in Si MBE for my PhD and found that dopants do not incorporate directly from the flux but rather from a surface reservoir that builds up on the growth surface. Controlling the amount of dopant in this reservoir is key to how the doping profiles are controlled. We got a chance to use my dissertation work in a state-of-the-art system, and wonder of wonders, we got the same result. I finally felt that I had indeed earned my PhD!

The other person was a young whippersnapper PhD from Stanford who had worked on bipolar transistors for his dissertation. Enter Gary Patton. In those days, new employees were left to pretty much fend for themselves and find a project that interested them. Sort of a free-agent system! It did not take Gary long to decide that building a SiGe-base HBT was what he wanted to do and he quickly assumed the role of integrator and device designer. We also dragged in my good friend Sandip Tiwari, who had built many III–V HBTs over the years, to help advise us on the nuances of HBT design and fabrication. Hans Stork (Gary's manager) and George Sai Halasz (my manager) ran interference when needed.

Around this time, there were many rumors that the folks at Bell Labs had produced a SiGe HBT, but it was very difficult for us to validate their results. Gary, Sandip, Hans, and I thought this through and agreed that we would need to definitively prove that we had indeed produced a SiGe HBT when the time came, and the best way to do that would be to create a rock-solid control (an MBE-grown Si BJT – same structure, but no Ge). In short, we decided to build devices with different Ge mole fractions and different base thicknesses. Everything else, the collector doping and the emitter doping and their thicknesses, would remain the same for all samples. I remember Gary coming in with the split table and nervously asking if this would be too much to do and how long to would take to develop the processes needed to build these different devices. I told him we would have them ready in a week. In fact, we built all those samples in less than two days. We would not have been able to do this with the amazing support of Gerry Scilla, our resident SIMS engineer. We would give Gerry a sample at 10.00 a.m. and he would give us the doping profile by lunch time. We would tweak the process and get the next cycle of feedback within an hour. We could park up to 20 wafers ready for growth in our system and could whip the samples out in no time. The processes, the operation of the shutters for the dopant fluxes, and the Si and Ge fluxes were all controlled by computer. The IBM PC had been developed by that time and I suspect we were the first MBE system with automated control of the dopant and Ge profiles. We could hit GO and watch the structures being grown without any further manual intervention. Many thanks to Rich Thompson, who implemented this control system.

Our SiGe HBTs borrowed from the AlGaAs/GaAs HBT structures that Sandip was very familiar with. We built a mesa isolated transistor with exposed junctions passivated by PECVD (plasma-enhanced chemical vapor deposition) oxides. We kept all the processing to within 550 °C to avoid any potential relaxation of the SiGe layers. The lowest neutral base width we managed to build was 50 nm, which I believe was a record in those days. After a couple of weeks, Gary announced that the device was ready to be measured. We all headed to Sandip's lab where he had a probe station already set up. We probed the control devices and they looked like transistors. Things were looking good! We then probed one of the SiGe HBTs and they too looked good. Importantly, to first order the collector currents in the SiGe devices were higher than in the controls and increased with Ge content. We all heaved a sigh of relief. We had a bona fide SiGe HBT!

It had been a long time since I had that eventful lunch with John Armstrong. That was November of 1982. The system was ordered in 1984 and we spent 1985 designing and building it. We received the system in mid-1986 and got it installed and running in early 1987. We had fabricated our first working transistors in fall of 1987. So many events and key people had to line up precisely for this to go according to plan. In my mind, I thought we would have this device in three years, in 1985, but it took us almost two more years to get there. Amazingly, even though there were many other very capable teams working toward the same goal (see discussion below), we kept our heads down and stuck to our plan.

In the next couple of weeks, we had characterized the devices in gory detail. There was no doubt that we had a SiGe HBT. We could extract the bandgap of our SiGe layers from the increase in the collector current and those numbers agreed well with optical bandgap measurements made by others. We had also grown the first epitaxial narrow base Si BJT. All we needed now was to get the word out. We had about three days to get meet the late news deadline for IEDM in 1987, and we agreed that we would shoot for that top-tier venue. We wrote up the paper. It was short, about one and a half pages of loosely spaced text and it had only four figures. The acceptance criteria for late-news' papers at IEDM were very stringent, but we had no doubt in our mind that it would be accepted. Gary presented the paper at IEDM in December of 1987, since I had a personal emergency and had to travel to India. The rest is history. At the 50th anniversary of IEDM in 2004, this late-news paper was selected as one of the fifty seminal papers presented at IEDM (Figure 18.5). I ran into Gary at IEDM at this very meeting. It had been a long time. He had left IBM and was running Quality Control at Hitachi (hard drives). Pleasantly surprised, we took this picture (Figure 18.6).

18.4 MBE vs. UHV/CVD vs. APCVD for SiGe epi (SSI)

No discussion of SiGe can be complete without the great (read: contentious) debate of those remarkable times. What was the best way to make a SiGe HBT? Key question. We built the first SiGe-base NpN HBT using MBE, but that was not all. Using MBE, we also built almost every other device imaginable (a SiGe-base PnP HBT, the first SiGe drift-base HBTs, SiGe MOSFETs, SiGe photodetectors, SiGe resonant tunneling diodes, and even an esoteric device called a BiCFET). MBE was the exploratory device engineer's tool of choice at the beginning.

Soon after the demonstration of the first SiGe HBT, I was appointed a first-line manager at IBM research and I called my department "Exploratory Si Bandgap-Engineered Devices," and we pretty much wiped the slate clean. If you could think of a bandgap-engineered device in the morning, we would have built it by the evening! A late evening perhaps, but the same day, nevertheless. The extreme doping and growth flexibility of MBE was a serious plus. However, MBE is a physical deposition method and some people thought it was too slow, although ALD (atomic layer deposition) is even slower and is a standard manufacturing process in CMOS in 2022. CVD (chemical vapor deposition), on the other hand, offered many advantages, not the least of which were conformality, and in some cases selectivity. In CVD, chemical precursors need to dissociate into elemental species on the growth surface and induce growth.

It was generally felt early on that the dissociation required a high temperature that would render the growth of practical strained layers impossible. Fortunately, this turned out to be not the case. Greg Higashi, a chemist at Bell Labs, showed that a simple dip of a silicon wafer in dilute HF rendered the surface hydrogen passivated. Importantly, it stayed that way for a long time, even at moderate temperatures. Our IBM Research colleague, Bernie Meyerson, leveraged this phenomenon by loading these hydrogen-terminated wafers into a UHV tube and was able to keep them passivated by supplying hydrogen radicals that would quickly replace any hydrogen termination that was lost. Importantly, this could be done at relatively low temperatures of 450–600 °C. At these temperatures, precursor gases such as silane and germane can disassociate less frequently, but rapidly enough to produce growth, and the products of dissociation, which include hydrogen radicals, can keep the surface passivated during growth. By conducting these reactions at low pressures, where the mean free path of the species was comparable to the spacings between wafers, UHV/CVD was able to yield uniform growth across the wafer, in a configuration where many wafers could be processed together (in batch).

This turned out to be the method of choice employed at IBM (now GlobalFoundries) as SiGe moved toward manufacturing (refer to references in [9]). Another CVD method was pioneered by the late Tom Sedgwick in my group and was called Atmospheric Pressure CVD (AP/CVD). The basic principle is the same as for UHV/CVD, but instead of relying on an ultrahigh vacuum base pressure, he flooded the system with ultrapure hydrogen and nitrogen. Growth proceeded very much the same way as in UHV/CVD, but at an ambient pressure that was 760 Torr. Joachim Burghartz, Detlev Grutzmacher (then a post-doc in my group working on APCVD, but who now does MBE at University of Aachen), and Tom Sedgewick demonstrated excellent self-aligned SiGe HBTs using APCVD. Different groups developed other variants of CVD techniques, such as Rapid Thermal CVD (RT/CVD) at Stanford University. Many companies use APCVD today to manufacture their SiGe HBTs. Not only that, we explored the entire Group IV quaternary system (C, Si, Ge Sn) using MBE, making Si-based strain and band engineering a fact of life. After all these years, I have concluded that a properly and carefully designed CVD epi growth process is more appropriate for scaling-up to manufacturing once you have demonstrated the working principles and key design parameters of the device in question using MBE. Yes, it all starts with MBE!

18.5 Putting Physics to Work – The Properties of SiGe HBTs (JDC)

Si BJTs were king of the hill in the mid-1980s (sigh … those were the days). It was during the reign of the Si BJT when I began my career at IBM Research (1984), making better/faster/stronger Si BJTs for room-sized, world-beating mainframe computers that cost LOTS of money and burn LOTS of power to get those Giga-FLOPS. Alas, by the mid-1980s, speed/power/integration/cost constraints were beginning to create major growing pains for the purveyors of these monster systems. The net: by the mid-1980s the world of large-scale computers really, really needed faster Si BJTs that did **not** require using more power to reach those faster speeds. And importantly, this faster Si BJT needed to remain compatible with the existing massive fabrication infrastructure associated with Si manufacturing and deployment.

Enter IBM Research. The Si BJT teams (at the time we were divided into two small groups, one device oriented and one technology oriented) worked tirelessly to answer to this narrowing constraint bottleneck and craft a vision of the future for Si BJTs. We asked the following open-ended question: What is possible in power-delay (switching energy) performance a generation or two out in the world of Si BJTs? We set to work, reducing thermal cycles and honing profile designs, and exploring new device structures (then dominated globally by implanted double-poly, self-aligned structures with shallow implanted base and polysilicon emitter). We developed complementary (*npn*+*pnp*) Si BJT platforms to enable new faster, lower power emitter coupled logic (ECL)-based circuits and architectures. Fun times. To be sure, impressive results across the board were achieved for these newer "future" Si BJTs (e.g. 50 GHz f_T with sub-10 ps ECL gates delays).

Along the way, I grew interested in operation of these newer flavors of Si BJTs at cryogenic temperatures (low-temperature computers had been long-pursued at IBM Research, all the way back the Josephson Junction cryo-computer days in the 1970s), a realm classically forbidden to Si BJTs due to carrier freezeout effects (more on this a little later). Before I had my very own Blanz Cryogenic Probe Station, the legendary Paul Solomon kindly let me use his little primitive cryo-prober down in enemy territory (III–V land!). Having a pet side interest was almost a mandate in those days at IBM Research, and cryo-T operation of transistors has continued to serve me well over the years as quantum systems have come into vogue!

Progress was made, but still, at least in retrospect, graded-base SiGe HBTs were the more elegant solution, though at this stage of the game the technology was still miles away from being mainframe computer ready. That said, as I stop and think about it now, there are several good reasons, some obvious, some not, why hybridizing the conventional Si BJT with the addition of a carefully placed SiGe epi layer, was a logical step in the quest for the desired faster/lower power Si BJTs. As an aside, Si CMOS, below 90 nm nodes or so, also uses SiGe for optimizing pFETs, but they were late to the party! And the design of III–V HBTs, at least at that time, was approached very differently (lightly doped, wide bandgap emitter on top of a heavily doped narrow bandgap base). These III–V HBTs most definitely did not look like Si BJTs with a nanoscale film of SiGe epi tucked within, which is what the SiGe HBT was to quickly become. Yes, as might be anticipated, at early conferences in the late 1980s, SiGe proponents were routinely subjected to extreme derision for creating "pseudo-HBTs." Read: Not good enough to be called a "real" HBT. Alas, we got the last laugh!

Consider this compelling argument for using SiGe as an alternate path for improving Si BJTs:

1. Thermodynamic film stability constraints mandate using only very, very thin SiGe films (e.g. 100–200 nm), with only modest peak Ge fraction (e.g. 20–25%). Integrated Ge content is what matters to film stability. Given the fact that the speed of Si BJTs is limited by the base width (to first order), and the base, by definition, has to be very thin, using thin SiGe layers in the base to engineer a better Si BJT fits like a glove. Check!
2. By III–V standards, the valence band offset of strained SiGe on Si is, sorry to say, kind of wimpy (150 meV per 10% Ge, give or take). To make up for this, one can (courtesy of Kroemer's theory) compositionally grade the Ge across the base to add an additional drift field to speed up electron transport (read: faster devices). Check!
3. Unlike in a Si BJT, the epi (MBE or otherwise) is grown at low temperatures, and the base doping can be (i) large and abrupt – a.k.a. thin and heavily doped, like a III–V HBT, and (ii) is electrically active as grown. Meaning, low base resistance (faster gate delay) and no nasty base-widening thermal anneal required. Check!
4. Importantly, using SiGe epi gives substantial leverage in design margins, enabling the device designer to hammer on particular metrics one cares about. Need more current gain? Easy: put more Ge at the emitter-base junction. Need higher f_T? Easy: use more Ge grading. Have performance to burn? Great, then use that extra performance to save power. Are there limits for working in a non-lattice-matched systems? Of course, there are. Stability boundaries must be respected, else device yield and long-term reliability will suffer. Still, this broadened parameter trade-space is pivotal in the grand scheme of things, particularly as technology scales to more aggressive nodes. Witness the exceptional growth in SiGe performance over the past 30 years. To be sure, this same game had been long enjoyed by the III–V HBT camp, but here the game is played in Si, in all that that implies. A key discriminator. Refer to the rules at the beginning of the paper. Check!

A compelling list, by any stretch. And so, the SiGe HBT was born. Psst! There were competitors vying for the same prize at the same time as the IBM Research crew (Shhh, do not tell Subu!). The game was afoot! At least four independent research groups were simultaneously racing neck-and-neck to demonstrate the first functional SiGe HBT, all using the MBE growth technique (the detailed history is given in [9]): the IBM team [7], a Japanese team [11], a Bell Labs team [12], and a Linköping University team [13]. The IBM team is fairly credited with the victory since they presented (and published) their results in early December 1987 at the IEDM. Even for the published journal articles that followed, the IBM team [14] was the first to submit their paper for review (on 17 November 1987). All four papers appeared in print in the spring of 1988. Predictably, other groups soon followed with more SiGe HBT demonstrations. I was in the audience at the IEDM watching the landmark presentation, though admittedly, still slightly skeptical of the sort-of leaky DC characteristics of the first SiGe HBT I was being shown.

An interesting anecdote. IBM Research regularly held their famous research seminar, a prestigious venue for making a name for yourself in a sea of ridiculously bright and accomplished people. Seminal new results were shown, primarily as a way to seed the ideas and aspirations of others. Being a Si BJT devotee at that point, I recall attending, with members of the Si BJT club, the talk on this new SiGe HBT. We listened, deeply skeptical. My boss, the legendary Denny Tang, architect (with Tak Ning) of bipolar scaling theory, leaned over to me at the time and said, "That device is going to have serious breakdown issues. Won't be reliable." Me: "Why?" Denny: "You can't put a narrow bandgap in the collector-base (CB) junction without degrading BV_{CBO}." I nodded appreciatively. Well of course you cannot! As I recall he asked the presenter (Gary Patton) that question. No satisfying answer. Well, turns out, actually you can! The "dead-space" effect, discovered by Sandeep Tiwari and other III–V HBT researchers, ensures that the maximum carrier energy for initiating breakdown does not occur in the SiGe layer (inevitably in the center of the CB space charge region), but much deeper in the collector, where life is pure Si. A useful perk of high-field transport in semiconductors.

The race was on at this point, both within IBM Research and around the world. History unspooled quickly, making for fun times. As outlined above, different sub-teams with IBM explored different epi growth techniques, invented new structures in the quest to build not just a SiGe HBT, but something scalable and manufacturable (no small feat), and developed benchmarking circuits (that would be me!). The watershed moment happened two years later (1990), with the publication of remarkable AC results for a UHV/CVD SiGe HBT in a non-self-aligned structure (not especially good for high-speed digital due to excess parasitics) [15, 16]. Result: 75 GHz peak f_T, essentially **double** the best available Si BJT f_T at the time! Trust me, an overnight doubling of speed will get the world's attention! It did. Later that same year, the world's first self-aligned device debuted. My name was in that author list [17]. I was now officially aboard the SiGe train! Now 32 years and counting!

In short order, management of IBM Research abruptly shifted organizational gears, to essentially drop on-going Si BJT research and revector the people and dollars to develop SiGe HBT as a viable technology for … wait-for-it … . digital circuits. Wait, what?! Yup. Digital. You know, ECL. Mainframe computers. Was the abrupt change of direction painful for some of the folks involved? Indeed. Fortunately, politics was above my pay grade, so I did not witness much of the in-fighting, arm-twisting, or angst that surely went on, but in the end, exciting times commenced as IBM swiftly ran with SiGe in 100 different directions. Those few years were great fun, a highlight of my career. Always this rush of being the first to do something and claim it with a stake in the ground. Intoxicating.

So … what, then, does one get by working like a devil to put SiGe inside a fairly conventional looking Si BJT? Turns out, quite a lot! Some of these attributes were obvious at the get-go (high gain, high f_T), though most were nonobvious (low noise, high output resistance, wonderful cryogenic properties, and inherent radiation immunity). Features of a well-made SiGe HBT include:

- All of the numerous advantages a bipolar transistor enjoys (courtesy of an exponential I–V) over an FET (quadratic I–V), Si or otherwise. Sorry, could not resist!
- As much current gain as you want (bigger is not always better, from a breakdown voltage perspective), but importantly, gain is decoupled from the base profile design, so you get high gain with low base resistance, a wonderful combination for almost any circuit you can think of, and a major reason SiGe HBTs and analog circuits are so often put in the same sentence today.

- High transconductance per unit area (high g_m in tiny devices, a consequence of those exponentials). Lovely feature from a circuits perspective.
- High f_T and f_{max} and hence fast digital switching. Good for RF power gain at high frequencies too!
- High output resistance (Early voltage – V_A), and the V_A is decoupled via Ge from the base profile design, something prized by analog designers especially (a key analog metric is the βV_A product). A highly useful perk.
- Record 1/*f* noise corner frequency (the lowest out there, in fact – can be as low as a few 100 Hz at useful bias, typically 100 times smaller than CMOS). Not so important for digital, but pivotal for RF.
- Low broadband noise (aided by high f_T, high current gain, and low base resistance). Also good for RF. AND, the minimum noise occurs in the SiGe HBT about 10 times lower in current density than the peak f_T. Read: significant power savings!
- Surprisingly good distortion properties (especially for an exponentially nonlinear device!). Nice to have for RF circuits.
- Aside from the epi growth tool, the SiGe HBT can fabricated using existing Si fabrication techniques and leverage the manufacturing economy-of-scale of Si (large wafers, low cost, reliable, easy doping, good oxides, robust BEOL (back-end of the line), fine-line lithography, good heat conduction, etc., etc.). For all intents and purposes, SiGe is Si. Refer to the rules at the beginning of the paper.

Sound like the perfect transistor? Well, I am admittedly biased, but still, mighty darn close to a dream come true in transistor land! As you might glean from the above list, however, there are a lot of benefits to SiGe that sound decidedly non-digital in nature. Surprise! Remember, the entire reason for IBM to be interested in the SiGe HBT was to build better ECL gates for mainframe computers. Period. More on this in a moment.

18.6 SiGe BiCMOS: Devices to Circuits to Systems (JDC and SSI)

After the first SiGe HBT was demonstrated in 1987, our IBM colleagues, including David Harame, Gary Patton, Emmanuel Crabbé, Jim Comfort, Jack Sun, Bernie Meyerson, and others (Figure 18.7), initially began finessing the device design and crafting the technology towards a scalable, manufacturable process technology, integrating the SiGe epi base with polysilicon emitters and deep trench isolation, things already found in mature Si BJTs.

Sadly, history can sometimes be a merciless beast. Not long after the self-aligned SiGe HBT was demonstrated (1990) and the world was all abuzz, a radical decision at IBM Corporate was made "in the room where it happens." Hint, I was not invited. Guess what? No more Si BJTs for mainframe computers! None. Zippo. Zilch. Kaput. Too much power, too much cost, too many scaling issues, shrinking profits, no future. But wait, were the corporate execs not paying attention?! Here comes SiGe HBTs to save the day! By this time, we had made self-aligned *pnp* SiGe HBTs [18], had cooled those puppies down to cryogenic temperatures [19], and were building useful circuits galore. No dice. Not needed.

IBM experienced a financially triggered existential crisis, truly a near-death encounter, and for the first time in its history laid off a large number of its ever-devoted, seriously bright employees. The Si BJT manufacturing lines in East Fishkill (EF) were abandoned. Subu recalls going into the then-empty EF 300 building fab and seeing wafers on probe stations with the probes down and the microscope lights still on. Folks had been called in suddenly and told leave immediately. Total chaos ensued.

Instead of using Si BJTs, IBM had decided to turn to CMOS for help (gasp!) with its current and future mainframe needs. A reminder that in 1990, CMOS was at 0.5 µm gate length (read: dinosaur slow compared to Si BJTs for digital circuits). What to do about the major loss in system computing throughput? Well, do the obvious – gang up and parallelize processors to make up for the inevitable cut in FLOPS. Easier packaging, less cooling, smaller boxes; in short, a path forward for IBM mainframes .... It was a path that did not include SiGe.

Yep. Hard to believe, but there it is. The veritable rug had been pulled out from under the SiGe camp just when it was beginning to rev its engines and roar. I recall well the day of the announcement. Doom and gloom. Despair. We all felt it. The halls at IBM Research grew quiet.

At the time, Subu was already working on developing SOI wafers, and that project too was slated for the chopping block. Fortunately, his manager at that time, Paul Horn, helped him find external funding to do a startup to manufacture SOI wafers, which incidentally, used a boron-doped SiGe etch stop layer and the SiGe was grown by APCVD.

Fortunately for me, this was right about the time I received my calling to become a professor (I was teaching in the evenings as an adjunct and Western Connecticut State University and fell in love with teaching). After some serious handwringing and late-night conversations with my wife, I decided to leave IBM for Auburn University (back close to home), though not before IBM stuck me in a newly formed CMOS device group for a few months. Talk about a thumb in the eye! I still blanch when I think about it. Psst! Do not tell my students! In the end, IBM even denied me the cushy golden parachute exit package they were offering volunteers ready to leave IBM Research. It seems they wanted me to stay after all, but NOT to work on SiGe. After more discernment, I left anyway, and have never once, over the 30+ years since, regretted that decision.

In the years following 1994, IBM made a bold transition to CMOS. At that time CMOS technology was much slower than bipolar technology but possessed a power burn that was significantly lower. It was also clear that there was plenty of room to scale CMOS (which continues to scale even to this day). But with this, there was no need for SiGe HBTs! Or was there?

In what was perhaps one of the most innovative pivots of technology repurposing in somewhat trying times, the realization that the SiGe HBT was far friendlier to analog and RF circuits than they were to digital circuits gave people an idea. But, sadly, it was not something IBM needed. The fact that SiGe survived at all during that tumultuous time is due in no small measure to a single individual, one short of stature, but animated beyond measure, with an infectious, bellowing laugh, now turned into a tireless advocate for all-things-SiGe: one David Harame. Smile David! David had already added CMOS to SiGe (SiGe BiCMOS [20]) for an ideal division of labor (use the SiGe HBT where best suited, and CMOS where best suited), so after corralling a cadre of folks and securing smirky approval from management, he set his mind on transferring the new SiGe BiCMOS technology to manufacturing (since bipolar fabs were now empty!). As a side note, adding CMOS to SiGe HBTs to form SiGe BiCMOS is not cake, given the larger thermal cycles required for CMOS and the fact that CMOS standard cell IP libraries must be preserved at all costs for backward compatibility with standalone CMOS. SiGe HBTs were venturing into the territory where III–V HBTs ruled but with an incredibly powerful ally – Si CMOS, with its incredibly high level of integration and lower power. You could not lose!

David moved to Burlington, VT and with his colleagues (my second PhD student, Alvin Joseph, soon quickly became David's right-hand man) came up with a viable (manufacturable) process flow. But that was not all that was needed. Analog and RF circuits require additional elements such as precision capacitors, inductors, and resistors that digital technologies do not possess, and they had to be developed and seamlessly integrated as well. Other additions soon followed, including Thru-Si-Via (TSV), which is required for better AC grounding and noise suppression in analog/RF. (Subu now works in advanced packaging and TSVs are the way of life there).

Along the way to manufacturing, David worked tirelessly to attract new business and customers for IBM, not in digital, but in analog and RF (for which he was still a novice – sorry David!). And with SiGe BiCMOS in hand, mixed-signal ICs were an obvious target (analog + digital + RF on a chip). Thus was birthed, in late 1994, IBM 5HP SiGe BiCMOS technology (0.5 μm lithography generation, high performance), the world's first truly manufacturable SiGe BiCMOS technology [21]. The peak f_T

was a measly 50 GHz! Nothing by today's standards, but it was first. Gary Patton built the business end of things and made this into a viable business that continues to be the backbone of GlobalFoundries dominant position in the RF and mixed-signal business.

By this time, JDC was a fledgling young professor at Auburn University and, of course, a diehard SiGe fan. It turned out to be the perfect combination: David and colleagues were spending countless hours getting 5HP SiGe up in production and yielding, and I was provided hardware to help understand, model, and optimize these devices, especially in the context of emerging analog/RF applications. My students and I did so happily! Well, of course we did! Key results began to fall in place. The first 1/f noise measurements (great story), impact of Ge grading on analog metrics, temperature dependence in general (my old love), but also a landslide of interesting cryogenic physics down at 4 K, C-SiGe (*npn+pnp*) for analog apps, a variety of analog and RF circuit demos, TCAD simulations galore for understanding and profile optimization, especially RF noise and linearity (another interesting story), radiation effects (more on this below). On and on it went (see, for example, [22–27]). A shout-out to Guofu Niu, back then my post-doc extraordinaire, now professor, who was instrumental in much of that foundational SiGe HBT research. We pulled our aggregate findings (and more!) together to write the first serious SiGe HBT textbook [8]. It has endured nicely. Physics is physics, right?! We think of it as the SiGe "bible," and is known affectionately by my own students as "the blue book."

It was an exciting time. And, surprise, the analog and RF powerhouse companies began noticing SiGe, were putting dollars into IBM to make it happen, and of course they needed R&D. In fact, my very first research contract as a professor was with Analog Devices, for $20k (tiny but mighty!). Thanks, Dennis Buss! A copy of the check sits in a frame in my office to this day. Not much money, but it was my first, and I was off to the races!

By the mid-1990s, impactful analog and RF circuits were coming out by the bucketful. The first LSI SiGe HBT circuit, a 1.2 GS/s 12-bit digital-to-analog converter – DAC, was demonstrated in December of 1993 [28], and things began to roll forward quickly in the world of analog and RF circuit demonstrations (see [9] and references within).

A robust scaling path for any viable new IC technology is vital. SiGe managed that feat quite seamlessly over the past 30 years. While first generation SiGe BiCMOS began as 0.5 μm/50 GHz, second generation came quickly (180 nm/100 GHz), then third generation (130 nm/200 GHz), fourth generation (90 nm/300 GHz), and finally this year (2022), fifth generation (45 nm/600 GHz) (see references in [9]). And, we are not close to the end of scaling, at least theoretically (SiGe intrinsic performance is well above a THz). In addition to epi growth and structural innovations along the way, the serendipitous discovery that carbon doping of SiGe in 1996 [29] strongly suppresses boron diffusion, has proved its weight in gold (an interesting story to be told there, though I will save it for a rainy day). With carbon doping, increasingly aggressive base profiles could be kept in place as thermal cycles continued to shrink with each generational turn of the crank.

From the get-go, the vision for SiGe was BiCMOS, and not simply optimized for one type of circuit application (say RF), but instead for all envisioned applications – analog, RF through mm-wave, and yes, even digital. A one-size fits all sort of approach, with CMOS brought along for the ride (facilitating a revitalization of sorts for older CMOS nodes). That initial vision from 1994 has endured to this day. The world depends upon leading edge SiGe technology (now available in open-access foundries globally) to explore and help define the ever-evolving land of electronic circuits.

A particular focus these days is on mm-wave (30–300 GHz) for wireless. It is my firm belief that SiGe is quite capable to working across the entire 30–300 GHz band for robust emerging applications, which are many and varied, perfect for SiGe! Particular sweet spots in frequency, due largely to wireless atmospheric transmission properties, are found in V-band (40–75 GHz), W-band (75–110 GHz), D-band (110–170 GHz), and G-band (110–300). My research team has active work in all of these bands, and more. An example of a V-band (60 GHz) down-looking, space-based remote sensing SiGe radiometer is shown in Figure 18.8. Needless to say, state-of-the-art performance in Si!

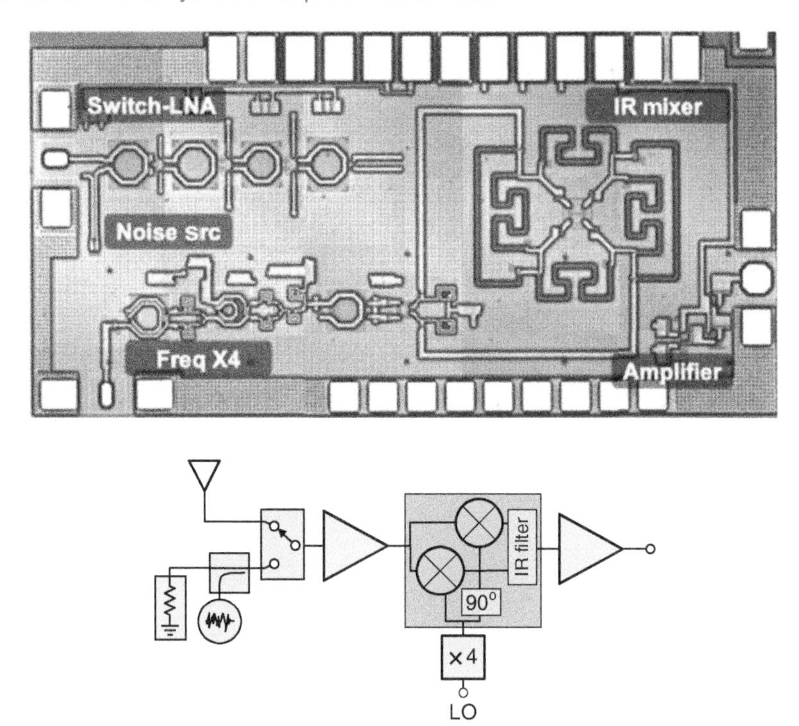

Figure 18.8 Example of a SiGe BiCMOS mm-wave integrated circuit built by Cressler's team, a V-band (60 GHz) SiGe radiometer for space-based remote sensing of the Earth from a CubeSat. The die size is less than 2 mm^2.

18.7 Using SiGe in Extreme Environments (JDC)

One of the most rewarding (and fun!) aspects of SiGe research over these subsequent decades has been the many opportunities afforded for breaking new ground, helping see new ways to doing old things, new ways that ultimately enable new applications which benefit all of us. I would like, finally, to talk briefly about using SiGe in "extreme environments" [30]. Extreme, in this case, means places where electronics has no business going! The rigors of outer space, which is bathed in nasty radiation, or down to extremely low temperatures (close to absolute zero, burr!), or even up to sweltering conditions (250 °C). Meaningful applications abound in all three contexts. It has long been my contention that SiGe is an ideal, perhaps ***the*** ideal solution, for all three scenarios (copious references for all three can be found in [9]).

The discovery that SiGe HBTs, as fabricated, can withstand exposure to wicked levels of ionizing radiation was serendipitous (like the transistor!). In 1994, Lew Cohn, then an influential project manager at the Defense Thread Reduction Agency (DTRA), visited me at Auburn as an external monitor on an Army contract that I had which was developing SiGe for RF applications. Go figure. At the end of the meeting, Lew was impressed. Afterward, he casually approached me and asked, "Have you ever irradiated these SiGe HBTs?" Hint: I knew nothing about radiation at the time, other than the existence of the van Allen belts! Me: "No, I have not." Lew: "You should. I can arrange some beam access if you are interested." Me: "I am. Let's do it." And so, we did.

Intrigued, I had my student, Jeff Babcock, take some 5HP SiGe HBTs to a gamma source and exposed them to beyond lethal doses of radiation (more than they would ever see in Earth orbit). Low and behold … nothing happened. Virtually no damage (Figures 18.9 and 18.10). Null result. And that

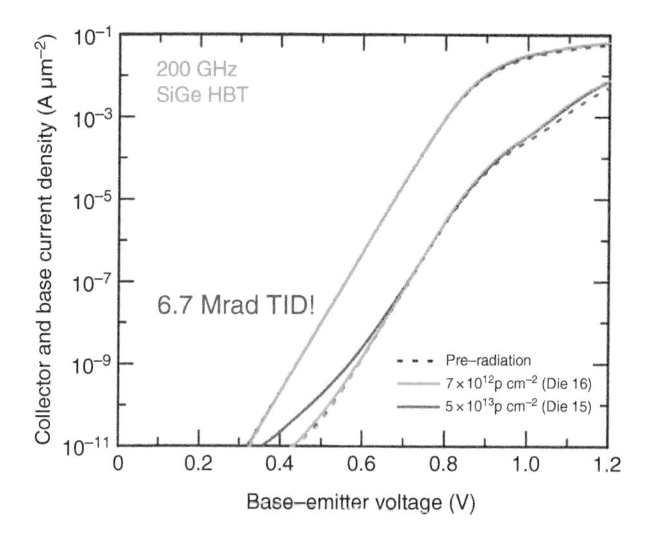

Figure 18.9 DC total ionizing dose (TID) radiation response of third-generation SiGe HBTs. Note that 5×10^{13} protons/cm^2 is equivalent to 6.7 Mrad of radiation. For reference, low Earth orbit (LEO) experiences about 100 krad over 10 years.

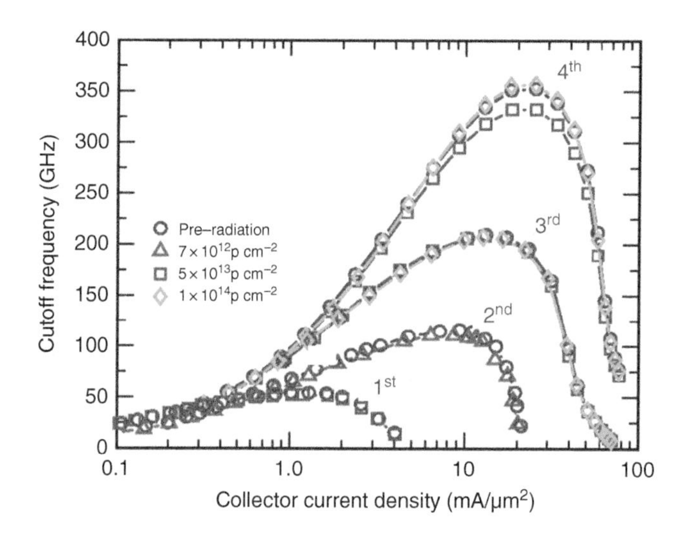

Figure 18.10 AC total ionizing dose (TID) radiation response of four generations of SiGe HBTs. Note that 5×10^{13} protons/cm^2 is equivalent to 6.7 Mrad of radiation. For reference, low Earth orbit (LEO) experiences about 100 krad over 10 years.

was the point! Take a SiGe HBT designed for terrestrial use and fly it space with no additional (expensive) total dose hardening required. Like magic! The holy grail! A pivotal result for the space community. Space electronics might sound niche, but alas, no, especially these days with commercial space amping up and a burgeoning space infrastructure rapidly evolving for everything from remote sensing in support of climate science, to communications, to eavesdropping. My team has aggressively pursued this field for the past 27 years and have published well over 200 papers on the topic (aimed at TID, ELDRS, SEU, SET, SEL, DD, in both SiGe devices and circuits, using gammas, X-rays, protons, electrons, neutrons, heavy ions, lasers, etc. – apologies for all the radiation acronyms – do not bother to look them up!) Essentially, everything you can think to do which intersects SiGe and radiation, we have done [31]! Suffice it to say that there is plenty of SiGe circuitry presently orbiting the Earth, and growing by the day. My team is working hard with NASA (National Aeronautics and Space Administration) to develop SiGe for the Moon (dark polar craters with water ice) and for Europa surface (liquid water ocean underneath an icecap)! SiGe is an enabling platform, as it turns out, since it can do both radiation and low temperatures at the same time so effortlessly. Go figure. It has been a fun ride.

I have alluded to my longstanding interest in opening bipolar transistors at very cold temperatures. This began with Si BJTs, but quickly morphed in earnest to SiGe HBTs, where, quite uncharacteristically, nature is actually on our side! That is, everything we do with SiGe device engineering to optimize it for analog/RF/digital terrestrial applications, is arrayed in such a way as to improve the device's properties, both DC and AC, as the temperature drops toward absolute zero! Remarkable, if you stop and think about it. Initially there was the legacy interest in operating large-scale computers at low temperatures to speed them up, but the physics turned out to be so interesting, and the performance so compelling, that cooling SiGe became pretty common, for applications as varied as precision instrumentation, to radio astronomy, to sensors of various kinds, and these days especially, to support quantum computing and quantum science as that field grows by leaps and bounds.

From a device engineering perspective, cooling SiGe HBTs is especially useful as a means to gauge the limits of device scaling. Meaning, cool a fourth generation device and you may get a glimpse at fifth or sixth generation performance. A look into the future, so to speak. The world's first half-THz (500 GHz f_{max} at 4.3 K for a 90 nm IBM SiGe HBT [32]) Si-based transistor was done this way, as was the world's fastest Si-based transistor (800 GHz f_{max} and 4.3 K for a 130 nm IHP SiGe HBT [33]), both measured in my lab at Georgia Tech (Figures 18.11 and 18.12).

Ever curious about the limits, we have even cooled SiGe HBTs down into the mK quantum regime, as a possible path to qubit interfaces, and well, just for the fun of it! Trust me, not an experiment for the faint of heart. Result? Advanced SiGe HBTs work just fine at mK temperatures, as surprising as that may sound [34]. To be sure, lots of nuanced physics gets brought to the table in the mK regime, and research continues in this arena.

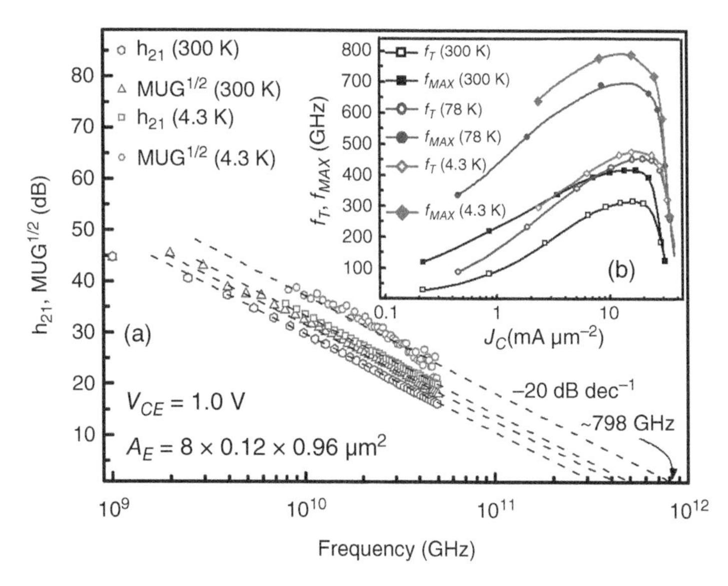

Figure 18.11 Frequency response of the world's fastest SiGe HBT, 800 GHz at 4.3 K. Source: Chakraborty et al. [33]/IEEE.

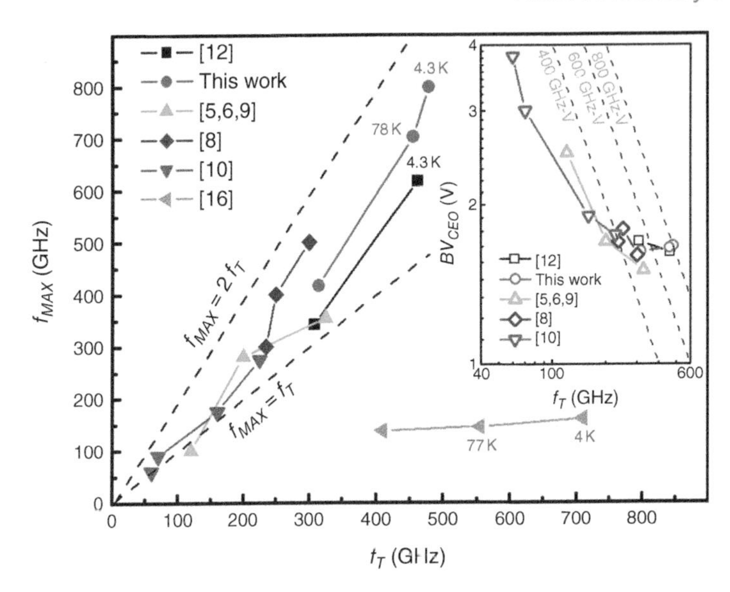

Figure 18.12 Global trends in f_T and f_{max} for SiGe HBTs. Source: Chakraborty et al. [33]/IEEE.

18.8 New Directions (JDC and SSI)

So where is SiGe technology headed? Great question! As mentioned above, practical manufacturing constraints aside, we are nowhere close to the ultimate speed limits of SiGe HBTs and will not be in the foreseeable future. The three leading commercial foundries (GlobalFoundries (GF), TowerJazz (TJ), and IHP (Germany)) all offer robust multi-hundred GHz SiGe BiCMOS platforms, and GF has recently announced a 45 nm, 600 GHz (at 300 K) SiGe HBT, combined with the 45 nm CMOS on SOI. Exciting stuff! Bring on the high-mm-wave apps! Dozens of companies across the globe have their own proprietary SiGe technologies. These types of ultrahigh-speed SiGe platforms offer circuit designers ample room to trade parameters off against each other (e.g. to trade f_{max} for much lower power), thereby broadening the design space dramatically. The future is very bright for SiGe BiCMOS, that much is obvious.

A new SiGe technology has come on the horizon in the past five years or so. Integrated Si photonics. The new-kid-on-the-block. In essence, integrated Si photonics brings light onto the Si wafer, letting light (photonics) and electrons (analog/RF electronic circuits) interact in clever ways, all the while being fabricated in a Si fab, using Si manufacturing techniques, in all that that means. Why call this a SiGe platform? Simple. While built in Si, you require a pure Ge photodetector, and you either have integrated SiGe HBTs (IHP) or CMOS (GF), whose pFETs use SiGe strain engineering. Call me pedantic, but that looks and smells like a SiGe platform to me! This vision of creating a SiGe "super-chip," which embodies photonics and electronics, dates back to Richard Soref of AFRL in the 1980s. Now, finally come to fruition. Lovely stuff. My team's research in this arena centers on how integrated Si/SiGe photonics behave in a radiation environment (No surpise!). We are already seeing lots of interesting effects [35–37] and are squarely focused in on the electronic–photonic interface (light circuits/electronic circuits) in microwave photonic components. We are also pushing on novel detectors for free-space optical systems and better photonic receivers [38]. Stay tuned on both.

We have come a long way from the original mesa-isolated SiGe HBT that we proposed back in 1982 and finally demonstrated in 1987. SiGe HBTs have evolved quickly over the years and so have the application needs. Frequencies are moving inevitably higher to support increasing data bandwidth, pushing into the near-THz regime (300–800 GHz). Thus far, SiGe has held its own against all comers, whether III–V (InP and GaN) or Si CMOS. That said, moving into the near-THz regime likely presents an inflexion point for business as usual. It is not that SiGe HBTs cannot operate in that regime. They can and they do, and we are likely to see THz levels of small-signal bandwidth soon enough. While SiGe receivers will likely be viable, efficiency will inevitably suffer at those frequencies for power generation (transmitters).

Fortunately, advanced packaging (Subu's current preoccupation) is coming to the rescue and will facilitate the intimate assembly of heterogenous chiplets (side-by-side or stacked or both), opening up new pathways for integrating Si, SiGe, and III–Vs, without compromising the unique benefits of any. For example, we could use SiGe HBTs for analog circuits, receivers, and for driving more efficient III–V power amplifiers, then bring deeply scaled CMOS/SOI or FinFETs to the table for microprocessors and memory. And heck, let's invite Si photonics to the party! In such a scenario, no longer would SiGe compete with III–V or Si, but rather all three would become synergistic, forming a team, all enabled by rapidly advancing heterogenous integration techniques. One thing remains clear: SiGe is here to stay, even in a post-Moore's Law era.

18.9 Some Parting Words (SSI)

It is worth noting that many of original IBM SiGe team have done really well for themselves: Bernie Meyerson, David Harame, Emmanuel Crabbé, and Subu Iyer were all appointed IBM Fellows. Gary Patton held several executive management positions at IBM, Hitachi, GlobalFoundries, and now Intel. Hans Stork held several executive management positions at TI, AMAT, and OnSemi, and Sylvain Delage heads the III–V effort at Thales in France. Detlev Grutzmacher, Joachim Burghartz, Sandip Tiwari, and John Cressler have had stellar academic careers in Germany and the United States. Subu finally made the transition to academia seven years ago following retirement from IBM after a very productive 30 years. And we all still keep in touch! It turns out that working on SiGe HBTs early in your career is not just fun, it is a sure ticket to success ... we think that every young engineer should try it! Thanks for reading. Happy 40th, SiGe HBT!

References

[1] Bardeen, J. and Brattain, W. (1947). The transistor, a semi-conductor triode. *Physical Review* 71: 230–231.

[2] Shockley, W., Sparks, M., and Teal, G. (1951). p-n junction transistors. *Physical Review* 83: 151–162.

[3] Shockley, W. (1950). *Semiconductor Amplifier.* US Patent 2,502,488 and 2,569,347.

[4] Kroemer, H. (1954). Zur theorie des diöusions und des drifttransistors, part III. *Arch. Elektr. Ubertragung* 8: 499–504.

[5] Kroemer, H. (1957). Quasielectric and quasimagnetic fields in nonuniform semiconductors. *RCA Review* 18: 332–342.

[6] Kroemer, H. (1957). Theory of a wide-gap emitter for transistors. *Proceedings of the IRE* 45: 1535–1537.

[7] Iyer, S.S., Patton, G.L., Delage, S.L. et al. (1987). Silicon-germanium base heterojunction bipolar transistors by molecular beam epitaxy. *Technical Digest of the IEEE International Electron Devices Meeting*, San Francisco (6–9 December 1987), pp. 874–876. IEEE.

[8] Cressler, J.D. and Niu, G. (2003). *Silicon-Germanium Heterojunction Bipolar Transistors.* Boston, MA: Artech House.

[9] Cressler, J.D. (ed.) (2006). *Silicon Heterostructure Handbook – Materials, Fabrication, Devices, Circuits, and Applications of SiGe and Si Strained-Layer Epitaxy.* Boca Raton, FL: CRC Press.

[10] Cressler, J.D. and Mantooth, A. (ed.) (2013). *Extreme Environment Electronics.* Boca Raton, FL: CRC Press.

[11] Tatsumi, T., Hirayama, H., and Aizaki, N. (1988). Si/$Ge_{0:3}Si_{0:7}$ heterojunction bipolar transistor made with Si molecular beam epitaxy. *Applied Physics Letters* 52: 895–897.

[12] Temkin, H., Bean, J.C., Antreasyan, A., and Leibenguth, R. (1988). Ge_xSi_{1-x} strained-layer heterostructure, bipolar transistors. *Applied Physics Letters* 52: 1089–1091.

[13] Xu, D.-X., Shen, G.-D., Willander, M. et al. (1988). n-Si/p-$Si_{1-x}Ge_x$/n-Si double-heterojunction bipolar transistors. *Applied Physics Letters* 52: 2239–2241.

[14] Patton, G.L., Iyer, S.S., Delage, S.L. et al. (1988). Silicon-germanium-base heterojunction bipolar transistors by molecular beam epitaxy. *IEEE Electron Device Letters* 9: 165–167.

[15] Patton, G.L., Comfort, J.H., Meyerson, B.S. et al. (1990). 63–75 GHz fT SiGe-base heterojunction-bipolar technology. *Technical Digest IEEE Symposium on VLSI Technology*, Honolulu, HI (4–7 June 1990), pp 49–50. IEEE.

[16] Patton, G.L., Comfort, J.H., Meyerson, B.S. et al. (1990). 75 GHz f_T SiGe base heterojunction bipolar transistors. *IEEE Electron Device Letters* 11: 171–173.

[17] Comfort, J.H., Patton, G.L., Cressler, J.D. et al. (1990). Profile leverage in a self-aligned epitaxial Si or SiGe-base bipolar technology. *Technical Digest IEEE International Electron Devices Meeting*, Washington (9–12 December 1990), pp 21–24. IEEE.

[18] Harame, D.L., Stork, J.M.C., Meyerson, B.S. et al. (1990). SiGe-base PNP transistors fabrication with n-type UHV/CVD LTE in a "NO DT" process. *Technical Digest IEEE Symposium on VLSI Technology*, Honolulu, pp 47–48.

[19] Crabbé, E.F., Patton, G.L., Stork, J.M.C. et al. (1990). Low temperature operation of Si and SiGe bipolar transistors. *Technical Digest IEEE International Electron Devices Meeting*, Washington (9–12 December 1990), pp 17–20. IEEE.

[20] Harame, D.L., Crabbé, E.F., Cressler, J.D. et al. (1992). A high-performance epitaxial SiGe-base ECL BiCMOS technology. *Technical Digest IEEE International Electron Devices Meeting*, Washington (13–16 December 1992), pp 19–22. IEEE.

[21] Harame, D.L., Schonenberg, K., Gilbert, M. et al. (1994). A 200 mm SiGe-HBT technology for wireless and mixed-signal applications. *Technical Digest IEEE International Electron Devices Meeting*, Washington (11–14 December 1994), pp 437–440. IEEE.

[22] Cressler, J.D., Comfort, J.H., Crabbé, E.F. et al. (1991). Sub-30-ps ECL circuit operation at liquid-nitrogen temperature using self-aligned epitaxial SiGe-base bipolar transistors. *IEEE Electron Device Letters* 12: 166–168.

[23] Cressler, J.D., Comfort, J.H., Crabbé, E.F. et al. (1993). On the profile design and optimization of epitaxial Si- and SiGe-base bipolar technology for 77 K applications – part I: transistor dc design considerations. *IEEE Transactions on Electron Devices* 40: 525–541.

[24] Cressler, J.D., Crabbé, E.F., Comfort, J.H. et al. (1993). On the profile design and optimization of epitaxial Si- and SiGe-base bipolar technology for 77 K applications – part II: circuit performance issues. *IEEE Transactions on Electron Devices* 40: 542–556.

[25] Cressler, J.D., Crabbé, E.F., Comfort, J.H. et al. (1994). An epitaxial emitter-cap SiGe-base bipolar technology optimized for liquid-nitrogen temperature operation. *IEEE Electron Device Letters* 15: 472–474.

[26] Babcock, J.A., Cressler, J.D., Vempati, L.S. et al. (1995). Ionizing radiation tolerance of high performance SiGe HBTs grown by UHV/CVD. *IEEE Transactions on Nuclear Science* 42: 1558–1566.

[27] Vempati, L.S., Cressler, J.D., Jaeger, R.C., and Harame, D.L. (1995). Low-frequency Noise in UHV/CVD Si- and SiGe-base Bipolar Transistors. *Proceedings of the IEEE Bipolar/BiCMOS Circuits and Technology Meeting*, Minnneapolis (2–3 October 1995), pp 129–132. IEEE.

[28] Harame, D.L., Stork, J.M.C., Meyerson, B.S. et al. (1993). Optimization of SiGe HBT technology for high speed analog and mixed-signal applications. *Technical Digest IEEE International Electron Devices Meeting*, San Francisco (5–8 December 1993), pp 71–74. IEEE.

[29] Lanzerotti, L., St Amour, A., Liu, C.W. et al. (1996). $Si/Si_{1-x-y}Ge_xC_y/Si$ heterojunction bipolar transistors. *IEEE Electron Device Letters* 17: 334–337.

[30] Cressler, J.D. (2005). On the potential of SiGe HBTs for extreme environment electronics. *Proceedings of the IEEE* 93: 1559–1582.

[31] Cressler, J.D. (2013). Radiation effects in SiGe technology. *IEEE Transactions on Nuclear Science* 60: 1992–2014.

[32] Krithivasan, R., Lu, Y., Cressler, J.D. et al. (2006). Half-TeraHertz operation of SiGe HBTs. *IEEE Electron Device Letters* 27: 567–569.

[33] Chakraborty, P.S., Cardoso, A.S., Wier, B.R. et al. (2014). 130 nm, 0.8 THz f_{max}, 1.6 V BV_{CEO} SiGe HBTs operating at 4.3 K. *IEEE Electron Device Letters* 35: 151–153.

[34] Ying, H., Dark, J., Ge, L. et al. (2017). Operation of SiGe HBTs down to 70 mK temperatures. *IEEE Electron Device Letters* 38: 12–15.

[35] Goley, P.S., Fleetwood, Z.E., and Cressler, J.D. (2018). Fundamental limitations on integrated silicon photonic waveguides operating in a heavy ion environment. *IEEE Transactions on Nuclear Science* 65 (1): 141–148.

[36] Tzintzarov, G.N., Ildefonso, A., Goley, P.S. et al. (2020). Electronic-to-photonic single-event transient propagation analysis in a segmented Mach-Zehnder modulator in a Si/SiGe integrated photonics platform. *IEEE Transactions on Nuclear Science* 67 (1): 260–267.

[37] Tzintzarov, G.N., Ildefonso, A., Teng, J.W. et al. (2021). Optical single-event transients induced in integrated silicon-photonic waveguides by two-photon absorption. *IEEE Transactions on Nuclear Science* 68 (5): 785–792.

[38] Frounchi, M., Tzintzarov, G.N., Ildefonso, A., and Cressler, J.D. (2021). High responsivity Ge phototransistor in commercial CMOS Si-photonics platform for monolithic optoelectronic receivers. *IEEE Electron Device Letters* 42 (2): 196–199. https://doi.org/10.1109/LED.2020.3042941.

Chapter 19

The 25-Year Disruptive Path of InP/GaAsSb Double Heterojunction Bipolar Transistors

Colombo R. Bolognesi

Department of Information Technology and Electrical Engineering (D-ITET), Laboratory for Millimeter-Wave Electronics, ETH Zurich, Zurich, Switzerland

19.1 Introduction

While reflecting on the source of my interest in transistors after receiving an invitation to contribute to this Special Commemorative Issue for the 75th anniversary of the invention of the transistor, I returned to my first undergraduate semester and its obligatory Electronics Circuits course at McGill University. It was the first time I was confronted to "transistors." It did not go well for me. While the course began with Ohm's Law and the mesh and loop circuit equations, diode and transistor first-order models were soon introduced: by the end of the semester, logic families such as RTL (resistor–transistor logic), DTL (diode–transistor logic), TTL (transistor–transistor logic), and eventually CMOS (complementary metal-oxide-semiconductor) were covered. We were told diodes conduct only if their forward voltage V_D is 0.7V, but not how this comes about nor why the forward voltage apparently could not much exceed 0.7V. This was very difficult to believe. I also had great difficulty with the "PI" small-signal transistor model, and in particular, accepting that a voltage at the input (the gate or base contact) could control the output (drain or collector) current through a "controlled-current source" could be controlled by a input potential with no direct model connection (*i.e.* a wire) to the drain or collector. I spent many hours in McGill's Rutherford Library looking through stacks of books on semiconductor electronics, hoping to find the insights I missed. Without the prerequisite modern physics that come later in engineering curricula it was of course a lost cause. The same scenario is undoubtedly experienced by today's students: it certainly contributes to the aversion many feel toward semiconductor electronics.

75th Anniversary of the Transistor, First Edition. Edited by Arokia Nathan, Samar K. Saha, and Ravi M. Todi.

In time, with the prerequisite modern physics background under my belt, the first semiconductor device course deeply interested me: for instance, the fact that charge carriers could move in a crystal as "free" quasi-particles in "energy bands" with an effective mass often less than the free electron rest mass m_o was captivating. I decided that solid-state electronics would be my specialization area. In those days personal computers with monochrome monitors were becoming popular: the McGill Engineering Student Cooperative began selling PC's with 8 MHz clock frequencies, and I reasoned that the world would always need transistors in a form or another, at least for my foreseeable employment lifetime. I took a lab course where we fabricated silicon diodes and metal-oxide-semiconductor field-effect transistors (MOSFETs) hands-on, with open windows and Montreal air in the fabrication area. The MOSFETs even worked, albeit with threshold voltages far from their intended value. The lithography was done with an ultraviolet tanning lamp: I was hooked when I saw my first pattern appear in the developer solution.

The next fateful step came during a summer employment at Northern Telecom where I was assigned by my manager the task to characterize the linearity of multistage GaAs MESFET power amplifiers (PAs) used in microwave line-of-sight digital communications. Northern Telecom (later known as the now defunct Nortel Networks) was a Canadian company that launched the digital communications revolution in 1976. The PAs were impressive: the heatsink was shoebox-sized and one could warm their hands over them. I came across a glossy brochure from Fujitsu in the laboratory: it described a new type of field-effect transistor: the so-called High Electron Mobility Transistor (HEMT) wherein a wider energy gap AlGaAs layer was selectively doped so as to supply electrons to an undoped GaAs channel. This technique physically separated donors from the channel electrons to enhance the mobility of a two-dimensional electron gas (2DEG) confined near the AlGaAs/GaAs interface. This "modulation-doped" 2DEG increased cutoff frequencies, and enabled lower noise figures which would shrink the size of parabolic dish antennas used in satellite TV. One could ENGINEER energy gaps! I was fascinated and could hardly wait to return to McGill in the fall to show this flyer to my favorite Professor in the semiconductors area! Upon returning to school, I brought him a copy of the brochure, barely able to contain my excitement. The Professor's response was disappointing: in his view, "HEMTs were a fad that would soon pass, but on the other hand, if I could grow a good Tellurium crystal, I could get a job anywhere." Upon leaving his office, I already knew that I had to leave McGill's hallowed halls if I ever was to get involved in heterojunction transistors.

Following graduation with a BEng (EE) and a summer position working on avalanche photodiodes receiver circuits at Bell Northern Research (BNR, effectively the "Bell Labs" of Nortel) in Ottawa, I joined Prof. Roy Boothroyd's group at Carleton University in the same city. This was a fortunate choice for me: Boothroyd was a pioneering contributor in silicon-based MOSFET and bipolar transistors (notably the drift base transistor). He was a quiet humble person and a benevolent supervisor. I discussed my interest in heterojunction transistors with him, and he promptly agreed to supervise a Masters project for me in this area. BNR was developing III–V compound transistor technologies in Ottawa, and we approached people there seeking a project on HBTs or HEMTs. Again, the quest for heterojunction transistors was foiled immediately: Paul Jay and Robert Surridge countered our HEMT/HBT proposal by intimating that "nobody will work on HEMTs or HBTs in a few years, but the future is in resonant tunneling diodes (RTDs)." So, I had no choice but to work on RTDs if I wanted to be involved in III–V device research. Boothroyd eventually observed that most RTDs are symmetrical and non-rectifying, and he suggested I try to design a rectifying RTD. My approach to achieve rectification was to use a multiple quantum well structure which would align quasi-bound states to allow resonant tunneling in one bias while suppressing tunneling in the reverse bias. The concept worked on the first pass in the lab, earning me a Masters degree. In the same time frame, I secured PhD studentships at a few US universities: there was a lot more III–V device research activity in the United States than in Canada. I finally chose a project focused on the development of InAs/AlSb HEMTs in Prof. Herbert Kroemer's group at UCSB – I had never heard of InAs/AlSb heterojunctions, but I was finally going to work on some heterojunction transistor.... Imagine that! In a phone interview, Kroemer made it clear he wanted an experimentalist since "the world only needs the best theorists" (*i.e.* not me). Developing HEMTs, growing the layers by molecular beam epitaxy (MBE), right next to the Pacific Ocean and under palm trees!

The UCSB learning/research environment was great, but my project work was not easy: working in a new material system never is. There are no ready answers to be found in textbooks or journals. Professors provide vision, PhD students perform the "pioneering" work, and often are the supervisor's coal mine canary in the lab. One needs to be resilient in the face of repeated failures, clear eyed, and *lucky*. For the first couple years, our HEMTs were showing no modulation at all, they were effectively gate-to-channel short-circuits. Meanwhile, my task was to characterize the MBE growth and transport properties of InAs/AlSb quantum wells [1] while another PhD student was in charge of device processing. I quickly managed to increase the 300 K mobility of InAs/AlSb quantum wells grown on GaAs from 20,000 to 30,000 $cm^2\ Vs^{-1}$ by applying MBE techniques I had read about during my Masters research. We had a couple well-separated lucky HEMT runs that yielded devices with high transconductances and enabled some publications. The processing student graduated, and I cumulated growth and device fabrication. Before departing, my colleague showed me his process once, and I went right back to making short-circuits! Around the same time, my supervisor grew bored with the InAs/AlSb HEMT project, and asked me to propose an alternate project. In our group, most people graduated based on their second or third "project," something quickly put together in the last year of their PhD. I thought up of some pulsed velocity-field measurement on InAs channels using antimonide-based photoconductive switches (which, 30 years later, still have not been demonstrated) – the alternate project had even less chance of being successful, but my supervisor liked it! All I needed to do to be successful with the InAs/AlSb HEMTs was to reduce the gate leakage, and I had by far not exhausted all my options on that front. I refused to abandon my InAs/AlSb HEMT work: this led to a very tense situation between me and my advisor for the rest of my PhD. I had come to California to work on heterojunction transistors: giving that up was never an option for me.

Then luck hit again: while processing Hall samples, the mask shifted in the aligner and the pattern was misaligned on my sample. After stripping the photoresist to start over, I noticed the Hall pattern remained visible on the sample surface, no matter how thoroughly I tried to clean it. There was only one conclusion: some of the process chemicals were attacking the sample surface. I modified the surface layers in my next MBE growth run and was able to obtain working InAs/AlSb HEMTs with good yields every time I processed. The path forward was not found in quantum mechanics, but in mere chemistry and unglamorous materials science. I could now *engineer* devices with confidence: I would modify the channel design and measure a change in device characteristics. I introduced delta-doping using Tellurium [2] (funny how Tellurium worked its way back into my life) in the top AlSb barrier, grew digital-alloy AlSbAs barriers to reduce gate leakage [3], varied the InAs channel thickness, and studied the RF performance of the devices. My work culminated with the demonstration of 0.5 µm gate devices with a current gain cutoff frequency $f_T = 95$ GHz [4], then the highest achieved in any material for that gate length. I would have liked to remain at UCSB for another year or so to make 0.1 µm gate devices: had they scaled properly, I expected them to be the fastest HEMTs ever! Surely, my supervisor would have been delighted – not so! He had clearly seen enough of me (apparently, one can be too much of an experimentalist), and stated: "nobody gives a sh!t about fast transistors" after I shared my plans to make a very fast HEMT [5]. So, 95 GHz was enough to earn me a PhD from UCSB and put an end to my days under palm trees.

Near the end of my PhD work, I interviewed at Hewlett Packard Research Labs in Palo Alto, CA. There I met Nick Moll, the son of John Moll (of the Ebers-Moll model found in all textbooks), and himself a leading III–V device scientist. In my eyes, the Moll name was effectively transistor royalty. The position was slated for the development of InP HBTs, a dream job for me even though I had no experience with bipolar devices. I felt a good rapport with Nick who would have been my direct supervisor, and really hoped to be hired. Having shown up for the interview in my wedding suit, the only one I owned, I was out of sort around the khaki-wearing members of the technical staff who quizzed me on semiconductors over a full day. Nick Moll wore jeans, and if my memory is right, he left the office to go ride his horse after taking me out to lunch. He vaguely reminded me of the Marlboro Man. Later that day, I met Nick's upper level manager, Jim Hollenhorst – I left with the distinct impression the man did not care a bit for me. The feeling was verified at a conference a few years later when a raging thunderstorm caught attendees outside: Hollenhorst and a small group from Agilent had car nearby when Hollenhorst triumphantly called out to me "Colombo, you can't come with us!" Rain and its consequences were familiar to me, and not a serious problem for an East-Coaster. I thanked my good fortune for sparing me work under his management. Many years later, I learnt from Nick Moll in his Barney Oliver Prize presentation that he was the *only one* at HP who thought they should hire me [6]. More on HP and its derivatives (Agilent/Keysight) later.

Following graduation, I chose a position as a BiCMOS Process Integration Engineer responsible for developing 0.5 μm single-polysilicon BJTs for 10 Gb s^{-1} fiber networks at Nortel Networks in Ottawa, Canada. I had turned down a position for GaAs HBTs in Thousand Oaks and an Assistant Professorship at a well-known Midwestern university. In principle, the plan at Nortel was to make SiGe HBTs and I thought that having a heterojunction to work with should keep me happy. That was not to be. The starting point was Nortel's 9 GHz process, and my Manager had me investigate whether or not SiGe was needed for 10 Gb s^{-1} applications for my very first engineering report. I judged that 10 Gb s^{-1} could be achieved with silicon-only BJTs and was able to improve device performance to 25 GHz within a year. There was quite a bit of uncertainty about the future at Nortel, and I was called for an interview for a faculty position at Simon Fraser University (SFU) in British Columbia, a residual of my post PhD job search of the previous year. I had really little interest in becoming a Professor having already declined such an opportunity. The Director of Engineering, Robert Hadaway, had told me of a raise after six months to make me accept a much lower salary than I could have had in the United States. When the six-month raise did not materialize, and with the uncertainty plaguing the company, I joined SFU in January 1995 to establish and lead its Compound Semiconductor Device Laboratory. The 51 weeks as a Nortel BiCMOS Engineer left a lasting impression. Whereas the InAs/AlSb HEMTs I made during my PhD had acceptable yields, one could see that each device had its own personality quirks. In contrast, one could probe row upon row of Si BJTs on 6-in. wafers with little or no variation in device characteristics: in Silicon, one really needs auto-prober data to superpose hundreds or thousands of device characteristics to assess variability. Manufacturability was clearly far better in the Silicon world, and its importance would play a key role in shaping my future research. Gordon Moore was still right: manufacturability is primordial.

19.2 Phase I: Simon Fraser Years (1995–2006)

After joining SFU, I set out to equip my laboratory space and launch some research activities. The first things I did was to resume work on InAs/AlSb HEMTs using layers grown at Hughes Research Labs (HRL) by David Chow whom I knew from interactions had while I was at UCSB. This healed my frustration with the early termination of my PhD work, and was helpful in building up my record for promotion. I was also looking for original research ideas of my own.

In 1995, I attended the Cornell High-Speed Conference where I saw former UCSB colleague and friend Chanh Nguyen (then from HRL) present Chirp Superlattice (CSL) Collector InP double heterojunction bipolar transistors (DHBTs) [7]. InP/GaInAs HBTs with a GaInAs collector showed high cutoff frequencies but had low breakdown voltages due to the low GaInAs energy gap of 0.75 eV. Using a wider gap InP (1.35 eV) collector was the obvious solution, but the band alignment between InP and GaInAs would block electrons in the GaInAs base if an abrupt heterojunction between GaInAs and the InP collector was used (Figure 19.1). The Japanese used a GaInAs spacer into the collector to allow electrons to overcome the blocking junction to InP [8]. Nguyen's approach was to use a digital alloy superlattice consisting of GaInAs and AlInAs layers with a shifting duty ratio to grade the average conduction band edge and overcome the collector blocking effect. Nguyen's work showed that the CSL

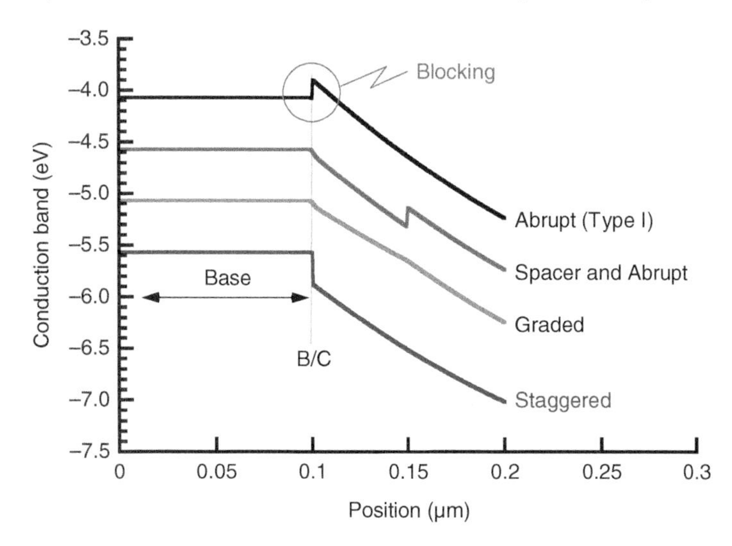

Figure 19.1 The four possible band conduction band profiles at the base/collector heterojunction of a DHBT. With a staggered ("Type-II") heterojunction, electrons are easily injected into the collector: there is no need to fight nature. Good injection is maintained down to low reverse (and even slight forward) biases. High dynamic performance is maintained down to low collector–emitter voltages, with great benefits in mixed-mode high-speed circuit applications.

period (equal to the thickness of a GaInAs and AlInAs pair of layers) needed to be quite thin, from 1.0 to 1.5 nm, to avoid regions of negative differential resistance in the transistor *I–V* characteristics. To put that in perspective, the lattice parameter on InP is 0.59 nm. I met Chanh following his talk and after the usual pleasantries our discussion turned to the challenges of performing such CSL growths: besides questions of compositional control and uniformity, he told me their MBE grower was never keen on performing such growth runs. Even if MBE had then been around for some 25 years, one still worried about machine wear from multiple shutter actuations, and the occasional shutter falling off during a growth run and leading to a major tool maintenance. Our conversation got me thinking about an alternate way of making InP DHBTs.

I recalled seeing a paper about GaAsSb/GaInAs superlattices from a Japanese group that determined the band alignment between $GaAs_{0.51}Sb_{0.49}$ and $Ga_{0.47}In_{0.53}As$ layers lattice-matched to InP: it turned out the alignment was "staggered" (also known as "Type-II") with the GaAsSb conduction band substantially higher than that in GaInAs. Knowing the "straddling" ("Type-I") conduction band discontinuity between InP and GaInAs was ~0.25 eV implied that it should be possible to build abrupt heterojunction NpN InP/GaAsSb/InP DHBTs. The expected band diagram was intriguing: the large valence band discontinuity made the GaAsSb base layer look like a quantum well for holes (shown in Figure 19.2 for a modern device). The use of abrupt junctions to InP on the emitter and collector sides suggested the HBTs could be used emitter- or collector-up. The low hole Schottky barrier height on antimonides should allow better p-Ohmic contacts than on GaInAs. A literature search showed some MBE and MOCVD growth studies on GaAsSb, but no works on the InP/GaAsSb heterojunction system. I had found my original research topic. The first technical step would then be to determine then actual band alignment at InP/GaAsSb heterojunctions. The first practical step would be to secure funding for research.

BNR in Ottawa had a solid GaInP/GaAs HBT process in place but no InP HBTs. I thought BNR would not take me seriously if I proposed work on GaAs: they already knew more about it than I could expect to contribute. But what about InP? I reached out to John Sitch to request a support letter for a potential application to the NSERC Strategic Grant Program (which had a 20–25% success rate).

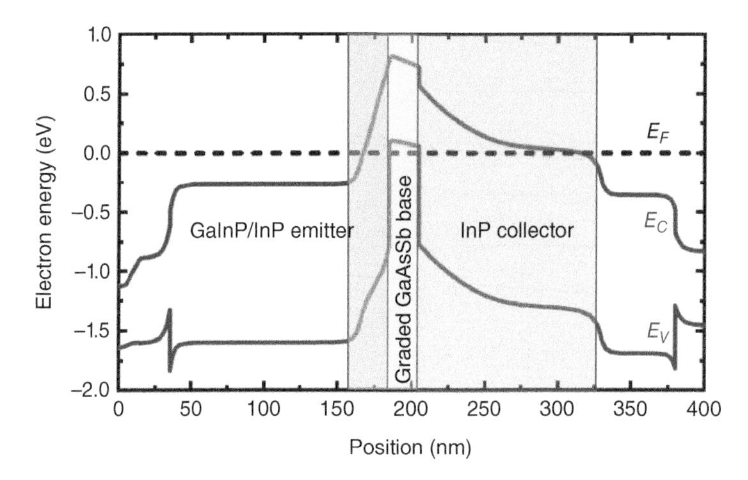

Figure 19.2 Equilibrium band diagram of a modern InP/GaAsSb DHBT featuring ETH-MWE innovations such as a composite graded InP/GaInP emitter for improved gain and scalability, and a mixed-group V Ga(A/s, Sb) graded base layer. The very large valence band discontinuity suppresses base push-out into the collector when the device operates with high currents and low collector voltages. No other bipolar transistor technology can replicate this effect while maintaining superior collector transport properties. The author credits this feature for the outstanding mixed-mode IC performances achieved in cooperation with III–V Lab.

Sitch was intrigued and agreed to support the research and provide some in-kind contributions in the form of chrome mask fabrication. Just before submitting my application, I came across a couple of articles from Bellcore [9] and Rockwell [10] demonstrating InP/GaAsSb DHBTs. In those days one had to physically go to the library and turn journal pages. Not only had we been scooped, but the results were not competitive with existing InP/GaInAs DHBTs. Doubly disappointed, I hesitated to submit the application: it did not seem like a very good idea. Having invested much time in the proposal, I finally sent it in thinking the odds of being funded were low. Surprisingly, in the absence of any track record for me in compound semiconductor HBTs, funding for three years was approved for a project targeting the fabrication of 150GHz InP/GaAsSb DHBTs for potential wireless communication applications [11]. My SFU co-applicant was Simon Watkins who would take care of the MOCVD growth of the InP/GaAsSb layers. Watkins had also secured another Strategic grant of his own, and we agreed to divide his growth time equally: six months per year on each project.

The NSERC InP/GaAsSb DHBT funding allowed for a Postdoctoral position. Noureddine Matine applied for the position having completed his PhD in France with work on the fabrication of InP/GaInAs HBTs. His expertise with the crystallographic etching properties of InP and with airbridge processes would prove very helpful in quickly launching InP HBT activities at SFU.

The first step was to master the growth of GaAsSb layers by MOCVD in Watkins' lab using a small vertical shower head reactor allowing growth on small samples (a quarter of a 2″ wafer). Initial efforts failed because even smaller pieces were used to save money, and poor morphologies resulted from edge effects. With bigger samples, good morphologies were achieved quickly. The next step was to develop the growth of adequate quality InP/GaAsSb heterojunctions [12], and to experimentally verify their band alignment [13]. As anticipated, the GaAsSb conduction was slightly higher than that of InP and thus in principle suitable for the realization of NpN InP/GaAsSb/InP DHBTs.

Around the same time period, I received a call from Nick Moll of HP Labs. He asked if my grant application had been successful, and informed me he had been one of the reviewers who evaluated the proposal. I answered affirmatively, and he announced that his organization would like to work with me on that project. That was amazing news! It also turned out the MIT graduate HP Labs hired over me had not made any InP HBTs. Over the next several years, Nick and I had frequent conversations which helped me refine my ideas. He was profoundly amused by the volume of material I had to prepare for the NSERC proposal in relation to the attributed funding. He told me HP Labs could provide me some support based on one-or two-page internal grant competitions that he justified by "access to the Professor." Sometimes it came as research funding, sometimes test equipment (*e.g.* semiconductor parameter analyzer, *C–V* meter, etc.). We agreed I should visit HP Labs and I traveled to Palo Alto. The night before my visit, my first PhD student working with Matine faxed me good and bad news. The good: the first large area InP/GaAsSb DHBTs from SFU had just been made. Their Gummel characteristics were ideal in that both the collector and base currents showed a slope of $60\,\mathrm{mV\,dec^{-1}}$. The bad: the current gain β was less than unity. I decided to show the results to the small number of HP engineers attending my presentation on the next day. It made the audience chuckle. Nick Moll was one who did not laugh. He stated "well, you will figure it out." And we did. It turned out that to start we had used a fairly regular InP/GaInAs HBT base design with a 60 nm C-doped GaAsSb doped at $2\times10^{19}\,\mathrm{cm^{-3}}$. I ventured that the electron mobility in GaAsSb was probably lower than in GaInAs (recall, there is little information on new materials), causing electrons to recombine before they could diffuse across the base into the InP collector. The current ideality suggested the InP-GaAsSb heterojunctions were of acceptable quality. We next used thinner GaAsSb bases: this brought up the gain to acceptable values but increased the base sheet resistance to values well above those achievable in GaAs or even GaInAs. If we had to use thin base layers, they would have to be doped more heavily. I convinced Watkins to increase the C-base doping level, and to our delight we found that GaAsSb really likes C-doping [14]. Very high C-concentrations could be achieved, without any sign of hydrogen passivation (a unique advantage of MOCVD-grown GaAsSb compared to GaInAs). Engineers from BNR with whom I discussed our work during a visit in Ottawa told me we really should not C-dope our bases above $4\times10^{19}\,\mathrm{cm^{-3}}$ because that caused reliability problems in GaAs HBTs. We had no choice: we increased the doping to $8\times10^{19}\,\mathrm{cm^{-3}}$ and more, and experienced none of the problems seen in GaAs. GaAsSb, unlike GaInAs, is not akin to GaAs.

The time came to make RF testable transistors in our airbridge process. The first run showed a current gain cutoff frequency of $f_T = 20\,\mathrm{GHz}$. In 1998, we first presented DC characteristics at the IEEE Device Research Conference held at UCSB. There I encountered Aiden Higgins whom I had first met while searching for employment at the Rockwell Science Center toward the end of my studies. Higgins asked me what I was working on. I told him of the InP/GaAsSb DHBT project before he replied "we did that … you're wasting your time." On a different note, Mark Rodwell a former professor of mine at UCSB, commented "if you can make better Ohmic contacts on GaAsSb, I will be very scared." It took a couple years but we did.… At the 1999 GaAs Integrated Circuit Symposium in Monterey, we reported $1\times24\,\mu\mathrm{m}^2$ DHBTs with $f_T/f_{MAX} = 80/90\,\mathrm{GHz}$ and an 8 V breakdown voltage [15], earning me my single Best Paper Award. Subsequent progress was swift: by 2000, we demonstrated the first bipolar transistors with $f_T/f_{MAX} = 300/300\,\mathrm{GHz}$ and $BV_{CEO} > 6\,\mathrm{V}$, a record for bipolar transistors in any material system [16, 17]. At the time, the best InP/GaInAs DHBT results offered cutoff frequencies of around 200 GHz with significantly lower breakdown voltages, and InP HEMTs were thought to be the only choice for 300 GHz and above. In order to confirm our results, we sent device to Tom MacElwee a former BiCMOS colleague of mine who had migrated to Nortel's Technology Access Group. MacElwee's measurements showed $f_T/f_{MAX} = 305/300\,\mathrm{GHz}$. Our paper [17] was initially rejected because a reviewer and the Editor in Charge felt mentioning independent f_T measurements of 300 and 305 GHz from different labs would cast an "*intolerable uncertainty*" on our claims! Our paper was rejected because we had our results independently cross-checked! In a mere three years (and effectively one year and a half of epitaxy) from measuring the InP/GaAsSb band alignment, we had transitioned from large-area DC device with $\beta < 1$ to state-of-the-art 300 GHz with record $f_T \times BV_{CEO}$ products of 1.8 THz-V.

Back at Nortel in Ottawa, MacElwee continued periodically testing our unpassivated InP/GaAsSb DHBTs. Their characteristics remained stable despite being kept in room air and on his bookshelf between measurement cycles. There was clearly something special about the sturdiness of these devices. Their raw performance generated growing interest from HP/Agilent and Nortel, with both organizations launching GaAsSb and GaInAs-based DHBT development efforts. In both companies, efforts to develop InP/GaInAs DHBTs proved frustrating.

It is interesting that the *only* early adopters of InP/GaAsSb technology (Agilent, and the defunct Nortel) did *not* have prior InP/GaInAs SHBT processes: they initiated InP/GaInAs and InP/GaAsSb DHBT development in *parallel*. Both organizations proofed their home-grown InP/GaAsSb layers by having high-speed DHBTs fabricated in the author's group before making their own to speed up development [18, 19]. A basic fabrication process was transferred from SFU to HP Labs in 2001. Their InP/GaAsSb DHBT process went into production in 2004, barely seven years after initiation in our group. In contrast, it took HP 10 years to put a GaInP/GaAs HBT process in place. Ironically, not being hired at HP Labs after my doctorate enabled me to initiate InP/GaAsSb DHBT research. With Martin Dvorak, the lone graduate student responsible for the 300 GHz graduating to take a position at Agilent in Santa Rosa, CA, to industrialize GaAsSb, he first had to train me on the latest process: as a Professor, I returned to the cleanroom to fabricate DHBTs using Agilent and Nortel epilayers until I could train new personnel.

19.3 Phase II: ETH Years (2006–2022)

Following the internet dot-com bubble of year 2000, Nortel's stock plunged and massive layouts followed. In a call with Robert Hadaway to see what I could do for continued collaboration and support his answer was "Nothing. Transistors like yours are a commodity." 300 GHz devices with 6 V breakdowns never were commodities. Things were not brighter on the HP/Agilent (now Keysight) front: Nick Moll fell gravely ill and went on indefinite leave before passing away in 2011. I had lost my champion in Palo Alto. His organization was also reeling from the internet bubble. Without industrial partners, I could no longer secure funding to keep my group and lab going. I had also reached the limit of what could be done with Watkins' epitaxy at SFU. I sought a Canadian Research Chair in "THz Electronics" that would fund my lab, including my own epitaxial growth system – it was denied. By 2004, my research's outlook was bleak. I was invited to apply to ETH-Zürich (Switzerland) to fill the position left vacant in 2005 by Prof. Werner Baechtold, a well-known GaAs MESFET pioneer. The interview process was in part cringeworthy: it was organized as a "mini-symposium" with four finalists simultaneously present. A further cringe factor was that one of the other three candidates was a former professor of mine turned competitor (he left the room during my research presentation). I was offered the position and secured my own MOCVD growth system as part of the start-up package. For the first time since my PhD days at UCSB, could I freely determine layers to be grown, year-round. As fate would have it, I would again leave Canada for heterojunction transistors, along with three members of my SFU group.

At ETH, I soon hired Dr. Olivier Ostinelli, a dedicated and talented MOCVD grower recently graduated under Baechtold's supervision: he had experience with the growth of mixed-Arsenide/Antimonide alloys. Luck was again on our side. Ostinelli also had some industrial experience: he understood a grower's role in device fabrication despite his background in Physics. Too many times in my academic career working with physicists had I seen them grow bored with transistor work (I count three times), with two eventually declaring that "growing transistors is not physics." With Ostinelli handling growth, progress resumed and a steady stream of innovations followed: InP/GaInP composite emitters were introduced to increase gain and improve scaling (reduced size-effects) [20]; mixed-group V GaAsSb compositionally graded bases to reduce base transit times and boost device gain [21]. This likely was the first use of mixed-group-V grading in a functional device – it could be done *automatically* by grading the C-doping precursor flow during growth to achieve simultaneous compositional and doping built-in drift fields. Ostinelli also was instrumental in developing GaInAsSb base layers at ETH, an idea we could not get to work at SFU.

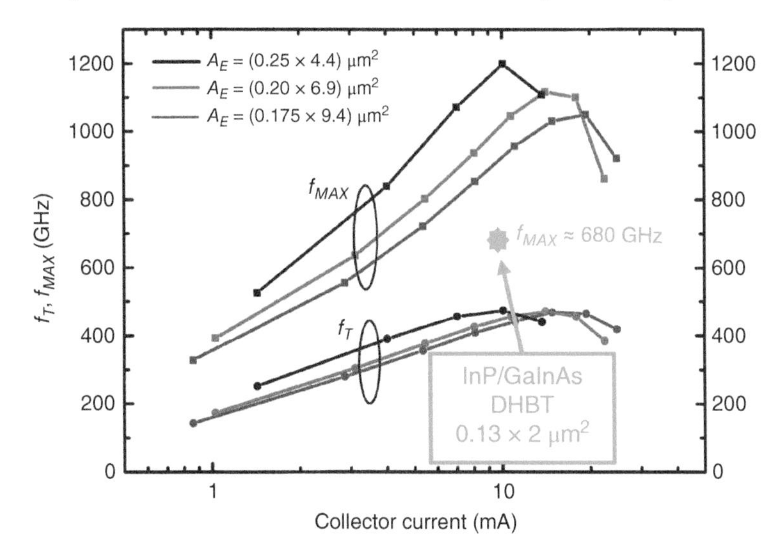

Figure 19.3 Scaling dependence of THz emitter-fin InP/GaAsSb DHBTs. InP/GaInAs DHBTs showing THz performance for a 2 μm long emitter clearly do not scale as well as "Type-II" DHBTs. Source: Adapted from Arabhavi et al. [24] and InP/GaInAs data from [26].

With the excellent nanofabrication facilities available at ETH, device performance and yields rose steadily: $f_T \times BV_{CEO}$ products of 2.8 THz-V were recently achieved for devices showing record output power and power density at 94 GHz [22]. As maximum oscillation frequencies rose to 800 GHz and above [23], our attempts to publish faced multiple rejections based on the very weak argument that InP/GaInAs DHBTs showed f_{MAX} values of ~1 THz, and that our work was thus nothing new (other than being based on a completely different alloy). For over two years, my PhD student Akshay Arabhavi (graduated in 2022) was dealt such rejections. His response was to devise a revolutionary emitter-fin process that would allow him to scale his InP/GaAsSb DHBTs to achieve record figures-of-merit and demonstrate the outstanding scaling properties of InP/GaAsSb technology. The emitter-fin process was inspired from the HEMT T-gate formation process used in our group to fabricate low-noise InP HEMTs used by the European Space Agency for Deep Space Communications. Arabhavi's work was first reported at the 2021 International Electron Device Meeting (IEDM) [24], and selected as a conference highlight for an extended special issue paper [25]. The emitter fin process demonstrated a record f_{MAX} = 1.2 THz in a (0.25×4.4) μm² InP/GaAsSb DHBT (Figure 19.3). The devices offer $f_{MAX} \times BV_{CEO}$ = 6.48 THz-V, the highest ever reported for a transistor. Thanks to thick base metal contacts, 9.4 μm long devices also achieve f_{MAX} > 1 THz for an emitter area of 1.645 μm². This represents a true breakthrough in device scaling. In comparison, the previous InP DHBT state-of-the-art reported f_{MAX} = 1.15 THz with BV_{CEO} = 3.5 V and a 0.26 μm² emitter area [26]. Beyond outstanding performance, the dissipated power density of InP/GaAsSb DHBTs is 40× lower than that of InP/GaInAs DHBTs and SiGe HBTs [25]. It could be distressing to the reviewers and editors who rejected Arabhavi's previous good works that he would have graduated and never developed his emitter-fin devices without their motivating interference (a nice demonstration of the Law of Unintended Consequences).

At about the same time Arabhavi was experiencing the rejection of his manuscripts, Virginie Nodjiadjim from III–V Lab in France contacted me to ask whether we could supply them InP/GaAsSb DHBT layers. Due to various circumstances, her organization faced delays in securing their usual InP/GaInAs DHBT layers for a French/Swiss project we were jointly involved in. Our standard layer design was slightly adapted and InP/GaAsSb DHBTs were grown at ETH on 3-in. InP substrates suitable for fabrication at III–V Lab. Multiple mixed-mode circuit blocks used in high-speed optical communications were fabricated. The circuit designs were based on III–V Lab's InP/GaInAs DHBT process, and thus *not* optimized for InP/GaAsSb DHBTs. The conservative 0.7 µm technology showed f_T/f_{MAX} = 360/420 GHz at J_C = 6 mA µm^{-2} on the ETH-grown InP/GaAsSb layers. In this technology, 100 Gb s^{-1} lumped linear drivers were measured with both with four-level pulse-amplitude modulation (PAM-4) and non-return-to-zero (NRZ) modulations [27]. The drivers delivered a 4.1-Vppd 100-Gb s^{-1} NRZ output signal eye diagram with a clear eye opening, emphasizing the InP/GaAsSb DHBT's ability to provide a large-output swing with high-symbol-rate operation while ensuring excellent signal integrity. A 3-V differential eye amplitude was obtained, with an eye signal-to-noise ratio (*S/N*) of 7 and an rms-jitter below 730 fs. To the best of our knowledge, this was the highest >3-Vppd 100-Gb s^{-1} NRZ eye S/N reported for a linear driver to date. Power-DAC circuits which can be used in single-carrier 1.0 Tb s^{-1} transmitters were also demonstrated [27]. PAM-4 operation of the DAC-drivers at 90 GBd (180 Gb s^{-1}) and 112 GBd (224 Gb s^{-1}) were achieved. The corresponding power consumption was 1.1 and 0.6 W for the two symbol-rates, respectively. The 90-GBd PAM-4 differential output eye diagram showed a record 5.5-Vppd swing, while 112 GBd PAM-4 operation also produced a record 3.35-Vppd output swing [27]. The corresponding E/O driver figure-of-merit (FoM) defined in terms of the PAM-4 symbol-rate D_S, the differential output swing V_{Opp}, the differential output matching impedance Z_0, and the circuit's DC power consumption P_{DC} is given as

$$FoM = D_S V_{Opp}^2 / 8 Z_0 P_{DC},$$

leading to record FoM values of 3.1 and 2.6 GBd at 180 and 224 Gb s^{-1}, respectively [27]. They are to be compared to the E/O FoM of 0.13 GBd achieved in a 256 Gbs PAM-4 generator IC implemented in a 0.25-µm InP/GaInAs DHBT technology with f_T/f_{MAX} = 460/480 GHz (J_C = 10–13 mA µm^{-2}) [28]. These results show the outstanding advantages of InP/GaAsSb DHBTs for high symbol-rate large output swing ICs: a favorable band structure is of central importance – raw transistor metrics (i.e. high f_T/f_{MAX} and J_C) are not the sole determinants of mixed-signal IC performance, regardless of published "roadmaps." Remarkably, these results were achieved without optimizing designs for GaAsSb, and with the same collector structure used by Arabhavi in THz emitter-fin transistor [24, 25], demonstrating the great versatility of InP/GaAsSb DHBT technology. Interestingly (and sadly for the high-speed electronics community), these outstanding results from III–V Lab also faced multiple rejections from journals and conferences over two years before they could successfully be published.

19.4 Response to Innovation

There are clear indications in the preceding text that disruptive innovations are often not received with open arms. In his 1966 book [29], MIT Professor Elting E. Morison examined the response to innovation beginning with naval artillery as an example. He found the response generally has three clear stages. In the first stage, there is *no* response: the development is ignored as if deemed not credible. In the second stage, *attempts* are made to formulate what appear as logical, rational rebuttals. Morison found such rebuttals tend to be factually wrong on important points ([29] p. 39). Morison's third stage is *argumentum ad hominem* – effectively, reputational attacks on the innovator. According to Morison, the three-stage response takes place when an innovative technology disrupts the "encrusted hierarchy" of a given community: the reaction is driven by "disbelief in the dramatic but substantiated improvements of the new process, and by protection of the existing devices and instruments with which they (the encrusted opposers) identified themselves." Morison notes that personal identification with a concept is a powerful barrier to otherwise easily acceptable innovation. Judged by the three-stage response it continues to receive, InP/GaAsSb DHBT technology seems to be remarkably disruptive. The "Information is Beautiful" website lists several innovators across multiple disciplines [30]: for example, Ohm's law needed 25 years before gaining acceptance while Galileo's ideas had to simmer for a solid 219 years. Surprisingly, Engineering is the slowest field to recognize innovation according to the website (40 average years until acceptance, though based on sparse data). If this proves accurate, one must wonder how much more technologically advanced our world could be. The supreme irony is that Morison's "encrusted hierarchy" largely drives the funding agencies tasked to stimulate innovation toward (at least partially) self-defeat. Technical diversification is advantageous, as diversification is in all other investment endeavors. The InP/GaAsSb DHBT work described here was carried out in small teams, without large-scale funding support from North American or European agencies.

Equally if not more disturbing, at the other range of the response spectrum, lie some early adopters whose strategy is to appropriate innovation and deliberately obscure its origin: as Agilent/Keysight's Don d'Avanzo put it, "we're happy to be the only ones working on it."

19.5 Final Words

The preceding personal account attempted to chronicle the surprising evolution of a powerfully disruptive technology initiated by an improbable contributor. I am thankful to Professor Arokia Nathan for inviting this contribution and providing the opportunity to tell the InP/GaAsSb DHBT story and show "how the sausage" was made. I know of no other technology developed in a single team drive, from band lineup measurements to THz transistor cutoff frequencies in a 25-year span. If luck favors a prepared mind, it acted as if InP/GaAsSb DHBTs were meant to be. I do hope it will inspire some readers.

Most of the work chronicled in this article was achieved through the work of PhD students. I am forever thankful to my PhD students for buying into their Professor's vision, for their dedication, cleverness, and resilience in moving it forward, regardless of project. To use Richard Feyman's words, the "pleasure of finding things out" with them is exhilarating.

In the specific context of this article focusing on InP/GaAsSb DHBTs, I am especially indebted to Dr. Olivier Ostinelli, and my past and present PhD students at ETH: Maria Alexandrova, Akshay Arabhavi, Filippo Ciabattini, Mojtaba Ebrahimi, Ralf Flückiger, Sara Hamzeloui, Rickard Lövblom, Wei Quan, and Yuping Zeng. Dr. Honggang Liu is credited for transferring the DHBT process from SFU to ETH, thus ensuring the continuity of our work. Sincere gratitude is also due to the entire collaborative team at III–V Lab for demonstrating the outstanding potential of InP/GaAsSb DHBTs in high-speed mixed circuit applications.

References

[1] Bolognesi, C.R., Kroemer, H., and English, J.H. (1992). Interface roughness scattering in InAs/AlSb quantum wells. *Appl. Phys. Lett.* 61: 213–215. https://doi.org/10.1063/1.108221.

[2] Werking, J.D., Bolognesi, C.R., Chang, L.-D. et al. (1992). High-transconductance InAs/AlSb heterojunction field-effect transistors with delta -doped AlSb upper barriers. *IEEE Electron Device Lett.* 13 (3): 164–166. https://doi.org/10.1109/55.144998.

[3] Bolognesi, C.R., Werking, J.D., Caine, E.J. et al. (1993). Microwave performance of a digital alloy barrier Al(Sb,As)/AlSb/InAs heterostructure field-effect transistor. *IEEE Electron Device Lett.* 14 (1): 13–15. https://doi.org/10.1109/55.215085.

[4] Bolognesi, C.R., Caine, E.J., and Kroemer, H. (1994). Improved charge control and frequency performance in InAs/AlSb-based heterostructure field-effect transistors. *IEEE Electron Device Lett.* 15 (1): 16–18. https://doi.org/10.1109/55.289476.

[5] Appropriately, his year 2000 Nobel Prize motivation was for "for developing semiconductor heterostructures used in high-speed- and opto-electronics."

[6] Admittedly, this puzzled me for years. Shortly before his passing, I visited Nick and his wife Barbara at their ranch with David DiSanto from Keysight. I pressed Nick on why I was not hired at HP Labs. He revealed that it was perceived that "my personality would not allow me to do things that do not make sense." This already had been demonstrated.

[7] Nguyen, C., Liu, T., Sun, H.-C. et al. (1995). Current transport in band-gap engineered AlInAs/GaInAs/InP double heterojunction bipolar transistor using chirped superlattice. *Proceedings IEEE/Cornell Conference on Advanced Concepts in High Speed Semiconductor Devices and Circuits*, Ithaca, NY, USA (7–9 August 1995), pp. 552–562. http://doi.org/10.1109/CORNEL.1995.482552.

[8] Fujihara, A., Ikenaga, Y., Takahashi, H. (2001). *High-speed InP/InGaAs DHBTs with ballistic collector launcher structure. International Electron Devices Meeting. Technical Digest*, Washington, DC, USA (2–5 December 2001), pp. 35.3.1–35.3.4. http://doi.org/10.1109/IEDM.2001.979629.

[9] Bhat, R., Hong, W-P., Caneau, C. et al. (1996). InP/GaAsSb/InP and InP/GaAsSb/InGaAsP double heterojunction bipolar transistors with a carbon-doped base grown by organometallic chemical vapor deposition. *Appl. Phys. Lett.* 68: 985–987. https://doi.org/10.1063/1.116120.

[10] McDermott, B.T., Gertner, E.R., Pittman, S. et al. (1996). Growth and doping of GaAsSb via metalorganic chemical vapor deposition for InP heterojunction bipolar transistors. *Appl. Phys. Lett.* 68: 1386–1388. https://doi.org/10.1063/1.116088.

[11] NSERC (1996). Awards database. https://www.nserc-crsng.gc.ca/ase-oro/Details-Detailles_eng.asp?id=1273.

[12] Xu, X.G., Hu, J., Watkins, S.P. et al. (1999). Metalorganic vapor phase epitaxy of high-quality GaAs0.5Sb0.5 and its application to heterostructure bipolar transistors. *Appl. Phys. Lett.* 74: 976–978. https://doi.org/10.1063/1.123428.

[13] Hu, J., Xu, X.G., Stotz, J.A.H. et al. (1998). Type II photoluminescence and conduction band offsets of GaAsSb/InGaAs and GaAsSb/InP heterostructures grown by metalorganic vapor phase epitaxy. *Appl. Phys. Lett.* 73: 2799–2801. https://doi.org/10.1063/1.122594.

[14] Watkins, S.P., Pitts, O.J., Dale, C. et al. (2000). Heavily carbon-doped GaAsSb grown on InP for HBT applications. *J. Cryst. Growth* 221: 59–65. https://doi.org/10.1016/S0022-0248(00)00649-7.

[15] Bolognesi, C.R., Matine, N., Dvorak, M.W. et al. (1999). InP/GaAsSb/InP DHBTs with high f_T and f_{MAX} for wireless communication applications. *GaAs IC Symposium. IEEE Gallium Arsenide Integrated Circuit Symposium. 21st Annual. Technical Digest*, Monterey, CA, USA (17–20 October 1999), pp. 63–66. http://doi.org/10.1109/GAAS.1999.803728.

[16] Dvorak, M.W., Pitts, O.J., Watkins, S.P., and Bolognesi, C.R. (2000). Abrupt junction InP/GaAsSb/InP double heterojunction bipolar transistors with F_T as high as 250 GHz and BV_{CEO}>6 V. *International Electron Devices Meeting. Technical Digest. IEDM*, San Francisco, CA, USA (10–13 December 2000), pp. 178–181. http//doi.org/10.1109/IEDM.2000.904287.

[17] Dvorak, M.W., Bolognesi, C.R., Pitts, O.J., and Watkins, S.P. (2001). 300 GHz InP/GaAsSb/InP double HBTs with high current capability and BV_{CEO}>6 V. *IEEE Electron Device Lett.* 22 (8): 361–363. https://doi.org/10.1109/55.936343.

[18] Matine, N., Dvorak, M.W., Lam, S. et al. (2000). Demonstration of GSMBE grown InP-$GaAs_{0.51}$ $As_{0.49}$/InP DHBTs. *Conference Proceedings. 2000 International Conference on Indium Phosphide and Related Materials*, Williamsburg, VA, USA (14–18 May 2000), pp. 239–242. http://doi.org/10.1109/ICIPRM.2000.850276.

[19] Liu, H.G., Wu, J.Q., Tao, N. et al. (2004). High-performance InP/GaAsSb/InP DHBTs grown by MOCVD on 100mm InP substrates using PH_3 and AsH_3. *J. Cryst. Growth* 267: 592–597. https://doi.org/10.1016/j.jcrysgro.2004.04.035.

[20] Liu, H.G., Ostinelli, O., Zeng, Y.P., and Bolognesi, C.R. (2008). Emitter-size effects and ultimate scalability of InP:GaInP/GaAsSb/InP DHBTs. *IEEE Electron Device Lett.* 29 (6): 546–548. https://doi.org/10.1109/LED.2008.920850.

[21] Liu, H.G., Ostinelli, O., Zeng, Y., and Bolognesi, C.R. (2007). 600 GHz InP/GaAsSb/InP DHBTs grown by MOCVD with a Ga(As, Sb) graded-base and $f_T \times BV_{CEO} \gg 2.5$ THz-V at room temperature. *2007 IEEE International Electron Devices Meeting*, Washington, DC, USA (10–12 December 2007), pp. 667–670. http://doi.org/10.1109/IEDM.2007.4419032.

[22] Hamzeloui, S., Arabhavi, A.M., Ciabattini, F. et al. (2022). High power InP/GaInAsSb DHBTs for Millimeter-wave PAs: 14.5 dBm output power and 10.4 mw/μm^2 power density at 94 GHz. *IEEE J. Microw.* https://doi.org/10.1109/JMW.2022.3202854.

[23] Arabhavi, A.M., Quan, W., Ostinelli, O., and Bolognesi, C.R. (2018). Scaling of InP/GaAsSb DHBTs: a simultaneous fT/fMAX 463/829 GHz in a 10 μm long emitter. *IEEE BiCMOS and Compound Semiconductor Integrated Circuits and Technology Symposium (BCICTS)*, San Diego, CA, USA (5–17 October 2018), pp. 132–135. http://doi.org/10.1109/BCICTS.2018.8551036.

[24] Arabhavi, A.M., Ciabattini, F., Hamzeloui, S. et al. (2021). InP/GaAsSb double heterojunction bipolar transistor technology with f_{MAX} = 1.2 THz. *2021 IEEE International Electron Devices Meeting (IEDM)*, San Francisco, CA, USA (11–16 December 2021), pp. 11.4.1–11.4.4. http://doi.org/10.1109/IEDM19574.2021.9720644.

[25] Arabhavi, A.M., Ciabattini, F., Hamzeloui, S. et al. (2022). InP/GaAsSb double heterojunction bipolar transistor emitter-fin Technology with f_{MAX} = 1.2 THz. *IEEE Trans. Electron Devices* 69 (4): 2122–2129. https://doi.org/10.1109/TED.2021.3138379.

[26] Urteaga, M., Pierson, R., Rowell, P. et al. (2011). 130nm InP DHBTs with f_t >0.52THz and f_{max} >1.1 THz. *69th Device Research Conference IEEE,* Santa Barbara, CA, USA (20–22 June 2011), pp. 281–282. http://doi/org/10.1109/DRC.2011.5994532.

[27] Konczykowska, A., Hersent, R., Jorge, F. et al. (2022). 112 GBaud (224 Gb/S) Large Output Swing InP DHBT PAM-4 DAC-driver. *24th International Microwave and Radar Conference MIKON'22*, Gdansk, Poland (12–14 September 2022).

[28] Nagatani, M., Wakita, H., Jyo, T. et al. (2018). A 256-Gbps PAM-4 signal generator IC in 0.25-μm InP DHBT technology. *2018 IEEE BiCMOS and Compound Semiconductor Integrated Circuits and Technology Symposium (BCICTS)*, San Diego, CA, USA (15–17 October 2018), pp. 28–31. http://doi.org/10.1109/BCICTS.2018.8550977.

[29] Morison, E.E. (1966 and 2016). *Men, Machines and Modern Times.* MIT Press.

[30] https://www.informationisbeautiful.net/visualizations/mavericks-and-heretics.

Chapter 20

The High Electron Mobility Transistor

40 Years of Excitement and Surprises

Jesús A. del Alamo

Department of Electrical Engineering and Computer Science, Massachusetts Institute of Technology, Cambridge, MA, USA

20.1 Introduction

THE High-Electron Mobility Transistor (HEMT) was first demonstrated by Mimura and colleagues at Fujitsu Labs in 1980 [1]. The invention of the HEMT represented the latest triumph of bandgap engineering and molecular beam epitaxy (MBE) that had earlier brought to the world the heterostructure laser, the heterojunction bipolar transistor, heterostructure avalanche photodiodes, and other electronic and photonic devices that would come to revolutionize communication systems.

The HEMT was based on the concept of *modulation doping* first demonstrated by Dingle and collaborators at Bell Labs in 1978 [2]. A modulation-doped structure creates a two-dimensional electron gas at the interface between two semiconductors of different bandgaps. The spatial separation that is created between dopants and electrons confers these with a mobility that exceeds the bulk value even at relatively high carrier concentrations (Figure 20.1).

The first demonstrations of modulation doping and the HEMT took place in the AlGaAs/GaAs system. These would not have been possible without the atomic layer precision growth capabilities of MBE. Through MBE, heterostructures approaching monolayer-level abruptness, nearly perfectly coherent interfaces, and precise dopant placement became possible. All these are essential ingredients for realizing modulation-doped structures with improved mobility over bulk values. The initial mobility enhancements that were observed were quite modest at room temperature but much more significant at

75th Anniversary of the Transistor, First Edition. Edited by Arokia Nathan, Samar K. Saha, and Ravi M. Todi.

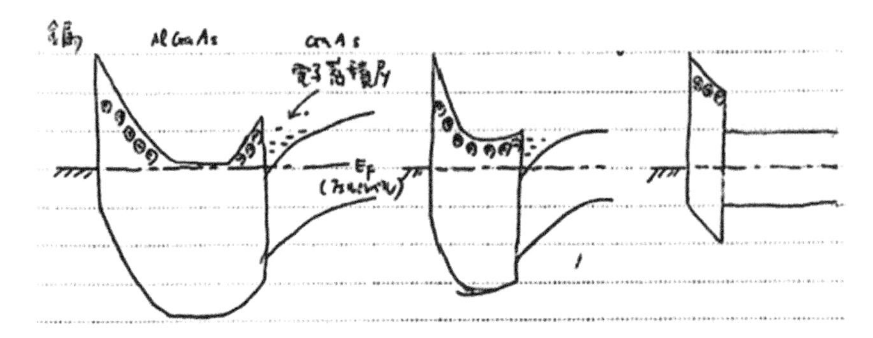

Figure 20.1 Energy band diagrams explaining field-effect modulation of two-dimensional electron gas. Source: From Mimura 3/IOP Publishing.

low temperatures. Over time, the low temperature mobility would dramatically improve leading to fundamental discoveries in solid-state physics [4].

This chapter traces the most significant steps of the evolution of the HEMT from its inception to its current state in various semiconductor systems with emphasis on III–V compound semiconductors based on arsenides and phosphides. It also outlines some of the new discoveries that the development of high crystalline quality AlGaAs/GaAs heterostructures enabled in solid-state physics. It finishes by sketching a future for new applications for HEMT-like structures in several domains.

20.2 HEMT Electronics

Right from its outset, the high mobility and highly confined nature of the two-dimensional electron gas suggested that modulation doping could be exploited to make high-speed field-effect transistors with excellent short-channel effects and great scaling potential. This spurred efforts to demonstrate high-speed logic circuits. Progress was spectacular. Within one year, Fujitsu had demonstrated the first HEMT integrated circuit featuring an enhancement/depletion-mode logic ring oscillator that showed a switching delay of 17.1 ps, "the lowest of all the semiconductor logic technologies reported thus far" [5]. Soon the advantages of HEMTs over contemporary GaAs MESFETs (Metal-Semiconductor Field-Effect Transistors) and Si ECL (Emitter-Coupled Logic) in terms of speed-power performance become evident and Fujitsu and others launched a race to demonstrate HEMT circuits for high-speed computers. Fujitsu achieved 1 kbit static random access memories (SRAMs) in 1984, 4 kb in 1987, and 64 kbit in 1991 [3]. Several companies realized many kinds of circuits with various levels of complexity. In the end, in spite of interesting demonstrations in supercomputers and other systems, this first commercial direction for HEMT ICs did not pan out. This technology, just as Si ECL later, could not compete with the increasing densities, much lower power consumption, and huge economies of scale of Si CMOS.

The first commercial application of a HEMT came from a totally unexpected area. In 1983, Fujitsu demonstrated a four-stage HEMT amplifier operating at 20 GHz [6]. Its gain and noise performance was observed to improve significantly as the operating temperature was reduced. This was attributed to an enhancement of mobility at low temperatures. These cryogenically cooled HEMTs significantly outperformed state-of-the-art GaAs MESFETs of the time. This result came to the attention of the Nobeyama Radio Observatory in Nagano (Japan) who ordered several low-noise amplifiers for their 45 m radio telescope. These were installed in 1985 and in 1986, this telescope discovered new interstellar molecules in the Taurus Molecular Cloud, about 400 light years away [3]. The great stability of the HEMT, compared with the then conventional parametric amplifiers, allowed prolonged observations, something that had great value to astronomers. Since then, HEMTs have been widely used in radio telescopes throughout the world [7].

The first mass-market applications of AlGaAs/GaAs HEMTs came in the communications arena after the recognition of their superior high-frequency noise characteristics over MESFETs. This was attributed to the excellent current-gain cut-off frequency, f_T, and the high aspect ratio of the channel. Direct Broadcasting Satellite receivers including low-noise HEMT amplifiers first went commercial in 1987. They enabled a reduction in antenna size by one half. By 1988, the yearly worldwide production of HEMT receivers had already become about 20 million [8].

Two key innovations in the mid-1980s would greatly expand the high-frequency capabilities of GaAs-based HEMTs and would open up many new application areas in communications and radar systems. 1985 saw the invention of the *Pseudomorphic HEMT*, or PHEMT [9, 10]. This is an AlGaAs/InGaAs/GaAs quantum-well structure where the enhanced transport properties of electrons in InGaAs coupled with the tight quantum-well confinement and large conduction band discontinuity between channel and barrier resulted in improved device scalability and performance. A second innovation was planar or *delta doping* of the AlGaAs barrier [11]. This allowed the thinning down of the barrier yielding improvements in transconductance, channel aspect ratio, and device scalability. The removal of dopants from directly underneath the gate also improved the breakdown voltage of the device. The combination of planar doping and pseudomorphic InGaAs channel propelled PHEMTs to the 0.1 μm gate length regime and the achievement of record noise and power performance up to very high frequencies (Figure 20.2).

The extraordinary capabilities of the PHEMT in terms of noise, power, and low-loss switching at very high frequencies have made this device technology a success in the commercial arena. Some of the most notable applications are cell phones, broadband wireless, satellite communications, instrumentation, cable TV, DBS, GPS, space, radar, fiber-optic communications, sensing, military, and many others.

PHEMT IC technology is now manufactured around the world on 6″ GaAs wafers and is available in foundry mode from several companies. Following a path similar to Si, new PHEMT IC technologies are emerging that monolithically integrate enhancement- and depletion-mode devices, HBTs, and high-quality passive elements to enable enhanced functionality single-chip systems [12]. These will make possible the development of highly integrated front-end modules for high-volume cellular phone applications.

In parallel with progress in GaAs HEMT technology, modulation doping and HEMTs were also demonstrated in other material systems, most notably, InAlAs/InGaAs [13, 14], AlSb/InAs [15, 16], AlGaN/GaN [17, 18], and even SiGe/Si [19, 20]. Similarly, modulation-doped two-dimensional hole gases and *High Hole Mobility Transistors* have also been realized in several heterostructure systems such as AlGaAs/GaAs [21, 22], AlGaAs/InGaAs [23, 24], InAlAs/InGaAs [25, 26], Si/SiGe [27, 28], AlAsSb/GaSb [29, 30], AlGaSb/InGaSb [31], and others. This spurred the co-integration of n-channel

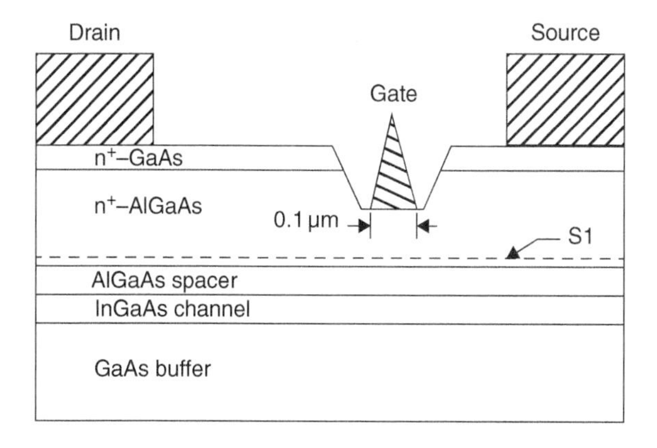

Figure 20.2 Cross-sectional schematic of delta-doped pseudomorphic HEMT ca. 1987. Source: Adapted from Ketterson et al. 10.

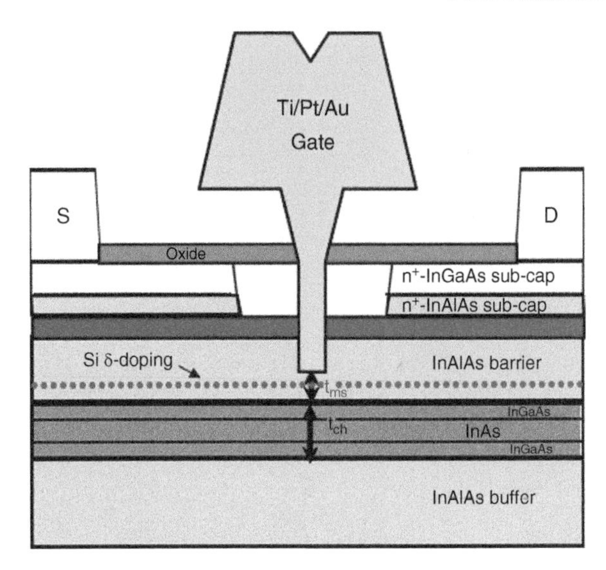

Figure 20.3 Schematic cross-section of state-of-the-art InAlAs/InGaAs HEMT on InP ("InP HEMT"). Source: Adapted from Kim et al. 35.

and p-channel devices which was achieved in the AlGaAs/GaAs [32] and InAlAs/InGaAs systems [33] and the demonstration of complementary circuits. Some of these circuits attained amazing levels of integration, in excess of 10^5 transistors [34].

InAlAs/InGaAs HEMTs on InP (also known as "InP HEMTs") have particularly excelled at the extremes of noise and frequency. Increasing the InAs content of InGaAs rapidly improves electron transport properties and enlarges the conduction band discontinuity which results in enhanced carrier confinement. This unique set of attributes confers ultra-scaled pseudomorphic InAlAs/InGaAs HEMTs on InP with outstanding low noise and high-frequency characteristics. These are in fact the first transistors of any kind in any material system to simultaneously exhibit f_T and f_{max} values in excess of 640 GHz [35] (Figure 20.3). The record f_T stands today at 738 GHz [36]. Particularly remarkable is the fact that these results were obtained at an operating voltage of 0.5 V, showcasing the outstanding low power potential of this technology. Separately, f_{max} values in excess of 1 THz [37, 38] and a noise figure as low as 0.71 dB at 95 GHz [39] have also been demonstrated in this material system. Ultralow noise characteristics are reported on InAlAs/InGaAs HEMTs at cryogenic temperatures [40]. Figure 20.4 shows the evolution of the current-gain cut-off frequency in GaAs MESFETs, GaAs PHEMTs, and InP HEMTs as a function of time.

InGaAs HEMTs on InP are uniquely suited for mm and sub-mm wave applications in radar, radio astronomy, space communications, high-resolution imaging arrays, high-capacity wireless communications, high data rate optical fiber systems, and sub-mm wave spectroscopy. A variety of sub-mm wave amplifiers and other circuits have been demonstrated operating at frequencies approaching and exceeding 1 THz [41–43]. A 120 Gb s^{-1} wireless link operating in the 300 GHz band has also been demonstrated [44].

Similar InAlAs/InGaAs HEMTs with high InAs channel content have also been metamorphically grown on GaAs substrates [45, 46]. These *metamorphic HEMTs*, or "MHEMTs," provide comparable performance with improved manufacturability (e.g. larger wafer size, improved reproducibility, greater ease of wafer handling, GaAs backside process flow), lower cost, and better-established packaging technology. InAlAs/InGaAs MHEMT processes have been available in foundry mode since 2003 [47]. A wide range of circuits with performance that rivals the best InP HEMT technologies have been demonstrated, including sub-mm wave amplifiers that operate at 460 GHz [48].

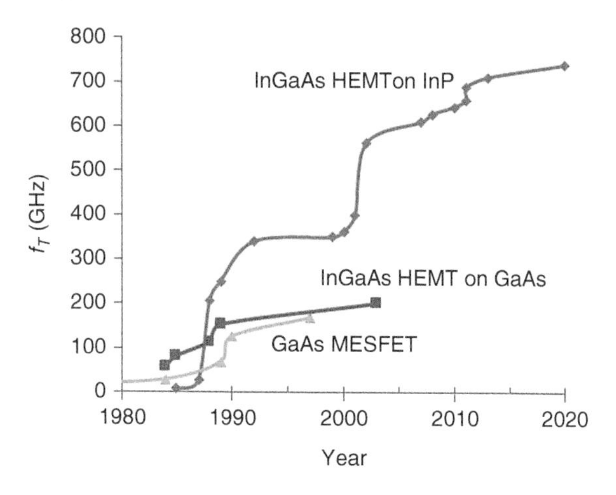

Figure 20.4 Record short-circuit current-gain cut-off frequency (f_T) for GaAs MESFETs, GaAs PHEMTs, and InP HEMTs as a function of year of demonstration.

Remarkable as the performance of GaAs- and InP-based HEMTs is, it is clear from looking at Figure 20.4 that these devices have reached their scaling limit. Performance improvements over the years were obtained through harmonious scaling of the vertical and lateral dimensions and by increasing the InAs composition in the channel. These improvements have essentially reached the end of the road and little benefit is to be expected without a radical device structure overhaul. This is because, further scaling of the traditional planar HEMT structure has hit a limit due to exponentially increasing gate leakage current as the wide bandgap barrier is thinned down [49]. Options for future performance enhancements include the introduction of a dielectric in the gate stack followed by further gate length scaling by adopting a FinFET configuration or a lateral or vertical nanowire geometry [50].

One of the most exciting device structures at the present time is the AlGaN/GaN HEMT. Modulation doping in this system was first demonstrated in 1992 [17] and a working transistor was first reported in 1993 [18], in both cases by Khan and collaborators at APA Optics. The enormous current interest in this system stems from several factors. First, the wide bandgap of GaN yields a high breakdown voltage which enables the high-voltage operation of transistors. Also, the saturation velocity of electrons in GaN is over a factor of two higher than Si, which is valuable for high-frequency operation. In addition, in a HEMT structure, a high electron concentration is induced at the AlGaN/GaN interface yielding high current. The confluence of all these factors results in devices with unique power amplification attributes at frequencies all the way up to the millimeter-wave regime [51]. Device designs suitable for power switching in electrical power management applications have been developed and they are many quick inroads in all manner of power management applications [52]. GaN HEMTs constitute a world in themselves and will not be treated any further in this review.

20.3 Modulation-Doped Structures in Physics

The AlGaAs/GaAs heterostructure is perhaps the most perfect crystalline interfacial system ever fabricated by humans. AlGaAs/GaAs structures can be grown with the atoms at the interface in near-perfect crystalline registry and therefore without dangling chemical bonds or interface traps. In the absence of disorder, electrons confined to the AlGaAs/GaAs interface constitute a two-dimensional electron gas and manifest striking quantum phenomena. In fact, the discovery of the *integral quantum-Hall effect*

(IQHE) in the Si/SiO_2 system spurred a great deal of interest in AlGaAs/GaAs modulation-doped structures in the quest for a more perfect interface where 2D electrons would exhibit extremely precise quantization of the Hall resistance.

Toward this goal, a great deal of effort has been devoted to eliminating all forms of scattering and improving the mobility of electrons at the AlGaAs/GaAs interface [53]. A continuous stream of innovations in MBE technology has brought the low temperature mobility from an early value of around $1.5\times10^4\,cm^2\,V^{-1}\,s^{-1}$ to a current record value of $5.7\times10^7\,cm^2\,V^{-1}\,s^{-1}$ [54]. This is the highest mobility on record in any semiconductor structure and attests to the extreme purity and perfection of the materials involved.

As the mobility increased, a veritable treasure trove of new and unexpected physics emerged that was previously hidden by disorder [4]. The *fractional quantum Hall effect* (FQHE) was discovered on a sample with a mobility of $9\times10^4\,cm^2\,V^{-1}\,s^{-1}$ [55]. Zero resistance states in magnetotransport induced by microwaves were discovered in a sample with a mobility of $3\times10^6\,cm^2\,V^{-1}\,s^{-1}$. Interactions between composite fermions (bound states of electrons and quantized vortices introduced to explain the FQHE) were first postulated on a sample with a mobility of $1\times10^7\,cm^2\,V^{-1}\,s^{-1}$. A very fragile even-denominator 5/2 quantum Hall state was discovered on a sample with a mobility of $3.1\times10^7\,cm^2\,V^{-1}\,s^{-1}$. This is speculated to possibly reveal the existence of "non-abelian quasi-particles" that hold potential for future quantum computers.

In parallel with this effort, using gates of various shapes on the surface of AlGaAs/GaAs heterostructures and applying appropriate potentials, the 2DEG can be carved into electron distributions of capricious geometries. In this way, 1DEG (*quantum wires*), 0DEG (*quantum dots*), and many other structures have been created where electrons are coherent and show prominent quantum phenomena over very long distances. In quantum wires, for example, quantized conductance has been observed [56]. Quantum dots exhibit a discrete set of energies, much like in an artificial atom. Field-effect-induced quantum dots can be made so small that they are able to contain just one electron [57]. Electrons can be induced to tunnel in and out of this artificial atom one at a time in what is called a *single-electron transistor* [58].

These findings already go beyond physics. The fact that the Hall resistance plateaus in the IQHE are only linked to fundamental constants (Planck's constant and the charge of the electron) and the extreme accuracy with which the Hall resistance can be measured has led to a new international standard for resistance. The Ohm today is defined through a measurement of the Hall resistance of the first plateau in an array of quantum Hall bars fabricated in AlGaAs/GaAs heterostructures [59]. A precision of a few parts in 10^9 is achieved.

20.4 Exciting Prospects

The future is bright for structures and devices based on two-dimensional carrier confinement in electronics, communications, physics, and other disciplines. GaAs, InP, and GaN HEMTs will continue their march toward greater levels of integration, higher frequency, higher power, higher power efficiency, lower noise, and lower cost. Society will derive great benefits from this. GaN, in particular, offers the promise of penetrating into the high-power high-frequency territory of vacuum tubes and lead to lighter, more efficient, and more reliable radar and communication systems on the ground and in space. The GaN system also is poised to revolutionize power electronics which will reach and deeply impact consumer, industrial, transportation, communications, and military systems. Here, an enhancement-mode device with very low leakage current is favored and MOS-HEMT or MISFET structures and p-GaN Gate HEMTs are likely to prevail. Material systems currently under research, such as the antimonides, might also emerge into the real world and enable a new generation of ultralow power and high-speed systems.

If the past serves as any guidance, it is reasonable to expect that continued progress in the quality of AlGaAs/GaAs structures will result in new insights into the often bizarre physics of quantized electrons. Similar improvements in other heterostructure material systems will also bring surprises. The FQHE, perhaps the greatest exponent for impeccable purity and atomic order, has also been observed in ZnO, SiGe, and GaN [53] and so the field is ripe for surprises.

The area of sensors is also pregnant with possibilities. The ultrahigh mobilities that are possible in InAlSb/InAsSb system enable high-sensitivity micro-Hall sensors for many applications including scanning Hall probe microscopy and biorecognition [60]. Three-axis Hall magnetic sensors have been demonstrated in micromachined AlGaAs/GaAs structures [61]. These devices have the potential of measuring the three vector components of a magnetic field and they may find use in future electronic compasses and navigation. 2DEG-based devices can also be used for THz detection, mixing, and frequency multiplication [62] as has been demonstrated in the AlGaAs/GaAs system [63]. GaN and related materials are chemically stable semiconductors with strong piezoelectric polarization. "Functionalized" GaN-based HEMT structures can be used as sensitive and robust chemical and biological sensors in gases and liquids [64]. More sophisticated sensors can be realized by combining functionalized GaN-based 2DEG structures and free-standing resonators [65]. These offer the possibility of simultaneous measurements of several properties such as viscosity, pH, and temperature.

20.5 Conclusions

The first 40 years of the HEMT, rich and exciting as they have been, represent just the opening act. The best for HEMT electronics is yet to come. *Kroemer's Lemma of New Technology* is appropriate here: "The principal applications of any sufficiently new and innovative technology have always been – and will continue to be – applications *created* by that technology" [66]. We are just now starting to see what 2DEG semiconductor heterostructures are capable of. From the cosmic to the personal, the next 40 years will bring us wonderful surprises and satisfactions.

Acknowledgments

The author is grateful to several colleagues for many valuable suggestions: Ray Ashoori (MIT), Brian Bennett (NRL), Bobby Brar (Teledyne), P. C. Chao (BAE Systems), Takatomo Enoki (NTT), Augusto Guttierez-Aitken (Northrop Grumman), Eric Higham (Strategy Analytics), Jose Jimenez (Qorvo), Marc Kastner (MIT), Richard Lai (Northrop Grumman), Takashi Mimura (Fujitsu), Tomas Palacios (MIT), Loren Pfeiffer (Princeton University), Philip Smith (BAE Systems), Tetsuya Suemitsu (Tohoku University), and Ling Xia (MIT).

References

[1] Mimura, T., Hiyamizu, S., Fujii, T., and Nambu, K. (1980). A new field-effect transistor with selectively doped GaAs/n-AlxGa1-xAs heterojunctions. *Japn. J. Appl. Phys.* 19: L225.

[2] Dingle, R., Stormer, H.L., Gossard, A.C., and Wiegmann, W. (1978). Electron mobilities in modulation-doped semiconductor heterojunction superlattices. *Appl. Phys. Lett.* 33: 665.

[3] Mimura, T. (2005). Development of high electron mobility transistor. *Japn. J. Appl. Phys.* 44: 8263.

[4] Pfeiffer, L. and West, K.W. (2003). The role of MBE in recent quantum hall effect physics discoveries. *Phys. E.* 20: 57.

[5] Joshin, T.M.K., Hiyamizu, S., Hikosaka, K. et al. (1981). High electron mobility transistor logic. *Japn. J. Appl. Phys.* 20: L598.

[6] Niori, M., Saito, T., Joshin, K., and Mimura, T. (1983). A 20 GHz high electron mobility transistor amplifier for satellite communiations. *1983 IEEE International Solid-State Circuits Conference*, New York, NY (23–25 February 1983). IEEE. p. 198.

[7] Weber, J.C. and Pospieszalski, M.W. (2002). Microwave instrumentation for radio astronomy. *IEEE Trans. Microw. Theory Techn.* 50: 986.

[8] Mimura, T. (1990). HEMT technology: impact on computer and communications systems. *Surf. Sci.* 228: 504.

[9] Rosenberg, J.J., Benlamri, M., Kirchner, P.D. et al. (1985). An $In_{0.15}$ $Ga_{0.85}$ As/GaAs pseudomorphic single quantum Well HEMT. *IEEE Electron Dev. Lett.* 6: 491.

[10] Ketterson, A., Moloney, M., Masselink, W.T. et al. (1985). High transconductance InGaAs/AlGaAs pseudomorphic modulation-doped field-effect transistors. *IEEE Electron Dev. Lett.* 6: 628.

[11] Chao, P.C., Smith, P.M., Duh, K.H.G. et al. (1987). High performance 0.1 um gate-length planar-doped HEMTs. *Proceedings., IEEE/Cornell Conference on Advanced Concepts in High Speed Semiconductor Devices and Circuits,* Ithaca, NY (7–9 August 1989). IEEE. p. 410.

[12] Henderson, T., Middleton, J., Mahoney, J. et al. (2007). High-performance BiHEMT HBT /E-D pHEMT integration. *CS MANTECH,* p. 247.

[13] Cheng, K.Y., Cho, A.Y., Drummond, T.J., and Morkoc, H. (1982). Electron mobilities in modulation doped $Ga_{0.47}$ $In_{0.53}$ $As/Al_{0.48}$ $In_{0.52}As$ heterojunctions grown by molecular beam epitaxy. *Appl. Phys. Lett.* 40: 147.

[14] Chen, C.Y., Cho, A.Y., Cheng, K.Y. et al. (1982). Depletion mode modulation doped $Al_{0.48}$ $In_{0.52}$ As – $Ga_{0.47}$ $In_{0.53}$ As heterojunction field effect transistors. *IEEE Electron Dev. Lett.* 3: 152.

[15] Chang, C.H., Chang, L.L., Mendez, E.E. et al. (1984). Electron densities in InAs/AlSb quantum wells. *J. Vac. Sci. Tech. B* 2: 214.

[16] Tuttle, T. and Kroemer, H. (1987). An AlSb/InAs/AlSb quantum well HFET. *IEEE Trans. Electron Dev.* 34: 2358.

[17] Asif Khan, M., Kuznia, J.N., Van Hove, J.M. et al. (1992). Observation of a two-dimensional electron gas in low pressure metalorganic chemical vapor deposited GaN-$Al_xGa_{1-x}N$ heterojunctions. *Appl. Phys. Lett.* 60: 3027.

[18] Asif Khan, M., Bhattarai, A., Kuznia, J.N., and Olson, D.T. (1993). High electron mobility transistor based on a GaN- Al_xGa_{1-x} N heterojunction. *Appl. Phys. Lett.* 63: 1214.

[19] Abstreiter, G., Brugger, H., Wolf, T. et al. (1985). Strain-induced two-dimensional electron gas in selectively doped Si/Si_xGe_{1-x} superlattices. *Phys. Rev. Lett.* 54: 2441.

[20] Daembkes, H., Herzog, H.-J., Jorke, H. et al. (1986). The n-channel SiGe/Si modulation-doped field-effect transistor. *IEEE Trans Electron. Dev.* 33: 633.

[21] Stormer, H.L. and Tsang, W.-T. (1980). Two-dimensional hole gas at a semiconductor heterojunction interface. *Appl. Phys. Lett.* 36: 685.

[22] Stormer, H.L., Baldwin, K., Gossard, A.C., and Wiegmann, W. (1984). Modulation-doped field-effect transistor based on a two-dimensional hole gas. *Appl. Phys. Lett.* 44: 1062.

[23] Fritz, I.J., Doyle, B.L., Schirber, J.E. et al. (1986). Influence of built-in strain on Hall effect in InGaAs/GaAs quantum well structures with p-type modulation doping. *Appl. Phys. Lett.* 49: 581.

[24] Lee, C.-P., Wang, H.T., Sullivan, G.J. et al. (1987). High-transconductance p-channel InGaAs/AlGaAs modulation-doped field effect transistors. *IEEE Electron Dev. Lett.* 8: 85.

[25] Razeghi, M., Maurel, P., Tardella, A. et al. (1986). First observation of two-dimensional hole gas in a $Ga_{0.47}$ $In_{0.53}$ As/InP heterojunction grown by metalorganic vapor deposition. *J. Appl. Phys.* 60: 2453.

[26] Feuer, D., He, Y., Tennant, D.M. et al. (1989). InP-based highfet's for complementary circuits. *IEEE Trans. Electron Dev.* 36: 2616.

[27] People, R., Bean, J.C., Lang, D.V. et al. (1984). Modulation doping in Ge_xSi_{1-x}/Si strained layer heterostructures. *Appl. Phys. Lett.* 45: 1231.

[28] Pearsall, T.P. and Bean, J.C. (1986). Enhancement- and depletion-mode p-channel Ge_xSi_{1-x} modulation-doped FET's. *IEEE Electron Dev. Lett.* 7: 308.

[29] Hansen, W., Smith, T.P. III, Piao, J. et al. (1990). Magnetoresistance measurements of doping symmetry and strain effects in GaSb-AlSb quantum wells. *Appl. Phys. Lett.* 56: 81.

[30] Luo, L., Longenbach, K.F., and Wang, W.I. (1990). P-channel modulation-doped field-effect transistors based on $AlSb_{0.9}As_{0.1}/GaSb$. *IEEE Electron Dev. Lett.* 11: 567.

[31] Klem, J.F., Lott, J.A., Schirber, J.E. et al. (1993). Strained quantum well modulation-doped InGaSb/AlGaSb structures grown by molecular beam epitaxy. *J. Electron. Mater.* 22: 315.

[32] Cirillo, N.C., Jr., Shur, M.S., Vold, P.J. et al. (1985). Complementar heterostructure insulated gate field effect transistors (HIGFETs). *1985 International Electron Devices Meeting*, Washington, DC (1–4 December 1985). IEEE. p. 317.

[33] Swirhun, S., Nohava, T., Huber, J. et al. (1991). P- and N-channel InAlAs/InGaAs heterojunction insulated gate FETs (HIGFETs) on InP. *Third International Conference Indium Phosphide and Related Materials*, Cardiff (8–11 April 1991). IEEE. p. 238.

[34] Brown, R.B., Bernhardt, B., LaMacchia, M. et al. (1998). Overview of complementary GaAs technology for high-speed VLSI circuits. *IEEE Trans. VLSI Syst.* 6: 47.

[35] Kim, D.-H. and del Alamo, J.A. (2010). 30 nm InAs PHEMTs with f_T=644 GHz and f_{max}=681 GHz. *IEEE Electron Dev. Lett.* 31: 806.

[36] Jo, H.-B., Yun, S.-W., Kim, J.-G. et al. (2020). L_g=19 nm $In_{0.8}Ga_{0.2}As$ composite-channel HEMTs with f_t=738 GHz and f_{max}=492 GHz. *2020 IEEE International Electron Devices Meeting (IEDM)*, San Francisco, CA (12–18 December 2020). IEEE. p. 167.

[37] Lai, R., Mei, X.B., Deal, W.R. et al. (2007). Sub 50 nm InP HEMT Device with f_{max} greater than 1 THz. *2007 IEEE International Electron Devices Meeting*, Washington, DC (10–12 December 2007). IEEE. p. 609.

[38] Kim, D.-H., Alamo, J. A. del., Chen, P.et al. (2010). 50-nm E-mode $In_{0.7}Ga_{0.3}As$ PHEMT on 100-mm InP substrate with f_{max}>1 THz. *2010 International Electron Devices Meeting*, San Francisco, CA (6–8 December 2010). IEEE. p. 692.

[39] Takahashi, T., Makiyama, K., Hara, N. et al. (2008). Improvement in high frequency and noise characteristics of InP-based HEMTs by reducing parasitic capacitance. *2008 20th International Conference on Indium Phosphide and Related Materials*, Versailles (25–29 May 2008). IEEE. Paper MoA3.4.

[40] Cha, E., Wadefalk, N., Moschetti, G. et al. (2020). In P HEMTs for Sub-mW cryogenic low-noise amplifiers. *EDL* 41: 1005.

[41] Deal, W.R., Mei, X.B., Radisic, V. et al. (2007). Demonstration of a S-MMIC LNA with 16-dB Gain at 340-GHz. *Compound Semiconductor IC Symposium*, Portland, OR (14–17 October 2007). IEEE.

[42] Deal, W.R., Mei, X.B., Radisic, V. et al. (2010). Demonstration of a 0.48 THz amplifier module using InP HEMT transistors. *IEEE Microw. Wireless Comp. Lett.* 20: 289.

[43] Deal, W.R., Leong, K., Yoshida, W. et al. (2016). InP HEMT integrated circuits operating above 1,000 GHz. *2016 IEEE International Electron Devices Meeting (IEDM)*, San Francisco, CA (3–7 December 2016). IEEE. p. 707.

[44] Hamada, H., Tsutsumi, T., Matsuzaki, H. et al. (2021). Ultra-high-speed 300-GHz InP IC technology for beyond 5G. *NTT Tech. Rev.* 19: 74.

[45] Wang, G.-W., Chen, Y.-K., Schaff, W.J. et al. (1988). A 0.1 um gate $Al_{0.5}$ $In_{0.5}$ As/$Ga_{0.5}$ $In_{0.5}$As MODFET fabricated on GaAs substrates. *IEEE Trans. Electron Dev.* 35: 818.

[46] Smith, P.M. Dugas, D., Chu, K. et al. (2003). Progress in GaAs Metamorphic HEMT technology for microwave applications. *25th Annual Technical Digest 2003. IEEE Gallium Arsenide Integrated Circuit (GaAs IC) Symposium*, San Diego, CA (9–12 November 2003). IEEE. p. 21.

[47] Chertouk, M. Chang, W.D., Yuan, C.G. et al. (2003). The first 0.15 um MHEMT 6" GaAs foundry service: highly reliable process for 3 V drain bias operations. *GaAs Symposium*, Munich. IEEE. p. 9.

[48] Tessmann, A., Leuther, A., Loesch, R. et al. (2010). A metamorphic HEMT S-MMIC amplifier with 16.1 dB gain at 460 GHz. *Comp. Semic. IC Symp.* Monterrey, CA (3–6 October 2010).

[49] Alamo, J. A. del. (2016). In As High-electron mobility transistors on the path to THz operation. *SSDM.*

[50] del Alamo, J.A. (2011). Nanometre-scale Electronics with III-V compound semiconductors. *Nature* 479: 317.

[51] Mishra, U., Shen, L., Kazior, T.E., and Wu, Y.-F. (2008). GaN-based RF power devices and amplifiers. *Proc. IEEE* 96: 287.

[52] Chen, K.J., Haberlen, O., Lido, A. et al. (2017). GaN-on-Si power technology: devices and applications. *IEEE Trans. Electron Dev.* 64: 779.

[53] Schlom, D.G. and Pfeiffer, L.N. (2010). Upward mobility rocks! *Nat. Mater.* 9: 881.

[54] Chung, Y.J., Gupta, A., Baldwin, K.W. et al. (2022). Understanding limits to mobility in ultrahigh-mobility GaAs two-dimensional electron systems: 100 million cm^2/V.s and beyond. *Phys. Rev. B* 106: 075134.

[55] Tsui, D.C., Stormer, H.L., Gossard, A.C. et al. (1982). Two-dimensional magnetotransport in the extreme quantum limit. *Phys. Rev. Lett.* 48: 1559.

[56] van Wees, B.J., van Houten, H., Beenakker, C.W.J. et al. (1988). Quantized conductance of point contacts in a two-dimensional electron gas. *Phys. Rev. Lett.* 60: 848.

[57] Ashoori, R. (1996). Electron in artificial atoms. *Nature* 379: 413.

[58] Kastner, M. (2000). The single electron transistor and artificial atoms. *Ann. Phys.* 11: 885.

[59] Poirier, W. and Schopfer, F. (2009). Resistance metrology based on the quantum hall effect. *Eur. Phys. J. Spec. Top.* 172: 207.

[60] Bando, M., Ohashi, T., Dede, M. et al. (2009). High sensitivity and multifunctional micro-Hall sensors fabricated using InAlSb/InAsSb/InAlSb heterostructures. *J. Appl. Phys.* 105: 07E909.

[61] Todaro, M.T., Sileo, L., Epifani, G. et al. (2010). A fully integrated GaAs-based three-axis Hall magnetic sensor exploiting self-positioned strain released structures. *J. Micromech. Microeng.* 20: 105013.

[62] Dyakonov, M. and Shur, M. (1996). Detection, mixing, and frequency multiplication of Teraherz radiation by two-dimensional electronic fluid. *IEEE Trans. Electron Dev.* 43: 380.

[63] Kawano, Y. and Ishibashi, K. (2010). On-chip near-field terahertz detection based on a two-dimensional electron gas. *Phys. E.* 42: 1188.

[64] Pearton, S.J., Kang, B.S., Kim, S. et al. (2004). GaN-based diodes and transistors for chemical, gas, bilological and pressure sensing. *J. Phys. Condens. Matter* 16: R961.

[65] Niebelschutz, F., Cimalla, V., Tonisch, K. et al. (2008). AlGaN/GaN-based MEMS with two-dimensional electron gas for novel sensor applications. *Phys. Stat.Sol. C* 5: 1914.

[66] Kroemer, H. (2000). Nobel lecture: quasielectric field and band offsets: teaching electrons new tricks. *Rev. Mod. Phys.* 73: 783.

Chapter 21

The Thin Film Transistor and Emergence of Large Area, Flexible Electronics and Beyond

Yue Kuo[1], Jin Jang[2], and Arokia Nathan[3,4]

[1]Thin Film Nano & Microelectronics Research Laboratory, Texas A&M University, College Station, TX, USA
[2]Advanced Display Research Center, Kyung Hee University, Seoul, Korea
[3]Darwin College, Cambridge, UK
[4]School of Information Sciences and Engineering, Shandong University, Qingdao, China

21.1 Birth of Large Area Electronics

The integrated circuit (IC), or chip technology, and thin film transistor (TFT)-based active-matrix flat panel displays (AMFPDs) are presently the two largest electronic industries. Both the IC and the TFT backplane are made by a similar procedure of repeated deposition, patterning, etching, and annealing steps [1]. The substrate for ICs is the single-crystal silicon wafer and that of the TFT backplane is the glass panel. The concept and operating principle of the TFT are similar to those of the metal-oxide-semiconductor field-effect transistor (MOSFET), as shown in Figure 21.1 [modified from ref. 2]. The first functional TFT was reported by Weimer in 1962 [3]. It has a staggered structure with a microcrystalline CdS semiconductor layer, as shown in Figure 21.2a. The output characteristics are shown in Figure 21.2b. In the subsequent year, the tin oxide TFT was reported [4].

Originally, the application of the TFT was intended for complex circuits, which was never realized due to the emergence of the single-crystal silicon-based MOSFETs. The quest for new applications continued taking into consideration the no-requirement-advantage of crystalline substrates. In 1973, Brody et al. [5] demonstrated the CdSe TFT AMLCD that provides the gray-scale switching capability video image of the LC in high-resolution pixels. This opened the window to high-performance AMFPDs and human–machine interactivity, whose requirements are in stark contrast to the world of computational CMOS ICs which thrive on high speed and dense integration. Subsequently, TFTs made of various inorganic and organic semiconductors were reported. However, the progress toward the mass production of TFT LCDs was very slow due to issues in manufacturing, reliability, and lifetime.

75th Anniversary of the Transistor, First Edition. Edited by Arokia Nathan, Samar K. Saha, and Ravi M. Todi.

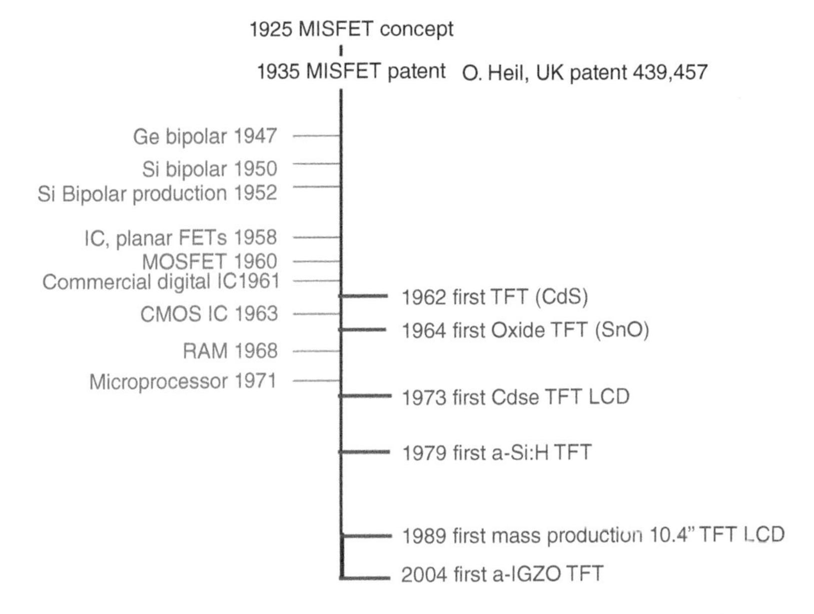

Figure 21.1 History of TFT and IC development. Source: Adapted from Kuo [2].

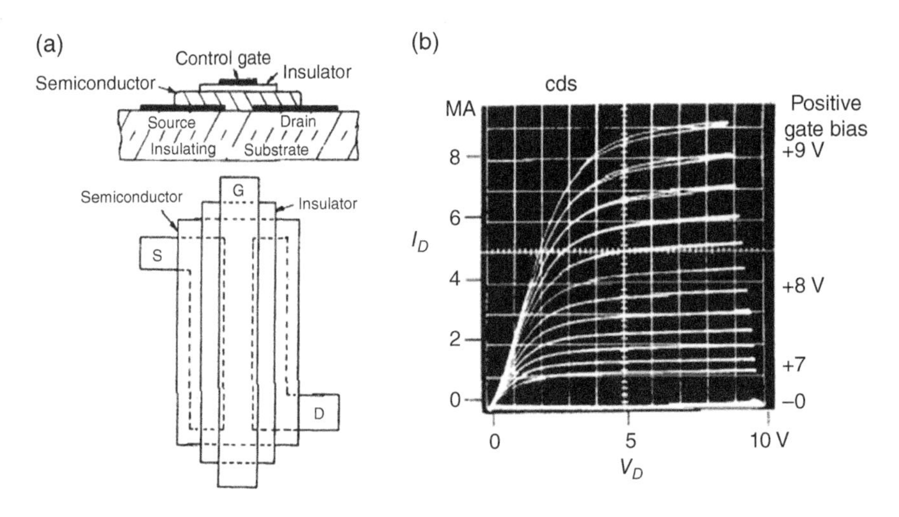

Figure 21.2 (a) Structure and (b) transfer characteristics of the first TFT. Source: Weimer [3]/IEEE.

In 1979, LeComber et al. [6] reported the hydrogenated amorphous silicon (a-Si:H) TFT made by the plasma-enhanced chemical vapor deposition (PECVD) method. This kind of TFT could be fabricated into a large-area array at a low temperature, e.g. ≤300 °C, with high throughput. Although the a-Si:H TFT has a low field-effect mobility of ≤1 $cm^2 V^{-1} s$, it is high enough to switch the LC in a pixel at video rates. Therefore, it immediately attracted widespread attention and stimulated vigorous R&D efforts in flat panel semiconductor companies. Eventually, the commercial production of high-quality, direct-view TFT LCDs was realized. The largest size display was 10.4 inches in 1990, which cost more than $2000, but was increased to >86 inches in 2023, which cost less than $1000, as shown in Figure 21.3. At the same time, the a-Si:H TFT-based flat panel X-ray medical imagers were proposed and translated to commercial products [7]. Instead of driving the LC pixel, the TFT reads out charges stored in an

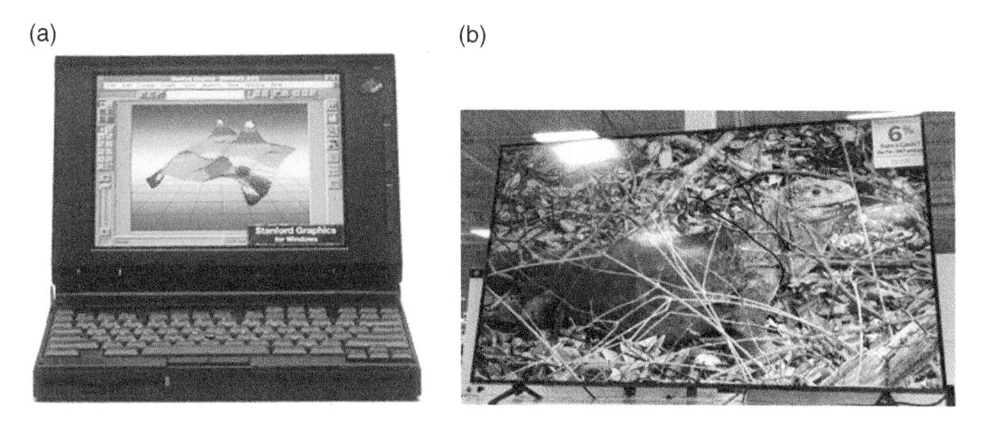

Figure 21.3 (a) A 10.4-inch TFT LCD in a 1990 IBM ThinkPad PC and (b) a LG 86-inch TFT LCD TV sold in a US supermarket in 2023. Source: IBM.

a-Si:H pin diode-connected-capacitor, above which a scintillation layer converts the X-rays into visible light. There are also many reports on a-Si:H TFT applications in gas-, pH-, photo-, and magnetic-sensing [2, 8].

21.2 Polycrystalline Silicon and Oxide Thin Film Transistor

Although the a-Si:H TFT technology has dominated the flat panel display market for over three decades, the low mobility greatly limits applications in products that require high-speed and complex operation, such as high-driving current capacity circuits for organic light-emitting diodes (OLEDs). Polycrystalline silicon (poly-Si) TFTs with field-effect mobilities larger than $100\,cm^2\,V^{-1}\,s$ have been reported for many years [9]. However, currently, poly-Si TFTs are mostly used in small-size displays in which the driving circuit and pixel transistors can be fabricated at the same time. The lack of large-area display or circuit applications is due to the absence of a proper fabrication process. To achieve high mobility, the poly-Si film in the TFT needs to contain low defects within grains and at grain boundaries, which can be achieved with high-temperature annealing or laser recrystallization of an a-Si film. The former is not compatible with the TFT glass substrates and the latter suffers from high manufacturing costs.

The first metal oxide TFT, which was made of the evaporated SnO_2 semiconductor layer, was reported in 1964 [4]. Later on, other oxide TFTs have been studied. This kind of TFT can have mobilities larger than that of the a-Si:H TFT. However, they had not been used in commercial products probably due to the high leakage current, e.g. several orders of magnitude larger than 10^{-12} A, and difficulty in device-grade film quality control. In 2004, Hosono's group reported room-temperature deposited amorphous indium gallium zinc oxide (a-IGZO) TFT, which enjoyed mobilities several times higher than that of the a-Si:H TFT [10]. Subsequently, TFTs with channels made of various types of metal oxide materials with mobilities larger than $>10\,cm^2\,V^{-1}\,s$ were reported. The transfer characteristics of a typical a-IGZO TFT fall between those of the a-Si:H and poly-Si TFT, as shown in Figure 21.4 [11]. The oxide semiconductor layer can be amorphous or polycrystalline. Typically, the amorphous oxide TFT has mobilities around $10\,cm^2\,V^{-1}\,s$ or higher mobility. Since the oxide TFT can have ultralow leakage current, e.g. of the order of $10^{-20}\,A\,\mu m^{-1}$, and can be fabricated using conventional low-temperature deposition processes, such as sputtering or PECVD, it can replace the a-Si:H TFT in AMLCDs. The high mobility enables applications for products that require a high driving current, such as micro-LEDs, or certain signal processing circuits. However, the fabrication process of the multicomponent oxide TFT has evolved to be more complex than that of the a-Si:H TFT, which can be a challenge for low-cost, large-area display production. Taking advantage of the high mobility of the amorphous oxide TFT, many new applications have

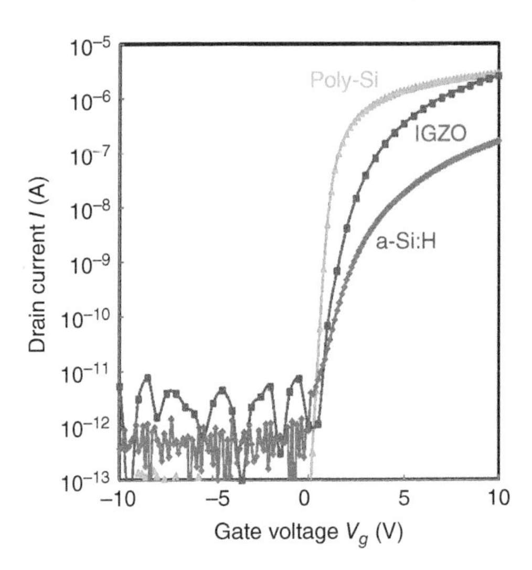

Figure 21.4 Transfer characteristics of a-Si:H, poly-Si, and a-IGZO TFTs. Source: Kuo and Chang [11]/IOP Publishing.

Table 21.1 Examples of applications of oxide TFTs.

Category	Specifics
A driving or reading unit for the attached device	OLEDs, LCDs, micro-LEDs, flexible displays, imagers, in-pixel memories, X-ray images, etc.
A switching unit in circuits	Inverters, ring oscillators, display gate drivers, RF IDs, nonvolatile memories, high-voltage transistors, etc.
A sensing device	CMOS image sensors, APS image sensors, UV sensors, near-IR sensors, temperature sensors, pressure sensors, biosensors, chemical sensors, pH sensors, etc.

been developed in the last decade. For example, Table 21.1 lists a broad range of oxide TFT applications from high-performance displays to circuits, sensors, and many novel devices.

Polycrystalline oxide TFT can have mobility several times higher than that of the amorphous oxide TFT but is often disadvantaged by a high leakage current. There are reports on low leakage current polycrystalline oxide TFTs [12, 13], which can open the door to more complex circuit applications. However, a greater understanding of reliability, manufacturability, and process compatibility with other devices and the substrate is required before acceptance by the industry.

21.3 Trends in TFT Development

21.3.1 Quest for CMOS Functionality

Although the current TFTs provide for proper functions for many products, there are numerous research efforts to improve some fundamental device characteristics. For example, high mobility enables more complicated applications, such as high-speed circuits or large driving currents. Mobility much higher than that of the commonly reported value of around 10 $cm^2 V^{-1}$ s for the amorphous oxide TFT to values larger than 100 $cm^2 V^{-1}$ s is the target of many researchers. This involves new materials or improved fabrication processes.

In addition, complimentary CMOS-type functions are needed for the TFT-based high-performance circuits. CMOS circuits made of n-type oxide and p-type low-temperature poly-Si TFTs have been demonstrated on mobile phones [14, 15]. Due to the complexities in the fabrication process, they are not used in large-area products. On the other hand, the amorphous oxide TFT array is easily fabricated on a low-temperature, large-area substrates but there is a lack of a properly functional p-channel oxide TFT, e.g. high mobility, low off-current, and thermodynamic stability. Currently, the p-channel poly-Si TFT is used in combination with the n-channel oxide TFT to integrate CMOS circuits in the commercial, small-size display products [16]. In the long run, circuits made of the same type of semiconductor material are necessary. The p-channel TFT needs to be developed, requiring new p-type oxide semiconductor materials to be discovered.

21.3.2 Low Power Consumption

The quest for low current, low voltage operation remains compelling in applications, such as analog front-end interfaces for high-resolution sensory signal acquisition, wearable or medical devices with inherently limited battery lifetime. It can be met with the TFTs operating in deep-subthreshold or near-OFF state regimes. Subthreshold operation of silicon CMOS devices had been intensively researched in the 1970s with great success in the watch industry [17]. In the case of TFTs, it has been demonstrated that biasing a Schottky-barrier TFT in the deep subthreshold region can improve the small signal gain and significantly reduce power consumption [18, 19]. Here, the transistor works in the leakage or so-called sleep mode where the drain-source (I_{DS}) current is extremely low [20, 21]. The Schottky-barrier TFT shows significant advantages in terms of high output resistance and high intrinsic gain. The Schottky barrier can be easily realized at the source or drain contacts with any semiconductor channel material. In IGZO TFTs, the Schottky contact at the source/drain contact can be formed by modulating the degree of compensation of oxygen vacancies/defects (V_{ox}), which in turn modulates the conductivity in the channel layer. Since the oxygen vacancies serve as electron donors, compensation can limit the injection of charge carriers at the source side forming a Schottky barrier [18]. The TFT has a much larger output resistance with no channel length dependence. The device shows bipolar behavior with geometry-independent electrical characteristics thus accommodating large dimensional variations in device features. This makes the subthreshold operational regime naturally suited for printed TFTs, offering a more tolerant design/fabrication window in analog circuit design. It should be noted that because current levels are low in subthreshold operation, so are the operating frequencies which are not a limitation since there is a plethora of low-frequency bio-related applications.

21.3.3 Flexible Electronics

For flexible electronic applications, the TFTs need to be fabricated on substrates that can be bent or twisted on repeated cycles without deterioration of material or device properties [22]. Polyimide (PI), polyethylene terephthalate (PET), polyethersulfone (PES), polycarbonate (PC), and silicone membranes are common flexible substrates for TFTs. Among them, PI is most used because it can withstand a temperature higher than 400 °C [23]. There are challenges in the substrate material. For example, the coefficient of thermal expansion (CTE), gas permeability, chemical stability, etc., especially during the plasma deposition or thermal treatment process, have to match with those of the TFT composing layers [24, 25]. It is often necessary to coat the substrate with an inorganic barrier to mitigate problems with the diffusion of moisture or contaminants through it. To achieve high flexibility, the substrates have to be very thin. A rigid supporting substrate is needed when TFTs are fabricated on top of it.

Subsequently, the thin substrate has to be separated from the supporting material with a mechanical or photo-blast method [26, 27]. The resistance of the TFT to mechanical bending is a major issue in flexible electronics. There are reports that electrical properties of the inorganic TFT deteriorate after the bending test, e.g. the shift of the threshold voltage (V_{th}), an increase of the leakage current, and increase in the subthreshold slope (SS) [28, 29]. The high-temperature-resistant stainless steel foil has been used as the substrate for poly-Si TFT AMOLEDs [30]. However, the manufacturability of this kind of product needs further research and development. On the other hand, organic TFTs (OTFTs) with polymer semiconductor channels are suitable for flexible electronic applications. They have been demonstrated in E-Ink displays and rollable OLEDs [31, 32]. The main advantage of OTFTs is that they can be made by non-vacuum, low-cost processes, such as 3D printing or solution-coating. However, OTFTs usually have a low resistance to electrical, thermal, and environmental stresses.

21.3.4 Biological Systems and Human–Machine Interactivity

There are many reports on TFT-based gas, liquid, and solid monitors that exploit the change of electrical characteristics from variation of the environmental condition or properties of attached devices [2, 8, 33]. OTFTs are especially useful in sensing chemical and biological systems due to the advantage of the easy attachment of functional groups to the polymer chain. Another advantage is that they can be fabricated with a low-cost, low-temperature solution coating or printing method.

Figure 21.5a shows an example of human–machine interactivity in which a human eye electro-oculogram signal (EOG) monitor made of two Schottky-barrier OTFTs [19]. The corresponding transfer characteristics and amplifier gain are shown in Figure 21.5b. The circuit has very low

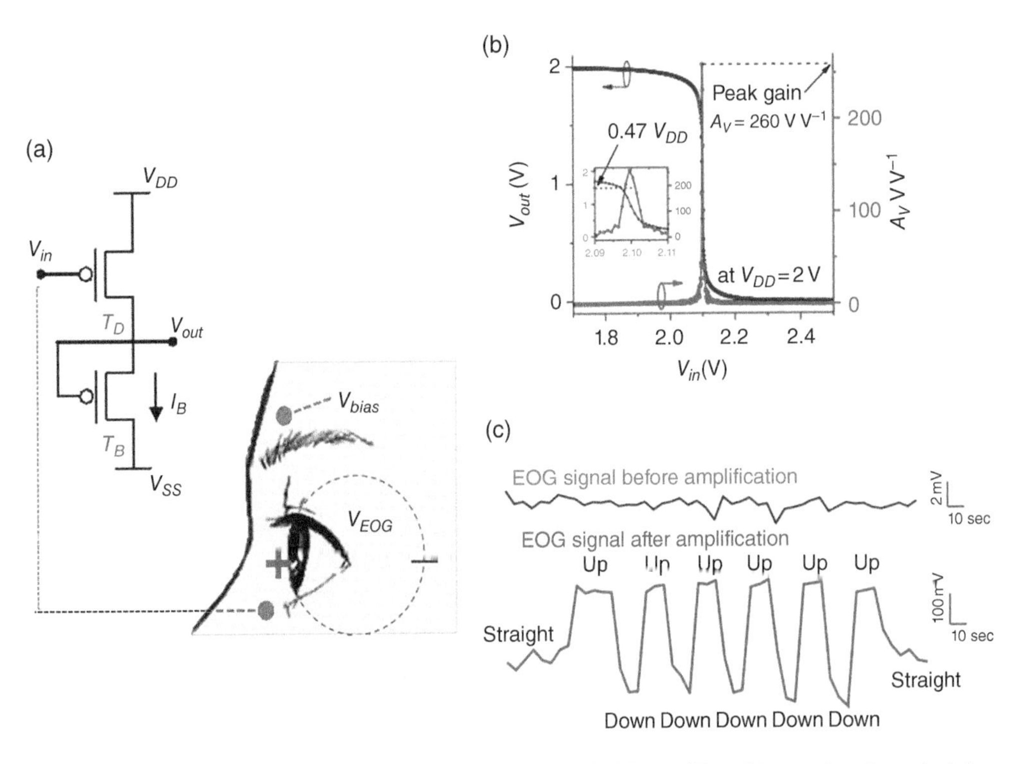

Figure 21.5 (a) An EOG human eye monitor made of a 2-OTFT amplifier, (b) transfer characteristics and amplifier gain, and (c) signals before and after amplification. Source: Jiang et al. [19]/AAAS.

power consumption of <1 nW and a very large peak voltage gain of 260 V V^{-1} when operated in the subthreshold regime with a bias current of 342 pA. The low power is due to the operation of the Schottky-barrier OTFT in the deep subthreshold region, which also can be electrically stable for an extended period [18, 19]. The large signal difference before and after amplification is obvious, as shown in Figure 21.5c, which can detect subtle eye movements for a better depiction of the virtual environment.

Currently, most OTFTs are made of p-type semiconductors and high mobility is usually achieved from the crystalline polymer channel [34, 35]. For practical applications, control over the grain distribution in the channel region is challenging. Also, reliability and environmental contamination are important issues to be addressed.

21.3.5 Low-Cost Solution Processing

OTFTs are suitable for flexible displays because of the flexibility of the organic thin film including polymer semiconductors. The main advantage of OTFTs is that they can be made by non-vacuum, solution processes such as spin coating, inkjet printing, dip coating, and spray coating. It can reduce the complicated processes involved in a conventional vacuum process. Pentacene is a typical organic semiconductor for the OTFT [36] but it cannot be deposited using a solution process. Therefore, 6,13-bis (tri-isopropylsilylethynyl) pentacene (TIPS pentacene) semiconductor has been developed for solution processing of TFTs [37]. OTFTs with mobility higher than 5 cm^2 V^{-1} s and prepared by solution processes have been developed using small molecule and polymer organic semiconductors [38].

It is noted that the silane-based liquid precursor has been developed and used for poly-Si TFT, exhibiting mobilities over 100 cm^2 V^{-1} s using spin coating [39]. But liquid Si is not easy to manufacture and has the demerit of easy oxidation during the thin film process.

Oxide semiconductors such as IGZO are deposited by vacuum process for industrial display products. To reduce the manufacturing cost, oxide thin films can be deposited by inkjet printing, spin coating, and spray pyrolysis [40]. Spin coating is the most popular technique, but film uniformity over a large area, important in display applications, is a concern. Inkjet printing also can be used for oxide TFT but it is seldom studied except in the printed OLED display case. On the other hand, spray pyrolysis can be used for large area deposition but the TFT's performance does not match that of the vacuum-processed counterpart [41]. It has been reported that bubble-free, dense a-IGZO thin films could be achieved by spray pyrolysis at a high substrate temperature of over 350 °C [42]. In order to compete with the vacuum-processed oxide TFTs, the process integration and device architecture should be further studied especially from the standpoint of achieving high mobility and excellent stability. Additionally, solution-processed gate insulators, especially high-κ dielectrics such as Al_2O_3, MgO, Zr_2O_3, Ta_2O_5, and HfO_2, have been studied in an attempt to reduce the operating voltage and power consumption. These requirements are of particular significance for mobile electronics [43]. Eventually, the realization of a complete oxide TFT, including gate dielectric, semiconductor channel, and passivation layers by solution processing on flexible substrates will open up a new generation of low cost, flexible electronics.

Acknowledgments

The authors would like to thank Dr. Chen Jiang, Tsinghua University, China, for technical assistance with the example of human–machine interaction using fully printed OTFTs.

References

[1] Kuo, Y. (2008). Thin-film transistor and ultra-large-scale integrated circuit: competition or collaboration. *Jpn. J. Appl. Phys.* 47 (3): 1845–1852.

[2] Kuo, Y. (2013). Thin film transistor technology – past, present, and future. *Electrochem. Soc. Interface* 22: 55–61.

[3] Weimer, P.K. (1962). The TFT a new thin-film transistor. *Proc. IRE* 50 (6): 1462–1469.

[4] Klasens, H.A. and Koelmans, H. (1964). A tin oxide field-effect transistor. *Solid State Electron.* 7 (9): 701–702.

[5] Brody, T.P., Asars, J.A., and Dixon, G.D. (1973). A 6 X6 inch 20 lines-per-inch liquid crystal display panel. *IEEE Trans. Electron Devices* ED-20 (11): 995–1001.

[6] LeComber, P.G., Spear, W.E., and Ghaith, A. (1979). Amorphous-silicon field-effect device and possible application. *Electron. Lett.* 15: 179–181.

[7] Moy, J.P. (1999). Large area X-ray detectors based on amorphous silicon technology. *Thin Solid Films* 337: 213–221.

[8] Kuo, Y. (ed.) (2004). Chapt. 11, non-LCD applications of a-Si:H TFTs. In: *Amorphous Silicon Thin Film Transistors*, 485–505. Norwell, MA: Kluwer Academic Publishers.

[9] Sameshima, T. (2004). Chapt. 5, poly-Si TFTs by laser crystallization methods. In: *Polycrystalline Silicon Thin Film Transistors* (ed. Kuo, Y.), 176–219. Norwell, MA: Kluwer Academic Publishers.

[10] Nomura, K., Ohta, H., Takagi, A. et al. (2004). Room-temperature fabrication of transparent flexible thin-film transistors using amorphous oxide semiconductors. *Nature* 432: 488–492.

[11] Kuo, Y. and Chang, G.-W. (2014). Thin film transistors as driving devices for attached devices. *ECS Trans.* 64 (10): 145–153.

[12] Yang, H.J., Seul, H.J., Kim, M.J. et al. (2020). High-performance thin-film transistors with an atomic-layer deposited indium gallium oxide channel: a cation combinatorial approach. *ACS Appl. Mater. Interfaces* 12: 52937–52951.

[13] Magari, Y., Kataoka, T., Yeh, W., and Furuta, M. (2022). High-mobility hydrogenated polycrystalline In_2O_3 (In_2O_3:H) thin-film transistors. *Nat. Commun.* 13: 1078.

[14] Chang, T.-K., Lin, C.-W., and Chang, S. (2022). LTPO TFT technology for AMOLEDs. *J. Soc. Inf. Disp.* 30 (3): 175–197.

[15] Kim, J. and Jang, J. (2022). Gate driver with low-temperature polycrystalline silicon and oxide thin-film-transistor circuits for low power consumption and narrow bezel AMOLED displays. *J. Soc. Inf. Disp.* 30: 482–488.

[16] Chen, C.D., Yang, B.-R., Liu, C. et al. (2017). Integrating poly-silicon and InGaZnO thin-film transistors for CMOS inverters. *IEEE Trans. Elec. Dev.* 64 (9): 3668–3671.

[17] Vittoz, E. and Fellrath, J. (1977). CMOS analog integrated circuits based on weak inversion operations. *IEEE J. Solid State Circuits* SSC-12 (3): 224–231.

[18] Lee, S. and Nathan, A. (2016). Subthreshold Schottky-barrier thin film transistors with ultralow power and high intrinsic gain. *Science* 354: 302–304.

[19] Jiang, C., Choi, H.W., Cheng, X. et al. (2019). Printed subthreshold organic transistors operating at high gain and ultralow power. *Science* 363: 719–723.

[20] Hosono, H. and Kumomi, H. (ed.) (2022). *Amorphous Oxide Semiconductors: IGZO and Related Materials for Display and Memory*. Wiley.

[21] Sekine, Y., Furutani, K., Shionoiri, Y. et al. (2011). Success in measurement the lowest off-state current of transistor in the world. *ECS Trans.* 37: 77–88.

[22] Nathan, A., Ahnood, A., Cole, M.T. et al. (2012). Flexible electronics: the next ubiquitous platform. *Proc. IEEE.* (Centennial Issue) 100: 1479–1510.

[23] Glescova, H. and Wagner, S. (1999). Amorphous silicon thin-film transistors on compliant polyimide foil substrates. *IEEE Elec. Dev. Lett.* 20: 473–475.

[24] Jang, J., Choi, M.H., and Cheon, J.H. (2010). *TFT Technologies for Flexible Displays*, 1143. SID 10 DIGEST.

[25] Sazonov, A., Nathan, A., and Striakhilev, D. (2000). Materials optimization for thin film transistors fabricated at low temperature on a plastic substrate. *J. Non-Cryst. Sol.* 266–269: 1329–1334.

[26] Miyasaka, M. (2007). *Suftla Flexible Microelectronics on their Way to Business*, 1673–1676. SID 07 Digest.

[27] Pecora, A., Maiolo, L., Cuscunà, M. et al. (2008). Low-temperature polysilicon thin film transistors on polyimide substrates for electronics on plastic. *Solid State Electron.* 52: 348–352.

[28] Billah, M.M., Hasan, M.M., and Jang, J. (2017). Effect of tensile and compressive bending stress on electrical performance of flexible a-IGZO TFTs. *IEEE Elec. Dev. Lett.* 38 (7): 890.

[29] Hasan, M.M., Billah, M.M., Naik, M.N. et al. (2017). Bending stress induced performance change in plastic oxide thin-film transistor and recovery by annealing at 300 °C. *IEEE Elec. Dev. Lett.* 38 (8): 1035.

[30] Ma, E.Y. and Wagner, S. (1999). Amorphous silicon transistors on ultrathin steel foil substrates. *Appl. Phys. Lett.* 74: 2661–2662.

[31] Song, D.H., Choi, M.H., Kim, J.Y. et al. (2007). Process optimization of the organic thin-film transistor by ink-jet printing of DH4T on plastic. *Appl. Phys. Lett.* 90: 53504–53507.

[32] Sou, A., Jung, S., Gili, E. et al. (2014). Programmable logic circuits for functional integrated smart plastic systems. *Org. Electron.* 15: 3111–3119.

[33] Kuo, Y. (2004). Chapt. 13, poly-Si TFTs for non-LCD applications. In: *Polycrystalline Silicon Thin Film Transistors*, 464–496. Norwell, MA: Kluwer Academic Publishers.

[34] Newman, C.R., Frisbie, C.D., da Silva Filho, D.A. et al. (2004). Introduction to organic thin film transistors and design of n-channel organic semiconductors. *Chem. Mater.* 16: 4436–4451.

[35] Nelson, S.F., Lin, Y.Y., Gundlach, D.J., and Jackson, T.N. (1998). Temperature-independent transport in high-mobility pentacene transistors. *Appl. Phys. Lett.* 72: 1854–1856.

[36] Campbell, R.B., Robertson, J.M., and Trotter, J. (1961). The crystal and molecular structure of pentacene. *Acta Crystallogr.* 14: 705–711.

[37] Park, S.K., Anthony, J.E., and Jackson, T.N. (2007). Solution-processed TIPS-pentacene organic thin-film-transistor circuits. *IEEE Elec. Dev. Lett.* 28: 877–879.

[38] Li, J., Zhao, Y., Tan, H.S. et al. (2012). A stable solution-processed polymer semiconductor with record high-mobility for printed transistors. *Sci. Rep.* 2: 754–762.

[39] Shimoda, T., Matsuki, Y., Furusawa, M. et al. (2006). Solution-processed silicon films and transistors. *Nature* 440: 783–786.

[40] Lee, D.-H., Chang, Y.-J., Herman, G.S., and Chang, C.-H. (2007). A general route to printable high-mobility transparent amorphous oxide semiconductors. *Adv. Mater.* 19: 843–847.

[41] Saha, J.K., Bukke, R.N., Mude, N.N., and Jang, J. (2020). Signifcant improvement of spray pyrolyzed ZnO thin flm by precursor optimization for high mobility thin film transistors. *Sci. Rep.* 10: 8999.

[42] Bae, J., Ali, A., and Jang, J. (2023). Spray pyrolyzed amorphous InGaZnO for high performance, self-aligned coplanar thin-film transistor backplanes. *Adv. Mater. Technol.* 8: 22007.

[43] Liu, A., Zhu, H., Sun, H. et al. (2018). Solution processed metal oxide high-κ dielectrics for emerging transistors and circuits. *Adv. Mater.* 30: 1706364.

Chapter 22

Imaging Inventions

Charge-Coupled Devices

Michael F. Tompsett

22.1 Setting the Stage for the Invention of the Charge-Coupled Device (CCD)

Every new idea is a product of its time and place and I have been lucky enough to make three pioneering imaging technology inventions by being in the right places at the right times and asking the right question. These inventions include uncooled thermal imaging, charge-coupled imaging devices, and a CMOS video analog-to-digital converter chip.

Working at EEV plc, a manufacturer of camera tubes in England, in 1968, I was shown an early proposal for a Hg-Cd-Te IR sensor operating at 77 °K, which required an undesirable power-hungry cooling mechanism. Reference [1] reviews a proposal by Hadni in 1963 in France for a ferroelectric bolometric camera tube that would operate in an impractical dielectric mode. My solution to these problems was the invention of a pyroelectric camera tube [2] operating at room temperature. This tube used the sensitivity of an electron beam to read-out the small charge variations caused by infra-red radiation on the surface of a pyroelectric crystal, and then restored the surface potential. This tube became highly valued by fire-fighters because it could see through smoke. At the same time, I also invented a solid-state version, which was developed 20 years later and followed by dramatic improvements, reviewed in [1]. These inventions created a new field of "uncooled thermal imaging," which has led to today's highly sensitive uncooled thermal imagers being very widely used in firefighting, search-and-rescue, medical applications, etc.

75th Anniversary of the Transistor, First Edition. Edited by Arokia Nathan, Samar K. Saha, and Ravi M. Todi.

In 1969, I wanted to develop silicon image sensor chips and I moved to Bell Labs, where there was clearly a need to reduce the size and cost of a proposed audio-visual Picturephone®. However, when I arrived at Bell Labs, I was given the task of solving a serious problem with a new MOS silicon vidicon camera tube [3]. After only a few hours of operation, an inexplicable bright area appeared in the video output, which rendered the tube useless. Fortunately, I soon figured out what was happening. Some electrons from the read-out beam were hitting the 200 V mesh in front of the MOS silicon wafer and generating 200 eV X-rays. These X-rays created positive charge at the silicon/silicon-oxide interface and formed an inversion layer, which allowed electrons to diffuse freely along this interface, causing the bright area on read out. What I did not realize at the time was that by explaining this to George Smith, my department head, I was probably contributing to the invention of CCDs. Additionally, I had stumbled on radiation damage in MOS devices, which later became a major problem in space electronics.

The importance of having the right technology available at the time of an invention is clearly shown in the following examples. The Field-Effect Transistor was invented by Lilienfeld in 1925 and a structure resembling an MOS transistor was also proposed by the inventors of the bipolar transistor 75 years ago. In neither case was a demonstration of an MOS transistor possible until Kahng and Atalla [4] were able to do this in 1960. I did not know Dawon Kahng until later, but he was also in George Smith's department and subsequently contributed ideas for CCDs. It was not only the availability of suitable MOS samples that allowed CCDs to be demonstrated so quickly, but it was also the lack of fine-line photolithography that gave CCDs a 25-year window of opportunity before CMOS imagers became viable.

As the inventors of the bipolar transistor discovered at Bell Labs in 1947, and so did I, Bell Labs was a wonderful place for invention with top-notch facilities and an informal, creative workplace. One rubbed shoulders with a remarkable number of very talented scientists and engineers, who had been attracted by its reputation.

22.2 The Invention of the CCD

As I said, every new idea is a product of its time and place and none more so than that of Charge Coupled Devices (CCDs). The following probably contributed to the invention.

- Willard Boyle and George Smith were together one afternoon apparently to discuss a demand from a very forceful vice-president, Jack Morton, to his executive director Willard Boyle. Morton wanted a silicon version of "magnetic bubbles," which were being developed at Bell Labs and were being touted with a lot of publicity for magnetic memory [5]. The "bubbles" or magnetic domains in certain rare earth garnet crystals would move across the crystal surface guided by metal chevrons in a rotating magnetic field and could be seen under a microscope.
- MOS transistors were well known, but my work with the silicon vidicon clearly showed that charge could move freely along the silicon/silicon-oxide interface.
- A supervisor in Smith's Department, Neil Berglund, was working on MOS Bucket Brigade devices, which had been invented at Dutch Philips [6]. These devices moved charge from diode to diode with a two-phase clock, but similarly to the operation of a CCD.
- Gene Gordon, Smith's boss, had received a pamphlet from Burroughs Corporation announcing a new flat-panel display, in which a plasma was moved across the display using electrodes clocked in a three-phase sequence. Gene had very recently given this pamphlet to Smith with the suggestion that he use the idea to make a display device.

The CCD invention was to put three sets of interleaved metal electrodes on an MOS wafer. Biasing would then be such as to collect packets of charge in potential wells in the silicon under every third electrode and then clock the electrodes in a three-phase sequence to move the packets of charge sequentially along the surface under the electrodes.

22.3 Verifying the CCD Concept

The following shows how we were able to immediately verify the CCD principle [7] with a simple experiment. We bonded wires onto a row of MOS test capacitors, which happened to be available and shown in Figure 22.1.

We created charge at the MOS interface under one end capacitor using avalanche breakdown, pulsed the individual capacitors in a three-phase sequence to transfer the charge along the row, and then repelled it into the substrate, where it was detected. Fortunately, the gaps were small enough to enable electrostatic coupling in the silicon under the gaps.

After that, I designed an 8-bit pixel^{-1} CCD [8]. In 1970, the design rule was 5 μm although 3 μm gaps between the metal fingers were realizable. I designed diodes at both ends of the device, so that charge could be injected at one end and more importantly detected at the other. George Smith was adamantly against this, because he thought CCDs should be pure MOS devices without any diffusions, but I ignored him anyway. The device is shown in Figure 22.2 and was demonstrated with an electrical input as an 8-sample analog delay line, or as an 8-bit memory, or as an 8 pixel linear imaging device with an optical input.

Figure 22.1 Capacitor array test for CCD concept.

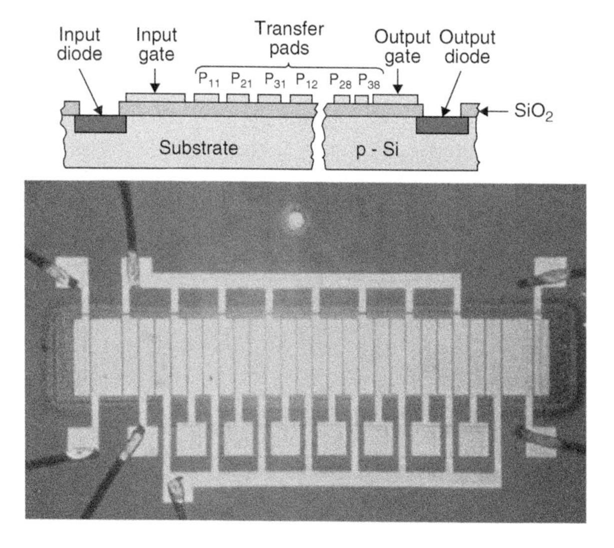

Figure 22.2 First 8-bit pixel^{-1} CCD. Source: Elsevier B.V.

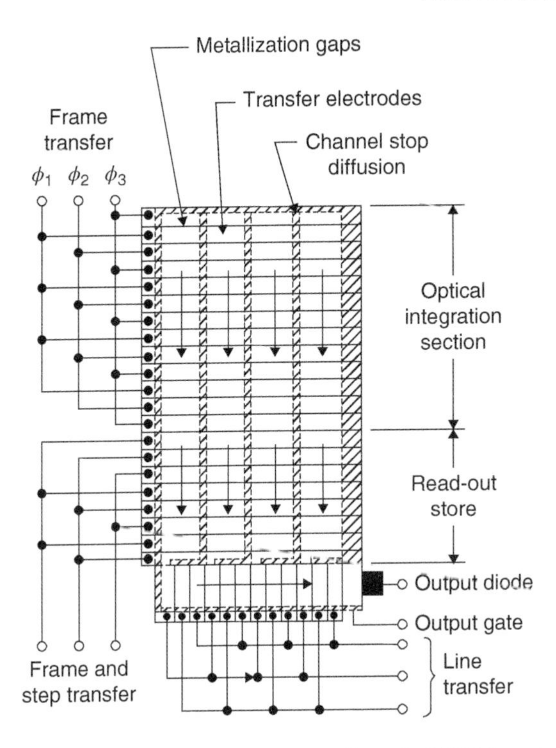

Figure 22.3 Readout principle of "Frame-Transfer" CCD imaging array.

22.4 The Invention of CCD Imagers

The Smith–Boyle patent application for CCDs did not cite any application to imaging, and the patent for both linear imaging devices and "Frame-Transfer" CCD imaging arrays is in my name alone [9]. My background in camera tubes for television, led me to design a CCD imager shown schematically in Figure 22.3 [9] with a serial output that exactly matched a standard television raster scan. Interlacing was obtained by integrating under different electrodes.

A three-level metallization scheme for the three-phase electrodes was proposed by Bertram and adopted, making the devices very tolerant to intra-level short-circuits. Since each layer was at the same voltage, a short-circuit across a gap was not fatal. This allowed exceptionally large chips to be made with finite yield using the 5 μm design rule of the time. I realized that the capacitance of a small silicon diode would be much smaller than the output capacitance of a vidicon camera tube, and this would lead to CCDs having much higher signal-to-noise ratios than camera tubes.

22.5 The First Solid-State Color TV Camera

My focus was on television cameras so the first CCD camera, which Ed Zimany and I demonstrated was a color camera (Figure 22.4) and not black and white, which would have been much simpler. The camera used three 106 × 128 pixel CCD imaging arrays mounted on *x*–*y* manipulators (Figure 22.5) and a large color splitting prism. The first ever color images from a solid-state camera taken with this camera were of my wife (Figure 22.6) and were published on the cover of Electronics Magazine [10].

We created a much smaller camera [11] about two years later using three 202 × 256 pixel imaging arrays with a custom-designed color splitting prism. A 525-line television black and white camera [12] was demonstrated in1976. However, by then many excellent new CCD ideas had been announced by researchers both inside and outside Bell Labs, commercial laboratories were also developing CCD

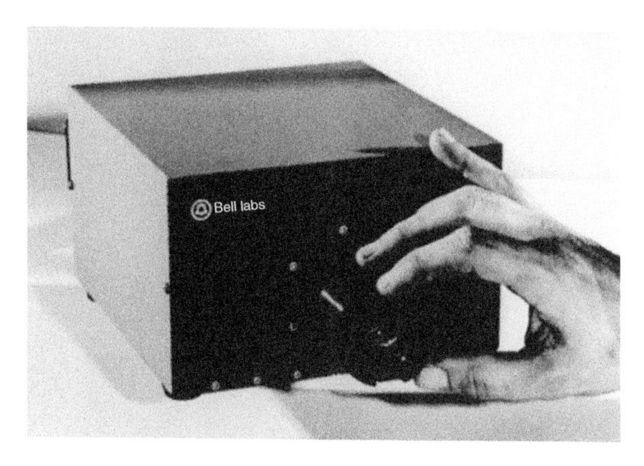

Figure 22.4 First CCD camera.

Figure 22.5 First CCD camera with prism removed showing 3 CCD imagers mounted on *x–y* positioners with associated control, amplifiers and power supplies.

imagers, I had coauthored a book on Charge Transfer Devices [13], and I had achieved my goal of demonstrating a solid-state camera. Picturephone had been dropped, and AT&T was not allowed to sell outside the telephone system, so continued development of imaging at Bell Labs did not seem to make sense. In addition, I correctly surmised that it would be years before CCD arrays would be developed to produce television or photographic quality imagers with yield, so I looked around for new challenges. I had the idea of integrating an analog-to-digital converter onto a CCD imaging chip, but I had to wait for the following technology development.

Figure 22.6 Image from first CCD camera.

22.6 Mixed Analog Design Modem Chip

In the mid-1970s, data modems were shoebox sized, so I decided to integrate both the analog circuits and the digital logic. With help from the modem design and filter design groups at the Bell Labs Holmdel location, my group developed two chips, a digital chip and a mixed analog–digital chip. Ed Zimany and I solved a dynamic range show-stopper problem by inventing an integrated MOS Analog Gain Control circuit. The analog chip [14] was successful and became the first mixed analog–digital MOS chip to go into manufacture worldwide.

This mixed analog–digital capability set the stage for me finally to invent a high-resolution video analog-to-digital converter chip in 1987. The chip was designed by Bang-Sup Song [15]. It was essential for any digital imaging application and was manufactured by Analog Devices Inc. in very large quantities.

The invention of CCD imagers was actually preceded by work on a thin-film-transistor array in 1963 by Paul Weimer at RCA Labs, and simultaneous proposals for *x–y* addressed MOS imagers in 1967 by Peter Noble at Plessey in England and Gene Weckler at Fairchild in California. However MOS technology was too primitive at the time and 25 years of MOS technology development were required before these proposals were picked up again by Eric Fossum and yielded viable devices. Today's CMOS imagers are made with incredible quality and yield and have replaced CCD imagers for other than space and astronomical work.

Acknowledgement

I have to thank all the members of my group and my colleagues at Bell Labs, who contributed to our successes as a team effort.

References

[1] Buser, R.G. and Tompsett, M.F. (1997). Semiconductors and semimetals. In: *Uncooled Infrared Imaging Arrays and Systems*, vol. 7 (ed. D. Skatrud and P. Kruse), 1–14. Academic Press.

[2] Tompsett, M.F. (1971). Magnetic Bubble Memory. *IEEE Transactions on Electron Devices* ED-18: 1070–1074.

[3] Crowell, M.H. and Labuda, E.F. (1969). The silicon diode array camera tube. *Bell System Technical Journal* 48: 1481–1528.

[4] Kahng, D. and Atalla, M.M. (1960). Silicon-silicon dioxide field induced surface devices. *IRE-AIEE Solid-State Device Res. Conf.*, University of Pennsylvania, Philadelphia (18–19 February).

[5] Bobeck, A.H. (1967). Magnetic Bubble Memory. *Bell System Technical Journal* 46: 1901–1925.

[6] Sangster, F.L.J. and Teer, K. (1969). Bucket brigade electronics – new possibilities for delay time-axis conversion and scanning. *IEEE Journal of Solid-State Circuits* SC-4: 131–136.

[7] Amelio, G.F., Tompsett, M.F., and Smith, G.E. (1970). Experimental verification of the charge coupled device concept. *Bell System Technical Journal* 49: 593–600.

[8] Tompsett, M.F., Amelio, G.F., and Smith, G.E. (1970). Charge coupled 8-bit shift register. *Applied Physics Letters* 17: 111–115.

[9] Tompsett, M.F., Amelio, G.F., Bertram, W.J. et al. (1971). Charge-coupled imaging devices: experimental results. *IEEE Transactions on Electron Devices* ED-18: 992–996.

[10] Tompsett, M.F., Bertram, W.J., Sealer, D.A., and Sequin, C.H. (1973). Charge-coupling improves its image challenging video camera tubes. *Electronics* 162–169.

[11] Sequin, C.H., Morris, F.J., Shankoff, T.A. et al. (1974). Charge-coupled area image sensor using three levels of polysilicon. *IEEE Transactions on Electron Devices* ED-21: 712–720.

[12] Sequin, C.H., Zimany, E.J., Tompsett, M.F. et al. (1976). All solid-state camera for the 525-line television format. *IEEE Journal of Solid-State Circuits* SC-11: 115–121.

[13] Sequin, C.H. and Tompsett, M.F. (1975). *Charge Transfer Devices*, 1978. Academic Press. Russian and Japanese translations.

[14] Daubert, S.J., Tompsett, M.F. et al. (1989). A CMOS modem analog processor for V.22bis modems. *IEEE Journal of Solid-State Circuits* SC-24: 281–291.

[15] Song, B.-S., Lee, S.H., and Tompsett, M.F. (1990). A 10b 15MHz CMOS recycling two-step A/D converter. *IEEE Journal of Solid-State Circuits* SC-25: 1328–1338.

Chapter 23

The Invention and Development of CMOS Image Sensors

A Camera in Every Pocket

Eric R. Fossum

Thayer School of Engineering, Dartmouth College, Hanover, NH, USA

23.1 Introduction

CMOS (complementary metal-oxide-semiconductor) image sensors (CIS) in 2022 are produced, tested, and sold at the rate of about 6 billion sensors per year. Assuming each goes into a camera, that translates to about 200 cameras per second. Most of these cameras wind up in smartphones since smartphones became the "killer application" for the technology. They are also widely used in nearly all other visible light sensor applications, such as in automobiles, medicine, security, webcams, body cams, and drones, and have also spawned social media giants and aided in social justice. All these picture or video-taking cameras use the active pixel sensor with intra-pixel charge transfer technology invented 30 years ago, around 1992 at the NASA Jet Propulsion Laboratory, to solve problems with Charge-Coupled-Device (CCD) cameras in interplanetary spacecraft cameras. However, that pivotal invention, like most inventions, stood on the shoulders of prior inventions, and has also been further improved and complemented by a host of other inventive technologies since then. In this paper, we will take a short swing through some of the key developments in the invention and development of present-day CMOS image sensors, as well discuss recent developments in photon-counting image sensors.

75th Anniversary of the Transistor, First Edition. Edited by Arokia Nathan, Samar K. Saha, and Ravi M. Todi.

23.2 Underlying Technology

At the heart of every camera, where the conversion from light to electronic signal takes place, is the image sensor – an integrated circuit (Figure 23.1). The optics of the camera focus the light as an image on the pixel-array portion of the surface of the sensor, which also has its own optical structures [2]. The absorption of visible-light photons creates electron–hole pairs in the semiconductor. The quantum efficiency (QE) is the ratio of useful photoelectrons to incident photons and often ranges from 50 to 80%. The photoelectrons are collected by diffusion and drift into an electrostatic "potential well" created by selective semiconductor doping via ion implantation. Photoelectrons are integrated in this storage well during the exposure (Figure 23.2). The pinned photodiode structure, invented by Teranishi, was used in CCDs and now in CMOS image sensors [3, 4]. At the end of the exposure integration period, the charge-domain signal is then converted to voltage by transferring the charge to a sense capacitance in the pixel. The change in the charge on this capacitive sense node causes a change in voltage, thus converting the signal from charge domain to voltage domain (Figure 23.3).

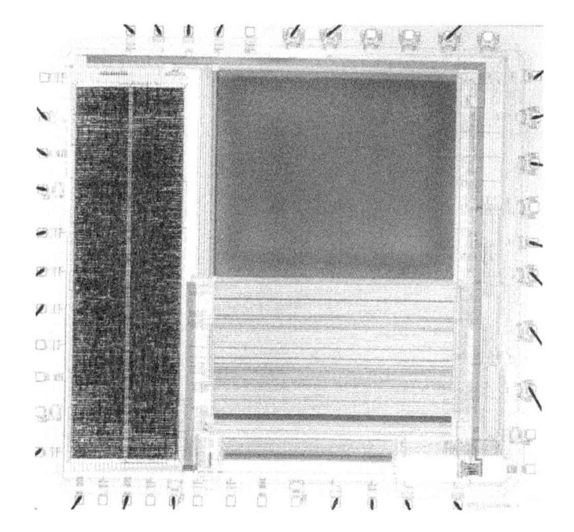

Figure 23.1 Early CMOS image sensor chip for webcams. Source: From [1] PHOTOBIT CORP. Left is digital logic for control and I/O. Top right is the pixel array. Bottom right are the column-parallel analog signal processing and analog-to-digital converter (ADC) circuits.

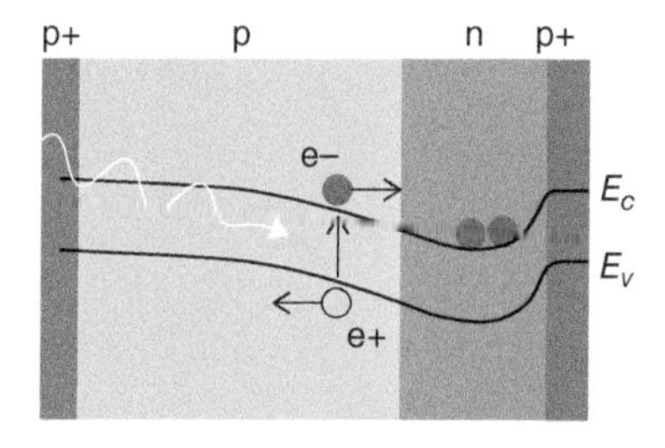

Figure 23.2 Photoelectrons are collected in an electrostatic storage well (N-region) created by selectively doping the silicon. In this case, photons enter the detector layer from the nominal substrate side, or backside, in so-called backside illumination (BSI) to increase pixel quantum efficiency.

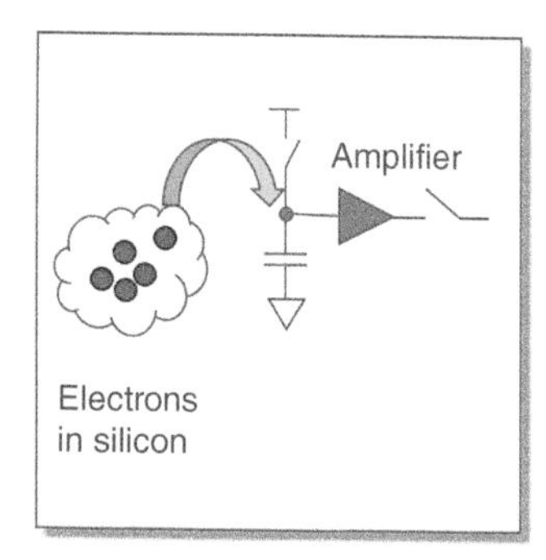

Figure 23.3 During readout of a CCD or CMOS image sensor, the signal charge is transferred onto a sense capacitance whose change in voltage is measured by an amplifier. For CDS, the amplifier voltage is sampled twice, once just before charge transfer, and once right after charge transfer, and then taking the difference to suppress reset noise and amplifier offset.

The change in voltage is usually buffered by a near-unity-gain source-follower metal-oxide-semiconductor field-effect transistor (MOSFET). This "floating diffusion amplifier" was first proposed by Kosonocky for a CCD and the signal may be further amplified by conventional electronics. The relationship between the change in output voltage and the number of photoelectrons is called the conversion gain (CG) with units of volts/electron. Typical CGs today are in the 30–75 $\mu V/e^-$ range.

Photon arrival rates are well-described by Poisson statistics, with a variance equal to the mean. Thus, repeated measurements of the same average photon flux will yield different results each time, and the standard deviation is referred to as photon shot noise. The signal-to-noise ratio (SNR) thus varies as the square root of the signal level, leading to low SNR at low signal levels. Thus, in low light, the signal often appears relatively more noisy (or "grainy"). The SNR can be improved by increasing the exposure time to increase the signal, but is limited according to the maximum number of photoelectrons that can be stored in the storage well (so-called "full well capacity" or FWC). An FWC of several thousand electrons is typical, leading to a maximum SNR of perhaps 50–70 or so.

Noise is also introduced by the readout process due to various noise sources in the readout circuit transistors and reset noise on capacitors. The use of correlated double sampling (CDS), invented by White [5], is one way to reduce reset noise and $1/f$ transistor noise. Correlated multiple sampling (CMS) may also be used. For larger signals, photon shot noise is the dominant noise. For low-light levels, the readout noise becomes critical in determining SNR. The readout noise is often referred to the equivalent number of input photoelectrons. For example, 150 μV rms of read noise with a CG of 50 $\mu V/e^-$ would be input-referred as 3 e^- rms. These days, the read noise of image sensors is typically in the 2–5 e^- rms range and is often dominated by residual $1/f$ noise from the first source-follower transistor.

23.3 Early Solid-State Image Sensors

In the 1960s, several groups pioneered building solid-state image sensors [6]. Weckler at Reticon realized that one could integrate photoelectrons on the built-in capacitance of a floating PN photodiode. This charge could then be readout by connecting the PN junction via a MOSFET switch to a readout circuit. This is now called a passive-pixel MOS image sensor. Noble at Plessey worked contemporaneously and built on Weckler's idea and used a source-follower as a readout amplifier. Adding a switch to reset the photodiode and a switch to connect the source-follower to some additional readout sequencing circuits, Noble's image sensor was the first three-transistor (3T) active pixel sensor. Unfortunately, these early approaches yielded neither good nor stable image quality compared to image-pickup tubes due to limitations of the circuit configuration and semiconductor technology of the mid-1960s. Images suffered from large temporal noise and fixed-pattern noise. The circuit was susceptible to drifts in operating voltage and manufacturing yield. Reticon and Hitachi later continued to explore these passive pixel and 3T active pixel approaches but none achieved viable commercial success due, in part, to the above but also due to the invention and rapid development of CCDs.

In 1969, the CCD concept was invented by Boyle and Smith at Bell Labs as a solid-state equivalent to magnetic bubble memory. The CCD was based on the MOS gate structure using a series of adjacent gates. By pulsing each gate in sequence, minority carriers in the semiconductor (e.g. electrons) can be dragged along in the semiconductor due to electrostatic attraction [7]. The electron signal charge can either be electrically injected or created by light. In the former, the CCD acts like a delay line, and in the latter, the charge (or voltage) that is read out is indicative of the spatial distribution of light. When the CCD is configured as a 2D array of MOS gates, an image can be focused on the device, and the charge from that light pattern read out as a digital image (Figure 23.4). The first practical CCD image sensor was invented by Tompsett [8] and many significant improvements quickly followed.

The CCD was immune from many of the manufacturing issues that plagued the early passive and active pixel sensors, and efforts in the United States, Europe, and especially Japan rapidly developed the CCD into a workhorse image sensor [7]. The CCD image sensor became the basis for consumer camcorders from Japanese manufacturers and later broadcast TV cameras, and by the 1990s, digital still cameras.

Despite their success, CCD image sensors had many drawbacks due to the thousands of charge transfer steps required to readout each pixel in the image. The charge transfer requires considerable energy or power, especially in driving the higher voltage and higher capacitance multitude of transfer gates to achieve a desired charge transfer efficiency of at least 99.999%. Whenever the pixel count in the image sensor is increased, even more energy and power is required for readout. The intrinsic performance of the CCD also must improve since more transfer steps are required, and the charge transfer speed must increase for the same frame rate. CCDs were thus hard to scale to larger sizes. The cost of CCDs was fairly high because the manufacturing process to make a CCD was a very different recipe from that used by mainstream CMOS electronics (e.g. for computers and memory), and essentially every generation of CCD had to have a new and improved process whose development cost was amortized across the sale of those CCDs. All the drive and readout electronics had to be on additional chips and further increased power dissipation and camera form-factor. While highly dwarfed in volume by CMOS image sensors, CCDs still find niche (albeit shrinking) applications in some machine vision and scientific cameras.

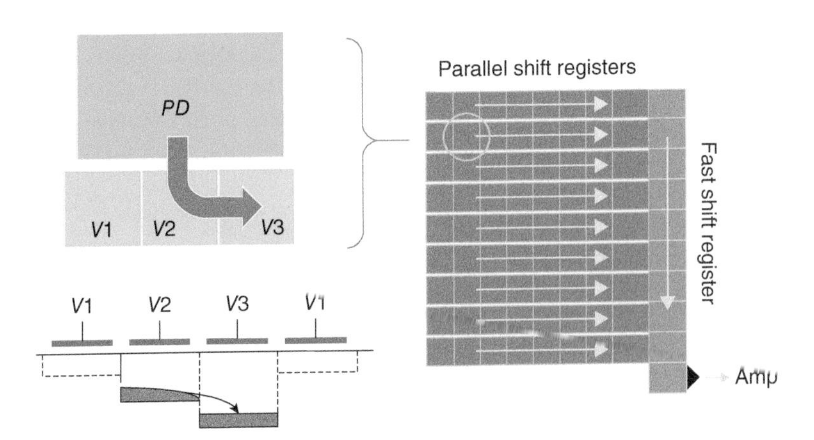

Figure 23.4 Illustration of a CCD interline transfer (ILT) pixel, top view (top left), charge-transfer shift register (bottom left), and a CCD ILT array (right).

23.4 Invention of CMOS Image Sensors

NASA had an additional problem with CCDs used in interplanetary spacecraft. Those CCD cameras were exposed to radiation in space that caused microscopic defects in the image sensor chip. The defects resulted in a continuous degradation in performance – both in charge transfer efficiency and dark current (like junction leakage current). Also, the power dissipation of CCD cameras was high – a problem when operating from a battery (CCD camcorder batteries were large, for example, and were depleted after perhaps 30 minutes of use) or a solar panel power supply or a radioisotope thermoelectric generator (RTG) aboard a spacecraft. Additionally, the electronics required to provide timing and clocking signals to drive the CCD, and to process the output signals including analog-to-digital conversion (ADC), were bulky and power-hungry.

In 1992, the author, working at the NASA Jet Propulsion Laboratory at Caltech, proposed a new approach for image sensors that would use an active pixel sensor with intra-pixel charge transfer, and use a mainstream CMOS microelectronics process flow as the baseline [9, 10] (see Figure 23.5). By using CMOS, one could not only capture the image, but one could also put all the timing and control circuits and all the signal processing circuits, including ADC, on the same chip. This would allow miniaturization of the spacecraft camera and reduce power consumption significantly (100×) as well as avoid many of the charge-transfer degradation issues associated with exposure to radiation in space [11] (see Figure 23.6).

Each pixel is composed of a tiny CCD with its own readout and selection electronics (see Figures 23.7 and 23.8). Using intra-pixel charge transfer allows the use of noise reduction techniques like CDS, thus enabling imaging performance comparable to CCDs. Incorporation of pinned photodiodes improved QE. By using CMOS as the baseline, the integration of additional signal processing and on-chip ADC to further improve performance of the active pixel sensor was also possible. A high-quality "camera-on-a-chip" became feasible. Both rolling shutter and global snapshot shutter implementations were conceived of in the early days, with rolling shutters yielding smaller, more competitive pixels. Global shutters have continued to improve and are desired when rolling shutter artifacts due to certain types of motion need to be suppressed [12].

The early sensors made in 1993–1995 showed great promise and it became clear that the technology was not only useful for space but also for consumers on planet Earth. However, the deeply entrenched CCD industry was very slow to recognize the advantages of the CMOS active pixel sensor camera-on-a-chip. US industry was similarly reluctant to engage with the new technology despite many attempts to evangelize and transfer the technology (with some notable exceptions). In 1995, the author and

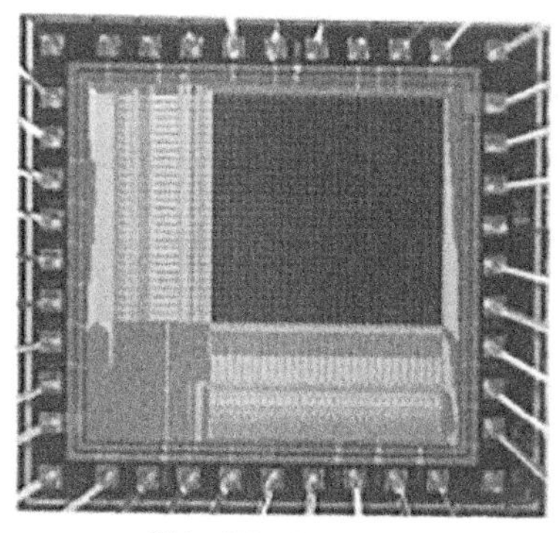

Chip (28 × 28 pixels)

"George"

Figure 23.5 First CMOS image sensor chip and acquired 28 × 28 pixel image of a one-dollar bill from 1993 at NASA's JPL at Caltech. Source: Courtesy of Eric R. Fossum.

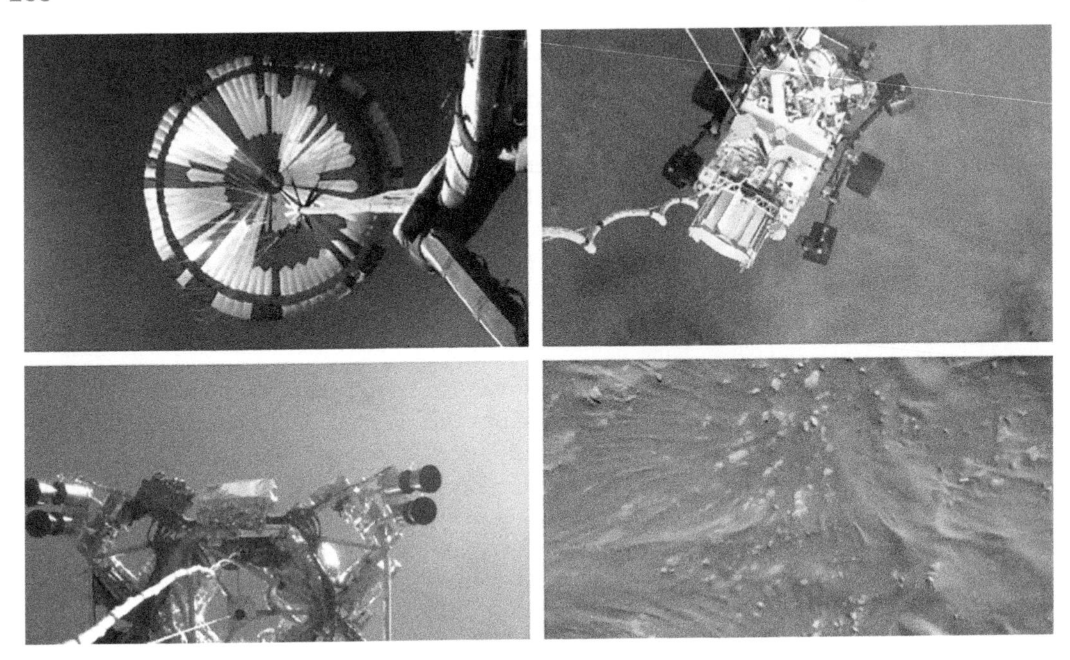

Figure 23.6 Video grabs of the Mars 2020 Perseverance rover landing. Perseverance has about 20 CMOS cameras on board including one on the Mars helicopter. Top left: Looking up at parachute. Bottom left: Looking up at "sky crane" descent vehicle. Top right: Looking down at the 1000 kg rover on the sky crane tether before landing: Bottom right: Mars surface just before landing. Source: NASA.

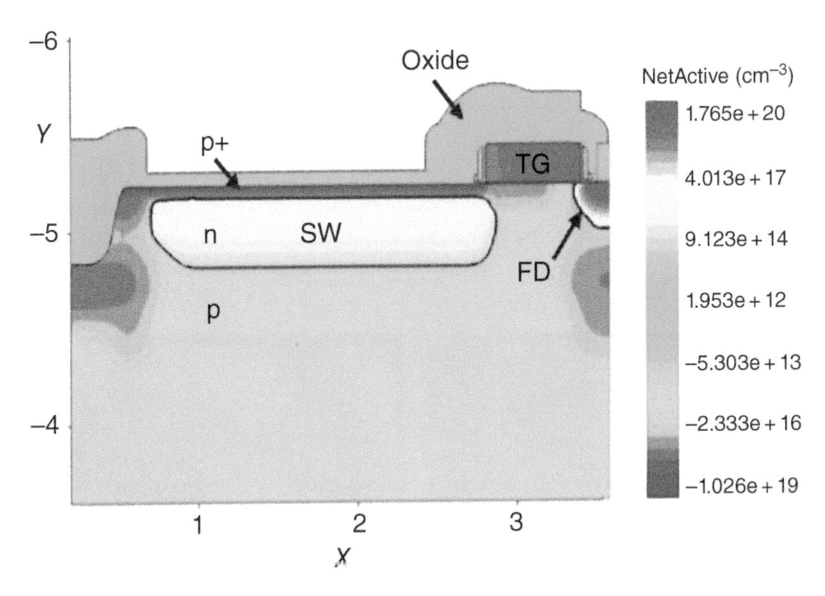

Figure 23.7 TCAD simulation of a pinned photodiode (PPD) charge storage region and transfer gate (TG) and sense node (FD). Source: From Fossum and Hondongwa [4]/IEEE.

several members of the team at JPL cofounded a spinoff company, Photobit, to commercialize the technology. Photobit grew to about 135 people before being acquired by Micron in 2001 and produced sensors for web cams, dental X-rays, swallowable pill cameras, high-speed machine vision systems, automotive rain wiper and high beam control, drowsy driver warning, star trackers, and other applications.

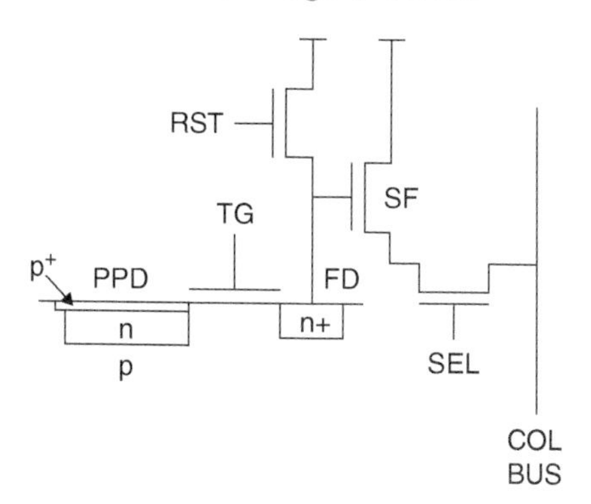

Figure 23.8 Circuit schematic of CMOS image sensor pixel including reset gate (RST), source-follower (SF), and row select switch (SEL). Source: From Fossum and Hondongwa [4]/IEEE.

By 2001, the "killer-app" of cellphone cameras was very visible on the horizon and Micron became a world leader in this technology [11].

Around the same time, some other large companies such as Sony and Samsung started to make large capital investments and seriously develop CMOS image sensors. Today, these two vertically integrated companies dominate the image sensor market place. Foundries like TSMC, X-Fab, and Tower-Jazz are also playing a major role for fabless companies. Micron later spun-off the image sensor business as Aptina, and later Aptina was acquired by ON-Semi (which grew out of Motorola).

In 2023, it is estimated that over 6.5 billion image sensors (or the same number of cameras) will be shipped worldwide corresponding to a semiconductor component revenue of $19.3B [13] as shown in Figure 23.9. While nearly all of these CMOS image sensors still utilize active pixels with intra-pixel charge transfer, many improvements have been made since the 1992 invention. The use of 3-D stacked structures and back-side illumination (BSI) with deep-trench isolation (DTI) has resulted in significantly improved performance and capability [14–16]. Shared readout [4] results in pixel pitch reduction, as well as a plummeting cost per pixel. Older CCD "VGA" sensors with 0.3 megapixels once cost about US$100 in the mid-1990s. Today, a VGA CMOS image sensor might cost more than 200× less or US$0.50.

CMOS image sensors today are ubiquitous, and several image sensors are typically found in every smartphone (e.g. front selfie camera and one or more rear-facing cameras) which in turn enabled

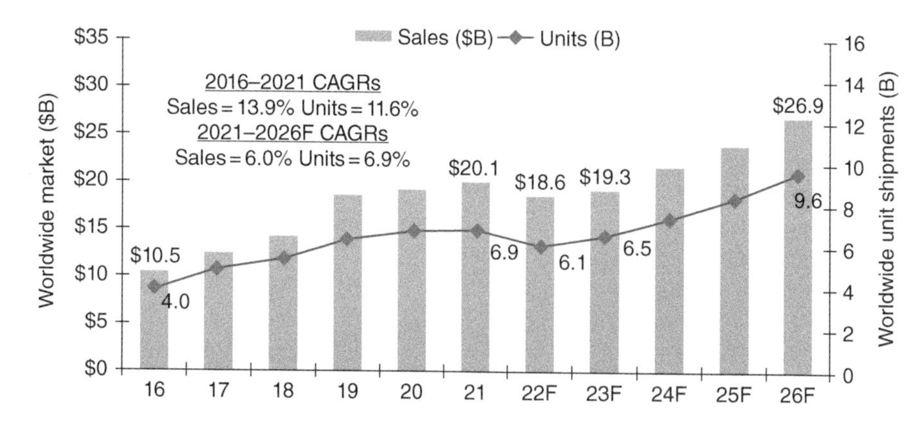

Figure 23.9 CMOS image sensor sales growth according to IC Insights. Source: Adapted from [13]. The Covid-19 pandemic and semiconductor supply chain issues cause a downward forecast for 2021–2023 with growth expected to recover after that.

Figure 23.10 2017 Queen Elizabeth Prize for Engineering presentation at Buckingham Palace. (L-R, HRH then-Prince-now King Charles III, Fossum, Tompsett, Teranishi). Source: Queen Elizabeth Prize for Engineering Foundation.

industry giants like Facebook, Instagram, Tiktok, and YouTube. They are also used in nearly every other camera application from automobiles to swallowable pill cameras to smart doorbells to police bodycams. While CCD unit volume was always small compared to today's CMOS image sensor unit volume, CCD manufacturing has come to an end for most applications.

In 2017, the importance of digital image sensors in everyday life was recognized through the awarding of the Queen Elizabeth Prize for Engineering to Smith and Tompsett (CCD image sensors), Teranishi (pinned photodiode), and Fossum (CMOS image sensor) (Figure 23.10).

23.5 Photon-Counting CMOS Image Sensors

In 2005, a different approach for image sensors was proposed [17]. In this proposed device, single photons would be detected and counted by a large number of specialized yet tiny pixels (called jots) operating at a high frame rate. Detection would be essentially binary, either 0 for no photon, or 1 for a photon. Frames of binary data could be used to recreate a gray-scale image, and image in the dimmest possible light. First called a digital film sensor, the concept was later renamed a Quanta Image Sensor (QIS) and extended to multi-bit operation [18].

Other groups began to demonstrate the QIS concept and prove the imaging characteristics model using single photon avalanche detector (SPAD) arrays. SPAD devices have been in development for about 26 years with recent rapid progress [19]. In 2021, a 3.2 Mpixel SPAD array was reported for the first time with a 6.4 μm pixel pitch [20]. Since the SPAD relies on avalanche multiplication for signal gain, it requires high internal electric fields and relatively large spacing between pixels to ensure isolation, and they may also typically have high dark count rates (dark current). Despite these issues, SPADs have been proven very useful for fast photon arrival timing applications such as 3-D imaging.

In 2012, work on realizing a CMOS QIS began at the Thayer School of Engineering at Dartmouth. Instead of using avalanche gain to detect single photoelectrons, the gain comes from using a very small sense node capacitance yielding CG in the range of 300–500 μV/e⁻. Using intra-pixel charge transfer, a single electron transferred to that capacitance can produce a discernible signal that is well above the noise floor (e.g. 0.2 e⁻ rms noise floor) and thus give a low error rate for detection of single photoelectrons. The detection process is slower than for SPADs, but sub-microsecond timing is achievable. Further, avoiding the high electric fields of SPADs enables smaller pixels or jots, improved manufacturability, and thus lower cost per pixel and smaller optics. Power dissipation is also considerably smaller. In 2017, Dartmouth reported a room-temperature 1 Mpixel QIS device implemented in a nearly standard BSI CIS 3D stacked process with 1.1 μm pixel pitch, operating at 1000 fps and dissipating about 20 mW total power [21] (see Figures 23.11–23.13). The 1 Mpixel QIS was demonstrated more than two years earlier than the first 1 Mpixel SPAD array and with much less development time and with much smaller pixels. About 34 CMOS QIS 1.1 μm pixels can fit into the area of one SPAD 6.4 μm pixel. This is the strength of working in a nearly standard CIS processes. SPAD technology has also been exploiting technologies used for CIS such as 3-D stacking and low dark current

Figure 23.11 QIS test chip containing 20 different 1 Mpixel arrays designed by Dartmouth and fabricated by TSMC in a 65 nm stacked BSI CIS process. Source: Eric R. Fossum.

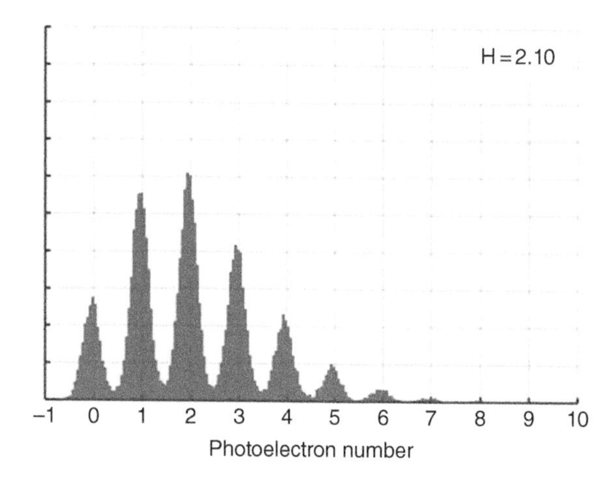

Figure 23.12 Photon-counting histogram (# occurrences vs. normalized readout voltage) showing clear quantization of photoelectrons. The peak heights correspond to the Poisson distribution for an average photoelectron arrival rate of 2.1 e⁻/sample.

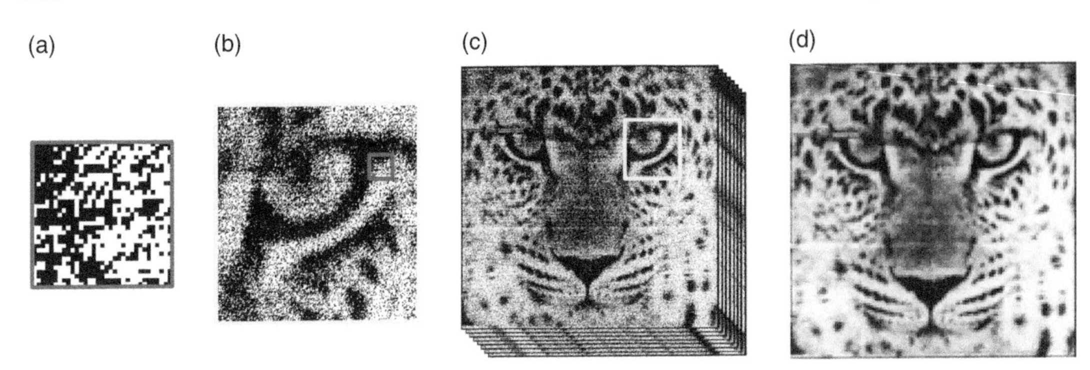

Figure 23.13 Data taken from a 1 Mpixel array on the QIS test chip. (a) Close-up of binary image data in one frame, (b) zoomed out version of (a), (c) further zoomed out and illustrating multiple binary frames that are combined to achieve the gray scale image in (d). Source: Optical Society of America.

structures. Both CMOS-image-sensor QIS technology and SPAD-QIS technology occupy important application areas [22].

Recently, CMOS image sensors that use QIS photon-counting technology have achieved 163 Mpixels in resolution and high dynamic range [23, 24]. The applications of QIS technology are currently being explored but include low-light imaging for security, defense, science, and possibly consumer devices.

23.6 Conclusion

CMOS image sensors continue to be used in an ever-increasing variety of applications. Some of these applications are useful in everyday life, some are for fun (photography, social media), some are for safety, and some continue to test the age-old balance between security and privacy. After 30 years, the future still looks bright for CMOS image sensors.

Acknowledgments

The author gratefully acknowledges the manifold contributions of his former and current students, colleagues at JPL, Photobit and Gigajot, and the support of NASA and DoD contracts and SBIR programs and corporate sponsors such as Gentex, Basler, Schick, Given Imaging, and Rambus, over many years. CMOS image sensors are only ubiquitous today because of the important contributions of thousands of engineers around the world, as well as the work done by many early pioneers in solid-state image sensors.

References

[1] (2022). https://spectrum.ieee.org/chip-hall-of-fame-photobit-pb100.

[2] Teranishi, N., Watanabe, H., Ueda, T., and Sengoku, N. (2012). Evolution of optical structure in image sensors. *2012 International Electron Devices Meeting*, San Francisco, CA (10–13 December 2012), pp. 24.1.1–24.1.4. IEEE. https://doi.org/10.1109/IEDM.2012.6479092.

[3] Teranishi, N., Kohono, A., Ishihara, Y. et al. (1982). No image lag photodiode structure in the interline CCD image sensor. *International Electron Devices Meeting*, San Francisco, CA (13–15 December 1982), pp. 324–327. IEEE. https://doi.org/10.1109/IEDM.1982.190285.

[4] Fossum, E.R. and Hondongwa, D.B. (2014). A review of the pinned photodiode for CCD and CMOS image sensors. *IEEE Journal of the Electron Devices Society* 2 (3): 33–43. https://doi.org/10.1109/JEDS.2014.2306412.

[5] White, M.H., Lampe, D.R., Mack, I.A., and Blaha, F.C. (1973). Characterization of charge-coupled device line and area-array imaging at low light levels. *Digest of Technical Papers – IEEE International Solid-State Circuits Conference* 16: https://doi.org/10.1109/ISSCC.1973.1155159.
[6] Fossum, E.R. (1997). CMOS image sensors: electronic camera-on-a-chip. *IEEE Transactions on Electron Devices* 44 (10): 1689–1698. https://doi.org/10.1109/16.628824.
[7] Theuwissen, A.J.P. (1995). *Solid-State Imaging with Charge-Coupled Devices*. Kluwer Academic Publishers. ISBN: ISBN 0-792-33456-6.
[8] Tompsett, M.F. (1978). Charge transfer imaging devices. US Patent No. 4,085,456, issued 1978.
[9] Fossum, E.R., Mendis, S., and Kemeny, S.E. (1995). Active pixel sensor with intra-pixel charge transfer. US Patent No. 5,471,515 issued 1995.
[10] Fossum, E.R. (1993). Active pixel sensors: are CCDs dinosaurs? *Proc. SPIE 1900, Charge-Coupled Devices and Solid State Optical Sensors III*, San Jose, CA (12 July 1993). SPIE. https://doi.org/10.1117/12.148585.
[11] Fossum, E.R. (2013). Camera-on-a-chip: technology transfer from saturn to your cell phone. *Technology and Innovation* 15 (3): 197–209. https://doi.org/10.3727/194982413X13790020921744.
[12] Velichko, S. (2022). Overview of CMOS global shutter pixels. *IEEE Transactions on Electron Devices* 69 (6): 2806–2814. https://doi.org/10.1109/TED.2021.3136148.
[13] (2022). https://www.icinsights.com/news/bulletins/CMOS-Image-Sensors-Stall-In-Perfect-Storm-Of-2022/.
[14] Fontaine, R. (2019). The state-of-the-art of smartphone imagers. *Proc. 2019 Int. Image Sensor Workshop*, Snowbird, Utah (23–27 June 2019). www.imagesensors.org.
[15] Oike, Y. (2022). Evolution of image sensor architectures with stacked device technologies. *IEEE Transactions on Electron Devices* 69 (6): 2757–2765. https://doi.org/10.1109/TED.2021.3097983.
[16] Wuu, S.G., Chen, H.-L., Chien, H.-C. et al. (2022). A review of 3-dimensional wafer level stacked backside illuminated CMOS image sensor process technologies. *IEEE Transactions on Electron Devices* 69 (6): 2766–2778. https://doi.org/10.1109/TED.2022.3152977.
[17] Fossum, E.R. (2005). What to do with sub-diffraction-limit (SDL) pixels? – a proposal for a gigapixel digital film sensor (DFS). *Proceedings of the 2005 IEEE Workshop on Charge-Coupled Devices and Advanced Image Sensors*, Karuizawa (9–11 June 2005). IEEE. www.imagesensors.org.
[18] Fossum, E.R., Ma, J., Masoodian, S. et al. (2016). The quanta image sensor: every photon counts. *Sensors* 16: 1260. https://doi.org/10.3390/s16081260.
[19] Gyongy, I., Dutton, N.A.W., and Henderson, R.K. (2022). Direct time-of-flight single-photon imaging. *IEEE Transactions on Electron Devices* 69 (6): 2794–2805. https://doi.org/10.1109/TED.2021.3131430.
[20] Morimoto, K., Iwata, J., Shinohara, M. et al. (2021). 3.2 megapixel 3D-stacked charge focusing SPAD for low-light imaging and depth sensing. *2021 IEEE International Electron Devices Meeting (IEDM)*, San Francisco, CA (11–16 December 2021), pp. 20.2.1–20.2.4. IEEE. https://doi.org/10.1109/IEDM19574.2021.9720605.
[21] Ma, J., Masoodian, S., Starkey, D.A., and Fossum, E.R. (2017). Photon-number-resolving megapixel image sensor at room temperature without avalanche gain. *Optica* 4: 1474–1481. https://doi.org/10.1364/OPTICA.4.001474.
[22] Ma, J., Chan, S., and Fossum, E.R. (2022). Review of quanta image sensors for ultralow-light imaging. *IEEE Transactions on Electron Devices* 69 (6): 2824–2839. https://doi.org/10.1109/TED.2022.3166716.
[23] Ma, J., Zhang, D., Elgendy, O., and Masoodian, S. (2021). A photon-counting 4Mpixel stacked BSI quanta image sensor with 0.3e- read noise and 100dB single-exposure dynamic range. *2021 Symposium on VLSI Circuits*, Kyoto (13–19 June 2021), pp. 1–2. IEEE. https://doi.org/10.23919/VLSICircuits52068.2021.9492410.
[24] Ma, J., Zhang, D., Robledo, D. et al. (2022). Ultra-high-resolution quanta image sensor with reliable photon-number-resolving and high dynamic range capabilities. *Scientific Reports* 12 (1): 1–9. https://doi.org/10.1038/s41598-022-17952-z.

Chapter 24

From Transistors to Microsensors

A Memoir

Henry Baltes

Micro and Nano Systems, D-MAVT, ETH Zurich, Zurich, Switzerland

24.1 Early Encounters

A long time after its invention in 1947, I met my first transistor. It happened in an advanced laboratory class I attended as a master student of physics at ETH Zurich around 1964. There were indeed three transistors connected to form a ring oscillator. Those were discrete bipolar devices. The dawn of integration had not reached us yet.

Next as a doctoral student, I inherited an impressive lock-in amplifier based on several big vacuum tubes. It was used to amplify weak signals from a far-infrared detector in a vacuum spectrometer. The amplifier worked quite well once burnt in, but that could take weeks. The vacuum tubes had to be replaced from time to time, but finally were not manufactured anymore.

When I ran out of spare tubes, I looked for a replacement and convinced my supervisor to buy a new commercial amplifier system based on discrete transistors. This helped me to get my DSc (doctor of science) degree in 1971 and soon to become a professor in Germany.

24.2 Integration

Later, I worked for the Swiss company Landis & Gyr in Zug, producing energy and currency management equipment such as electricity meters and optical banknote readers. Besides, I taught and had a research collaboration in statistical optics at École Polytechnique Fédérale de Lausanne (EPFL).

75th Anniversary of the Transistor, First Edition. Edited by Arokia Nathan, Samar K. Saha, and Ravi M. Todi.

In 1981, I started to explore the potential of integrated circuitry for instrumentation. To this end, I attended an advanced course on *Large Scale Integrated Circuits Technology: State of the Art and Prospects* held at Ettore Majorana Centre for Scientific Culture, Erice, Italy, in July 1981. The proceedings [1] edited by Leo Esaki and Giovanni Soncini appeared in 1982.

Within a week and to my utter amazement, I learnt a lot about the recent innovations in semiconductor electronics that lead to reducing the cost of transmitting, storing, and processing information with high performance based on large-scale integration. The critical dimension then was about five microns.

On the airplane back from Palermo to Zurich, I decided to leave optics behind me and move to microelectronics. To this end, I wanted to gain hands-on experience. Hans Melchior was so kind to arrange a crash lab class for me at ETH Zurich. I started with a small lot of five 1.25-in. wafers. My yield was low. Only one of them survived all the processing steps.

24.3 Silicon Sensors

Sensors convert physical (thermal, mechanical, magnetic, and radiant) or chemical signals to electrical signals. In the same year 1981, I got to know about silicon microsensor research at Delft University of Technology initiated by Simon Middelhoek (1931–2020).

Upon my request, he invited me to his laboratory to take a look at their ongoing research. Once more I was very impressed by what I saw. Silicon microsensors or integrated circuit (IC) sensors have the great potential to be co-integrated on the same chip together with signal conditioning and processing integrated circuitry.

I returned to Switzerland with a generous load of reprints, preprints, reports, and lecture notes. All of this inspired me to move my teaching at EPFL to integrated sensors (capteurs intégrés) and to start exploring on-chip silicon magnetic field sensors for current sensing in future integrated electricity meters.

Luckily, I had the chance to join this new research area early on. Its first conference had been held in 1980 with chairs Piet Bergveld and Jay Zemel. The new journal *Sensors and Actuators* started in 1981 with Simon Middelhoek as founding editor. He soon became my mentor.

24.4 Transistor Sensors

Transistors (bipolar or field effect) play two quite distinct roles in the context of microsensors. (i) With judicious design modifications they can act as sensing devices integrated on a silicon chip. (ii) On the other hand, transistors such as the metal-oxide-semiconductor field-effect transistor (MOSFET) are crucial components of the co-integrated circuitry enhancing the action of the sensing devices, including of course those not based on transistors.

The bipolar magneto-transistor and the field-effect transistor MAGFET are designed to detect magnetic fields. Another example is the ion sensitive field-effect transistor or ISFET for measuring ion concentration. Fabrication of the latter requires dedicated post-processing after finishing a standard CMOS IC process [2–4]; see Figures 24.1 and 24.2.

A unique "resonant gate transistor" has been demonstrated by Harvey Nathanson (1936–2019) et al. in 1967, where a vibrating post-processed gold beam uses a field-effect transistor (FET) read out [4]; see Figure 24.3. Before he retired in 2001, I had the good luck to meet him when I was invited to talk at Westinghouse Research Laboratories, Pittsburgh, PA.

Integrated magnetic sensors can be obtained just by design, compatible with the rules of a custom IC provider, and without post-processing. That is the point where I started my work in the area, first at the Swiss company, and since 1983 when I continued as a professor at the University of Alberta in Edmonton, Canada, and from 1988 to 2006 at ETH Zurich [6]. Our research included the development of computer aided design (CAD) tools for microsensors [7, 8]; see Figure 24.4.

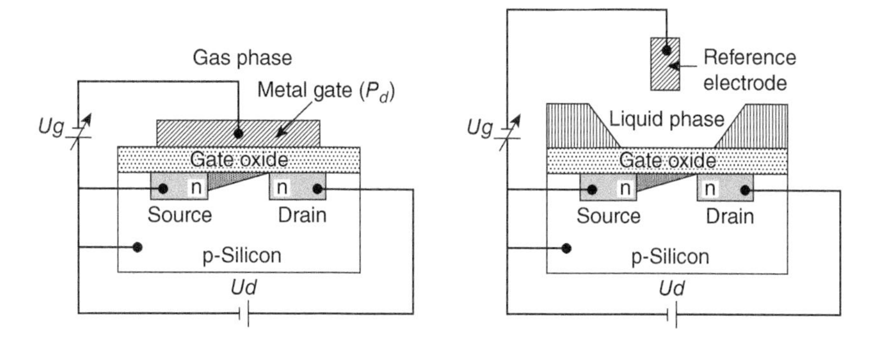

Figure 24.1 Schematic representation of MOSFET (left) and ISFET (right) structure for chemical sensing. *Ug* denotes the gate voltage, *Ud* the drain voltage. By replacing the metal gate of the MOSFET with an ionic solution and a reference electrode immersed into this solution, the ISFET has been developed [5]. Source: Reproduced with permission from Hierlemann et al. [5]. Copyright (2021) IEEE.

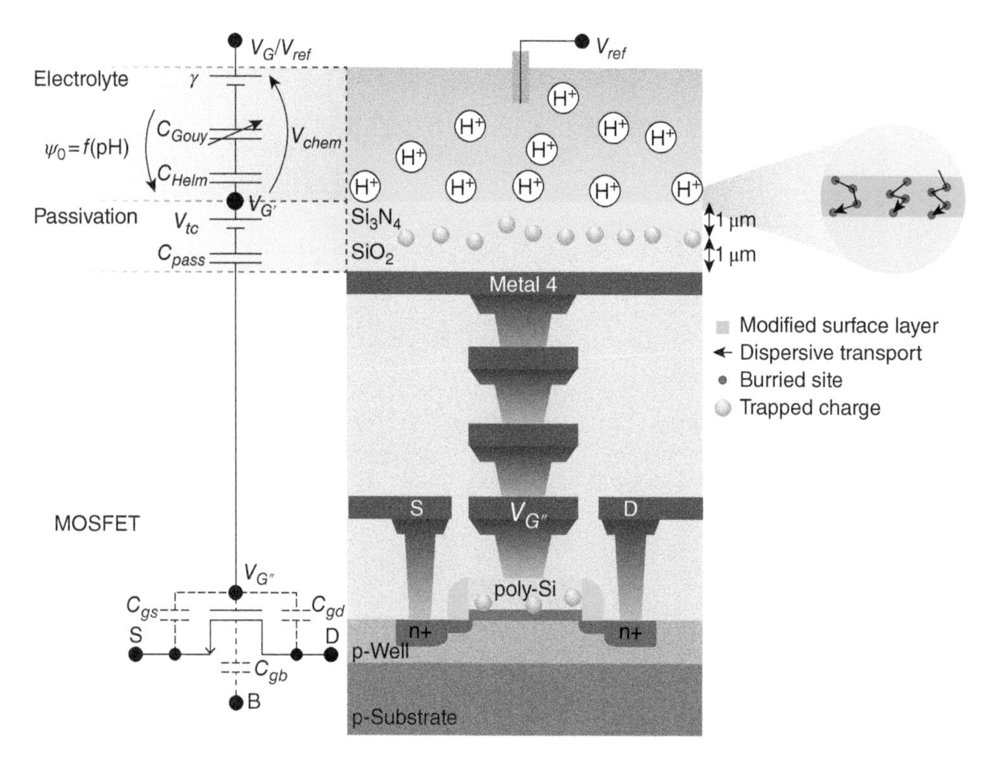

Figure 24.2 CMOS ISFET structure in 0.35 μm technology schematic and macro-model circuit (left). Inset on the right enlarges the electrolyte passivation layers interface, indicating (dark green) the layer modification by drift molecules [3]. Source: Reproduced with permission from Panteli et al. [3]. Copyright (2021) IEEE.

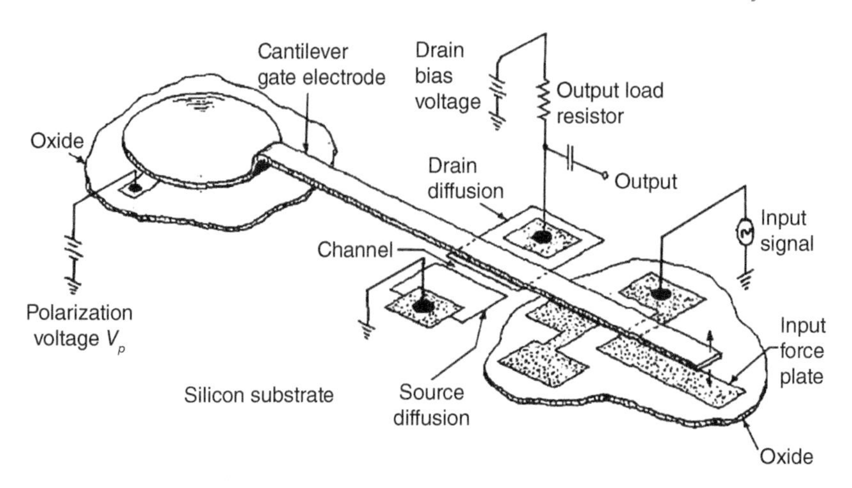

Figure 24.3 Schematic diagram of a resonant gate transistor consisting of a gold beam which is 0.1 mm in length, and 5–10 μm in thickness, resonating at 5 kHz with a quality factor of 500 [4]. Source: Reproduced with permission from Nathanson et al. [4]. Copyright (1967) IEEE.

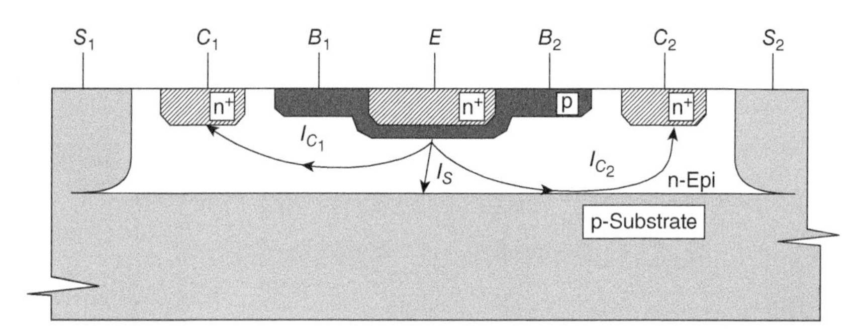

Figure 24.4 Basic structure and currents of a vertical pnp magneto-transistor with two collectors. For zero magnetic field the current paths are symmetric. For non-zero magnetic field the current paths are rotated due to the galvanomagnetic effect [7]. Source: Reproduced with permission from Riccobene et al. [7]. Copyright (1994) IEEE.

24.5 CMOS End Fabrication

During my active years in Canada, I enjoyed an exciting collaboration with a Silicon Valley company active in custom chip design and fabrication, LSI Logic Corporation, Milpitas, California, founded in 1981. With their help and a grant from the Alberta Government, a commercial CMOS end fabrication was set up at the University of Alberta. My research group did benefit from that facility for CMOS sensor post-processing.

Preparing my classes using the latest original literature, I found an amazing review article on silicon as a mechanical material [9] by Kurt Petersen. I learnt that silicon has excellent mechanical properties and concluded that selected mechanical structures could be achieved by post-processing on a CMOS wafer. This led to numerous mechanical, thermal, and chemical IC sensors pursued further at ETH Zurich [5], where my research group set up a post-processing lab dedicated to that purpose. Sensor circuits were developed in collaboration with Qiuting Huang. In 1998, our efforts led to the foundation of the spin-off company SENSIRION. A few years later, we developed a medical microsensor in close collaboration with the University Hospital Zurich [10].

24.6 Outlook

Further system miniaturization will demand for continuous down-scaling of sensor functions toward the nanoscale. New sensor device concepts will emerge to improve performance, e.g. sensitivity, or to utilize unique functional properties of nanoscale structures. Suspended carbon nanotube field-effect transistors (CNT FET) fabricated with a dry transfer technique demonstrate strong promise as ultralow-power, hysteresis-free gas sensors [11].

Acknowledgments

Looking back, I do not remember all details of CMOS sensor design, modeling, and fabrication. But I do remember the people who supported me in many ways since I embarked on the field in 1981 as mentors, sponsors, project partners, team leaders, technicians, secretaries, and providers of CMOS prototypes.

In Switzerland: Hans Melchior, Heinz Rüegg, Qiuting Huang, Christofer Hierold, David Moser, Oliver Paul, Andreas Hierlemann, Felix Mayer, Gerhard Wachutka, Kai-Uwe Kirstein, Cosmin Roman, Heike Hall, Christiaan Abbas, Donat Scheiwiller, Erna Hug, Wolfgang Fichtner, Ralph Hütter, Olaf Kübler, Heinrich Ursprung, Nico de Roij, Antonio Quattropani, René Hartmann, Jean-Frédéric Moser, Radivoje Popovic, Jon Grand.

Else in Europe: Simon Middelhoek, Lina Sarro, Massimo Rudan, Giorgio Baccarani, Giovanni Soncini, Günter Zimmer, Wolfgang Göpel, Franz Laermer, Volker Kempe, Siegfried Selberherr.

In Canada: Günter Schmidt-Weinmar, Igor Filanovsky, Walter Allegretto, Jed Harrison, Savvas Chamberlain, Bob James, Ken Broadfoot, Hugh Planche, Phillip Haswell, June Swanson.

In the United States: Kurt Petersen, Thomas Kenny, Mark Allen, Robert Muller, Greg Kovacs, Kensall Wise, Khalil Najafi, Jim Knutti, Robert Dennard, Denice Denton, Antonio Ricco, Jay Zemel, Mitchell Bohn, Charles Jungo.

In Japan: Shoei Kataoka, Tetsuo Nakamura, Susumu Sugiyama, Osamu Tabata, Shoji Kawahito, Harumi Yanagidaira.

Always and everywhere: Arokia Nathan, Oliver Brand, Jan Korvink, Sadik Hafizovic, Gabriele Baltes.

References

[1] L. Esaki and G. Soncini (eds.) (1981). Large scale integrated circuits technology: state of the art and prospects. *Proc. the NATO Advanced Study Institute*, Erice, Italy (15–27 July 1981). E. Nijhoff (1982). Vol. 55 of NATO Advanced Study Institutes Series.

[2] Bergveld, P. (1970). Development of an ion-sensitive solid-state device for neurophysiological measurements. *IEEE Transactions on Biomedical Engineering* BME 17: 70–71.

[3] Panteli, C., Georgiou, P., and Fobelets, K. (2021). Reduced drift of CMOS ISFET pH Sensors using graphene sheets. *IEEE Sensors Journal* 21: 14609–14618.

[4] Nathanson, H.C., Newell, W.E., Wickstrom, R.A., and Davis, J.R. (1967). The resonant gate transistor. *IEEE Transactions on Electron Devices* 14: 117–133.

[5] Hierlemann, A., Brand, O., Hagleitner, C., and Baltes, H. (2003). Microfabrication techniques for chemical/biosensors. *Proceedings of the IEEE* 91: 839–863.

[6] Baltes, H. and Castagnetti, R. (1994). Magnetic sensors. In: *Semiconductor Sensors* (ed. S.M. Sze), 205–269. New York: Wiley.

[7] Riccobene, C., Wachutka, G., Bürgler, J., and Baltes, H. (1994). Operating principle of dual collector magnetotransistors studied by two-dimensional simulation. *IEEE Transactions on Electron Devices* 41: 32–43.

[8] Nathan, A. and Baltes, H. (1999). *Microtransducer CAD Physical and Computational Aspects*. Wien: Springer.

[9] Petersen, K.E. (1982). Silicon as a mechanical material. *Proceedings of the IEEE* 70: 420–457.

[10] Salo, T., Kirstein, K., Sedivy, J. et al. (2004). Continuous blood pressure monitoring utilizing a CMOS tactile sensor. *The 26th Annual International Conference of the IEEE-Engineering in Medicine and Biology Society* 2: 2326–2329.

[11] Jung, S., Hauert, R., Haluska, M. et al. (2021). Understanding and improving carbon nanotube-electrode contact in bottom-contacted nanotube gas sensors. *Sensors and Actuators B: Chemical* 331: 1–7.

Chapter 25

Creation of the Insulated Gate Bipolar Transistor

B. Jayant Baliga

Electrical and Computer Engineering Department, North Carolina State University, Raleigh, NC, USA

25.1 Introduction

Invention, although an essential component in the creation of a new technology, is often not of paramount importance in its becoming a successful innovation. Most inventions do not become successful products due to either lack of follow-through by the inventor or recognition of its merit by corporations. This article describes the circumstances that led to the Insulated Gate Bipolar Transistor (IGBT) becoming widely accepted as the power device of choice for medium and high power applications where the circuit operating voltages exceed 200V. The most important driving force for the creation of the IGBT was recognition by General Electric (GE) Company of value of this idea and its potential impact on a diverse array of products. The strong support by the enlightened management at GE at the highest level allowed its rapid commercialization. This enabled the development of highly efficient adjustable speed motor drives for industrial applications, compact fluorescent lamps for lighting, and the electronic ignition system for transportation. The benefits of this technology to society have been quantified in previous publications [1, 2]. Independent power semiconductor manufacturers would not have embarked up on introducing the IGBT as a product due to an uncertain market. It was only after the value proposition for the IGBT was demonstrated at GE for a diverse array of applications that worldwide commercialization of the technology was undertaken.

75th Anniversary of the Transistor, First Edition. Edited by Arokia Nathan, Samar K. Saha, and Ravi M. Todi.

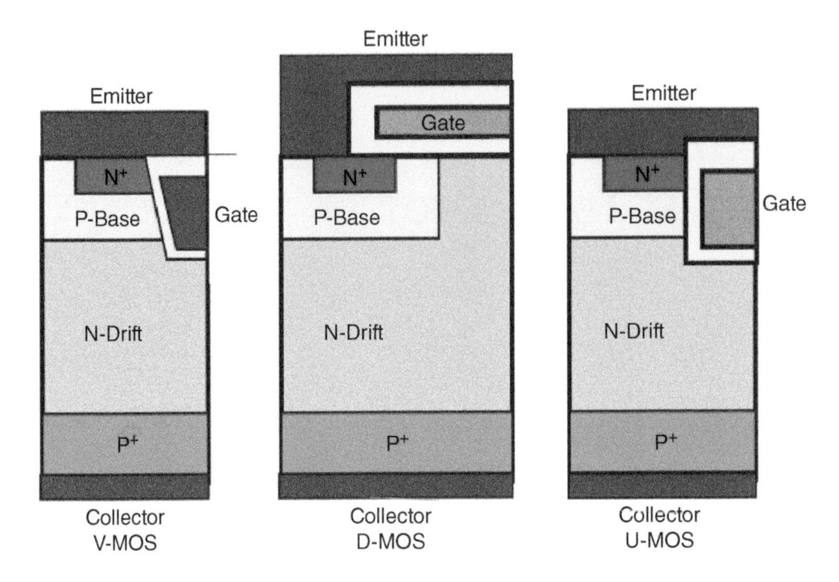

Figure 25.1 Cross sections of basic IGBT device structures.

The basic IGBT cross sections are illustrated in Figure 25.1. The V-groove MOS (V-MOS) structure was popular during early days of power metal-oxide-semiconductor field-effect transistor (MOSFET) development. First experimental evidence of the IGBT-mode was experimentally demonstrated at GE using this structure. The commercialization of the IGBT at GE was accomplished using the double-diffused MOS (D-MOS) structure. It became the dominant technology for manufacturing the device in the 1980s. The U-MOS (U-shaped vertical MOS) structure began to be used for products after 1990.

25.2 Historical Context

I joined the General Electric Corporate Research and Development Center (GE-CRD) in 1974. My first task after joining the GE-CRD was development of thyristors with 3 kV blocking voltage capability from 2 in. diameter wafers. At that time, a trade-off between the blocking voltage capability and gate drive current for thyristors was necessary. Large gate triggering currents were required for devices that could tolerate high *dV/dt* (time rate of change of voltage), and the involute gate structure was required for distributing the gate current across the large device area. Conception of the IGBT structure occurred when I tried to overcome these limitations.

I submitted a patent disclosure at GE on 26 July 1977 on using an MOS gate structure for thyristors [3]. A vertical f layer structure was described with the V-MOS gate structure shown in Figure 25.1. Setting up and optimizing a potassium hydroxide (KOH)-based silicon etching process to form the (truncated) V-grove region was first needed for fabrication of the device structure. The fabrication of the MOS-gated thyristors was started on 9 November 1978 and completed on 28 July 1979 under my supervision with Margaret Lazeri serving as my lead process technician [4]. My first experimental measurements on 30 July 1979 clearly showed what is now called the IGBT-mode [5]. This refers to MOS-gate bias-controlled current saturation in a 4-layer semiconductor structure. In addition, I discovered that the 4-layer structure could be turned-off by application of a negative gate bias to the MOS gate electrode. This revealed non-latch-up current transport in a vertical 4-layer semiconductor device for the first time, i.e. the IGBT-mode. The conception and experimental demonstration of the IGBT at GE can therefore be directly traced to my patent disclosure on 26 July 1977. The results were published on 27 September 1979 in Electronics Letters [6] showing the IGBT-mode of operation for a MOS-gated 4-layer semiconductor.

25.3 The Brock Effect

Tom Brock, the Vice-President of a new GE Product Division set up to create high-efficiency adjustable speed motor drives for air-conditioning, visited the GE-CRD in early September 1980. He was frustrated with using the available Darlington bipolar power transistors for this application due to their bulky and expensive gate drive and snubber circuits. He challenged my group to either deliver a better device technology or seek other employment. I realized that a device with an expanded IGBT-mode that I had discovered could be the answer. My patent disclosure [7] on 29 September 1980 projected the following characteristics for the device: (i) both forward and reverse blocking capability; (ii) forward drop similar to a p–i–n rectifier; (iii) turn-on and turn-off using a small gate voltage with low gate current; (iv) very high turn-off gain; (v) high *dV/dt* and *dI/dt* (time rate of change of current) capability; (vi) operating at elevated temperatures; (vii) tolerance to radiation. I named the device "Gate Enhanced Rectifier (GERECT)" to emphasize its p–i–n rectifier like on-state characteristics projected based up on my Field Controlled Thyristor (FCT) work in the 1970s [8–12].

25.4 My IGBT Proposal

My GERECT proposal in September 1980 was met with skepticism by colleagues at GE. They firstly pointed out that previous efforts at MOS-gating of 4-layer structures showed latch-up of the thyristors at low current levels [13, 14]. They also pointed out that my proposed IGBT structure consisted of an n-channel MOSFET driving a *wide-base p–n–p bipolar transistor*. Prevailing wisdom based on decades of work on power bipolar transistors recommended using a *narrow-base n–p–n structure* to get good current gain. Based up on this, the critics said that my proposed device can be expected to operate at a low on-state current density (below $20\,\text{A}\,\text{cm}^{-2}$) making it cost prohibitive. My projections of an on-state current density of $100–200\,\text{A}\,\text{cm}^{-2}$ for the IGBT based up on a p–i–n diode/MOSFET model were considered unrealistic. It was also argued that latch-up of the 4-layer structure could not be avoided leading to destructive failure of the IGBT in applications.

At the same time frame, Victor Temple at GE proposed the MOS-Controlled Thyristor (MCT) structure [15]. His concept was based up on a 4-layer thyristor operating with latch-up in the on-state. An integrated MOSFET was used in the MCT to short-circuit the junction between the cathode and the base region to interrupt the regenerative action and turn-off the device. Temple made convincing arguments that the MCT would have lower on-state voltage drop and larger on-state current density than my proposed IGBT making it superior for all applications. I pointed out that the IGBT was more suitable for replacement of bipolar transistors due its good forward biased safe operating area (FBSOA) which the MCT did not exhibit; and I argued that the MCT was more suitable for replacement of gate turn-off thyristors (GTOs) because it required snubbers to control the turn-on and turn-off process. In addition, the MCT was much more complex for fabrication extending its development time. Unfortunately, the management at GE-CRD began to view the IGBT as a temporary solution while awaiting the realization of the MCT. Moreover, Temple's arguments convinced U.S. funding agencies at the Electric Power Research Institute (EPRI) and the Department of Defense (DoD) to form a consortium to exclusively support the development of the MCT in the United States [16]. This decision resulted in shifting the eventual production of the IGBT to overseas manufacturers who recognized the true merits of the device.

25.5 The Welch Edict

Tom Brock returned to the GE Research Center in October 1980 to hear a response to his challenge. Due to lack of any practical alternatives, I proposed the IGBT to him although it had not yet been fully developed. My presentation showed that, in addition to his adjustable speed drives for heat pumps, the IGBT could be used for a wide variety of products made by GE in the small appliance, large appliance, industrial drives, and lighting divisions. Most importantly, I pointed out that I could engineer the process for my IGBT so that it could be manufactured in the existing power MOSFET GE production line. This would allow rapid availability of IGBTs in production quantities without going through the usual prolonged development phase required for most innovations with unique device structures, such as the MCT.

Brock found my arguments so compelling that he briefed GE Chairman Jack Welch of the potential company-wide impact. In response, Welch decided to get a personal briefing about the IGBT by making a trip to Schenectady in November 1980. I was asked to personally brief Welch about my IGBT idea and its impact on GE products. This was a highly risky proposition because I had not yet made a fully functional device at that time. Based on my pitch, Welch declared his support for the rapid development of the IGBT but declared a moratorium on public disclosure to allow GE to capitalize on this technology for its products. On the one hand, this was a desirable outcome because I got access to the GE power MOSFET production line. On the other hand, it delayed my getting recognition outside GE for this important innovation for several years. This edict also put a spotlight on me from the highest levels of the company. My career at GE would have been terminated if my chip and process design effort for fabricating the IGBT in the power MOSFET production line had failed. I was fortunately able to flawlessly create a chip design and a modified production process that enabled manufacturing the device on first pass at the production line.

The filing of the patent application for the IGBT was undertaken at the corporate level because of the interest shown by Welch. The patent application was filed on 2 December 1980 and issued only in 1990 after vigorous prosecution at the patent office [17]. GE obtained patent protection for the basic IGBT concept with broad claims until 2009 due to this delay. Meanwhile, a dozen patents related to the IGBT were issued to me in the 1980s [18].

25.6 Manufacturing the First IGBT Product

In January 1981, I flew from Schenectady to Cupertino, CA, to discuss the fabrication of the IGBT at the power MOSFET production facility. The manager of the production line, Nathan Zommer, was reticent to interrupt the production line but was compelled to do so due to the Welch edict. I described the IGBT concept to him and proposed my fabrication process with one additional mask to form a deep P^+ region for thyristor latch-up suppression. Zommer agreed to fabricate the IGBT if I provided the starting Si wafers, the chip mask design, and the revised process flow.

A team was assigned to work under my leadership as the acknowledged inventor and principle developer of the device at GE-CRD [19]. The team that I supervised consisted of Mike Adler to perform numerical simulations to confirm its p–i–n rectifier like on-state characteristics and the ability to suppress thyristor latch-up using the deep P^+ region; and Peter Gray to perform the mask layout I designed with the square-cell IGBTs and the 600-V multiple floating field ring edge termination. I determined the N-drift region doping and thickness to obtain the desired 600-V symmetric blocking capability after taking open-base transistor breakdown into account. I then ordered and procured the starting material (N-drift region on P^+ substrates) from vendors and supplied them to Zommer for manufacturing the IGBT.

The first IGBT wafers were completed by the production line in August 1981, less than one year after my meeting with Welch. My chip design and process produced fully functional IGBTs with 600V forward blocking capability. Most importantly, the IGBT chips had p-i-n rectifier like on-state characteristics with an on-state voltage drop of only 1V at 100–200 A cm^{-2} vindicating my projected performance and silencing the critics at GE. The IGBTs switched on and off under control of a voltage pulse applied to the gate as predicted. Thyristor latch-up was observed only at very high current densities (over 1000 A cm^{-2}) providing excellent margin in applications, a major achievement because critics had considered this to be impossible.

However, turn-off of the as-fabricated devices occurred with a current tail making switching losses too high for the intended adjustable speed motor drives. In anticipation of this issue, I had already worked on a lifetime control process using electron irradiation for power MOSFETs in 1980. Prior electron irradiation studies on power MOSFETs had showed large negative threshold voltage shifts due to charges produced in the gate oxide. I discovered that this threshold voltage shift could be removed by annealing the power MOSFETs at 140 °C in a nitrogen environment while retaining the lifetime reducing defects in the bulk. These results were withheld from publication by GE until 1982 [20]. Applying the electron irradiation process in 1981 to the fabricated 600-V IGBTs allowed adjusting the turn-off times from 17 to 0.2 μs. I was eventually allowed to reveal this breakthrough in 1983 [21]. It is worth highlighting this achievement as a critical step for making the IGBT a practical innovation with widespread applications. Without this capability, the IGBT may have been restricted to circuits operating only at low frequencies limiting its utility.

My early measurements also demonstrated that the IGBT had excellent high-temperature characteristics. These results were withheld from publication by GE until 1985 [22]. Meanwhile, this discovery encouraged GE application engineers to use the IGBT for controlling steam irons, space heaters, and developing the Triad-PAR lamp [23] among other products with adverse ambient temperatures.

The manufactured IGBTs treated with electron irradiation allowed GE to launch a 5 hp. "Smart Switch" adjustable speed motor drive for heat pumps on 3 October 1983. My leadership in inventing, championing, developing, and producing the IGBT in a manufacturing line within a remarkable short period of time was recognized by GE with the Coolidge Fellow Award given to me on 28 June 1983. I was the youngest scientist accorded this highest distinction within the company. During the award ceremony, Roland Schmitt, Senior Vice-President and Head of the GE Research Center stated: *"We have identified nearly $ 2 Billion in markets that will be impacted by the IGBT already."* [24].

The above description demonstrates that making the IGBT a viable power semiconductor device required overcoming many obstacles by me during 1981–1983. The creation of the IGBT took far more than just the invention of a 4-layer semiconductor device with MOS-gate structure. It was certainly much more challenging than merely replacing the N^+ drain of a power MOSFET with the P^+ collector region of the IGBT as sometimes stated in the literature.

25.7 First IGBT Product Release

I was chagrined in September 1982 to find out that a paper titled "The Insulated Gate Transistor" will be presented by Marvin Smith, from the Marketing Group at the GE Semiconductor Products Division, at the IEEE Industrial Applications Society Annual Meeting. Welch's embargo that prevented my publications on the IGBT had not been communicated to him. This let the cat out of the bag! When I complained about the unfairness of the situation, GE management responded by making me first author of the IAS paper [25] with focus on the IGBT as a commercial semiconductor device product. I was also allowed to submit a technical paper to the IEEE International Electron Devices Meeting (IEDM) describing the device physics [26].

GE Semiconductor Products Department released the first IGBT product on 20 June 1983: Power-MOS IGT D94FQ4,R4 with ratings of 400, 500-V, and 18-A. It utilized my chip design, manufacturing process flow, and electron irradiation for switching speed enhancement. I then accompanied Dan Schwab from the GE Marketing Department on a press tour to New York City and Boston to promote the product. This effort established GE as the leader in the invention, development, and commercialization of the IGBT in 1983. GE received a *"Product of the Year"* award from Electronic Products Magazine and an *IR-100 Award* from the Industrial Research Magazine for the IGT in 1983. I was featured in the GE 1983/84 Winter Monogram magazine among the six *"Edisons and Steinmetzes of our time"* [27] for my invention and development of the IGBT. The article states: *"The device is expected to have applications in almost every General Electric business."* The introduction of the IGBT as a product by GE in 1983 precedes any products announced by companies in Japan by two to three years. The IGBT was therefore not "born in Japan in 1985" as claimed in 2017 [28] and was already a product by 1983 that had been used for many applications at GE after solving the thyristor latch-up problem.

Figure 25.2 B. Jayant Baliga examining photomasks for manufacturing the first IGBT product at GE.

The first technical device paper on the IGBT device was presented by me at IEDM in December 1982 [26]. It included a 600-V symmetric blocking structure and a 600V asymmetric blocking structure with N-buffer layer. I was prohibited by GE from showing latch-up-free IGBT operation above $100 A cm^{-2}$. This restriction was later removed allowing publication of IGBT device characteristics at high current densities [29].

In conjunction with the IEDM publication, GE put out a press release containing my photograph shown in Figure 25.2 with the caption: *"GE scientists announced development of a new power semiconductor switch called an insulated gate rectifier, capable of operating at high current densities while requiring lower drive power. The device's inventor, B. Jayant Baliga, manager of GE's High Voltage Device Unit examines one of the photo masks used to create the devices intricate circuitry."* According to Bill Austin from GE [30], this photo and caption was distributed by United Press International (UPI) and appeared in some 750 daily newspapers across the United States in 1982.

25.8 IGBT Technology Enhancement

After creating the first IGBT product in 1983 at GE, I was keen on expanding its utility by developing higher blocking voltage and complementary structures. In 1984, I recruited Paul Chow from the VLSI group at GE to work under me in the Power Device Group on the IGBT. Our collaboration demonstrated that: (i) p-channel IGBTs have similar performance to n-channel devices [31] resulting in the Genius I/O product release by GE in 1986 for factory automation; (ii) excellent blocking voltage scaling is possible for IGBTs [32]; (iii) thinner gate oxide will greatly improve on-state and latch-up performance of IGBTs [33]. Under my supervision, Hsueh-Rong Chang demonstrated the first U-MOS IGBTs [34] shown in Figure 25.1. This idea was capitalized later by many Japanese companies to make the "injection-enhanced" IGBT [35].

After June 1983, the GE-CRD routinely made presentations on the IGBT to visitors from across the globe as one of the important innovations created there. This included visitors from Japanese power semiconductor manufacturers, including Fuji Electric, Hitachi, Mitsubishi, and Toshiba. The knowledge about the excellent performance achieved at GE for the IGBT convinced these companies to undertake a concerted effort in Japan to develop and manufacture IGBTs [36]. Toshiba commercialized the IGBT in 1985, followed by Fuji Electric in 1986, and Mitsubishi Electric in 1987. European and

U.S. power semiconductors followed suit once the burgeoning market for the IGBT in electronic ignition systems, adjustable speed motor drives, and compact florescent lamps became evident.

25.9 IGBT Evolution

As already discussed, the IGBT structures with the D-MOS and U-MOS gate structures were demonstrated in the 1980s at GE during the creation of the device. This included the symmetric and asymmetric blocking structures as shown in Figure 25.3. Further progress in improving the performance of the IGBT, particularly for higher blocking voltages needed in traction applications, was achieved in Europe and Japan with the transparent emitter and deep trench approaches shown in Figure 25.4. The dummy-cell concept was introduced in 2016 to improve the short-circuit withstand time [37].

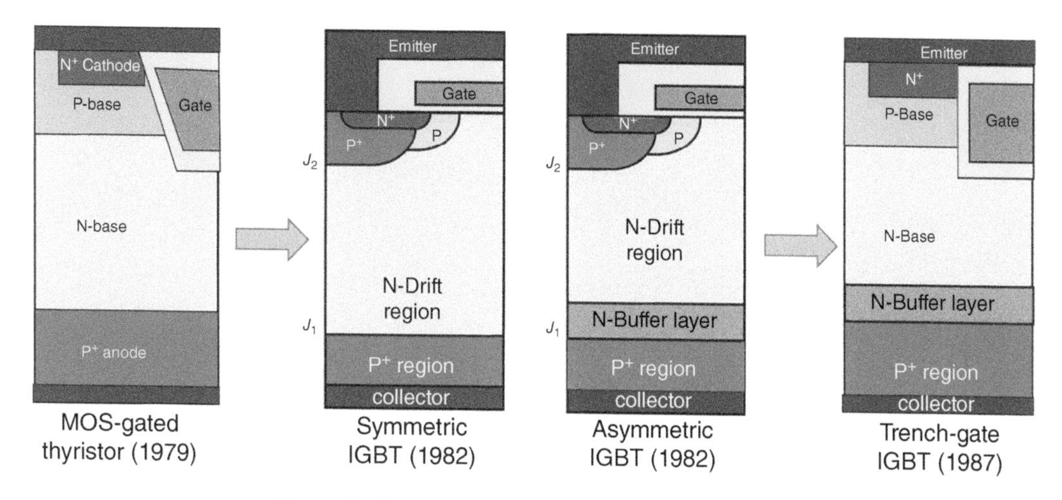

Figure 25.3 Evolution of the IGBT during the 1980s.

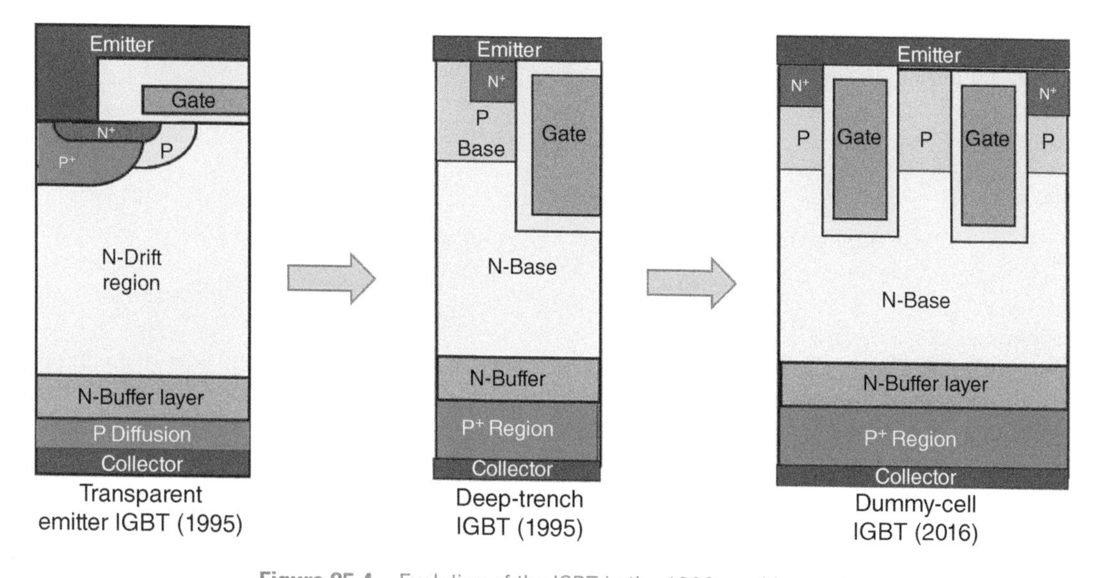

Figure 25.4 Evolution of the IGBT in the 1990s and beyond.

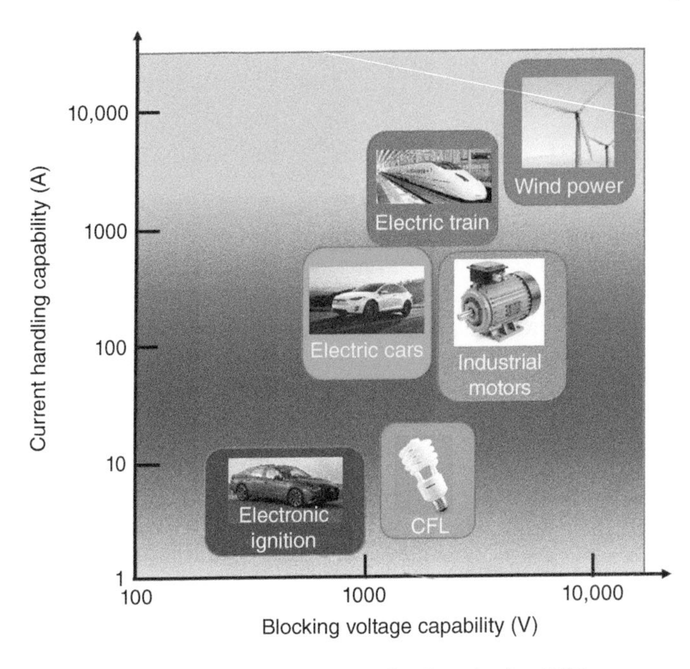

Figure 25.5 Prominent applications for the IGBT.

25.10 IGBT Applications

The IGBT is now utilized in a myriad of applications spanning all sectors of the economy. They have been described in a book to show its widespread adoption in consumer, industrial, lighting, transportation, and other sectors [38]. Some prominent applications are highlighted in Figure 25.5. They include: the electronic ignition system for all gasoline-powered vehicles, the compact florescent lamps for lighting, adjustable speed drives for industrial motors, and traction drives for high-speed urban and long-distance electric locomotives. In addition, the IGBT is used by electric vehicle manufacturers to drive the electric motors making it an essential component for migrating away from fossil fuels. All renewable energy generation, such as hydroelectric, photovoltaic, and wind-power, require the IGBT in inverters to convert the electricity to the well-regulated power delivered to homes and factories.

25.11 IGBT Social Impact

The IGBT enables huge gains in electrical energy efficiency leading to power conservation: adjustable speed motor drives improve efficiency by 40%, while compact florescent lamps improve efficiency by 75%. In addition, the electronic ignition system improved the fuel efficiency by 10% reducing the consumption of gasoline [2].

About 600 billion gallons of petroleum are consumed by automobiles and trucks each year around the world. The cumulative reduction in gasoline consumption from 1990 to 2020 due to the IGBT-enabled electronic ignition systems is 1.79 Trillion gallons due to the 10% improved fuel efficiency.

Two-thirds of the generated electricity around the world runs motors in consumer and industrial applications. An additional, one-fifth of the generated electricity is used for lighting. The improved electrical efficiency derived using the IGBT-enabled adjustable speed drives and compact florescent lamps has saved 133,165 Terra-Watt-Hours of electrical energy during the 1990–2020 time frame.

The above gasoline and electricity efficiency improvements saved worldwide consumers \$36.5 Trillion while eliminating 180 Trillion pounds of carbon dioxide emissions. This is equivalent to offsetting CO_2 emissions from human activity over a three-year time frame.

Even greater impact from the IGBT will be derived in the future with replacement of gasoline power cars with electric vehicles and replacement of coal-fired power plants with wind and solar power generation. These benefits have also been recently quantified [39].

25.12 My Sentiments

In this article, I have set forth the unique set of circumstances that allowed me to invent, develop, and commercialize the IGBT device in a remarkable short time frame. My invention of the idea provided the starting point but my overcoming many critical barriers was equally, if not more important, to transforming the idea into a practical product that could be used for many applications. The culture at GE inculcated by a new Chairman provided the support structure needed by me for its rapid manufacturing and internal adoption. After the success of the IGBT, there have been claims that a similar idea was postulated by others. However, they failed to recognize its value and take it from a concept to a working device with performance useful to the power electronics community. My ability to solve the thyristor latch-up problem and formulate an electron irradiation-based switching speed control process were crucial to its success. My fabrication of the device from the inception in a power MOSFET manufacturing line within 10 months was an unprecedented accomplishment allowing its rapid adoption for developing products in the consumer, industrial, lighting, and transportation sectors.

References

[1] Baliga, B.J. (2013). The role of power semiconductor devices in creating a sustainable society. *IEEE Applied Power Electronics Conference, Plenary Paper*, Long Beach, CA (March 2013).

[2] Baliga, B.J. (2014). Social impact of power semiconductor devices. *IEEE International Electron Devices Meeting, Invited Talk*, San Francisco, CA (December 2014).

[3] Baliga, B.J. (1977). MOS gate structure for thyristors. GE Patent Disclosure Letter RD-10,243, submitted 26 July 1977.

[4] Baliga, B.J. (1978). *GE Process Ticket MOSCR-1* (Start date 9 November 1978; Completion date 28 July 1979).

[5] Baliga, B.J. (1979). "Reduction to practice" for Patent Disclosure Letter RD-10,243, 30 July 1979.

[6] Baliga, B.J. (1979). Enhancement and depletion mode vertical channel MOS gated thyristors. *Electronics Letters* 15: 645–647, Submitted 28 August 1979, Published 27 September 1979.

[7] Baliga, B.J. (1980). Gate enhanced rectifier (GERECT). GE Patent Disclosure Letter RD-13,112, 29 September 1980.

[8] Baliga, B.J. (1976). Electric field controlled semiconductor device. US Patent 4,132,996, filed 8 November 1976, issued 2 January 1979.

[9] Baliga, B.J. (1979). Grid depth dependence of the characteristics of vertical channel field controlled thyristors. *Solid State Electronics* 22: 237–239.

[10] Baliga, B.J. (1977). Asymmetrical field controlled thyristor. GE Patent Disclosure Letter, 9 December 1977.

[11] Baliga, B.J. (1980). The asymmetrical field controlled thyristor. *IEEE Transactions on Electron Devices* ED-27: 1262–1268.

[12] Baliga, B.J. (1982). High gain power switching using field controlled thyristors. *Solid State Electronics* 25: 345–353.

[13] Plummer, J.D. and Scharf, B.W. (1980). Insulated-gate planar thyristors: I – structure and basic operation. *IEEE Transactions on Electron Devices* ED-27: 380–387.

[14] Plummer, J.D. and Scharf, B.W. (1980). Insulated-gate planar thyristors: II – quantitative modeling. *IEEE Transactions on Electron Devices* ED-27: 387–394.

[15] Temple, V.A.K. (1984). MOS controlled thyristors. *IEEE International Electron Devices Meeting, Abstract* 10 (7): 282–285.
[16] Hingorani, N.G., Mehta, H., Levy, S. et al. (1989). Research coordination for power semiconductor technology. *Proceeding of the IEEE* 77: 1376–1389.
[17] Baliga, B.J. (1980). Gate enhanced rectifier. U.S. Patent 4,969,028, filed 2 December 1980, issued 6 November 1990.
[18] Twelve IGBT related US Patents issued to GE with B. J. Baliga as inventor: (1) #4443931, issued 4/14/1984; (2) #4618872, issued 21 October 1986; (3) #4620211, issued 28 October 1986; (4) #4717679, issued 5 January 1988; (5) #4782379, issued 1 November 1998; (6) #4801986, issued 31 January 1989; (7) #4823176, issued 18 April 1989; (8) #4883767, issued 28 November 1989; (9) #4901127, issued 13 February 1990; (10) #4933740, issued 12 June 1990; (11) #4980740, issued 25 December 1990; (12) #4994871, issued 19 February 1991.
[19] GE Corporate Research and Development Post, No. 936, 28 September 1983 (as one example among many).
[20] Baliga, B.J. (1982). Power MOSFET integral diode reverse recovery tailoring using electron irradiation. *IEEE Device Research Conference, Paper* IVA-7 (June 1982).
[21] Baliga, B.J. (1983). Fast switching insulated gate transistors. *IEEE Electron Device Letters* EDL-4: 452–454.
[22] Baliga, B.J. (1985). Temperature behavior of insulated gate transistor characteristics. *Solid State Electronics* 28: 289–297.
[23] Bloomer, M.D., Laughton, W.J., and Watrous, D.L. (1986). Triad Low-Voltage Converter: Final Report. GE Class 3 Report 86CRD011 (February 1986).
[24] Baliga, B.J. (1983). Brown win Coolidge Awards – Top GE Research Center Honor. Schenectady Gazette (28 June 1983).
[25] Baliga, B.J., Chang, M., Schafer, P., and Smith, M.W. (1983). The insulated gate transistor. *IEEE Industry Applications Society Meeting*, pp. 794–803.
[26] Baliga, B.J. et al. (1982). The insulated gate rectifier. *IEEE International Electron Devices Meeting, Abstract* 10 (6): 264–267.
[27] Carpenter, D.R. (1983–1984). A patent. *GE Monogram* 61 (4).
[28] Iwamuro, N. and Laska, T. (2017). IGBT history, state-of-the-art, and future prospects. *IEEE Transactions on Electron Devices* 64: 741–752.
[29] Baliga, B.J. (1982). MOS controlled bipolar devices. *IEEE Electro*, paper 7/1, pp. 1–8 (19 April 1983).
[30] Bill Austin Memo to B. J. Baliga dated 14 December 1982.
[31] Baliga, B.J., et al. (1984). Comparison of n and p channel IGTs. *IEEE International Electron Devices Meeting, Abstract* 10.6, pp. 278–281, December 1984.
[32] Chow, T.P. and Baliga, B.J. (1985). Comparison of 300-, 600-, and 1200-V n-channel IGTs. *IEEE Electron Device Letters* EDL-6: 161–163.
[33] Chow, T.P., Baliga, B.J., and Chang, M.F. (1985). The effect of channel length and gate oxide thickness on the performance of insulated gate transistors. *IEEE Transactions on Electron Devices* ED-32: 2554.
[34] Chang, H.R., Baliga, B.J., Kretchmer, J.W., et al. Insulated gate bipolar transistor with a trench gate structure. *IEEE International Electron Devices Meeting, Abstract* 29.5, pp. 674–677, December 1987.
[35] Kitigawa, M. et al. (1993). A 4500V injection enhanced insulated gate bipolar transistor. *IEEE International Electron Devices Meeting, Abstract* 28.3.1: 679–682.
[36] Uchida, Y., retired Member of the Board of Directors, Fuji Electric Company, 'Personal Communication' during innauguration of the PSRC consortium at NCSU in 1991 at Raleigh, NC.
[37] Eikyu, K. et al. (2016). On the scaling limit of the Si-IGBTs with very narrow mesa structure. *IEEE International Symposium on Power Semiconductor Devices and ICs*, pp. 211–214.
[38] Baliga, B.J. (2015). *The IGBT Device*. Elsevier Press.
[39] Baliga, B.J. (2021). Enabling green and renewable energy solutions with power semiconductor technology. *Plenary/Keynote Talk, Virtual Global Renewable Energy Researchers Meet*, Australia (7–8 May 2021).

Chapter 26

The History of Noise in Metal-Oxide-Semiconductor Field-Effect Transistors

Renuka P. Jindal

VanderZiel Institute of Science and Technology, LLC, Princeton, NJ, USA

26.1 Introduction

Almost a hundred years ago, the first shot was fired to trigger the *transistor race* by a Polish-American physicist Julius Lilienfeld [1]. His patent disclosed how to achieve conductivity modulation at the surface of a solid by the application of an electric field. A follow-up patent was filed by Oskar Heil [2]. However, due to an electrically dirty surface, a practical implementation of such a device eluded researchers for another quarter of a century years. Finally, a little known fact, in 1958, this silence was broken by Ray Warner [3] of Bell Labs, disclosing the construction of a planar FET using substrate as a gate electrode which could be called the OSFET. Following close behind, Dawon Kahng and Mohamed Atalla [4], also of Bell Labs, published their first realization of a MOSFET using a top metal plate as the gate electrode. Ray's colleagues used to tease him with the quip that he invented 6-UP and not 7-UP. Years later, I had the opportunity of working with Ray as a co-advisor with Aldert vanderZiel for my PhD thesis at the University of Minnesota and Dawon as a senior colleague at Bell Labs. This Kahng–Atalla device had a channel length of 10 µm, about the diameter of a human hair. Since then, over the last six decades, advances in lithography, processing techniques, T-CAD, and device modeling have resulted in a 2000-fold shrinkage [5] in channel length. This phenomenal reduction has fundamentally altered the nature of carrier transport in these devices with far-reaching consequences on the device performance, exacerbating noise. To overcome the noise limitations, the system demanded higher power, compromising battery life, system reliability, and ultimately resulting in a higher cost to the consumer. In this paper, we will track how a careful peeling of the proverbial onion,

75th Anniversary of the Transistor, First Edition. Edited by Arokia Nathan, Samar K. Saha, and Ravi M. Todi.

layer by layer, led to almost an order of magnitude [6] reduction in noise, making the bulk MOSFET and its progenies the technology of choice for most low-cost, high-volume modern-day electronics, enjoyed by humanity. Although it took a century to reach this level of proliferation, for Lilienfeld [1], it would be a dream come true.

In common parlance, anything unwanted is referred to as noise. In the present context, however, noise has a very specific meaning. Temporal variability in the magnitude and/or direction of any measurable quantity is fundamentally present in all physical systems. Its existence is guaranteed by thermodynamics. This stochastic temporal variability is referred to as noise. Spatial variability of *I–V* characteristics of otherwise identical devices across a wafer is an important topic discussed elsewhere [7].

26.2 MOSFET Noise Time Line

After the demonstration of the MOSFET in 1960 [4], its noise was first analyzed by Jordan and Jordan [8]. Based on these findings, MOSFET appeared to be a promising competitor to its solid-state sibling, the bipolar junction transistor, which had been demonstrated [9] at Bell Labs 13 years earlier. When I joined Bell Labs, the now ubiquitous wireless communication was in its infancy and long-haul lightwave communication was in the developmental stage. The leading candidate for lightwave application was bipolar technology. Nonetheless, MOS effort was being pursued aggressively in the Advanced Silicon Lab under the leadership of Martin Lepselter. There was stiff competition from the MESFET folks using Gallium Arsenide as the semiconducting material. The goal of Marty's lab with 100+ PhDs was to blow Gallium Arsenide out of the water. As he used to say "Gallium Arsenide is the technology of the future, and we will keep it that way" professing "Even God made this earth out of Silicon."

A rude awakening to this MOSFET euphoria dawned upon us when we translated our IC chip designs into working silicon [10] with an unexplained severe noise penalty. Since I had worked under the noise pioneer van der Ziel, fondly referred to as the "Father of Noise," I was given the charter to develop a fundamental understanding of noise in MOS devices for ultralow-noise lightwave and wireless applications. It was then when my *love affair* with noise took firm roots, continuing for more than 40 years and counting. In the Bell Labs style of "Give them the direction and get out of their way" [9], I was given full support by Marty Lepselter, a MEMS pioneer, George Smith, a Nobel Laureate and Harry Boll, a DRAM pioneer and Bell Labs Fellow, to cut the Gordian knot. The first item of business was to develop an ultrasensitive, yet automated, noise measurement system. Next, collaborating with colleagues in the IC FAB line, referred by us as the "Blue Zoo," I designed and fabricated noise testers with metallurgical channel lengths down to 250 nm. This was in the early 1980s. Iterating between design, layout, fabrication, packaging, and testing, I spent the next several years at Murray Hill understanding this unexplained noise, leveraging measurements, theoretical analyses, and simulations. Living close by in Berkeley Heights, my research often spilled into evenings and weekends. George Smith aptly captured my efforts in a limerick that he read out aloud at my 10th anniversary luncheon.

When Renuka first came to Ma Bell;
His work went on jolly and well
Then he carefully found
Those strange IGFET sounds
That now blast out louder than Hell.

Contrary to the rest of the world, we at Bell Labs, have used the acronym IGFET instead of a MOSFET, highlighting an important characteristic of the gate electrode. In the rest of this work, the history of the understanding of noise in MOS devices, as written by the global noise community, will be elucidated, skipping prior details [11]. I coined the terms *intrinsic* and *extrinsic* noise. *Intrinsic noise* is expected of an ideal device, independent of technology constraints. The two *intrinsic* noise mechanisms in a MOSFET are channel thermal noise and induced-gate noise. *Extrinsic noise* mechanisms are not fundamental to device operation and yet can be dominant.

26.3 Channel Thermal Noise

The first contribution to the intrinsic noise comes from channel thermal noise guaranteed by thermodynamics [12]. Jordan and Jordan [8], under the implicit assumptions of drift-diffusion (DD) carrier transport and quasi-static (QS) approximation, gave the following equivalent expression for drain current noise current spectral density:

$$S_{I_d}(f) = 4k_B T \gamma_{lc} g_{d0} \tag{26.1}$$

where k_B is the Boltzmann constant, T is the absolute sample temperature, and g_{d0} is the drain conductance at drain-to-source voltage $V_{DS} = 0$. Here, γ_{lc} is a bias-dependent noise parameter for long-channel devices. As expected, $\gamma_{lc} \rightarrow 1$ as $V_{DS} \rightarrow 0$ and monotonically decreases to 2/3 as V_{DS} is increased and the device enters the saturation regime and is flat thereafter.

The saturation value of γ_{lc} = 2/3 is valid only for long-channel devices built on lightly doped substrates. The effect of the fixed bulk-charge on channel thermal noise was modeled by Sah et al. [13], extended by Klaassen et al. [14], and further extended by Paasschens et al. [15].

Under saturation, in the non-quasistatic (NQS) regime, Deshpande and Jindal [16] updated (26.1) to give the following expression:

$$S_{I_d}(f) = 4k_B T g_{d0}\left[\frac{2}{3} + \frac{76}{4725}\left(\frac{\omega}{\omega_T}\right)^2\right] \tag{26.2}$$

where ω is the operating angular frequency and ω_T is the angular transit frequency.

26.4 Induced Gate and Substrate Current Noise

The second intrinsic noise mechanism arises when channel thermal noise voltage fluctuations couple to the FET gate resulting in an induced gate current. Shoji [17], a former student of van der Ziel and my Bell Labs colleague, developed an expression for induced gate noise in the QS regime. Deshpande and Jindal [16] extended Shoji's treatment to the NQS operation. These results were further enhanced by Vallur and Jindal [18], taking electron-temperature and velocity-saturation into account keeping the NQS regime intact. The results are given below.

$$S_{I_g}(f)_{NQS} = \frac{64}{135} k_B T g_{d0}\left(1 + \frac{E}{E_C}\right)^2\left(\frac{\omega}{\omega_T}\right)^2 \left[1 - \frac{43}{2970}\left(1 + \frac{E}{E_C}\right)^2\left(\frac{\omega}{\omega_T}\right)^2\right]\Delta f \tag{26.3}$$

Here, E is the channel electric field, E_C the critical electric field, and other symbols have been defined earlier. As a check, if we substitute $E_C = \infty$, we recover [16] and if, in addition, we substitute $\omega_T = \infty$, we recover [17]. The above results show an explicit frequency dependence and an increase in the induced gate noise as a function of the electric field E in the unpinched portion of the channel. One factor $(1 + E/E_C)$ arises due to electron temperature and another factor $(1 + E/E_C)$ arises due to mobility degradation.

Analogous to the induction of gate current, one expects induced substrate-current noise. This was first analyzed by Pu and Tsividis [19]. As pointed out by Jindal [20], in connection with substrate-resistance noise, application of substrate reverse bias helps suppress this noise. In the context of a surrounding gate and other gate structures, the analysis is straightforward.

26.5 Gate–Drain Current Noise Cross Correlation

Since the induced gate noise is a direct consequence of the channel thermal fluctuations, it is expected that the two are correlated. An expression for the cross-correlation spectrum, in saturation and under QS approximation, was derived by Shoji [17]. Extending his results to the NQS regime, Deshpande and Jindal [16] showed that the cross-correlation current spectral density is given by

$$S_{I_g^* I_d}(f) = j\frac{4}{9}k_B T g_{d0}\left(\frac{\omega}{\omega_T}\right) + \frac{176}{945}k_B T g_{d0}\left(\frac{\omega}{\omega_T}\right)^2 \tag{26.4}$$

Here, the first term corresponds to Shoji's results and is purely imaginary with a linear frequency dependence. The new second term, with a quadratic frequency dependence, is real. This real component was alluded to by Jungemann et al. [21] based on Monte-Carlo hydrodynamic simulations of carrier transport in the FET channel. Its analytical formulation given above now lends itself to important LNA design optimization [22].

26.6 Equilibrium Noise

At $V_{DS} = 0$, all gate interactions are invisible across the source–drain terminals of the MOSFET due to vanishing gate and substrate transconductances g_m and g_{mb}, respectively. However, Jindal [23] has shown that due to the induced gate current, there are fluctuations in the inversion layer charge Q giving rise to channel conductance fluctuations given by

$$S_Q(f) = C^2 4k_B T\left(\frac{1/g_{do}}{12}\right) \tag{26.5}$$

where C is the total gate capacitance. However, to sense these fluctuations, a current must be injected. This topology is leveraged in the design of transimpedance amplifiers [24] using a MOSFET as a tunable resistive feedback element.

Nonetheless, as early as 1979, Takagi and van der Ziel [25] reported excess channel noise at $V_{DS} = 0$ for devices with gate oxide thickness of 1000 Å. They ascribed it to a strong transverse electric field in the channel, decreasing with a decrease in sample temperature. This proves the absence of hot-carrier effects. Later, at a gate oxide thickness of 18 Å, Vadyala [26] also observed a similar effect. A possible explanation for this effect can be provided by the earlier work of Jindal and van der Ziel [27].

26.7 Bulk Charge Effects

Considering a FET as a 4-terminal black-box, it can be argued [28] that $g_s = g_d + g_m + g_{mb}$. Here, g_s and g_d are the source and drain conductances, respectively. At $V_{DS} = 0$, $g_m = g_{mb} = 0$. Then, $g_s = g_d \equiv g_{d0}$. Now for typical MOSFET operation with the pinch-off occurring on the drain side of the channel, g_s is fairly independent of the drain voltage. Thus, in general, $g_{d0} = g_m + g_{mb} + g_d$. Under saturation, since $g_d << g_m$ or g_{mb}, we can write, $g_{d0} \approx g_m + g_{mb}$.

For a given g_{d0}, since a higher substrate doping gives rise to a higher g_{mb} at the expense of g_m, a higher substrate doping will result in poorer noise performance [20]. However, a mild substrate reverse bias [20] eliminates this penalty. Surrounding gate devices avoid this penalty.

Next, we will focus on the *extrinsic* noise mechanisms which are amenable to significant reduction by a suitable changes in the device structure.

26.8 Gate Resistance Noise

Any thermal voltage fluctuations in a resistive gate should generate channel noise. Today, this may appear to be a trivial assertion. However, I must admit that 40 years ago when this concept was first enunciated, it took two theoreticians to finally nail this down. Karvel Thornber, a Caltech graduate and Bell Labs colleague, modeled [29] this for a single-stripe gate. However, in a typical IC layout, the gate consists of multiple stripes over thin-oxide, producing transistor action, and over thick-oxide, serving only as an interconnect, along with resistive contact windows. Jindal [30] developed a generalized analytical formulation to quantify the above, taking all phase relationships into account. The results are quite complex. Assuming a constant device channel length, the equivalent noise resistance simplifies to the following expression:

$$R_n = \frac{1}{g_{m_T}^2}\left[\sum_{j=1}^{p}\left(\sum_{i=1}^{p} g_{m_i} A_{v_{ij}}\right)^2 R_j + \sum_{i=1}^{p}\frac{g_{m_i}^2}{12} R_i\right] \tag{26.6}$$

Here, g_{m_i} is the transconductance associated with the ith resistor R_i, and g_{m_T} is the total transconductance of the device. $A_{v_{ij}}$ are the average voltage fluctuation ratios calculated either by inspection or using standard circuit simulation tools. When the signal is fed to a single-stripe gate of resistance R from one end, the effective noise resistance is $R/3$. This reduces to $R/12$ when the signal is fed from both ends. These findings have been adopted as rules of thumb with widespread usage in the global circuit design community.

Large W/L ratio devices with nanoscale channel lengths especially require a careful evaluation of this noise source. For an ultralow-noise design, the effective gate resistance can be minimized by using multi-finger gates, larger contact area, and placing them strategically throughout the layout. Using lower resistivity gate materials helps at the same time reducing the interfacial contact resistance.

Since the publication of Jindal's results [30], researchers worldwide [31–54] have extensively used these concepts to design RF CMOS front-ends for lightwave and wireless applications, minimizing cost to the consumer.

26.9 Substrate Resistance Noise

As pointed out by Jindal [54], substrate resistance also produces a fluctuating drain current through the substrate transconductance with spectral density

$$S_{I_d}(f) = 4k_B T R_{sub} g_{mb}^2 \tag{26.7}$$

Here, R_{sub} is the distributed resistance between the FET channel and the substrate contact. A reduction in g_{mb} by applying a mild reverse bias minimizes [20, 54] this noise, improving LNA performance. In a surrounding gate device, there is no substrate contribution. Due to the importance of substrate noise in LNA design, it has drawn interest from the device [34, 35, 55, 56], compact modeling [39], and circuit-design [44, 51, 57–61] communities.

It should be noted that in some later literature, the phrase "epi-noise" [62] is being used which can be misleading since it is the same noise mechanism.

26.10 Substrate and Gate Current Noise

As channel lengths shrink, carriers traveling from the source to the drain acquire enough energy to cause impact ionization, generating position-dependent, gate, substrate, source and/or drain currents. For long-channel MOSFETs, Kim [63] reported this mechanism giving rise to full shot noise. Rucker and van der Ziel [64] also reported full shot noise associated with gate current in a JFET. For submicrometer MOSFETs, Jindal [65] observed full shot noise at low V_{DS} transforming into supershot noise at higher V_{DS}. While shot noise can be easily explained, supershot noise is more complex. This supershot noise was modeled by Jindal [66] as a multistep ionization process involving both carrier types, resulting in an avalanche gain. This is effectively altering the charge quantum, hence generating supershot noise. For most of this regime, the equivalent saturated diode noise current is given by the expression

$$I_{eq} = I_{sub}\sqrt{\frac{I_{sub}}{I_{sub0}}} \tag{26.8}$$

where I_{sub0} is the highest value of substrate current for which it exhibits full shot noise. This noisy I_{sub} generates a fluctuating substrate potential coupling to the FET channel through the substrate transconductance as follows:

$$S_{I_d}\left(f\right) = 2qI_{eq}R_{sub}^2 g_{mb}^2 \tag{26.9}$$

where q is the electronic charge. Mild substrate reverse bias essentially decouples [20] this noise from the channel. Also, impact ionization can be minimized by careful device design. This noise mechanism has drawn the interest of device [31, 67, 68] and circuit designers, *too many to cite*.

Paradoxically, it was shown by Jindal [69], how this *orderly* carrier multiplication can be used for ultralow-noise amplification of photocurrent in a lightwave preamplifier leading to the principle of random multiplication [70] which leverages device speed to quench noise.

As mentioned above, a gate current can also be generated due to the high-energy carrier transport. Alternatively, Fowler–Nordheim tunneling through the thin gate oxide can also produce gate current. This becomes important as gate oxide thicknesses approach 10s of Angstroms. Both such mechanisms are expected to generate only full shot noise.

26.11 Short-Channel Effects

Incessant scaling over the last 62 years has resulted in several phenomena having important bearing on the noise performance of MOSFETs. These include channel-length modulation (CLM) [28], carrier heating [71], velocity saturation [72], and nonlocal quasi-ballistic and ballistic transport [73]. While directly or indirectly influencing all noise sources discussed above, they have a profound effect on channel thermal noise.

Carrier heating can be reasonably accounted for by assuming a Maxwellian-like carrier energy distribution with an effective electron temperature given by [71]

$$T_e = T\left[1 + \frac{E(x)}{E_C}\right] \tag{26.10}$$

Velocity saturation can be incorporated by the following mobility expression [72]:

$$\mu_n\left(x\right) = \frac{\mu_{n0}}{1 + E\left(x\right)/E_C} \tag{26.11}$$

where the symbols have their usual meaning.

For lack of a tractable analytical expression, nonlocal quasi-ballistic effects [73] must be addressed numerically.

26.12 Effect on Channel Thermal Noise

Studies on micrometer-scale MOSFETs [74–76] yielded only a small increase in noise at room temperature becoming more pronounced at 77 K. From early theory [8], the expectation was for the channel thermal noise parameter γ to equal unity at $V_{DS} = 0$ and dip to 2/3 as the device nears saturation. However, at a metallurgical channel length of 500 and 250 nm, Jindal [6, 20] found that this dip was never observed. In actuality, γ kept increasing monotonically with increasing V_{DS}. These were the first ever *room-temperature* experimental observations of this noise escalation [6, 20, 77, 78]. The initial increase in γ in the linear region of MOSFET operation is believed to be due to carrier heating [78]. Approaching saturation, γ continues to increase, albeit at a slower rate. Since the initial observation of these results [6], this topic has been vigorously pursued by the device community worldwide [21, 34, 51, 78–111] with similar attention paid by circuit designers *too many to cite*. Most later investigations [34, 42, 84, 90, 96, 98, 106] have largely confirmed Jindal's earliest findings [20, 77, 78].

As channel lengths continue to shrink, attempts to develop a physical understanding of this excess noise have yet to converge, partly since it is a moving target. The research efforts can be classified into the "no excess noise," "shot noise," and the "hot carrier noise" groups. We will next discuss each of these schools of thought.

The "no excess noise" group [15, 42, 107, 112] has explained channel noise in MOSFET with gate lengths down to 80 nm without invoking any excess noise mechanism with 20% unexplained noise at shorter channel lengths.

Under the shot noise umbrella, there are three approaches. Obrecht et al. [91, 97] simulate a significant diffusion current component in the FET channel near the source end of the device as a possible source of shot noise. Andersson et al. [99] experimentally verified this approach.

Navid and Dutton [100] ascribed the shot noise to nonequilibrium transport due to few collisions of the carriers in their journey from the source to the drain. Their theory was able to explain the experimental results reported by Jindal [20] for 750 nm channel length devices but not for 500 and 250 nm devices.

Later, Sirohi [101] and Navid [102] ascribed shot noise generation due to random carrier emission over source–channel barrier. Using a semiempirical approach, Sirohi [101] explained the noise measurements [20] down to 250 nm. Later, Devulapalli [103] was able analytically account for excess noise at the sub-100 nm scale.

Next, we discuss the hot-carrier line of thought. Baechtold [75] claimed that the noise in n-silicon layers increases due to an increase in the carrier temperature supported by later theories [34, 84, 88, 94]. However, carrier heating and a reduction in mobility go hand in hand, making hot carriers less responsive to electric fields, implying a reduction in noise. Under velocity saturation, this noise must vanish. Therefore, following Chen and Deen [96], the pinched-off portion of the channel cannot contribute to the channel thermal. Using the transmission line analysis, Vallur and Jindal [18] have shown that in the unpinched off portion of the channel, for a specific dependence of the carrier temperature (26.10) and mobility (26.11) on electric field, the two effects cancel each other out in the *QS* portion of the noise. The channel noise is then given by

$$\overline{i_d i_d^*}_{NQS} = 4k_B T g_{d0}\left\{\frac{2}{3}+\frac{76}{4725}\left[\frac{\omega\left(1+E/E_C\right)}{\omega_T}\right]^2\right\}\Delta f \tag{26.12}$$

This cancelation has perhaps led to significant confusion in the literature. To make the above relationship consistent with CLM, g_{d0} is now the drain-to-source conductance at $V_{DS} = 0$ for a device with channel length equal to CLM-driven electrical channel length. There was a sincere effort [42, 96, 105, 110] to incorporate these effects in models. However, the details were not in agreement [108]. Later, Patalay et al. [106] unambiguously demonstrated that for channel lengths in the sub-100 nm regime, the above theories cannot explain the measured noise.

With a physics-based numerical investigation, using DD and hydrodynamic transport models incorporating velocity saturation with and without carrier heating, Mahajan et al. [111] have compared simulations with experimental data [106]. They demonstrate that in the presence of mobility degradation, neither CLM nor carrier heating is enough to explain the measured results. Thus, a new noise mechanism is needed to explain measurements.

In view of this gap in understanding, it is not surprising that TCAD models are grossly inadequate in simulating channel thermal noise in nanoscale MOS devices. For PD-SOI, at 40 nm channel length, Wadje et al. [113] found the simulated noise was 2.6 times lower than experimental data. As carrier transport in FET channels becomes near-ballistic, this excess channel noise will continue to morph itself and be an active area of research.

26.13 1/*f* Noise

In any dissipative system under equilibrium, Nyquist theory [12] precludes the observation of 1/*f* noise across its terminals. The system must depart from equilibrium to observe these conductivity fluctuations in the sample. However, the conductivity is a product of the number of carriers and their mobility. Thus, the question whether it is the number of mobility fluctuations is not easily answered.

The number fluctuation model goes back to 1950 as proposed by van der Ziel [114], Du Pre [115], and McWhorter [116]. Noise is generated by trapping and de-trapping of carriers at the oxide–semiconductor interface. A superposition of Lorentzians corresponding to each time constant gives rise to the celebrated 1/*f* spectrum. In 1969, based on experimental data, Hooge [117] claimed that 1/*f* noise is related to the number of carriers in the bulk of the sample. Due to the absence of bulk traps, he suggested that the source of this noise is tied to mobility fluctuations. However, no physical mechanism was suggested. Subsequently, Handel [118] proposed a quantum 1/*f* noise model to explain the effect. Jindal and van der Ziel [119] put forth a phonon fluctuation model ascribing these mobility fluctuations to the scattering of the carriers by the crystal lattice.

Working with MOSFETs with ultra-small dimensions, Ralls et al. [120] observed random telegraph signal due to single electron trapping and de-trapping events with a characteristic time constant. For larger devices, collectively these fluctuations would give rise to a 1/*f* spectrum. Currently there is consensus that number fluctuations are the source of 1/*f* noise in n-channel MOS devices. For p-channel devices, the situation is still not entirely clear.

26.14 Conclusions

The need for a better understanding of noise behavior of MOSFETs arose in the early 1980s at Bell Labs to enable the design of front-end electronics for lightwave systems [109] and is now leveraged by the ubiquitous wireless handset industry. Over this 40-year period, the understanding of intrinsic noise mechanisms has been refined and several extrinsic noise sources have been found, understood, and eliminated. As a result, MOSFET has emerged as the technology leader with unprecedented levels of integration for most modern-day high-volume low-cost applications. The symbiosis of technological breakthroughs, theoretical analyses, and numerical experiments will continue to drive the innovation engine in nanoelectronics toward lower cost and ever-increasing performance.

Acknowledgments

This work is dedicated to Aldert van der Ziel whose fond memory continues to inspire my activities. Early on, this research was launched by the vision of the famous trio Harry Boll, George Smith, and Marty Lepselter at Bell Labs. Later, support from University of Louisiana at Lafayette, as William and Mary Hansen Hall, Louisiana Board of Regents, Eminent Scholar Endowed Chair, where the later research was carried out, is gratefully acknowledged. The dedicated effort of my graduate students pursuing successful careers in the telecom industry is rightfully remembered. Technical help from Shelby Williams in supporting these efforts was key their effectiveness. My current involvement in bio-related noise research at VIST as Eminent Scientist and Chief Technology Officer continues to fire up my neurons.

On the personal side, I would like to acknowledge the support of my beloved wife Cynthia Lucille. She was instrumental in nurturing my perennial thirst for knowledge, sometimes at the expense of family time. Her memory and support from our three children will continue to enrich me.

References

[1] Lilienfeld, J.E. (1930). Method and apparatus for controlling electric currents, U.S. Patent 1,745,175, filed 8 October 1926 (22 October 1925 in Canada), issued 28 January 1930.

[2] Heil, O. (1935). Improvements in or relating to electrical amplifiers and other control arrangements and devices, British Patent 439,457, filed 5 March 1935, issued 6 December 1935.

[3] Warner, R.M. (2001). Microelectronics: its unusual origin and personality. *IEEE Trans. Electron Devices* 48 (11): 2457–2467.

[4] Kahng, D. and Atalla, M.M. (1960). Silicon–Silicon dioxide field induced surface devices. *Proc. IRE Solid-State Devices Res. Conf.* Pittsburgh, PA: Carnegie Inst. Technol.

[5] Yeap, G., Lin, S.S., Chen, Y.M. et al. (2019). 5nm CMOS production technology platform featuring full-fledged EUV and high-mobility channel FinFETs with Densest 0.021μm2 SRAM cells for mobile SoC and high-performance computing applications. *IEDM Tech. Dig.*, pp. 36.7.1–36.7.4.

[6] Jindal, R.P. (1984). *Internal Technical Memorandum.* Murray Hill, NJ: AT&T Bell Laboratories.

[7] Jindal, R.P. (2012). Harnessing a few to do the work of a million: the technological challenge. *2012 Int. Conf. on Emerging Electronics*, Mumbai, India. pp. 1–4. http://doi.org/10.1109/ICEmElec.2012.6636219.

[8] Jordan, A.G. and Jordan, N.A. (1965). Theory of noise in metal oxide semiconductor devices. *IEEE Trans. Electron Devices* 12 (3): 148–156.

[9] Jindal, R.P. (2001). Editorial – Hell's bells laboratory. *IEEE Trans. Electron Devices* 48 (11): 2453–2454.

[10] Fraser, D.L. Jr., Williams, G.F, Jindal, R.P. et al. (1983). A single chip NMOS preamplifier for optical fiber receivers. *IEEE Int. Solid-State Circuits Conf.* 26: 80–81.

[11] Jindal, R.P. (2010). Physics of noise performance of nanoscale bulk MOS transistors. In: *Compact Modeling, Principles, Techniques and Applications* (ed. G. Gildenblat), 137–159. Springer.

[12] Nyquist, H. (1928). Thermal agitation of electric charge in conductors. *Phys. Rev.* 32 (1): 110–113.

[13] Sah, C.T., Wu, S.Y., and Hielschler, F.H. (1966). The effects of fixed bulk charge on the thermal noise in metal-oxide-semiconductor transistors. *IEEE Trans. Electron Devices* 13 (4): 410–414.

[14] Klaassen, F.M. and Prins, J. (1967). Thermal noise in MOS transistors. *Philips Res. Rep.* 22: 504–514.

[15] Paasschens, J.C.J., Scholten, A.J., van Langevelde, R. et al. (2005). Generalizations of the Klaassen–Prins equation for calculating the noise of semiconductor devices. *IEEE Trans. Electron Devices* 52 (11): 2463–2472.

[16] Deshpande, A. and Jindal, R.P. (2008). Modeling non-quasi-static effects in thermal noise and induced gate noise in MOS field effect transistors. *Solid State Electron.* 52 (5): 771–774.

[17] Shoji, M. (Jun. 1966). Analysis of high frequency thermal noise of enhancement mode MOS field-effect transistors. *IEEE Trans. Electron Devices* 13, 6 (6): 520–524.

[18] Vallur, S. and Jindal, R.P. (2009). Modeling short-channel effects in channel thermal noise and induced-gate noise in MOSFETs in the NQS regime. *Solid State Electron.* 53 (1): 36–41.

[19] Pu, L.-J. and Tsividis, Y. (1990). Small-signal parameters and thermal noise of the four-terminal MOSFET in non-quasi-static operation. *Solid State Electron.* 33 (5): 513–521.

[20] Jindal, R.P. (1985). High frequency noise in fine line NMOS field effect transistors. *Invited Paper in IEDM Tech. Dig.*, pp. 68–71.

[21] Jungemann, C., Neinhus, B., Nguyen, C.D. et al. (2003). Hydrodynamic modeling of RF noise in CMOS devices. *IEDM Tech. Dig.*, pp. 36.3.1–36.3.4.

[22] Patel, A. (2007). Design optimization of MOS amplifiers for high frequency ultra-low-noise wireless communication applications. Master's Thesis. Univ. of Louisiana at Lafayette, Lafayette.

[23] Jindal, R.P. (2005). Effect of induced gate noise at zero drain bias in field-effect transistors. *IEEE Trans. Electron Devices* 52 (3): 432–434.

[24] Jindal, R.P. (1988). Transimpedance preamplifier with 70dB AGC range in fine line NMOS. *IEEE J. Solid State Circuits* 23 (2): 867–869.

[25] Takagi, K. and van der Ziel, A. (1979). Drain noise in MOSFETs at zero drain bias as a function of temperature. *Solid State Electron.* 22: 87–88.

[26] Vadyala, R. (2007). Experiments on the excess thermal noise in MOSFETs at zero drain bias. Master's Thesis. Univ. of Louisiana at Lafayette, Lafayette.

[27] Jindal, R.P. and van der Ziel, A. (1981). Effect of transverse electric field on Nyquist noise. *Solid State Electron.* 24 (10): 905–906.

[28] Tsividis, Y. (1999). *Operation and Modeling of the MOS Transistor*, 2e. New York: McGraw Hill.

[29] Thornber, K.K. (1981). Resistive-gate-induced thermal noise in IGFET's. *IEEE J. Solid State Circuits* 16 (4): 414–415.

[30] Jindal, R.P. (1984). Noise associated with distributed resistance of MOSFET gate structures in integrated circuits. *IEEE Trans. Electron Devices* 31 (10): 1505–1509.

[31] Shuang, X. and Conn, D.R. (1989). A low-noise gate structure for DMOS monolithic devices. *IEEE Trans. Electron Devices* 36 (7): 1393–1396.

[32] Manku, T. (1999). Microwave CMOS – device physics and design. *IEEE J. Solid State Circuits* 34 (3): 277–285.

[33] Tedja, S., Williams, H.H., van der Spiegel, A. et al. (1992). Noise spectral-density measurements of a radiation hardened CMOS process in the weak and moderate inversion. *IEEE Trans. Nucl. Sci.* 39 (4): 804–808.

[34] Tedja, S., van der Speigel, J., and Williams, H.H. (1994). Analytical and experimental studies of thermal noise in MOSFETs. *IEEE Trans. Electron Devices* 41 (11): 2069–2075.

[35] Re, V.I., Bietti, I., Castello, R. et al. (2001). Experimental study and modeling of the white noise sources in submicron P- and N-MOSFETs. *IEEE Trans. Nucl. Sci.* 48 (4): 1577–1586.

[36] Anelli, G., Faccio, F., and Florian, S. (2001). Noise characterization of a 0.25 μm CMOS technology for the LHC experiments. *Nucl. Instr. Methods Phys. Res. Sect. A – Accel. Spectrom. Detect. Assoc. Equip.* 457: 361–368.

[37] Abou-Allam, E. and Manku, T. (1997). A small-signal MOSFET model for radio frequency IC applications. *IEEE Trans. Comput. Aided Des. Integr. Circuits Syst.* 16 (5): 437–447.

[38] Abou-Allam, E. and Manku, T. (1999). An improved transmission-line model for MOS transistors. *IEEE Trans. Circuits Syst. II-Analog Digit. Signal Process.* 46 (11): 1380–1387.

[39] Tin, S.F., Osman, A.A., Mayaram, K. et al. (2000). A simple sub-circuit extension of the BSIM3v3 model for CMOS RF design. *IEEE J. Solid State Circuits* 35 (4): 612–624.

[40] Tsakas, E.F. and Birbas, A.N. (2000). Noise associated with interdigitated gate structures in RF submicron MOSFETs. *IEEE Trans. Electron Devices* 47 (9): 1745–1750.

[41] Goo, J.-S., Choi, C.-H., Danneville, F. et al. (2000). An accurate and efficient high frequency noise simulation technique for deep submicron MOSFETs. *IEEE Trans. Electron Devices* 47 (12): 2410–2419.

[42] Scholten, A.J., Tiemeijer, L.F., van Langevelde, R. et al. (2003). Noise modelling for RF CMOS circuit simulation. *IEEE Trans. Electron Devices* 50 (3): 618–632.

[43] Steyaert, M. and Chang, Z. (1990). Low voltage BIMOS AM front-end amplifier. *IEE Proc.-G Circuits Devices Syst.* 137: 57–60.

[44] Chang, Z., and Sansen, W. (1990). Low-noise, low-distortion CMOS AM wide-band amplifiers matching a capacitive source. *IEEE J. Solid State Circuits* 25 (3): 833–840.

[45] Shaeffer, D.K. and Lee, T.H. (1997). A 1.5-V, 1.5-GHz CMOS low noise amplifier. *IEEE J. Solid State Circuits* 32 (5): 745–759.

[46] Zhou, J. and Allstot, D. (1999). Addition to monolithic transformers and their application in a differential CMOS RF low-noise amplifier. *IEEE J. Solid State Circuits* 34: 1176.

[47] Christoforou, Y. and Rossetto, O. (1999). GaAs preamplifier and LED driver for use in cryogenic and highly irradiated environments. *Nucl. Instrum. Methods in Phys. Res. Sect. A – Accel. Spectrom. Detect. Assoc. Equip.* 425 (1): 347–356.

[48] Terrovitis, M.T. and Meyer, R.G. (1999). Noise in current-commutating CMOS mixers. *IEEE J. Solid State Circuits* 34: 772–783.

[49] Tsakas, E.F. and Birbas, A. (2000). Noise optimization for the design of a reliable high speed X-ray readout integrated circuit. *Microelectron. Reliab.* 40 (11): 1937–1942.

[50] Tsakas, E.F., Birbas, A.N., Manthos, N. et al. (2001). Low noise high-speed X-ray readout IC for imaging applications. *Nucl. Instrum. Methods Phys. Res. Sect. A – Accel. Spectrom. Detect. Assoc. Equip.* 469 (1): 106–115.

[51] Maschera, D., Simoni, A., Gottardi, M. et al. (2004). An automatically compensated readout channel for rotary encoder system. *IEEE Trans. Instrum. Meas.* 50 (6): 1801–1807.
[52] Goo, J.-S., Ahn, H.-T., Ladwig, D.J. et al. (2002). A noise optimization technique for integrated low-noise amplifiers. *IEEE J. Solid State Circuits* 37 (8): 994–1002.
[53] Scholten, A.J., Tiemeijer, L.F., van Langevelde, R. et al. (2004). Compact modelling of noise for RF CMOS circuit design. *IEE Proc.-Circuits Devices Syst.* 151 (2): 167–174.
[54] Jindal, R.P. (1985). Distributed substrate resistance noise in fine line NMOS field effect transistors. *IEEE Trans. Electron Devices* 32 (11): 2450–2453.
[55] Chang, C.Y., Su, J.G., Wong, S.C. et al. (2002). RF CMOS technology for MMIC. *Microelectron. Reliab.* 42 (4–5): 721–733.
[56] Re, V. and Svelto, F. (1995). High accuracy measurement of low-frequency noise in front-end P-channel FETs. *Nucl. Phys. B Proc. Suppl.* 44: 607–612.
[57] Kim, C.S., Park, J.-W., Yu, H.K. et al. (2000). Gate layout and bonding pad structure of a RF n-MOSFET for low noise performance. *IEEE Electron Device Lett.* 21 (12): 607–609.
[58] Binkley, D.M., Rochelle, J.M., Paulus, M.J. et al. (1990). Amplex, a low-noise, low-power analog CMOS signal processor for multi-element silicon particle detectors. *Nucl. Instrum. Methods Phys. Res. Sect. A – Accel. Spectrum. Detect. Assoc. Equip.* 288 (1): 157–167.
[59] Shaeffer, D.K., Shahani, A.R., Mohan, S.S. et al. (1998). A 115-mW, 0.5-μm CMOS GPS receiver with wide dynamic-range active filters. *IEEE J. Solid State Circuits* 33 (12): 2219–2231.
[60] Beuville, E., Borer, K., Chesi, E. et al. (1990). Amplex, a low-noise, low-power analog CMOS signal processor for multi-element silicon particle detectors. *Nucl. Instrum. Methods Phys. Res. Sect. A – Accel. Spectrum. Detect. Assoc. Equip.* 288 (1): 157–167.
[61] Gottardi, M., Gonzo, L., Gregori, S. et al. (2003). An integrated CMOS front-end for optical absolute rotary encoders. *Analog Integr. Circ. Sig. Process* 34 (2): 143–154.
[62] Lee, T.H. (1998). *The Design of CMOS Radio-Frequency Integrated Circuits*. Cambridge, UK.: Cambridge Univ. Press.
[63] Kim, C.S. (1971). Avalanche multiplication and related noise in silicon MOSFETs. Ph.D. dissertation. Univ. of Florida.
[64] Rucker, L.M. and van der Ziel, A. (1978). Noise associated with JFET gate current resulting from avalanching in the channel. *Solid State Electron.* 21: 798–799.
[65] Jindal, R.P. (1984). Noise associated with substrate current in fine line NMOS field effect transistors. *Presented at the 42nd Device Research Conference*, pp. 18–20. See *IEEE Trans. Electron Devices* 31(12): 1971.
[66] Jindal, R.P. (1985). Noise associated with substrate current in fine line NMOS field effect transistors. *IEEE Trans. Electron Devices* 32 (6): 1047–1052.
[67] Jin, W., Chan, P.C.H., Fung, S.K.H. et al. (1999). Shot-noise-induced excess low-frequency noise in floating-body partially depleted SOI MOSFETs. *IEEE Trans. Electron Devices* 46 (6): 1180–1185.
[68] Workman, G.O. and Fossum, J.G. (2000). Physical noise modeling of SOI MOSFETs with analysis of the Lorentzian component in the low-frequency noise spectrum. *IEEE Trans. Electron Devices* 47 (6): 1192–1201.
[69] Jindal, R.P. (1987). A scheme for ultra-low-noise avalanche multiplication of fiber-optics signals. *IEEE Trans. Electron Devices* 34 (2): 301–304.
[70] Jindal, R.P. (1989). A physical phenomenon to implement a random (noninstantaneous) multiplication photodetector. *IEEE Electron Device Lett.* 10 (2): 49–51.
[71] van der Ziel, A. (1986). *Noise in Solid-State Devices and Circuits*. New York: John Wiley.
[72] Caughey, D.M. and Thomas, R.E. (1967). Carrier mobilities in silicon empirically related to doping and field. *Proc. IEEE* 55 (12): 2192–2193.
[73] Lundstrom, M. (2000). *Fundamentals of Carrier Transport*, 2e. Cambridge: Cambridge University Press.
[74] Klaassen, F.M. (1970). On the influence of hot carrier effects on the thermal noise of field-effect transistors. *IEEE Trans. Electron Devices* 17 (10): 858–862.
[75] Baechtold, W. (1971). Noise temperature in silicon in the hot electron region. *IEEE Trans. Electron Devices* 18 (2): 1186–1187.
[76] Takagi, K. and Matsumoto, K. (1977). Noise in silicon and FET's at high electric fields. *Solid State Electron.* 20: 1–3.
[77] Jindal, R.P. (1985). Noise phenomena in submicron channel length silicon NMOS transistors. *8th Int. Conf. Noise Phys. Syst. 4th Int. Conf. 1/f Noise Conf.*, Rome, Italy, pp. 199–202.

[78] Jindal, R.P. (1986). Hot electron effects on channel thermal noise in fine line NMOS field effect transistors. *IEEE Trans. Electron Devices* 33 (9): 1395–1397.
[79] Sheu, B.J., Scharfetter, D.L., Ko, P.K. et al. (1987). BSIM: Berkeley Short-Channel IGFET model for MOS transistors. *IEEE J. Solid State Circuits* 22 (4): 558–566.
[80] Cappy, A. and Wolfgang, H. (1989). High-frequency FET noise performance: a new approach. *IEEE Trans. Electron Devices* 36 (2): 403–409.
[81] Danneville, F. (1994). Microscopic noise Modeling and macroscopic noise models: how good a connection? *IEEE Trans. Electron Devices* 41 (5): 779–786.
[82] Wang, B., Hellums, J.R., and Sodini, C.D. (1994). MOSFET thermal noise modeling for analog integrated circuits. *IEEE J. Solid State Circuits* 29 (7): 833–835.
[83] Yasuhisa, O. (1995). An improved analytical solution of energy balance equation for short-channel SOI MOSFET's and transverse-field-induced carrier heating. *IEEE Trans. Electron Devices* 42 (2): 301–306.
[84] Triantis, D.P., Birbas, A.N., and Kondis, D. (1996). Thermal noise modeling for short-channel MOSFETs. *IEEE Trans. Electron Devices* 43 (11): 1950–1955.
[85] Triantis, D.P. and Birbas, A.N. (1997). Optimal current for minimum thermal noise operation of submicrometer MOS transistors. *IEEE Trans. Electron Devices* 44 (11): 1990–1995.
[86] Bonani, F., Ghione, G., Pinto, M.R. et al. (1998). An efficient approach to noise analysis through multidimensional physics-based models. *IEEE Trans. Electron Devices* 45 (1): 261–269.
[87] Svelto, F. (1998). Noise analysis of submicron PMOS in NWELL. *Nucl. Phys. B* 61: 539–544.
[88] Klien, P. (1999). An analytical thermal noise model of deep submicron MOSFETs. *IEEE Electron Device Lett.* 20 (8): 399–401.
[89] Scholten, A.J., Tromp, H.J., Tiemeijer, L.F. et al. (1999). Accurate thermal noise model for deep-submicron CMOS. *IEDM Tech. Dig*., pp. 155–158.
[90] Ou, J.J., Jin, X., Hu, C. et al. (1999). Submicron CMOS thermal noise modeling from an RF perspective. *Symp. VLSI Tech. Dig.,* pp. 151–152.
[91] Obrecht, M.S., Manku, T., Elmasry, M.I. et al. (2000). Simulation of temperature dependence of microwave noise in metal-oxide-semiconductor-field-effect transistors. *Jpn. J. Appl. Phys.* 39: 1690–1693.
[92] Jin, W., Chan, P.C.H., Lau, J. et al. (2000). A physical thermal noise model for SOI MOSFET. *IEEE Trans. Electron Devices* 47 (4): 768–773.
[93] Goo, J.-S., Choi, C.-H., Abramo, A. et al. (2001). Physical origin of the excess thermal noise in short channel MOSFETs. *IEEE Electron Device Lett.* 22 (2): 101–103.
[94] Knoblinger, G., Klein, P., Tiebout, M. et al. (2001). A new model for thermal channel noise of deep-submicron MOSFETs and its application in RF-CMOS design. *IEEE J. Solid State Circuits* 36 (5): 831–837.
[95] Rengel, R., Mateos, J., Pardo, D. et al. (2001). Monte Carlo analysis of dynamic and noise performance of submicron MOSFETs at RF and microwave frequencies. *Semicond. Sci. Technol.* 16 (11): 939–946.
[96] Chen, C.-H. and Deen, M.J. (2002). Channel noise modeling of deep submicron MOSFETs. *IEEE Trans. Electron Devices* 49 (8): 1484–1487.
[97] Obrecht, M.S., Abou-Allam, E., Manku, T. et al. (2002). Diffusion current and its effect on noise in submicron MOSFETs. *IEEE Trans. Electron Devices* 49 (3): 524–526.
[98] Han, K., Abou-Allam, E., Manku, T. et al. (2003). Thermal noise modeling for short-channel MOSFETs. *Int. Conf. SISPAD* 79–82.
[99] Andersson, S. and Svensson, C. (2005). Direct experimental verification of shot noise in short channel MOS transistors. *Electron. Lett.* 41 (15): 869–871.
[100] Navid, R. and Dutton, R.W. (2002). The physical phenomena responsible for excess noise in short-channel MOS devices. *Proc. Int. Conf. Simul. of Semicond. Processes and Devices*, Kobe, Japan, pp. 75–78.
[101] Sirohi, S. (2005). Experiments and modeling of high-frequency noise in deep submicron MOSFETs. Master's Thesis. Univ. of Louisiana at Lafayette, Lafayette.
[102] Navid, R. (2005). Amplitude and phase noise in modern CMOS circuits. Ph.D. Thesis. Stanford Univ, Stanford.
[103] Devulapalli, V. (2008). Modeling Drain-current Noise in Short-channel MOS Field-Effect Transistors. Master's Thesis. Univ. of Louisiana at Lafayette, Lafayette.
[104] Teng, H.F. and Jang, S.L. (2003). A non-local channel thermal noise model for nMOSFETs. *Solid State Electron.* 47 (5): 815–819.
[105] Han, K., Shin, H., Lee, K. et al. (2004). Analytical drain thermal noise current model valid for deep submicron MOSFETs. *IEEE Trans. Electron Devices* 51 (2): 261–269.

[106] Patalay, P.R., Jindal, R.P., Shichijo, H. et al. (2009). High-frequency noise measurements on MOSFETs with channel-lengths in Sub-100 nm regime. *Proc. 2nd IEEE Int. Workshop on Elect. Devices and Semicond. Tech.*, Mumbai, India.

[107] Scholten, A.J., Tiemeijer, L.F., Zegers-van Duijnhoven, A.T.A. et al. (2005). Modeling and characterization of noise in 90-nm RF CMOS technology. *Proc. 18th ICNF* (ed. T. González, J. Mateos, and D. Pardo), pp. 735–740.

[108] Roy, A.S. and Enz, C.C. (2005). Compact modeling of thermal noise in the MOS transistor. *IEEE Trans. Electron Devices* 52 (4): 611–614.

[109] Jindal, R.P. (2009). From Millibits to terabits and beyond- over 60 years of innovation. *Invited Paper, Proc. 2nd IEEE Int. Workshop on Elect. Devices and Semicond. Tech.*, Mumbai, India.

[110] Asgaran, A.S., Deen, M.J., Chen, C.-H. et al. (2004). Analytical Modeling of MOSFETs channel noise and noise parameters. *IEEE Trans. Electron Devices* 51 (12): 2109–2114.

[111] Mahajan, V.M., Jindal, R.P., Shichijo, H. et al. (2009). Numerical investigation of excess RF channel noise in sub-100 nm MOSFETs. *Proc. 2nd IEEE Int. Workshop on Elect. Devices and Semicond. Tech.*, Mumbai, India.

[112] van Langevelde, R., Paasschens, J.C.J., Scholten, A.J. et al. (2003). New compact model for induced gate current noise. *IEDM Tech. Dig.*, pp. 36.2.1–36.2.4.

[113] Wadje, N.S., Neeli, V.B., Jindal, R.P. et al. (2010). Investigation of kink-induced excess RF channel noise in sub – 50 nm PD-SOI MOSFETs. *IEEE International SOI Conference*, pp. 1–2.

[114] van der Ziel, A. (1950). On the noise spectra of semiconductor noise and of flicker effect. *Physica 16* (4): 359–372.

[115] Du Pre, F.K. (1950). A suggestion regarding the spectral density of flicker noise. *Phys. Rev.* 78 (5): 615.

[116] McWhorter, A.L. (1957). 1/f noise and germanium surface properties. In: *in Semiconductor Surface Physics*. Philadelphia, PA: Univ. of Pennsylvania Press.

[117] Hooge, F.N. (1969). 1/f noise is no surface effect. *Phys. Lett.* 29A (3): 139–140.

[118] Handel, P.H. (1975). 1/f noise – an infrared phenomena. *Phys. Rev. Lett.* 34 (24): 1492–1495.

[119] Jindal, R.P. and van der Ziel, A. (1981). Phonon fluctuation model for flicker noise in elemental semiconductors. *J. Appl. Phys.* 52 (4): 2884–2888.

[120] Ralls, K.S., Skocpol, W.J., Jackel, L.D. et al. (1984). Discrete resistance switching in submicrometer silicon inversion layers: individual interface traps and low-frequency (1/f) noise. *Phys. Rev. Lett.* 52 (3): 228–231.

Chapter 27

A Miraculously Reliable Transistor

A Short History

Muhammad Ashraful Alam[1] and Ahmed Ehteshamul Islam[2]

[1]Elmore Family School of Electrical and Computer Engineering, Purdue University, West Lafayette, IN, USA
[2]Air Force Research Laboratory, Wright-Patterson Air Force Base, Dayton, OH, USA

27.1 Introduction: A Transistor is Born

The history of metal-oxide-semiconductor field-effect transistor (MOSFET), as illustrated by Moore's law, can be divided into several phases. These include Proto-scaling (1959–1971), Geometric-scaling (1971–2003) [1], and Equivalent-scaling (2004–2020) [1] eras, while the Hyper-scaling (2020–) [1] and the Functional-scaling [2] eras are in the horizon (Figure 27.1). The Proto-scaling era spanned almost a quarter century between the invention of point-contact transistors on 16 December 1947, and the introduction of the Intel 4004 microprocessor on 15 November 1971. Atalla's discovery of thermal oxidation, originally intended for the surface passivation of high-voltage diodes, was quickly adapted to create the gate-oxide of the first MOSFET in 1959 [3–5]. In the 1960s, bipolar junction transistor (BJT) and MOSFET were being developed in parallel: despite poorer reliability, metal-oxide-semiconductor (MOS) integrated circuits (ICs) offered higher density and lower manufacturing costs. In 1962, Hofstein and Heiman at RCA built the first 16-transistor multipurpose MOS logic circuit [6]. The CMOS (complementary metal-oxide-semiconductor) transistors, first demonstrated in 1963 [7] would, by 1965, find applications in static random-access memory (SRAM) circuits [8]. Between 1967 and 1971, a series of MOS-based systems arrived in quick succession [9]: Autonetics created D200, a general-purpose 24-chip MOS computer for aviation and navigation, such as the guidance system for Poseidon submarine-launched ballistic missile. Honeywell followed up with the flight controller for F-14 fighter jets, Four-phase Systems created System-IV chipset, and Viatron Computer Systems

75th Anniversary of the Transistor, First Edition. Edited by Arokia Nathan, Samar K. Saha, and Ravi M. Todi.

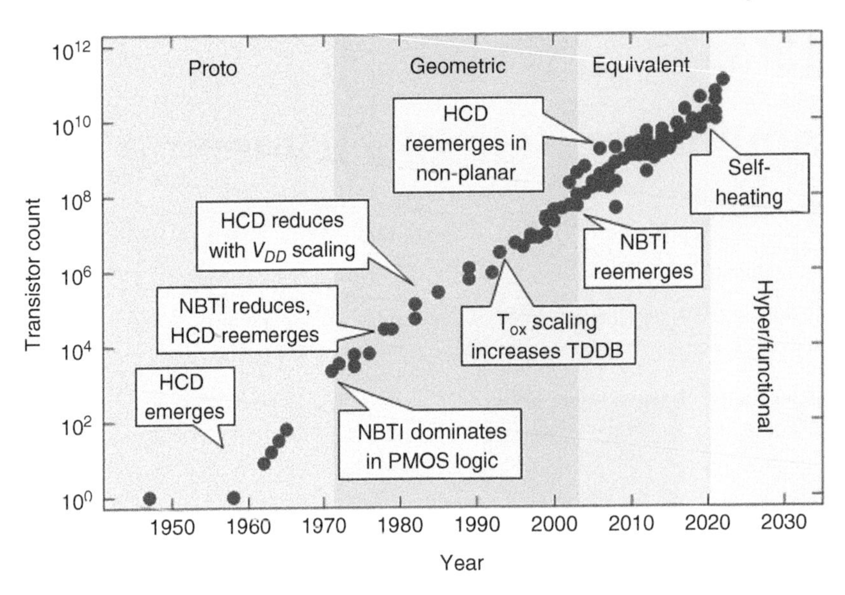

Figure 27.1 The increasing transistor count in ICs fabricated during the Proto-, Geometric-, Equivalent-, Hyper-/ Functional-scaling eras. Data for this plot is obtained from Wikipedia for Intel, AMD, and Apple technologies. The emergence, disappearance, and recurrence of different reliability issues (e.g., HCD: Hot carrier degradation; NBTI: Negative bias temperature instability; TDDB: Time-dependent dielectric breakdown, etc.) are highlighted.

introduced System 21, a 16-bit mini-computer. These early systems used both MOSFET-based logic and memory components (e.g. SRAM, DRAM [dynamic random-access memory], and FGMOS [floating-gate MOS]).

Intel 4004 ICs launched the Geometric-scaling era (1971–2003) where successive design innovations would lead to, over the next 30 years, 100-fold scaling of the transistor dimensions from 10 μm to 90 nm, and the oxide thickness from 200 nm to 2–3 nm. A number of process innovations made this Dennard scaling [10] possible: self-aligned poly-silicon gate, surface-channel transistors, lightly doped drain (LDD), halo implant, channel strain, multilevel interconnects, transition to Cu-interconnects, among many others. These process innovations either solved a reliability challenge or created new ones.

The Equivalent scaling (2003–2020) allowed continued improvement in transistor density and performance, with the introduction of strain, high-κ/metal gate stack technologies, successive generation of fin field-effect transistor (FinFET), ultrathin body silicon-on-insulator (SOI), and track-height reduction techniques. Today, the 2–5 nm technologies are approaching 2D density limits. Future improvement in the Hyper-scaling era (2020–) will rely on packaging innovation involving 3D heterogeneous integrated systems of various types of vertically stacked logic and memory technologies connected laterally by interposers [1, 2]. Indeed, 3D monolithic stacking of flash memories has inspired a corresponding drive to integrate indium gallium zinc oxide (IGZO)-based thin-film transistors between successive backend metal levels. The Hyper scaling of IC will be ultimately limited by phonon-bottleneck. Eventually in the future, Functional scaling will focus on the quality of data/insight generated by sensors and edge processors by focusing on actionable information, rather than raw computational speed [2].

Moore's law is often framed in terms of process integration challenges of increasing transistor density and computational speed. However, less obvious are the stability/reliability challenges that the industry needed (and will need) to overcome in order to make progress possible. In this chapter, we will discuss in Section 27.2 the key stability challenges associated with the Proto-scaling phase that allowed MOSFET technology to achieve, by early 1970s, reliability metrics comparable to more mature bipolar transistor technology. The emergence of packaged CMOS and memory ICs led to the definition of degradation modes (e.g. bias temperature instability or bias-temperature instability (BTI), negative BTI or NBTI, hot carrier degradation or HCD, time-dependent dielectric breakdown or TDDB, electromigration, stress migration, corrosion/intermetallic diffusion) that have structured the reliability discussion over the last 50 years (see Figure 27.1). In Section 27.3, we will provide a brief history of the degradation modes. In Section 27.4, we will conclude by summarizing the emerging reliability challenges of the Hyper-scaling and Functional-scaling eras. As an aside, any historical review is necessarily partial and colored by the authors' own evolving understanding of the field. The chapter should be viewed as an introduction to the vast literature on the topic of Transistor Reliability.

27.2 Transistor Reliability in the Proto-Scaling Era

The invention of the point-contact transistor was triggered by the failure to solve a key challenge associated with Lilienfeld/Shockley's original idea of field effect transistors [5]. Bardeen attributed this failure to surface states and Fermi-level pinning that frustrate the effort to create a gate-controlled inversion layer in semiconductors [11]. Atalla's solved this reliability challenge by innovative surface-cleaning and thermal oxidation of the Si substrate to reduce the interface density by several orders of magnitude, which led to the first demonstration of Si MOSFET [4, 5]. The surface passivation and the planar technology developed by Atalla, Hoernei, and Noyce dramatically improved the reliability of planar ICs compared to systems based on discrete components [3, 5, 6, 8]. Indeed, planar IC was a critical component in developing a reasonably reliable 6-BJT ICs that would support Apollo missions. The support from the space/military programs was responsible for the rapid rise in transistor density reported in the Moore's original paper published in Electronics Magazine in 1965 [8]. Moore emphasized that, apart from the benefit of size, weight, and power, high reliability was a key driver for the emergence of the IC technology.

A series of MOSFET reliability challenges had to solve before Geometric scaling would be possible. In early 1960s, the MOS threshold voltage drifted rapidly at high bias and elevated temperature. Industry-wide research efforts at Bell Labs, Fairchild, Westinghouse, GTE, RCA, and Sylvania concluded that BTI originates from contamination/accumulation of relatively mobile ions such as sodium [12] at the Si/SiO_2 interface. Cleaner environment and processing techniques (e.g. gettering) reduced BTI significantly. Further, the discovery of forming gas anneal and optimum oxidation temperature minimized interface defect density. The rapid improvement in stability/reliability allowed Bell Labs to introduce, in 1967, the dual-dielectric MOS IC with more than 1000 transistors, providing the data point anticipated by Moore's law. By 1971, many MOS ICs (e.g. D200, System IV, System 21 [9]) would have comparable density/reliability, and the industry was ready for the era of Geometric (Dennard) scaling [10].

27.3 Reliability of Geometric- and Equivalent-Scaling Eras

The consideration of very long-term (~10 years) reliability challenges of MOS-ICs began to come to the fore after Intel introduced its 4004 microprocessors in 1971. The subsequent development of CMOS technology, the economic motivation to continue Moore's law scaling, the introduction of Erasable programmable read-only memory (EPROM), etc., initiated an urgent need to develop a foundational theory of reliability of modern MOS ICs. Indeed, the progress of Moore's law reflects ingenuity, creativity, and ever-deepening understanding of the performance and reliability of MOSFET technology. We will discuss a few of these long-term reliability challenges related to CMOS technology that have, at various points, threatened the Moore's law (see Figure 27.1) and the innovations needed to address these challenges.

27.3.1 NBTI: A Persistent Reliability Challenge

NBTI in PMOS transistors arises from the dissociation of Si—H bonds at the Si-SiO_2 interface (giving rise to interface traps) and due to trapping of holes into pre-existing and/or stress-generated oxide traps. The dissociation of Si-H as the origin of NBTI has been known since the Proto-scaling era [13]. Till the early period of the Geometric-scaling era, NBTI persisted as the major reliability concern in PMOS logic and memory devices [14–16]. As NMOS devices became operational and replaced PMOS-based systems, NBTI disappeared in the late 1970s. Even after the introduction of CMOS logic in the 1980s, the use of buried channels in the PMOS device held back NBTI concerns till the 130 nm technology node. However, the use of dual poly-process with surface channel PMOS devices, the increase in oxide electric field (due to slower scaling of operating voltages compared to more aggressive scaling of oxide thickness), and the introduction of nitrogen and hafnium into the gate oxide (to reduce gate leakage and to inhibit boron penetration), reintroduced NBTI as a major reliability concern in the Equivalent- and Hyper-scaling eras [17–20].

The mechanism of NBTI has been studied since 1977 [16]. The hole-assisted field-driven thermal dissociation (during NBTI stress) and reformation (during NBTI relaxation) of Si—H bonds (explained using the Reaction-Diffusion (R-D) theory [16, 18, 19]) dictates the key features of NBTI such as the universal existence of a long-term power law time exponent (i.e. $\Delta V_{IT} \sim t^n$ with $n \sim 1/6$), the frequency independence, and the duty cycle dependence [18–20]. The R-D model can also explain the variation of NBTI with mechanical strain resulting from the introduction of Ge and nonplanar geometries in Si transistors in the Equivalent-scaling era [21, 22]. The variation in strain changes the energy band-structure of the valence band (hence, the probability of hole capture by the Si—H bond) and the Si—H bond density – thereby changing the magnitude of NBTI at a particular time without affecting the time exponent. The use of deuterium instead of hydrogen during oxide preparation (thereby replacing Si—H bonds with Si—D bonds at the Si–SiO_2 interface) can reduce NBTI concerns [23], but this approach is not cost-effective.

In the Equivalent-scaling era, the trapping of holes into pre-existing oxide traps (i.e. hole trapping) of nitride oxides and high-κ gate dielectric significantly affected NBTI [24–26]. Hole trapping reduced n below 1/6 (mostly during the initial phases of NBTI degradation) and the activation energy to values below 0.1 eV measured during the NBTI stress phase. Hole trapping also caused stronger relaxation, stronger duty cycle dependence near 80–90% duty cycle values, and frequency dependence during NBTI relaxation and AC stress measurements. NBTI stress also generates oxide defects, especially at higher stress bias and thus increases n to values above 1/6.

In the 3D-HI FinFET and gate all-around transistors of the Hyper-scaling era, the self-heating is likely to increase NBTI degradation in operational circuits. Therefore, it is clear that NBTI will persist as a key reliability concern for all future technologies. Improved interface engineering, careful management of channel strain, and NBTI-aware circuit and system design [27] would keep NBTI a persistent but manageable reliability concern for the foreseeable future.

27.3.2 The Rise, Fall, and Rise Again of HCD

HCD describes the breaking of atomic bonds (and corresponding performance degradation) by energetic (hot) electrons. The electrons are accelerated (heated) by the terminal voltages in a solid-state device. Hot electrons create hot holes through impact ionization; and their trapping/recombination leads to bond dissociation. By the 1960s, HCD was well known to reduce BJT current gain [28]. These BJTs had to be very carefully qualified for the Apollo program because impact ionization (and hence HCD) increases at the *lower* operating temperature.

Early in the Geometric-scaling era, n-MOSFET HCD emerged as an important reliability issue both for logic and memory transistors [29–31] The electrons/holes accelerated by the high drain (or substrate) field broke the Si—H bonds at the Si/SiO_2 interface or Si—O bonds in the bulk of the oxide. The charges trapped in the broken bonds scattered channel electrons and increased the threshold voltage, with a corresponding reduction in the drain current. The earlier observations of channel hot-electron HCD [29, 30] used gate current as an indicator for the electron injection efficiency into the gate (creating defects) to understand the efficiency of HCD and therefore used $V_{GS} = V_{DS}$ (where gate current had a maxima) as the HCD test condition.

By the early 1980s, it became increasingly important to develop a theory that would correlate HCD to the geometry/voltages relevant for the scaled transistor technologies. Based on the early work on the substrate hot-electron effect [32, 33], Hu et al. developed the phenomenological lucky electron model [34] for long-channel transistors. The model correlated gate current and substrate current during HCD in terms of transistor parameters (e.g. drain-current, mean-free path, oxide barrier height, and impact ionization threshold). The model is very popular for HCD because substrate current is easily measurable and serves as a predictor of the relative HCD robustness of a technology, without having to do long-term accelerated tests. For short-channel transistors being operated at BJT-compatible higher voltages, multiple groups [35–37] developed full-band Boltzmann equation solvers by direct or Monte Carlo methods. In addition, process innovations, such as the use of LDD structures for high-power operation, reduced HCD concerns [38]. By the late 1980s, charge-pumping technique demonstrated that $V_{GS} \sim V_{DS}/2$ bias maximizes HCD [39].

Fortunately, as MOS ICs proliferated in late 1980s and early 1990s, it was no longer necessary to use BJT-compatible voltage. The reduced operating voltage reduced HCD dramatically, and by late 1990s, HCD was no longer considered a major reliability issue for MOS ICs.

Since the introduction of ultra-scaled transistors high-κ dielectric and bulk FinFET, two things have changed. First, the dramatic decrease in thermal conductivity (related to confined geometry of Fins, surrounded by lower thermal high-κ dielectric) has increased self-heating significantly. Second, *for scaled transistors operated at low* V_{DD}, HCD increased with increasing *lattice temperature*. The increased self-heating and the positive temperature coefficient have led to a renewed concern for the HCD [40]. The effect is somewhat counterbalanced by the fact that, HCD in FinFETs reduces somewhat with the increasing operating frequency, because the fins cannot respond thermally within the short timescale.

An important development of HCD theory is the (re)discovery of universal scaling theory, namely, that the HCD degradation curves, for devices taken from multiple technology generations and stressed at various voltages/temperatures can all be scaled to a single degradation vs. time curve [40–42]. The universal scaling approach provides a new approach to accelerated testing of modern transistors, where long-term HCD performance can be predicted by short-term test of a few transistors. It can be shown that the scaling theory generalizes the power-law based ($\sim t^n$) HCD time-kinetics models developed by Hess et al. (explained using distributed Si—O bond energy model [43]) and by Alam et al. based on reaction–diffusion analysis of Si—H bond dissociation [44].

HCD will remain an important concern in the Hyper-scaling era, because self-heating will be increasingly important for gate-all-around transistors confined with 3D-HI systems. The size of the scaled transistors also makes the statistical distribution of HCD an increasingly important consideration [45].

27.3.3 A Short History of TDDB

A MOSFET must have a high-quality gate dielectric (to electrically isolate the channel from the gate). The leakage through the oxide should be negligible, the bulk trapping should be low, and the dielectric breakdown must be avoided. In the Proto-scaling era, the gate oxides were sufficiently thick (~100 nm) so that tunneling and dielectric breakdown were not significant concerns. As mentioned in Section 27.2, the key concerns were BTI related to mobile alkali ion redistribution or the de-passivation of surface states at the Si/SiO_2 interface.

At the onset of the Geometric-scaling era in the early 1970s, thinner gate oxides were needed to control shorter channel transistors [10]. The operating voltage for memory transistors was large: the short-circuit failure of DRAM and the read–write failure of E^2PROM were attributed to TDDB [46]. The anomalously low breakdown field of 50–100 nm oxides was attributed to localized oxide "thinning" (e.g. stacking faults in crystalline oxides, ion/particulate-related field enhancement in amorphous oxides). Enhanced injection through the localized hotspots led to premature breakdown [47, 48].

Improved processing reduced alkali ions and local weak spots so that new theories of intrinsic breakdown were needed to explain the time, temperature, and field-dependent TDDB lifetime (T_{BD}) of sub-50 nm oxides. In the 1980s, Hu et al. developed the 1/E model [49] based on positive feedback (the subsequent electro-thermal runaway) between electron injection into the conduction band of oxides and trapping of holes created by impact ionization. This resulted in the $\log(T_{BD}) \propto 1/E$ relationship, because both Fowler-Nordheim injection and impact ionization have (exponential) $1/E$ dependencies. It is easy to see that the lucky-electron model for HCI [34] and the $1/E$ model for TDDB [49] are based on the same conceptual framework (i.e. carrier heating and impact-ionization feedback).

By late 1980s and early 1990s, it became evident that reduced operating voltage makes oxide impact ionization impossible for sub-10 nm dielectrics [50]. Thinner oxides, however, have higher tunneling current. The tunneling electrons are ballistic, i.e. they arrive at the anode with no loss of energy. These energetic electrons can impact ionize in the anode to produce hot holes, which in turn can be injected back and trapped into the oxide (i.e. Anode Hole Injection, AHI). The subsequent electron–hole recombination breaks bonds and creates defects (at localized hot spots). Accumulation of these defects shorts the substrate to the gate, i.e. the dielectric is broken. The AHI model predicts $\log(T_{BD}) \propto 1/E$ relationship with a different proportionality constant, because now the impact ionization occurs in polysilicon gate (for NMOS) or silicon substrate (for PMOS). Another model, that focuses on field-induced reduction of bond-strength as the rate-limiting factor, interprets TDDB in terms of the E-model, i.e. $\log(T_{BD}) \propto -E$ [51]. The relevance of the two models was hotly debated and a synthesis has been proposed [52–54].

In the mid-1990s, an unusually wide distribution of failure times captured the attention of the industry for 3–7 nm gate oxides [55]. Initially presumed to be related to extrinsic defects (reminiscent of the oxide thinning in the Proto-scaling era), it was soon realized to be an intrinsic property of thin oxides and interpreted by the percolation model. The model predicted that if the defect generation is uncorrelated, the failure time would be Weibull distributed. The experimental validation of uncorrelated defect generation and Weibull distribution were important contributions to the TDDB literature.

Unfortunately, the anode hole injection model and statistical percolation theory led many to conclude that oxides may not scale below 3 nm thickness [56]. Two conceptual advances allowed scaling to continue, namely, minority carrier anode hole injection model [52] and theory of soft breakdown [57–59]. Using detailed numerical and analytical models, Alam et al. explained that defect generation in gate oxides depends on voltage (i.e. energy of the ballistic electrons), not the electric field across the oxide. At lower voltages, the defect generation reduces much faster, and thus, T_{BD} is much longer than previously presumed. A related multi-vibrational hydrogen-release model reached the similar conclusion of reduced defect generation, but focused on the energy needed to break hydrogen bonds [60]. Even with the increased T_{BD}, however, the first oxide breakdown in a typical IC would still occur within unacceptably short time. Fortunately, at typical operating voltages, the first few breakdowns are always soft [58, 61–63]. Despite multiple oxide breakdowns, the IC would not fail. The transistor could be scaled down to 1 nm, so long the circuit can manage the tunneling and leakage current due to soft breakdown. A fortunate confluence of several phenomena sustained oxide scaling, and Moore's law.

By the early 2000s, high-κ gate oxides were introduced at the onset of the Equivalent-scaling era. Percolation theory was generalized to include different in defect generation rates in high-κ vs. SiO_2 interface layers [64]. The trapping in high-κ is still important, however, thin oxide ensures that the threshold voltage shift is insignificant.

Although equivalent oxide thickness has unchanged since 2005, TDDB remains an important concern. For FinFET or gate-all-around transistors in 3D-HI system, the lifetime could reduce due to self-heating-induced correlated defect generation and oxide breakdown at nonplanar edges [65]. Despite the long history and significant work, there are still open questions. While the voltage, temperature, and thickness dependencies are well understood, the time-kinetics is not [66, 67]. A theory that unifies NBTI, HCI, and TDDB time-kinetics would be helpful.

Finally, the Hyper-scaling era anticipates the integration of ferroelectric oxides, which could create new TDDB challenges. For example, the oxide/ferroelectric interface is susceptible to trapping. Hot atom damage in ferroelectric oxide could be a concern. Details are discussed in the recent review article [68].

27.3.4 Application-Specific Reliability: Radiation

Although not intrinsic to the operation of the CMOS circuits, application-specific robustness is essential in the rapid evolution of IC technology over the last 75 years. We will consider two examples related to radiation and corrosion.

In the 1960s, ICs in satellites, space-crafts, and missiles had to survive high-radiation environments. After all, Starfish nuclear test in the exo-atmosphere destroyed the first AT&T satellite (Telestar) only after eight months of operation. The physics of radiation-induced electron–hole pair generation and the trapped holes causing threshold voltage instability was already well understood by early 1970s [69]. It became increasingly clear that cosmic and solar radiations are important concerns. Binder found signature of soft-errors in the satellite data; Intel attributed "random, non-recurring, single-bit" failures in DRAM to soft-errors; and Ziegler et al. reported that terrestrial failures will increase exponentially with altitude.

Throughout the Geometric-scaling era, transistor scaling and the commensurate decrease in charge stored led to an increased sensitivity single-event upset (SEU). The universality of this trend was noted by Petersen and Marshall [70] who observed a power law dependence of critical charge to upset as a function of technology feature size for a wide variety of technologies. The critical-charge reduced from 5 pC to 5 fC as the channel length scaled from 10 to 0.5 μm and below. Voltage scaling has been shown to have an exponential impact on SEU rates. The increasing SEU sensitivity of advanced commercial technologies is making even terrestrial upset a primary reliability concern in ICs, especially those employed in very large datacenters. As CubeSats and communication satellites (based on ultra-scaled transistors) proliferate, radiation hardness will remain an important concern for technology development.

Careful selection of the materials used in IC and packaging are also important. The radiation-induced failures of IBM Supercomputers in 1987 were traced to uranium contamination of the ceramic package. The package factory had been built along a river, downstream from an old uranium mine. Waste from the mine had contaminated the water and, indirectly, the packages. Careful sourcing of the packages solved the issue. In 1995, Baumann et al. showed that boron compounds (e.g. boron phosphosilicate glass [BPSG] used to improve interconnect step-coverage) could be important sources of soft-errors [71]. The subsequent introduction of chemical mechanical polishing (CMP) obviated BPSG and the related reliability concern for ICs produced in the Equivalent-scaling era. One must guard against accidental cross-contamination in the 3D-HI systems.

Finally, as the oxides have thinned, permanent faults related to single event gate rupture and single event burnout have been reported. These hard errors must be managed at the system and algorithm levels for the safe function of the ICs.

27.3.5 Application-Specific Reliability: Corrosion

Reliability of plastic packages has been the subject of intense interest since the 1960s [72]. In the 1970s and 1980s, over 70% of the IC package in the market employed plastic encapsulation [73, 74]. However, ensuring the reliability of plastic package is challenging, especially at high humidity and high temperature setting. Unlike hermetic packages, Al metallization in plastic packages are susceptible to electrochemical failure [73, 75]. Reich and Hakim [73] used both a small set of test and field data to posit that the failure rate of a plastic package is given by $\lambda = \exp(A + B(T + RH))$, where $(T + RH)$ are the sum of temperature and relative humidity. In 1986, Peck offered a more general empirical model for time-to-failure, $t_f \propto (RH)^n \exp(E_a/k_B T)$, using large datasets [76, 77]. Peck's equation has been used to qualify packaged technologies in the 1990s and early 2000s. However, as self-heating became increasingly important in the Equivalent-scaling era, it was clear that the temperature-humidity-bias testing and corresponding lifetime extrapolation using Peck's equations is overly pessimistic due to the excessive moisture loading during the entire test period [78, 79]. The only time moisture can get into the ICs is when they are turned off [79], hence testing requirements for packaged ICs should be determined based on the IC constant longest off duration, such as seven days per year. Ref. [80] has reported a generalized Peck's equation suitable for lifetime extrapolation under use condition intermittent IC self-heating, i.e. time to failure, $t_f = A(\alpha \cdot RH)^n \exp(E_a/\beta \cdot k_B \cdot T)$ that explicitly account for the correlated dynamics of IC bondwire electrochemical failures, moisture ingress/egress, and IC intermittent self-heating. According to this model, the increased IC lifetime is not as significant as the reduced RH due to self-heating may imply. This is because the corrosion of the bond-wire is not defined by the average RH, but by the short timespan at the OFF-to-ON transition during which the coexistence of local moisture (before it can escape) and high T (defined by the IC thermal mass) leads to a "quantized" increase in corrosion. Packaging reliability will become increasingly important in the Hyper-scaling and Functional-scaling eras, and there exists significant research opportunities in these topics.

27.4 Conclusions: Reliability Challenges for the Hyper-Scaling and Functional-Scaling Eras

The characterization and resolution of reliability challenges (e.g. NBTI, HCI, TDDB, etc.) have been essential prerequisite for sustaining the Moore's law over the last 75 years. The similarity of terminology sometimes hides the evolving complexity of the underlying phenomena. For example, the ion-related TDDB challenges in the 1970s are distinctly different from those in the 1990s related to soft-breakdown and Anode Hole injection. The history over the past 75 years teaches us several lessons:

- The discovery of thermal oxidation and the passivation of surface states were important for the development of Si-based MOS transistors. The success and sustainability for other semiconductor materials such as SiC [82], GaN [83], Ga_2O_3 [84] in transistor applications will also rely on the handling of these surface states.
- It is important to distinguish between extrinsic (e.g. mobile ions in oxides, uranium ion in packages) vs. intrinsic effects (e.g. percolation model, self-heating, and interface defects). Extrinsic effects can be solved by process improvement; intrinsic effects (e.g. tunneling limit in thin oxides) requires fundamental shift in technology scaling (e.g. Geometric scaling to Equivalent scaling). In this regard, thermal limit would be a key reliability challenge for the Hyper-scaling era based on 3D-packaging.
- Simple phenomenological reliability models (e.g. $1/E$ model, R-D model) could be as helpful as detailed modeling (e.g. full band Monte Carlo model for HCD). A model should be judged based on the useful predictions it makes, rather than the complexity of the physics it includes.
- Multi-scale and heterogeneous reliability modeling will be increasingly important for the Hyper-scaling era. After all, the packaged IC will contain a variety of technologies. Reliability CAD models should be able to capture the spatiotemporal interaction and reliability implications of the subsystems.
- It will be important to create new qualification protocols that consider IC (rather than transistors) as the unit of reliability physics. In a feedback loop, an effective use field-data would help assure long-term viability of the ICs.
- Finally, over the last 75 years, we have focused on transistor reliability in reasonably controlled environments (e.g. computers in the datacenter) as the key driver for Moore's law. Future technologies will be deployed in a wide variety of applications (e.g. smart agriculture, smart health [81]). Being able to predict the reliability of edge-deployed systems operating in "extreme" and unpredictable environment would be a key challenge for the Functional-scaling era of Moore's law.

References

[1] Datta, S., Chakraborty, W., and Radosavljevic, M. (2022). Toward attojoule switching energy in logic transistors. *Science* 378 (6621): 733–740. https://doi.org/10.1126/science.ade7656.

[2] Lundstrom, M.S. and Alam, M.A. (2022). Moore's law: the journey ahead. *Science* 378 (6621): 722–723. https://doi.org/10.1126/science.ade2191.

[3] Atalla, M.M., Tannenbaum, E., and Scheibner, E.J. (1959). Stabilization of silicon surfaces by thermally grown oxides. *Bell System Technical Journal* 38 (3): 749–783. https://doi.org/10.1002/j.1538-7305.1959.tb03907.x.

[4] Kahng, D. and Atalla, M.M. (1960). Silicon-silicon dioxide field field induced surface devices. *IRE-AIEE Solid-State Device Res. Conf.*, Pittsburgh, PA (June 1960).

[5] Sah, C.-T. (1988). Evolution of the MOS transistor-from conception to VLSI. *Proceedings of the IEEE* 76 (10): 1280–1326. https://doi.org/10.1109/5.16328.

[6] Riordan, M. (2007). The silicon dioxide solution. *IEEE Spectrum* 44 (12): 51–56. https://doi.org/10.1109/MSPEC.2007.4390023.

[7] Moore, G.E., Sah, C.T., and Wanlass, F.M. (1964). Metal-oxide-semiconductor field effect devices for micropower logic circuitry in *Micropower Electronics*, Keonjian, E. (ed.), Pergamon, pp. 41–55.

[8] Moore, G.E. (1965). Cramming more components onto integrated circuits. *Electronics* 38: 114–117.

[9] Shirriff, K. (2016). The surprising story of the first microprocessors. *IEEE Spectrum* 53 (9): 48–54. https://doi.org/10.1109/MSPEC.2016.7551353.

[10] Dennard, R.H., Gaensslen, F.H., Yu, H.N. et al. (1974). Design of ion-implanted MOSFET's with very small physical dimensions. *IEEE Journal of Solid-State Circuits* 9 (5): 256–268. https://doi.org/10.1109/JSSC.1974.1050511.

[11] Bardeen, J. (1956). Surface states on semiconductors. *Journal of the Electrochemical Society* 103 (3): C59–C59.

[12] Snow, E.H., Grove, A.S., Deal, B.E., and Sah, C.T. (1965). Ion transport phenomena in insulating films. *Journal of Applied Physics* 36 (5): 1664–1673. https://doi.org/10.1063/1.1703105.

[13] Deal, B.E., Sklar, M., Grove, A.S., and Snow, E.H. (1967). Characteristics of the surface-state charge (Qss) of thermally oxidized silicon. *Journal of the Electrochemical Society* 114 (3): 266. https://doi.org/10.1149/1.2426565.

[14] Goetzberger, A., Lopez, A.D., and Strain, R.J. (1973). On the formation of surface states during stress aging of thermal Si-SiO2 interfaces. *Journal of the Electrochemical Society* 120 (1): 90. https://doi.org/10.1149/1.2403408.

[15] Frohman-Bentchkowsky, D. (1971). A fully decoded 2048-bit electrically programmable FAMOS read-only memory. *IEEE Journal of Solid-State Circuits* 6 (5): 301–306. https://doi.org/10.1109/JSSC.1971.1050191.

[16] Jeppson, K.O. and Svensson, C.M. (1977). Negative bias stress of MOS devices at high electric fields and degradation of MNOS devices. *Journal of Applied Physics* 48 (5): 2004–2014. https://doi.org/10.1063/1.323909.

[17] Schroder, D.K. and Babcock, J.A. (2003). Negative bias temperature instability: road to cross in deep submicron silicon semiconductor manufacturing. *Journal of Applied Physics* 94 (1): 1–18. https://doi.org/10.1063/1.1567461.

[18] Alam, M.A. and Mahapatra, S. (2005). A comprehensive model of PMOS NBTI degradation. *Microelectronics Reliability* 45 (1): 71–81. https://doi.org/10.1016/j.microrel.2004.03.019.

[19] Islam, A.E., Kufluoglu, H., Varghese, D. et al. (2007). Recent issues in negative-bias temperature instability: initial degradation, field dependence of interface trap generation, hole trapping effects, and relaxation. *IEEE Transactions on Electron Devices* 54 (9): 2143–2154. https://doi.org/10.1109/TED.2007.902883.

[20] Mahapatra, S. (ed.) (2022). *Recent Advances in PMOS Negative Bias Temperature Instability – Characterization and Modeling of Device Architecture, Material and Process Impact*. Springer Singapore.

[21] Tiwari, R., Parihar, N., Thakor K. et al. (2019). A 3-D TCAD framework for NBTI, Part-II: impact of mechanical strain, quantum effects, and FinFET dimension scaling. *IEEE Transactions on Electron Devices* 66 (5): 2093–2099. https://doi.org/10.1109/TED.2019.2906293.

[22] Tiwari, R., Parihar, N., Thakor K. et al. (2019). A 3-D TCAD framework for NBTI – part I: implementation details and FinFET channel material impact. *IEEE Transactions on Electron Devices* 66 (5): 2086–2092. https://doi.org/10.1109/TED.2019.2906339.

[23] Kimizuka, N., Yamamoto, T., Mogami, T. et al. (1999). The impact of bias temperature instability for direct-tunneling ultra-thin gate oxide on MOSFET scaling. *1999 Symposium on VLSI Technology. Digest of Technical Papers (IEEE Cat. No.99CH36325)*, Kyoto (14–16 June 1999), pp. 73–74. IEEE. https://doi.org/10.1109/VLSIT.1999.799346.

[24] Mahapatra, S., Ahmed, K., Varghese, D. et al. (2007). On the physical mechanism of NBTI in silicon oxynitride p-MOSFETs: can differences in insulator processing conditions resolve the interface trap generation versus hole trapping controversy? *2007 IEEE International Reliability Physics Symposium Proceedings. 45th Annual*, Phoenix, AZ (15–19 April 2007), pp. 1–9. IEEE. https://doi.org/10.1109/RELPHY.2007.369860.

[25] Zhang, J.F., Zhao, C.Z., Chang, M.H. et al. (2008). Impact of different defects on the kinetics of negative bias temperature instability of hafnium stacks. *Applied Physics Letters* 92 (1): 013501. https://doi.org/10.1063/1.2828697.

[26] Grasser, T. (2012). Stochastic charge trapping in oxides: from random telegraph noise to bias temperature instabilities. *Microelectronics Reliability* 52 (1): 39–70. https://doi.org/10.1016/j.microrel.2011.09.002.

[27] Alam, M. (2008). Reliability- and process-variation aware design of integrated circuits. *Microelectronics Reliability* 48 (8): 1114–1122. https://doi.org/10.1016/j.microrel.2008.07.039.

[28] Collins, D.R. (1968). Excess current generation due to reverse bias p-n junction stress. *Applied Physics Letters* 13 (8): 264–266. https://doi.org/10.1063/1.1652602.

[29] Abbas, S.A. and Dockerty, R.C. (1975). Hot-carrier instability in IGFET's. *Applied Physics Letters* 27 (3): 147–148. https://doi.org/10.1063/1.88387.

[30] Ning, T.H., Cook, P.W., Dennard, R.H. et al. (1979). 1 μm MOSFET VLSI technology: part IV – hot-electron design constraints. *IEEE Transactions on Electron Devices* 26 (4): 346–353. https://doi.org/10.1109/T-ED.1979.19433.

[31] Ning, T.H., Osburn, C.M., and Yu, H.N. (1976). Threshold instability in IGFET's due to emission of leakage electrons from silicon substrate into silicon dioxide. *Applied Physics Letters* 29 (3): 198–200. https://doi.org/10.1063/1.88991.

[32] Ning, T.H., Osburn, C.M., and Yu, H.N. (1977). Emission probability of hot electrons from silicon into silicon dioxide. *Journal of Applied Physics* 48 (1): 286–293. https://doi.org/10.1063/1.323374.

[33] Verwey, J.F., Kramer, R.P., and Maagt, B.J.D. (1975). Mean free path of hot electrons at the surface of boron-doped silicon. *Journal of Applied Physics* 46 (6): 2612–2619. https://doi.org/10.1063/1.321938.

[34] Simon, T., Ping-Keung, K., and Chenming, H. (1984). Lucky-electron model of channel hot-electron injection in MOSFET'S. *IEEE Transactions on Electron Devices* 31 (9): 1116–1125. https://doi.org/10.1109/T-ED.1984.21674.

[35] Hansch (1992). Modeling hot carrier reliability of MOSFET: what is necessary and what is possible? *1992 International Technical Digest on Electron Devices Meeting*, San Francisco, CA (13–16 December 1992), pp. 717–720. IEEE. https://doi.org/10.1109/IEDM.1992.307459.

[36] Sangiorgi, E., Ricco, B., and Venturi, F. (1988). MOS/sup 2/: an efficient MOnte carlo simulator for MOS devices. *IEEE Transactions on Computer-Aided Design of Integrated Circuits and Systems* 7 (2): 259–271. https://doi.org/10.1109/43.3157.

[37] Higman, J.M., Hess, K., Hwang, C.G., and Dutton, R.W. (1989). Coupled Monte Carlo-drift diffusion analysis of hot-electron effects in MOSFETs. *IEEE Transactions on Electron Devices* 36 (5): 930–937. https://doi.org/10.1109/16.299675.

[38] Sanchez, J.J., Hsueh, K.K., and DeMassa, T.A. (1989). Drain-engineered hot-electron-resistant device structures: a review. *IEEE Transactions on Electron Devices* 36 (6): 1125–1132. https://doi.org/10.1109/16.24357.

[39] Heremans, P., Maes, H.E., and Saks, N. (1986). Evaluation of hot carrier degradation of N-channel MOSFET's with the charge pumping technique. *IEEE Electron Device Letters* 7 (7): 428–430. https://doi.org/10.1109/EDL.1986.26425.

[40] Alam, M.A., Mahajan, B.K., Chen, Y.P. et al. (2019). A device-to-system perspective regarding self-heating enhanced hot carrier degradation in modern field-effect transistors: a topical review. *IEEE Transactions on Electron Devices* 66 (11): 4556–4565. https://doi.org/10.1109/TED.2019.2941445.

[41] Varghese, D., Kufluoglu, H., Reddy, V. et al. (2007). Off-state degradation in drain-extended NMOS transistors: interface damage and correlation to dielectric breakdown. *IEEE Transactions on Electron Devices* 54 (10): 2669–2678. https://doi.org/10.1109/TED.2007.904587.

[42] Mahapatra, S. and Saikia, R. (2018). On the universality of hot carrier degradation: multiple probes, various operating regimes, and different MOSFET architectures. *IEEE Transactions on Electron Devices* 65 (8): 3088–3094. https://doi.org/10.1109/TED.2018.2842129.

[43] Hess, K., Haggag, A., McMahon, W. et al. (2000). Simulation of Si-SiO/sub 2/ defect generation in CMOS chips: from atomistic structure to chip failure rates. *International Electron Devices Meeting 2000. Technical Digest. IEDM (Cat. No.00CH37138)*, San Francisco, CA (10–13 December 2000), pp. 93–96. IEEE. https://doi.org/10.1109/IEDM.2000.904266.

[44] Kufluoglu, H. and Alam, M.A. (2004). A geometrical unification of the theories of NBTI and HCI time-exponents and its implications for ultra-scaled planar and surround-gate MOSFETs. *IEDM Technical Digest. IEEE International Electron Devices Meeting, 2004.*, San Francisco, CA (13–15 December 2004), pp. 113–116. IEEE. https://doi.org/10.1109/IEDM.2004.1419081.

[45] Kerber, A. and Cartier, E. (2013). Application of VRS methodology for the statistical assessment of BTI in MG/HK CMOS devices. *IEEE Electron Device Letters* 34 (8): 960–962. https://doi.org/10.1109/LED.2013.2268050.

[46] Barrett, C.R. and Smith, R.C. (1976). Failure modes and reliability of dynamic RAMS. *1976 International Electron Devices Meeting*, Washington, DC (6–8 December 1976), pp. 319–322. IEEE. https://doi.org/10.1109/IEDM.1976.189047.

[47] Sing, Y.W. and Sudlow, B. (1980) Modeling and VLSI design constraints of substrate current. *1980 International Electron Devices Meeting*, Washington, DC (8–10 December 1980), pp. 732–735. IEEE. https://doi.org/10.1109/IEDM.1980.189941.

[48] Childs, P.A., Eccleston, W., and Stuart, R.A. Alternative mechanism for substrate minority carrier injection in MOS devices operating in low level avalanche. *Electronics Letters* 17 (8): 281–282.

[49] Chenming, H., Simon, C.T., Fu-Chieh, H. et al. (1985). Hot-electron-induced MOSFET degradation – model, monitor, and improvement. *IEEE Journal of Solid-State Circuits* 20 (1): 295–305. https://doi.org/10.1109/JSSC.1985.1052306.

[50] Schuegraf, K.F. and Chenming, H. (1994). Hole injection SiO/sub 2/ breakdown model for very low voltage lifetime extrapolation. *IEEE Transactions on Electron Devices* 41 (5): 761–767. https://doi.org/10.1109/16.285029.

[51] McPherson, J., Reddy, V., Banerjee, K., and Huy, L. (1998). Comparison of E and 1/E TDDB models for SiO/sub 2/ under long-term/low-field test conditions. *International Electron Devices Meeting 1998. Technical Digest (Cat. No.98CH36217)*, San Francisco, CA (6–9 December 1998), pp. 171–174. IEEE. https://doi.org/10.1109/IEDM.1998.746310.

[52] Alam, M.A., Bude, J., and Ghetti, A. (2000). Field acceleration for oxide breakdown-can an accurate anode hole injection model resolve the E vs. 1/E controversy? *2000 IEEE International Reliability Physics Symposium Proceedings. 38th Annual (Cat. No.00CH37059)*, San Jose, CA (10–13 April 2000), pp. 21–26. IEEE. https://doi.org/10.1109/RELPHY.2000.843886.

[53] McPherson, J.W., Khamankar, R.B., and Shanware, A. (2000). Complementary model for intrinsic time-dependent dielectric breakdown in SiO2 dielectrics. *Journal of Applied Physics* 88 (9): 5351–5359. https://doi.org/10.1063/1.1318369.

[54] McPherson, J.W. (2012). Time dependent dielectric breakdown physics – models revisited. *Microelectronics Reliability* 52 (9): 1753–1760. https://doi.org/10.1016/j.microrel.2012.06.007.

[55] Degraeve, R., Groeseneken, G., Bellens, R. et al. (1995). A consistent model for the thickness dependence of intrinsic breakdown in ultra-thin oxides. *Proceedings of International Electron Devices Meeting* 10–13 (1995): 863–866. https://doi.org/10.1109/IEDM.1995.499353.

[56] Stathis, J.H. and DiMaria, D.J. (1998). Reliability projection for ultra-thin oxides at low voltage. *International Electron Devices Meeting 1998. Technical Digest (Cat. No.98CH36217)*, San Francisco, CA (6–9 December 1998), pp. 167–170. IEEE. https://doi.org/10.1109/IEDM.1998.746309.

[57] Weir, B.E., Silverman, P.J., Monroe, D. et al. (1997). Ultra-thin gate dielectrics: they break down, but do they fail? *International Electron Devices Meeting IEDM Technical Digest*, Washington, DC (10–10 December 1997), pp. 73–76. IEEE. https://doi.org/10.1109/IEDM.1997.649463.

[58] Depas, M., Nigam, T., and Heyns, M.M. (1996). Soft breakdown of ultra-thin gate oxide layers. *IEEE Transactions on Electron Devices* 43 (9): 1499–1504. https://doi.org/10.1109/16.535341.

[59] Okada, K. and Kawasaki, S. (1995). New dielectric breakdown model of local wearout in ultra thin silicon dioxides. *Extended abstracts of the . . . Conference on Solid State Devices and Materials* 1995: 473–475.

[60] Suñé, J. and Wu, E.Y. (2004). Hydrogen-release mechanisms in the breakdown of thin SiO2 films. *Physical Review Letters* 92 (8): 087601. https://doi.org/10.1103/PhysRevLett.92.087601.

[61] Alam, M.A., Smith, R.K., Weir, B.E., and Silverman, P.J. (2002). Uncorrelated breakdown of integrated circuits. *Nature* 420 (6914): 378–378. https://doi.org/10.1038/420378a.

[62] Alam, M.A., Weir, B.E., and Silverman, P.J. (2002). A study of soft and hard breakdown – part I: analysis of statistical percolation conductance. *IEEE Transactions on Electron Devices* 49 (2): 232–238. https://doi.org/10.1109/16.981212.

[63] Alam, M.A., Weir, B.E., and Silverman, P.J. (2002). A study of soft and hard breakdown – part II: principles of area, thickness, and voltage scaling. *IEEE Transactions on Electron Devices* 49 (2): 239–246. https://doi.org/10.1109/16.981213.

[64] Nigam, T., Kerber, A., and Peumans, P. (2009). Accurate model for time-dependent dielectric breakdown of high-k metal gate stacks. *2009 IEEE International Reliability Physics Symposium*, Montreal, QC (26–30 April 2009), pp. 523–530. IEEE. https://doi.org/10.1109/IRPS.2009.5173307.

[65] Ranjan, R., Liu, Y., Nigam, T., et al. (2017). Impact of AC voltage stress on core NMOSFETs TDDB in FinFET and planar technologies. *2017 IEEE International Reliability Physics Symposium (IRPS)*, Monterey, CA (2–6 April 2017), pp. DG-10.1–DG-10.5. IEEE. https://doi.org/10.1109/IRPS.2017.7936367.

[66] Alam, M.A. (2002). SILC as a measure of trap generation and predictor of T/sub BD/ in ultrathin oxides. *IEEE Transactions on Electron Devices* 49 (2): 226–231. https://doi.org/10.1109/16.981211.

[67] Kumar, S., Anandkrishnan, R., Parihar, N., and Mahapatra, S. (2020). A stochastic framework for the time kinetics of interface and bulk oxide traps for BTI, SILC, and TDDB in MOSFETs. *IEEE Transactions on Electron Devices* 67 (11): 4741–4748. https://doi.org/10.1109/TED.2020.3020533.

[68] Zagni, N. and Alam, M.A. (2021). Reliability physics of ferroelectric/negative capacitance transistors for memory/logic applications: an integrative perspective. *Journal of Materials Research* 36 (24): 4908–4918. https://doi.org/10.1557/s43578-021-00420-1.

[69] Ma, T.P., Scoggan, G., and Leone, R. (1975). Comparison of interface-state generation by 25-keV electron beam irradiation in p-type and n-type MOS capacitors. *Applied Physics Letters* 27 (2): 61–63. https://doi.org/10.1063/1.88366.

[70] Petersen, E. and Marshall, P. (1988). Single event phenomena in the space and SDI arenas. *Journal of Radiation Effects* 6 (2): 1.

[71] Baumann, R., Hossain, T., Murata, S., and Kitagawa, H. (1995). Boron compounds as a dominant source of alpha particles in semiconductor devices. *Proceedings of 1995 IEEE International Reliability Physics Symposium*, Las Vegas, NV (4–6 April 1995), pp. 297–302. IEEE. https://doi.org/10.1109/RELPHY.1995.513695.

[72] Jayakumar, R. and Prabaharan, M. (ed.) (2021). *Advances in Polymer Science*. Springer. https://www.springer.com/series/12.

[73] Reich, B. and Hakim, E.B. (1972). Environmental factors governing field reliability of plastic transistors and integrated circuits. *10th Reliability Physics Symposium*, Las Vegas, NV (5–7 April 1972), pp. 82–87. IEEE. https://doi.org/10.1109/IRPS.1972.362532.

[74] Brauer, J.B., Kapfer, V.C., and Tamburrino, A.L. (1970). Can plastic encapsuiated microcircuits provide reliability with economy? *8th Reliability Physics Symposium*, Las Vegas, NV (7–10 April 1970), pp. 61–72. IEEE. https://doi.org/10.1109/IRPS.1970.362437.

[75] Peck, D.S. (1970). The design and evaluation of reliable plastic-encapsulated semiconductor devices. *8th Reliability Physics Symposium*, Las Vegas, NV (7–10 April 1970), pp. 81–93. IEEE. https://doi.org/10.1109/IRPS.1970.362439.

[76] Peck, D.S. (1986). Comprehensive model for humidity testing correlation. *24th International Reliability Physics Symposium*, Anaheim, CA (1–3 April 1986), pp. 44–50. IEEE. https://doi.org/10.1109/IRPS.1986.362110.

[77] Hallberg, Ö. and Peck, D.S. (1991). Recent humidity accelerations, a base for testing standards. *Quality and Reliability Engineering International* 7 (3): 169–180. https://doi.org/10.1002/qre.4680070308.

[78] Mavinkurve, A., Rongen, R., Soestbergen, M. Van. et al. (2015). Biased humidity degradation model for Cu-Al wire bonded contacts based on molding compound properties. *2015 European Microelectronics Packaging Conference (EMPC)*, Friedrichshafen (14–16 September 2015), pp. 1–10. IEEE.

[79] Rangaraj, S., Kwon D., Pei, M. et al. (2013). Accelerated stress testing methodology to risk assess silicon-package thermomechanical failure modes resulting from moisture exposure under use condition. *2013 IEEE International Reliability Physics Symposium (IRPS)*, Monterey, CA (14–18 April 2013), pp. 5C.3.1–5C.3.5. IEEE. https://doi.org/10.1109/IRPS.2013.6532032.

[80] Mamun, M.A.Z. and Alam, M.A. (2022). Reduced relative humidity (RH) enhances the corrosion-limited lifetime of self-heated ic: peck's equation generalized. *2022 IEEE International Reliability Physics Symposium (IRPS)*, Dallas, TX (27–31 March 2022), pp. 8A.5-1–8A.5-8. IEEE. https://doi.org/10.1109/IRPS48227.2022.9764577.

[81] Saha, A., Yermembetova, A., Mi, Y. et al. (2022). Temperature self-calibration of always-on, field-deployed ion-selective electrodes based on differential voltage measurement. *ACS Sensors* 7 (9): 2661–2670.

[82] Lelis, A. J., Green, R., and Habersat, D.B. (2018). SiC MOSFET threshold-stability issues. *Materials Science in Semiconductor Processing* 78: 32–37. https://doi.org/10.1016/j.mssp.2017.11.028.

[83] Meneghini, M., Santi, C.D., Abid, I. et al. (2021). GaN-based power devices: Physics, reliability, and perspectives. *Journal of Applied Physics* 130: 181101. https://doi.org/10.1063/5.0061354.

[84] Islam, A.E., Zhang, C., DeLello, K. et al. (2022). Defect Engineering at the Al_2O_3/(010) β-Ga_2O_3 Interface via Surface Treatments and Forming Gas Post-Deposition Anneals. *IEEE Transactions on Electron Device* 69 (10): 5656–5663. https://doi.org/10.1109/TED.2022.3200643.

Chapter 28

Technology Computer-Aided Design

A Key Component of Microelectronics' Development

Siegfried Selberherr[1] and Viktor Sverdlov[2]

[1] Institute for Microelectronics, TU Wien, Vienna, Austria
[2] Christian Doppler Laboratory for Nonvolatile Magnetoresistive Memory and Logic at the Institute for Microelectronics, TU Wien, Vienna, Austria

28.1 Introduction

The discovery of the transistor effect [1] and the invention of the transistor created an enormous impact on our modern society as it has undoubtfully revolutionized the way we deal with an ever-increasing information flow. Today, there are trillions of transistors on Earth. Computers built with transistors dramatically modified our approaches, how information is generated, processed, and stored. The breathtaking increase in computers' complexity, computational power, and speed has been supported by an amazing scalability of metal-oxide-semiconductor (MOS) field-effect transistors (FETs). The scaling is expected to continue for at least 30 years as is detailed by various roadmaps, e.g. [2]. With the 3 nm technology node expected by the end of the year 2022, new technology nodes are predicted to appear on the market every second year. The 3 nm technology node will probably be the last node based on fin field-effect transistors (FinFETs) which were introduced at the 22 nm node. The next 2 nm node is expected to be based on gate-all-around nanosheet transistors paving the way toward the Å (angstrom)-nodes. It is envisaged that the gate-all-around nanosheet transistors will evolve to forksheet architecture followed by complimentary FETs where p-type and n-type nanosheet transistors are stacked on top of each other. To proceed to angstrom-size nodes, or *A*-nodes, one must go beyond the current version of extreme ultraviolet lithography. Next generations will require new system platforms with improved optics to achieve 20 pm accuracy for about 1-m large lenses. To reduce the channel length and thickness, the use of novel atomically flat materials with advanced properties is foreseen.

75th Anniversary of the Transistor, First Edition. Edited by Arokia Nathan, Samar K. Saha, and Ravi M. Todi.

To assist the development and engineering of upcoming generations of emerging transistors at all stages from a concept to realization and reliability analysis, advanced Technology Computer-Aided Design (TCAD) tools are employed. TCAD allowed to reduce research and development-related costs by up to 40% [3]. To be predictive, TCAD tools must be physics-based. Concurrently, timely but perhaps less accurate results are of great value as multiple runs with a large set of parameters and parameters' values are often required. Therefore, TCAD tools usually contain several blocks based on models of diverse complexities.

Continuous scaling ensures extending the functionality of the MOSFET-based platform. However, as stated by the International Roadmap for Devices and Systems (IRDS) [4], novel computational paradigms are required to further boost the efficiency and the performance. It becomes increasingly difficult for the saturating MOSFETs' technologies to satisfy the rising demands. Beyond Complementary MOS (CMOS) technologies are therefore necessary for the upcoming era of novel computing paradigms.

In complement to the electron charge employed in modern CMOS-based integrated circuit, the electron spin and magnetization are viewed as promising state variable for emerging devices [4]. Although the efficiency of using the electron spin in logic devices is yet to be proven, the utilization of magnetization orientation is successfully employed in emerging magnetoresistive memories to store information [4]. The magnetization is switched between the two orientations by a spin-polarized electric current. The magnetization (and the information) is preserved, when the power is turned off. Magnetoresistive memory is therefore nonvolatile. As the data is written electrically and not by a magnetic field, the memory is also scalable. The memory is fast and is promising for embedded applications.

Below, partly based on our experience, we briefly review an earlier success of TCAD modeling. We begin with a short historic TCAD overview and outline modern developments. The drift-diffusion approach for charge transport complemented with the spin transport is well suited to model transport and torques in emerging magnetoresistive memories [5]. We conclude the overview with recent developments in TCAD with a vision that machine learning approaches combined with ab-initio calculations may become the main stream for parameter optimization and improving reliability.

28.2 Short History

TCAD tools are used to assist in development and engineering at practically all stages ranging from process simulation to device optimization. The first fully numerical solution of the transport equations was proposed in 1964 by Gummel [6] for a one-dimensional bipolar transistor. Coupled with the Poisson equation, the approach was further developed to perform transit-time simulations and applied to pn-junctions [7] and to avalanche diodes by [8]. First simulations combining current transport with a solution of the two-dimensional Poisson equation with an application to MOS transistors were performed by Loeb et al. [9] and Schroeder and Muller [10]. Simultaneous solutions of the coupled continuity and Poisson equations applied to FET [11] and to bipolar transistors [12] were reported in the late 1960s.

The superior scalability and lower power consumption, compared to bipolar transistors, made CMOS the dominant driver for integrated electronics in the mid-1980s. The monograph [13] was probably the first book comprehensively covering all aspects of discretization and numerical solutions of partial differential equations describing the physics and needed for modeling of the charge transport in semiconductor devices. The book was a great booster for the development of simulation tools which required new physics phenomena to be introduced in order to support the scaling of MOSFET devices. With personal computers broadly available, simulation tools became highly popular. It was realized that computer-based modeling and simulations help developing novel transistors with improved characteristics saving research and development-related costs and shortening the time from lab to market significantly.

The fundamental scaling laws of MOS devices were understood and implemented in TCAD tools. These achievements set the stage for TCAD becoming a key component to drive microelectronics' development through the very large and ultra-large-scale integration eras. With a transition to CMOS, the need for tightly coupled process and device simulations was realized [14]. The TCAD tools reached amazing maturity as they became capable of addressing in full the whole complexity of simulations and implementations of CMOS design rules [15].

Tighter requirements on control over the manufacturing process, electrical reliability, as well as business-related aspects (i.e. minimizing the time to market) guaranteed TCAD's growing impact on microelectronics' development. A modern TCAD suite is coupled with a circuit simulator and typically includes semiconductor process and device simulation blocks. To optimize the process and reduce the costs all leading companies including AT&T [16], IBM [17], Intel [18], and NEC [19] developed their own TCAD tools. In order to improve the quality of device modeling, the leading universities including Berkley [20] and Stanford [21] developed and introduced advanced methods in device and process simulations. Partly based on personal experience with MINIMOS [22], we highlight the development of the models involved in the simulations. The sophistication and the complexity of the models involved in the device simulations was dictated by the needs to resolve the technical problems and issues related to the development of advanced technology nodes.

28.3 Scaling and Model Complexity

From an engineering point of view, classical models like the drift-diffusion model have enjoyed an amazing success due to their relative simplicity, numerical robustness, and the ability to perform two- and three-dimensional simulations on large unstructured meshes [13]. Although the drift-diffusion model should not be applied to describe the charge transport in ultra-scaled MOSFETs, it turns out that results of more accurate quantum mechanical calculations can be reproduced with the drift-diffusion transport model, if the parameters of the model, in particular the carrier mobilities are properly adjusted and calibrated.

MINIMOS is a software tool for the numerical simulation of FETs such as silicon bulk and silicon on insulator (SOI) MOSFETs, and gallium arsenide MESFETs [22]. The fundamental semiconductor equations, consisting of the Poisson's equation and the transport equation, are solved numerically. The Poisson equation

$$\nabla\left(\varepsilon\nabla\varphi(\mathbf{r},t)\right)=-\rho(\mathbf{r},t) \tag{28.1}$$

establishes the relation between the position $\mathbf{r}$ and time t, total charge density $\rho(\mathbf{r}, t)$, and the electrostatic potential $\varphi(\mathbf{r}, t)$ in the device. Here, ε is the dielectric permittivity. The total charge density $\rho(\mathbf{r}, t)$ is determined by the corresponding contributions from the electron $n(\mathbf{r}, t)$ and hole $p(\mathbf{r}, t)$ concentrations and any material-dependent charges $C(\mathbf{r}, t)$.

$$\rho(\mathbf{r},t)=q\left(p(\mathbf{r},t)-n(\mathbf{r},t)+C(\mathbf{r},t)\right), \tag{28.2}$$

q is the absolute value of the electron charge and $C(\mathbf{r}, t)$ is the concentration of material dopants and various material defects. The carrier concentrations satisfy

$$q\frac{\partial n(\mathbf{r},t)}{\partial t}=\nabla \boldsymbol{j}_n(\mathbf{r},t)-qR(\mathbf{r},t), \tag{28.3a}$$

$$-q\frac{\partial p(\mathbf{r},t)}{\partial t}=\nabla \boldsymbol{j}_p(\mathbf{r},t)+qR(\mathbf{r},t), \tag{28.3b}$$

where $R(\mathbf{r}, t)$ is the electron–hole generation rate. The carrier (electron or hole) current density is in a simple manner modeled proportional to the electric field $\mathbf{E} = -\nabla\varphi(\mathbf{r}, t)$ (the drift term) and to the carrier density gradient (the diffusion term):

$$j_{n[p]}(\mathbf{r},t) = \mu_{n[p]}\mathbf{E} + [-]D_{n[p]}\nabla n[p](\mathbf{r},t), \tag{28.4}$$

Here, the symbols in the square brackets describe the current density due to holes. The coefficients of the proportionalities to the electric field or the density gradients are the electron [hole] mobilities $\mu_{n[p]}$ and the diffusion coefficients $D_{n[p]}$, correspondingly. The diffusion coefficient is related to the mobility via the Einstein relation, so only the mobilities must be determined. The carrier mobilities are determined by scattering processes and are material and temperature dependent. Physics-based mobility models facilitated the success of (1–4) to simulate even submicron MOSFETs. $C(\mathbf{r}, t)$ is obtained by process simulation and usually assumed to be constant, i.e. chargeable defects are neglected.

Initially, finite differences were employed for space discretization and the Backward Euler method for time discretization to solve the fundamental equations in a two-dimensional geometry only accounting for electrons [23, 24]. Although containing only a minimal set of models, MINIMOS boosted the development of particularly n-channel transistors considerably thanks to the source code and the program released to the public [25, 26]. Soon, MINIMOS was extended to its second version [27]. As the scaling continued, it was clear that the carriers' heating in the on-state must be included in the simulations. This was accomplished by extending the drift-diffusion model to include the energy balance equations [28]. To adequately account in the equations for the temperature dependence of the carrier mobility and for impact ionization required a complete redesign of the numerical methods to address the short-channel effects and hot-carrier phenomena at room and low ambient temperatures [29]. With the transistors' dimensions shrinking, three-dimensional capabilities to investigate the parasitic effects at the channel edges were introduced [30].

One of the main advantages of the drift-diffusion model of the current density is its flexibility to incorporate different physical phenomena affecting the transport via comprehensive mobility modeling [31]. With the downscaling continuing into a deep submicron region, a continuous improvement of the initial mobility model [23], which included lattice, ionized impurity, surface scattering, and also velocity saturation was paramount.

As the quality of the mobility models defined the predictivity of the TCAD tools, more accurate approaches addressing scattering mechanisms at a microscopic level became necessary. To model the carrier transport in the high and low electric field portion of the semiconductor channel, a Monte Carlo method of solving the kinetic Boltzmann equation was coupled to the drift-diffusion model [32]. It allowed to use the space-dependent parameters to address the nonlocal effects in high-field regions including velocity overshoot and ballistic transport [33].

MINIMOS was completely redesigned by making it more user friendly and adapt it for application demands introduced by the continuing MOSFET scaling. The simulator called from MINIMOS-NT[1] was completely rewritten with the C programming language [34]. The simulator was also extended to include new materials and physical effects. In [35], a ferroelectric memory cell was shown to increase the capabilities of integrated circuits. It was proposed that a similar approach could also work for ferromagnetic materials currently used in magnetoresistive memories. MINIMOS-NT was extended to model partially depleted silicon-on-insulator transistors considered as possible successors of bulk MOSFETs at this time [36]. A fully coupled electrothermal mixed-mode simulation of a circuit based on SiGe heterojunction bipolar transistors, which included self-heating and inter-device coupling effects was presented in [37]. For accurately simulating GaAs power heterojunction bipolar transistors, the drift-diffusion model was shown to be insufficient, and its generalization to a proper hydrodynamic formulation was a must. Different models for the thermal conductivity and the specific heat for relevant diamond and zinc-blende

[1] NT stands for New Technology alike Microsoft's Windows NT.

structure semiconductors were developed as functions of the temperature and the material compositions of semiconductor alloys [38]. This allowed to demonstrate that MINIMOS-NT reproduces accurately experimental results and RF data obtained for III–V compound semiconductors in heterojunction bipolar and high electron mobility transistors. Many of the existing physical models regarding the band gap, mobility, thermal conductivity, energy relaxation times, and specific heat were critically investigated, refined, replaced, and added to make MINIMOS-NT a generic device simulator for a variety of micro-materials, including group IV semiconductors, III–V compound semiconductors, and their alloys [39].

With reducing the channel length, the carriers accelerated by the source–drain voltage did not manage to relax their energy. Their energy distribution does not yield the Maxwell distribution. It became apparent that it was insufficient to describe the distribution function by their average carrier energy. Nevertheless, the drift-diffusion model and the energy-transport model rely on these assumptions. Therefore, it is necessary to include higher moments of the distribution function for more accurate description of the transport. A transport model based on six moments was derived and implemented [40]. The six-moments model was then generalized to include the band structure effects proven to be important for an accurate description [41]. The framework allows to model the scattering integral self-consistently without resorting to the relaxation time approximation [42] and to accurately describe effects like impact ionization, which is difficult and only approximate in lower order transport models.

While traditionally scaling the devices to put more transistors on a die, the semiconductor industry faced the problem of extensive heat generation per transistor already at the 90 nm technology node. To proceed to the 65 nm node, a modification of wafer processing was urgently required. An important engineering solution which allowed to continue scaling for the next node and to keep the heat generation under control was to introduce strain into the channel [43]. Strain improved the transport properties of carriers. The on-current through strained channel transistors is increased compared to that through relaxed ones, for the same V_{DD} voltage. At the same time, the off-current stays practically unchanged. Therefore, more speed is achieved at the same V_{DD}. Alternatively, similar speed is obtained at a lower V_{DD} in strained devices.

As the signs of strain needed to boost the mobilities of electrons and holes are opposite, strain must be controllably and independently delivered to each n- and p-MOSFET. This task is accomplished by mechanically stressing the transistor channels. The mechanical stress is process-induced and is created by capping layers alternatively, tensely, and compressively, stressing n- and p-channels, respectively. Additional stress to p-channels is achieved by replacing the Si source and drain contacts with epitaxially regrown SiGe contacts. As the lattice constant of Ge and therefore of SiGe is larger than that of Si, the SiGe source and drain squeeze the p-channel even more. Although a new technology of high permittivity dielectric/metal gate to reducing the gate leakage current was required to proceed to 45 nm technology node [44], stress remained the important booster of performance enhancement with scaling. At the 32 nm technology node, advanced channel stressors were introduced [45] allowing to get stress of 1.2–1.5 GPa.

The boost of the on-current by stress is due to the strain-induced mobility enhancement in the channel. Depending on the stress conditions, up to fourfold mobility enhancement for holes and nearly twofold for electrons were reported [46].

The mobility enhancement is mostly due to the strain-induced band structure modifications. Strain lifts the degeneracy between the light and heavy hole bands. Apart from that, a substantial band warping favorable for increasing the carriers' velocities at the same driving electric field appears in strained silicon. Practically relevant compressive strain along the [110] p-channel direction reduces the contribution from the heavy effective mass wings and substantially boosts the contribution from the light effective mass wings along the [110] direction [47]. As the light effective mass carriers with a higher mobility increase, the total channel mobility is enhanced.

Similarly, tensile strain along the [110] n-channel direction splits the degeneracy of the six conduction band valleys. The two valleys with the low effective masses in the channel plane move down in energy and contribute more to the transport. Induced by tensile stress in the [110] direction, shear strain additionally reduces the effective mass [48] along the [110] transport direction ensuring an even larger mobility enhancement [49].

Creating local strain on each transistor required additional steps in the fabrication process. However, extra process costs were negated by even larger gains due to performance enhancement. Strain remains the main mobility booster for all technology nodes beyond the 32 nm node including the most recent nodes.

28.4 MINIMOS Commercialization and Beyond

The introduction of a tri-gate transistor architecture at the 22 nm technology node by Intel [50] dramatically boosted the performance gain at low operating voltages and reduced the power, which allowed to outperform bulk, partly depleted and fully depleted silicon-on-insulator (FDSOI) devices. Most importantly, it added only 3% of the process costs, compared to at least 10% needed for FDSOI process. The tri-gate transistor technology survived for many device generations including the 3 nm technology node. At the same time, it became clear that the task of an ultimate refinement of the three-dimensional capability of efficient device modeling and simulation goes beyond the University's research as it requires a continuous interaction with potential customers and demands innovative computational solutions to deliver timely results. To address these challenges, a spin-off company was established by former PhD students of the Institute for Microelectronics, TU Wien. Because of the succession of its current commercial implementation [51], GTS MINIMOS-NT supports all previous developments and got enhanced by incorporating novel variability and reliability features including recent bias temperature instabilities models [52].

Integrated circuits development was dominated by CMOS technology for more than a quarter of the century. As the device sizes shrink and proceed to the nanoscale regime, their dimensions become comparable to an important physics parameter, the electron wavelength. At this point, the validity of the drift-diffusion model and also the higher moments' transport models based on the semiclassical Boltzmann transport equation is questionable as quantum effects due to the wave nature of the electrons propagating though the channel become important. However, a purely coherent quantum transport description based on the wave function being the solution of the Schrödinger equation with the corresponding boundary conditions is also not applicable as scattering with phonons at room temperature, impurities, and surface roughness remains strong and cannot be neglected. Therefore, a fully quantum mechanical transport formalism which consistently describes dissipative scattering processes must be employed.

Popular techniques to describe the dissipative quantum transport in nanoscale devices are the nonequilibrium Green's functions (NEGF) approach [53] and the approach based on the Wigner function [54]. The latter approach [55] represents the phase space formulation of quantum mechanics, as the Wigner function depends on both space and momentum. In the limit of the electrostatic potential slowly varying in space and time the Wigner equation reduces to the Boltzmann equation (without scattering) providing a seamless link to the semiclassical transport. By analogy to the Boltzmann transport equation, the Wigner equation can also be solved with stochastic methods [56]. However, in contrast to traditional particle-based Monte-Carlo methods used to compute the solution of the Boltzmann equation, particles with two different flavors, or weights of different signs [54], are required to capture the quantum mechanical nature of the solution of the Wigner equation.

The NEGF method allows lumping the electrodes into a surface Green's function, thus reducing the calculations to the device region only [57]. The knowledge of so-called self-energies is required in order to describe scattering. The self-energy depends on the Green's function and must be found self-consistently. The NEGF equations are mostly solved numerically by discretizing the simulation domain. In a stationary case, the NEGFs also depend on the energy. NEGFs are obtained by inverting a (huge) matrix in space and energies. The matrix becomes sparse, if the self-energy is local in space and energy, which simplifies its inversion [53].

The NEGF method allows to describe quantum transport through devices and molecules at an atomistic level, provided the electronic parameters are known. These electronic parameters of a material (i.e. sites energies, hopping integrals) can be obtained from first principles, or ab initio, calculations, e.g. [58–60]. The ab-initio method is designed to solve the Schrödinger equation [61]. On input, the method uses the positions of the nuclei and the number of electrons. As an output, the method provides electron densities, energies, and other useful properties. By combining ab-initio calculations with the NEGF method, the transport properties through a device are evaluated from first principles without adjustable parameters (provided the scattering self-energies are known). Therefore, major TCAD suppliers (e.g. Silvaco [62] and Synopsys [63]) include first-principle-based atomistic simulators in their TCAD design tools suites. Using ab-initio tools is paramount to investigate emerging devices beyond CMOS devices [4], which are expected to support novel computational paradigms.

28.5 Design Technology Co-Optimization at Advanced Nodes

The importance and the need of atomistic simulations for transistor design was recognized already 25 years ago [64]. Indeed, randomly distributed dopants in the channel affect the intrinsic MOSFETs characteristics and cause increasing parameter fluctuations with downscaling into the nano-meter regime [65]. Thus, predictive TCAD process and device simulations are mandatory to evaluate process-induced variability, statistical fluctuations, time-dependent variability resulting in device degradation, and their mutual correlations. These analyses must be performed at the earliest possible stage of the development of any new technology. Therefore, a development of TCAD-based Device-Technology Co-Optimization (DTCO) tools integrating a virtual fabrication platform with a TCAD-based device flow chain becomes necessary to reduce variability by calibrating results of the TCAD process and device simulation, extracting compact models and parameters, and performing statistical circuit simulations for advance nodes, e.g. [66–68].

28.6 Electron Spin for Microelectronics

In addition to the electron charge, the electron spin is considered as promising state variable for digital devices beyond CMOS transistors [4]. A spin field-effect transistor (SpinFET) [69] uses the spin polarization to enrich the transistor performance. A SpinFET is built of a semiconducting channel region placed between ferromagnetic source and drain contacts. The source contact serves as an injector of spin-polarized electrons which enter the channel. The drain magnetization orientation serves as a detector of the spin polarization of impinging electrons. The electrons with their spins parallel to the drain magnetization escape freely into the drain and the drain current is large. If the impinging spins are opposite to the drain magnetization, they cannot enter the drain, and the current is small. Importantly, the spin polarization of electrons propagating under the gate is controllably rotated by a spin–orbit field. The strength of this field is gate voltage dependent. This all-electric way to manipulate spins combined with the two relative source–drain magnetization orientations, parallel and anti-parallel, makes SpinFETs suitable for reconfigurable logic [70]. Although the spin injection, spin propagation, and spin manipulation in a semiconductor channel have been experimentally demonstrated [71], the overall characteristics were not impressive. In particular, the ratio of the current difference in parallel and anti-parallel configurations to the current in on-state in Si devices is still small [72], and the strength of the spin–orbit interaction is weak. A promising avenue is to explore unique properties of novel two-dimensional materials [73].

An alternative path is to explore devices with already large magnetoresistance, namely magnetic tunnel junctions (MTJs). An MTJ is composed of two ferromagnetic layers, the reference and the free layer, separated by a tunnel junction. Although an MTJ is quite different from a MOSFET, the charge drift-diffusion model (4) generalized to include the spin current is capable to properly describe the transport in ferromagnetic layers and spin valves [74]. However, to model the transport in MTJs, several additional features must be included. First, the dependence of the resistivity across the tunnel barrier on the relative magnetization, or the tunnel magnetoresistance, must be modeled by the magnetization-dependent resistivity [75]. To reproduce the spin transport through the tunnel barrier, and in particular its dependence on the relative magnetization, a special boundary condition for the spin current is required [76]. With these add-ons, the spin and charge drift-diffusion model is suitable to accurately evaluate the interface- and bulk-like spin-transfer torques in both MTJs and spin valves and their combinations. It is critically important to describe ultra-scaled memory cells with several tunnel barriers [77] or normal metal spacers [78] between elongated ferromagnetic parts with nonuniform magnetization.

By introducing the Hall component for the spin currents, the generalized spin and charge drift-diffusion approach also allows to describe spin-orbit torques (SOT) [79] in emerging SOT-MRAM cells.

28.7 Summary and Outlook

Based on a particular example of MINIMOS, TCAD was demonstrated to be a key component of microelectronics' development. The evolution of TCAD transport models from classical drift-diffusion-like to more rigorous models at the nanoscale level was synchronized with MOSFET scaling. It was shown that, regardless of its simplicity, the drift-diffusion-based transport models are appropriate to model emerging magnetoresistive memories. To proceed beyond CMOS into the nanoscale regime, the employment of ab-initio atomistic tools is attractive; however, the simulations are demanding heavy computational resources. Employing machine-learning methods to extend ab-initio calculations outside of the parameter range used for the network training may accelerate the search for new materials and devices with unique characteristics and, therefore, represents the most probable path of TCAD development in the future.

References

[1] Shokley, W.B., Bardeen, J., Brattain, W.H. (1956). Noble prize in physics. https://www.nobelprize.org/prizes/physics/1956/summary (accessed 8 May 2023).

[2] Van den Hove, L. (2022). 20 year roadmap: tearing down the walls. SEMICON West. https://www.imec-int.com/en/articles/20-year-roadmap-tearing-down-walls. Published on: 2 August 2022 (accessed 8 May 2023).

[3] International technology roadmap for semiconductors 2.0. (2015). Executive report. https://www.semiconductors.org/wp-content/uploads/2018/06/0_2015-ITRS-2.0-Executive-Report-1.pdf (accessed 8 May 2023).

[4] IEEE IRDS (2021). Beyond CMOS. https://irds.ieee.org/editions/2021/beyond-cmos (accessed 8 May 2023).

[5] Apalkov, D., Dieny, B., and Slaughter, J.M. (2016). Magnetoresistive random access memory. *Proc. IEEE* 104: 1796–1830. http://doi.org/10.1109/JPROC.2016.2590142.

[6] Gummel, H. (1964). A self-consistent iterative scheme for one-dimensional steady state transistor calculations. *IEEE Trans. Electron Devices* 11: 455–465. https://doi.org/10.1109/T-ED.1964.15364.

[7] DeMari, A. (1968). An accurate numerical steady-state one-dimensional solution of the P–N junction. *Solid State Electron.* 11: 33–58. https://doi.org/10.1016/0038-1101(68)90137-8.

[8] Scharfetter, D. and Gummel, H. (1969). Large-signal analysis of a silicon read diode oscillator. *IEEE Trans. Electron Devices* 16: 64–77. https://doi.org/10.1109/T-ED.1969.16566.

[9] Loeb, H., Andrew, R., and Love, W. (1969). Application of 2-dimensional solutions of the Shockley-Poisson equation to inversion-layer M.O.S.T. devices. *Electron. Lett.* 4: 352–354. https://doi.org/10.1049/el:19680277.

[10] Schroeder, J. and Muller, R. (1968). IGFET analysis through numerical solution of Poisson's equation. *IEEE Trans. Electron Devices* 15: 954–961. https://doi.org/10.1109/T-ED.1968.16545.

[11] Kennedy, D. (1969). On the ambipolar diffusion of impurities into silicon. *Proc. IEEE* 54: 1202–1203. https://doi.org/10.1109/PROC.1969.7194.

[12] Slotboom, J. (1973). Computer-aided two-dimensional analysis of bipolar transistors. *Electron. Lett.* 5: 677–678. https://doi.org/10.1109/T-ED.1973.17727.

[13] Selberherr, S. (1984). *Analysis and Simulation of Semiconductor Devices*. Springer. http://doi.org/10.1007/978-3-7091-8752-4.

[14] Dutton, R.W. and Hansen, S.E. (1981). Process modeling of integrated circuit device technology. *Proc. IEEE* 69: 1305–1320. https://doi.org/10.1109/PROC.1981.12168.

[15] Cham, K.M., Oh, S.-Y., Chin, D., and Moll, J.L. (1986). *Computer-Aided Design and VLSI Device Development*. Kluwer Academic Publishers. http://doi.org/10.1007/978-1-4613-2553-6.

[16] Lloyd, P., McAndrew, C.C., McLennan, M.J. et al. (1993). Technology CAD at AT&T. In: *Technology CAD Systems* (ed. F. Fasching, S. Halama, and S. Selberherr), 1–24. Vienna: Springer. http://doi.org/10.1007/978-3-7091-9315-0_1.

[17] Knepper, R.W., Johnson, J.B., Furkay, S. et al. (1993). *Technology CAD at IBM*, 25–62. Vienna: Springer. http://doi.org/10.1007/978-3-7091-9315-0_2.

[18] Mar, J. (1993). *Technology CAD at Intel*, 83–74. Vienna: Springer. http://doi.org/10.1007/978-3-7091-9315-0_3.

[19] Tanabe, N. (1993). *Technology CAD at NEC*, 237–253. Vienna: Springer. http://doi.org/10.1007/978-3-[7091]-9315-0_11.

[20] Neureuther, A., Wang, R., and Helmsen, J. (1993). *Perspectives on TCAD Integration at Berkley*, 75–81. Vienna: Springer. http://doi.org/10.1007/978-3-7091-9315-0_4.

[21] Dutton, R.W. and Goossens, R.J.G. (1993). *Technology CAD at Stanford University: Physics, Algorithms, Software, and Applications*, 113–130. Vienna: Springer. http://doi.org/10.1007/978-3-7091-9315-0_6.

[22] MINIMOS 6.1: A two- and three-dimensional MOS simulator. https://www.iue.tuwien.ac.at/software/minimos-61 (accessed 8 May 2023).

[23] Selberherr, S., Schutz, A., and Potzl, H.W. (1980). MINIMOS – a two-dimensional MOS transistor analyzer. *IEEE Trans. Electron Devices* 27: 1540–1550. https://doi.org/10.1109/T-ED.1980.20068.

[24] Selberherr, S., Schutz, A., and Potzl, H.W. (1980). MINIMOS – a two-dimensional MOS transistor analyzer. *IEEE J. Solid-State Circuits* 15: 605–615. https://doi.org/10.1109/JSSC.1980.1051444.

[25] Selberherr, S., Fichtner, W., and Pötzl, H. (1979). MINIMOS – a program package to facilitate MOS device design and analysis. *Proceedings Conf. on Numerical Analysis of Semiconductor Devices*, Dublin, Ireland, (27–29 June 1979), pp. 275–279. ISBN: 0-906783-00-3.

[26] Selberherr, S. (1981). Zweidimensionale Modellierung von MOS-Transistoren. Ph.D. Thesis. TU Wien. http://doi.org/10.34726/hss.1981.00317827.

[27] Schütz, A., Selberherr, S., and Pötzl, H.W. (1982). A two-dimensional model of the avalanche effect in MOS transistors. *Solid State Electron.* 25: 177–183. https://doi.org/10.1016/0038-1101(82)90105-8.

[28] Haensch, W. and Selberherr, S. (1987). MINIMOS 3: a MOSFET simulator that includes energy balance. *IEEE Trans. Electron Devices* 34: 1074–1078. https://doi.org/10.1109/T-ED.1987.23047.

[29] Selberherr, S. (1989). MOS device modeling at 77 K. *IEEE Trans. Comput.-Aided Design Integr. Circuits Syst.* 9: 1464–1474. https://doi.org/10.1109/16.30960.

[30] Thurner, M. and Selberherr, S. (1990). Three-dimensional effects due to the field oxide in MOS devices analyzed with MINIMOS 5. *IEEE Trans. Electron Devices* 9: 856–867. https://doi.org/10.1109/43.57786.

[31] Selberherr, S., Hänsch, W., Seavey, M., and Slotboom, J. (1990). The evolution of the MINIMOS mobility model. *Solid State Electron.* 33: 1425–1436. https://doi.org/10.1016/0038-1101(90)90117-W.

[32] Kosina, H. and Selberherr, S. (1990). Coupling of Monte Carlo and drift diffusion method with applications to metal oxide semiconductor field effect transistors. *Jpn. J. Appl. Phys.* 29: L2283–L2285. https://doi.org/10.1143/JJAP.29.L2283.

[33] Kosina, H. and Selberherr, S. (1994). A hybrid device simulator that combines Monte Carlo and drift-diffusion analysis. *IEEE Trans. Comput.-Aided Des. Integr. Circuits Syst.* 13: 201–210. https://doi.org/10.1109/43.259943.

[34] Grasser, T. and Selberherr, S. (2000). Mixed-mode device simulation. *Microelectron. J.* 31: 873–881. http://doi.org/10.1016/S0026-2692(00)00083-5.

[35] Dragosits, K. and Selberherr, S. (2001). Two-dimensional simulation of ferroelectric memory cells. *IEEE Trans. Electron Devices* 48: 316–322. https://doi.org/10.1109/16.902733.

[36] Gritsch, M., Kosina, H., Grasser, T., and Selberherr, S. (2001). Influence of generation/recombination effects in simulations of partially depleted SOI MOSFETs. *Solid State Electron.* 45: 621–627. https://doi.org/10.1016/S0038-1101(01)00080-6.

[37] Grasser, T. and Selberherr, S. (2001). Fully coupled electrothermal mixed-mode device simulation of SiGe HBT circuits. *IEEE Trans. Electron Devices* 48: 1421–1427. https://doi.org/10.1109/16.930661.

[38] Palankovski, V., Schultheis, R., and Selberherr, S. (2001). Simulation of power heterojunction bipolar transistors on gallium arsenide. *IEEE Trans. Electron Devices* 48: 1264–1269. https://doi.org/10.1109/16.925258.

[39] Palankovski, V. and Selberherr, S. (2001). Micro materials modeling in MINIMOS-NT. *Microsyst. Technol.* 7: 183–187. https://doi.org/10.1007/s005420000076.

[40] Grasser, T., Kosina, H., Gritsch, M., and Selberherr, S. (2001). Using six moments of Boltzmann's transport equation for device simulation. *J. Appl. Phys.* 90: 2389–2396. https://doi.org/10.1063/1.1389757.

[41] Grasser, T., Kosina, H., Heitzinger, C., and Selberherr, S. (2002). Characterization of the hot electron distribution function using six moments. *J. Appl. Phys.* 91: 3869–3879. https://doi.org/10.1063/1.1450257.

[42] Grasser, T., Kosina, H., and Selberherr, S. (2003). Hot carrier effects with macroscopic transport models. *Int. J. High Speed Electron. Syst.* 13: 873–890. https://doi.org/10.1142/S012915640300206X.

[43] Ghani, T., Armstrong, M., Auth, C. et al. (2003). A 90nm high volume manufacturing logic technology featuring novel 45nm gate length strained silicon CMOS transistors. *Intl. Electron Devices Meeting*, Washington, DC (08-10 December 2003), pp. 11.6.1–11.6.3. http://doi.org/10.1109/IEDM.2003.1269442.

[44] Mistry, K., Allen, C., Auth, C. et al. (2007). A 45nm logic technology with high-k+metal gate transistors, strained silicon, 9 Cu interconnect layers, 193nm dry patterning, and 100% Pb-free packaging. *Intl. Electron Devices Meeting*, Washington DC (10-12 December 2007), pp. 247–250. http://doi.org/10.1109/IEDM.2007.4418914.

[45] Natarajan, S., Armstrong, K., Bost, M. et al. (2008). A 32nm logic technology featuring 2nd-generation high-k + metal-gate transistors, enhanced channel strain and 0.171μm2 SRAM cell size in a 291Mb array. *Intl. Electron Devices Meeting*, San Francisco, USA (15–17 December 2008), pp. 1–3. http://doi.org/10.1109/IEDM.2008.4796777.

[46] Thompson, S.E., Suthram, S., Sun, Y. et al. (2006). Future of strained Si/semiconductors in nanoscale MOSFETs. *Intl. Electron Devices Meeting*, pp. 681–684. http://doi.org/10.1109/IEDM.2006.346877.

[47] Wang, E., Matagne, P., Shifren, L. et al. (2006). Physics of hole transport in strained silicon MOSFET inversion layers. *IEEE Trans. Electron Devices* 53: 1840–1851. https://doi.org/10.1109/TED.2006.877370.

[48] Ungersboeck, E., Dhar, S., Karlowatz, G. et al. (2007). The effect of general strain on band structure and electron mobility of silicon. *IEEE Trans. Electron Devices* 54: 2183–2190. https://doi.org/10.1109/TED.2007.902880.

[49] Sverdlov, V. (2011). *Strain-Induced Effects in Advanced MOSFETs* (ed. S. Selberherr). Wien, NY: Springer-Verlag, ISBN: 978-3-7091-0381-4, 252 pages. https://doi.org/10.1007/978-3-7091-0382-1.

[50] Bohr, M. (2011). The evolution of scaling from the homogeneous era to the heterogeneous era. *Intl. Electron Devices Meeting*, Washington, DC (5–7 December 2003), pp. 1.1.1–1.1.6. http://doi.org/10.1109/IEDM.2011.6131469.

[51] Classical semiconductor device and circuit simulator. https://www.globaltcad.com/products/gts-minimos-nt (accessed 8 May 2023).

[52] Grasser, T., Rott, K., Reisinger, H. et al. (2014). NBTI in nanoscale MOSFETs-the ultimate modeling benchmark. *IEEE Trans. Electron Devices* 61: 3586–3593. https://doi.org/10.1109/TED.2014.235357.

[53] Pourfath, M. (2014). *The Non-Equilibrium Green's Function Method for Nanoscale Device Simulation* (ed. S. Selberherr). Wien, NY: Springer-Verlag, ISBN: 978-3-7091-1800-9, 256 pages. https://doi.org/10.1007/978-3-7091-1800-9.

[54] Ferry, D.K., Weinbub, J., Nedjalkov, M., and Selberherr, S. (2022). A review of quantum transport in field-effect transistors. *Semicond. Sci. Technol.* 37: 043001-1–043001-32. https://doi.org/10.1088/1361-6641/ac4405.

[55] Wigner, E. (1932). On the quantum correction for thermodynamic equilibrium. *Phys. Rev.* 40: 749–759. https://doi.org/10.1103/PhysRev.40.749.

[56] Nedjalkov, M., Dimov, I., and Selberherr, S. (2021). *Stochastic Approaches to Electron Transport in Micro- and Nanostructures*. Birkhäuser Basel: https://doi.org/10.1007/978-3-030-67917-0.

[57] Datta, S. (2005). *Quantum Transport, Atom to Transistor*. Cambridge. https://doi.org/10.1017/CBO9781139164313.

[58] Kresse, G. and Furthmüller, J. (1999). Efficient iterative schemes for *ab initio* total-energy calculations using a plane-wave basis set. *Phys. Rev. B* 54: art. 11169.

[59] Giannozzi, P., Baroni, S., Bonini, N. et al. (2009). QUANTUM ESPRESSO: a modular and open-source software project for quantum simulations of materials. *J. Phys. Condens. Matter* 21: art. 395502. https://doi.org/10.1088/0953-8984/21/39/395502.

[60] Kühne, T.D., Iannuzzi, M., Del Ben, M. et al. (2020). CP2K: an electronic structure and molecular dynamics software package-quickstep: efficient and accurate electronic structure calculations. *J. Chem. Phys.* 152: art. 194103. https://doi.org/10.1063/5.0007045.

[61] Kohn, W. and Sham, L.J. (1965). Self-consistent equations including exchange and correlation effects. *Phys. Rev.* 140: A1133–A1138. https://doi.org/10.1103/PhysRev.140.A1133.

[62] Victory atomistic device and nanostructure simulator. https://silvaco.com/tcad/atomistic-simulation (accessed 8 May 2023).

[63] QuantumATK – atomistic simulation software. https://www.synopsys.com/silicon/quantumatk.html (accessed 8 May 2023).

[64] Asenov, A. (1988). Random dopant induced threshold voltage lowering and fluctuations in Sub-0.1 μm MOSFET's: a 3-D "atomistic" simulation study. *IEEE Trans. Electron Devices* 45: 2505–2513. https://doi.org/10.1109/16.735728.

[65] Asenov, A., Brown, A.R., Davies, J.H. et al. (2003). Simulation of intrinsic parameter fluctuations in Decananometer and Nanometer-scale MOSFETs. *IEEE Trans. Electron Devices* 50: 1837–1852. https://doi.org/10.1109/TED.2003.815862.

[66] Asenov, A., Cheng, B., Wang, X. et al. (2015). Variability aware simulation based design-technology Cooptimization (DTCO) flow in 14 nm FinFET/SRAM Cooptimization. *IEEE Trans. Electron Devices* 62: 1682–1690. https://doi.org/10.1109/TED.2014.2363117.

[67] Amoroso, S.M., Asenov, P., Lee, J. et al. (2020). Enabling variability-aware design-technology co-optimization for advanced memory technologies. *J. Microelectron. Manufact.* 3: 1–8. https://doi.org/10.33079/jomm.20030409.

[68] Lu, L.C. (2022). Semiconductor evolution for chip and system design-from 2D scaling to 3D heterogeneous integration. *International Symposium on VLSI Design, Automation and Test (VLSI-DAT)*, Hsinchu, Taiwan (18-21 April 2022), p. 1. http://doi.org/10.1109/VLSI-DAT54769.2022.9768086.

[69] Datta, S. and Das, B. (1990). Electronic analog of the electro-optic modulator. *Appl. Phys. Lett.* 56: 665–667. https://doi.org/10.1063/1.102730.

[70] Sugahara, S. and Nitta, J. (2010). Spin-transistor electronics: an overview and outlook. *Proc. IEEE* 98: 2124–2154. https://doi.org/10.1109/JPROC.2010.2064272.

[71] Chuang, P., Ho, S.-C., Smith, L.W. et al. (2015). All-electric all-semiconductor spin field-effect transistors. *Nat. Nanotechnol.* 10: 35–39. https://doi.org/10.1038/nnano.2014.296.

[72] Tahara, T., Koike, H., Kameno, M. et al. (2015). Room-temperature operation of Si spin MOSFET with high on/off spin signal ratio. *Appl. Phys. Express* 8: art. 11304. https://doi.org/10.7567/APEX.8.113004.

[73] Ingla-Aynes, J., Herling, F., Fabian, J. et al. (2021). Electrical control of valley-Zeeman spin-orbit-coupling–induced spin precession at room temperature. *Phys. Rev. Lett.* 127: art. 047202. https://doi.org/10.1103/PhysRevLett.127.047202.

[74] Abert, C., Ruggeri, M., Bruckner, F. et al. (2015). A three-dimensional spin-diffusion model for micromagnetics. *Sci. Rep.* 5: art. 14855. https://doi.org/10.1038/srep14855.

[75] Fiorentini, S., Ender, J., Selberherr, S. et al. (2021). Coupled spin and charge drift-diffusion approach applied to magnetic tunnel junctions. *Solid State Electron.* 186: art. 108103. https://doi.org/10.1016/j.sse.2021.108103.

[76] Fiorentini, S., Bendra, M., Ender, J. et al. (2022). Spin torques in ultra-scaled MRAM devices. ESSDERC (accepted).

[77] Jinnai, B., Igarashi, J, Watanabe, K. et al. (2020). High-performance shape-anisotropy magnetic tunnel junctions down to 2.3 nm. *Intl. Electron Devices Meeting*, San Francisco, USA (12-18 December 2020), pp. 24.6.1–24.6.4. http://doi.org/10.1109/IEDM13553.2020.9371972.

[78] Hu, G., Lauer, G., Sun, J.Z. et al. (2021). 2x reduction of STT-MRAM switching current using double spin-torque magnetic tunnel junction. *Intl. Electron Devices Meeting*, San Francisco, USA (11-16 December 2021), pp. 43–46. http://doi.org/10.1109/IEDM19574.2021.9720691.

[79] Jorstad, N.P., Fiorentini, S., Loch, W. et al. (2022). Finite element modeling of spin-orbit torques. *Solid State Electron.* 194: art. 108323. http://doi.org/10.1016/j.sse.2022.108323.

Chapter 29

Early Integrated Circuits

Willy Sansen

Faculty of Engineering Science, Katholieke Universiteit Leuven, Belgium

As my hobby is music, I wanted to build an audio power amplifier. I loved the sound from those huge bass-reflex boxes, with that ubiquitous Philips 9710M loudspeaker. That was in 1964. I had already read and digested my very first book on radio electronics by L. Feenstra in 1961. It showed how to build a 5-W HiFi amplifier with tubes EF86 at the input and EL84 at the output. That same tube EF86 was used in my Low-frequency Sine-square wave audio generator BEM 004, which I built up with components from MBLE in 1965. I still use it to produce shifting sounds for my course in experimental music today!

That same year, however, I attended a presentation by Philips people from Eindhoven, explaining how wonderful transistors could be as substitutes for tubes. They did not heat up so much and subsequently would replace all tubes in the future. That seminar was organized by the IEEE branch of the Katholieke Universiteit Leuven (KU) Leuven, Belgium. So, the first thing I did was to become a student member of the IEEE. I still had two years to go for my MSc degree and I thought I could use all help to be part of a future of such importance.

The first transistor I actually held in my hand was a OC71, renamed later AC125. It was a pnp bipolar transistor with Germanium as a material. It was a black little tube with three wires sticking out, and a dot on one side. It was used in a 10-W HiFi amplifier, as described in the April 1966 issue of the Kluwer journal, "Radio Electronica." The AC125 was used at the input and an AD139 at the output. The supply voltage was −34V (see Figure 29.1). A few months later, in the September issue of the same Journal, a 30-W HiFi amplifier was discussed with similar Germanium devices. I thought this is it. With 30W

75th Anniversary of the Transistor, First Edition. Edited by Arokia Nathan, Samar K. Saha, and Ravi M. Todi.

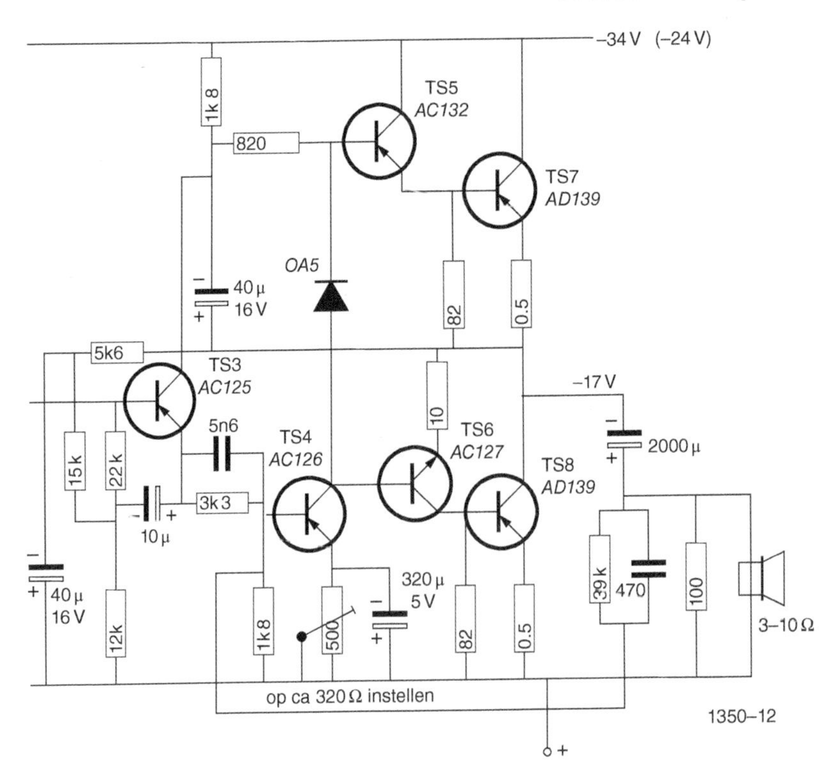

Figure 29.1 10-W HiFi amplifier with Germanium transistors (Radio Electronica, April 1966).

I can surely get HiFi, and focus again on my music. Unfortunately, these Germanium devices had too low output resistances to really provide low-distortion gain. How happy was I when I discovered in the same September issue a 30W amplifier with silicon devices. It had Motorola devices and the input and RCA power devices at the output. They were hard to find however. In the same September issue, we had thus amplifiers with Germanium devices and a few pages later amplifiers with silicon devices, all in that same year 1966.

One year later, in 1967, in the December issue of that same journal "Radio Electronica," I found another 10-W amplifier again with silicon transistors, however, different ones, and easily available. A BC 107 npn transistor was used at the input and 2N3055's at the output, driven by 2N3053's. This combination would be found in many HiFi amplifiers, even decades later. These bipolar silicon transistors were described in great detail in the Adzam Data handbook of October 1967. I am still fond of its quote "The BC107 is intended for a multitude of high gain low power applications, in particular for use in driver stages of hi-fi equipment and in the jungle of television receivers." The same Data handbook also showed a pMOS (p-channel metal-oxide-semiconductor) in combination with an nnp transistor, the TAA 320 from Philips. It offered a higher input impedance than any bipolar transistor amplifier. It also showed several operational amplifiers with bipolar transistors, all integrated in one single silicon chip.

In that same incredible year 1966, the first operational amplifiers with bipolar transistors appeared in the first Volume of the *IEEE Journal of Solid-State Circuits* [1]. For example, Jim Solomons paper in the September issue (on pages 19–28) clearly explained in detail how to go about this. I tried it myself when I took Don Pederson's class at UC Berkeley in 1969. It turned out a bit harder than explained, but surely, npn bipolar transistors were established as the technology to go and that was not expected to change all that much in the future. It was thus no surprise that, when I jumped into nonlinear circuits,

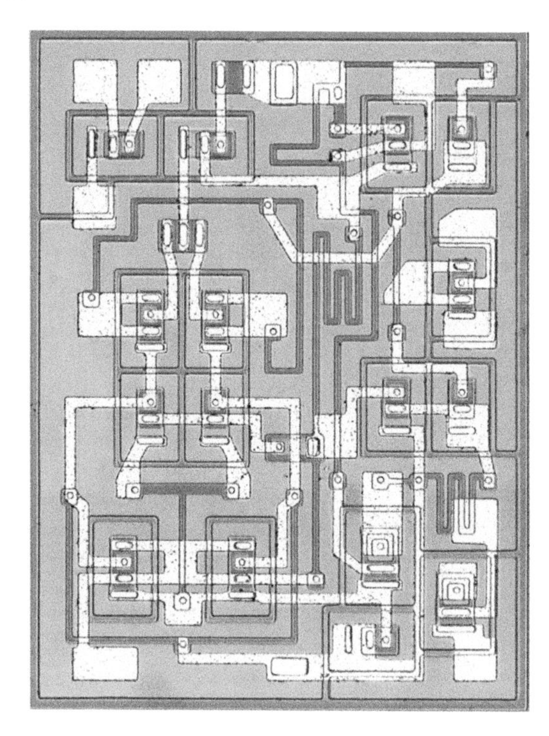

Figure 29.2 AGC amplifier in 10 μm bipolar technology (JSSC Sansen August 1974, pp. 159–166).

as explained in Bob Meyers class, I also used bipolar npn's. I am still proud of my automatic gain control (AGC) circuit (see Figure 29.2) [2] that lead to my PhD thesis at UC Berkeley in 1972. The npn bipolar transistors are easy to distinguish. That circuit had low noise and low-distortion, two topics which are still at the core of my international courses today. I had learned how to peel rubylith from Ian Getreu and Bruce Wooley. Especially those small contact windows were a headache. If your wife found one on your sleeve that evening, you knew that you had to back to Cory Hall to figure out which contact was missing!

The year 1968 was a stellar year for npn bipolar transistors. Barie Gilbert's multiplier was published in the *Journal of Solid-State Circuits* and so some of his wideband amplifiers [3]. Things changed all of a sudden however, when metal-oxide-semiconductor transistor (MOST) devices emerged. The first IC's I designed in pMOST technology were four chips for a private automatic branch exchange (PABX) communication system for GTE-ATEA in Leuven in 1973. A sketch of the clock generator is shown in Figure 29.3. The supply voltage was −12 V and the clocks at −24 V. The minimum line width was 10 μm. And then things went fast. After the pMOST transistors, n-channel MOSTs (nMOST's) emerged. This technology was used by Ricardo Suarez in a switched-capacitor ADC on International Solid-State Circuits Conference (ISSCC) 1974. Complementary MOS (CMOS) would follow several years later.

Nevertheless, our first paper on ISSCC 1976, a 4×4 crosspoint for communications for ITT-Bell Telephone, was again with bipolar transistors. It took us several more years to jump on the bandwagon of CMOS. In 1983, Marc Degrauwe published his first paper of a "Novel adaptive biasing amplifier" all in CMOS technology, with 10 μm minimum linewidth. On ISSCC 1983, he presented an opamp silicon compiler, in that same CMOS technology. Many other designs in CMOS saw light in parallel.

To cut the costs of masks however, Multi-Project Chips were initiated, in which all designs were put together. Our first one came out in January 1983, and was realized in a 5 μm p-well CMOS technology as shown in Figure 29.4. It was gracefully processed within the ESAT facilities at the KU Leuven by Gilbert Declerck and his team. The previous processing lab had been lost by fire in 1981. This allowed

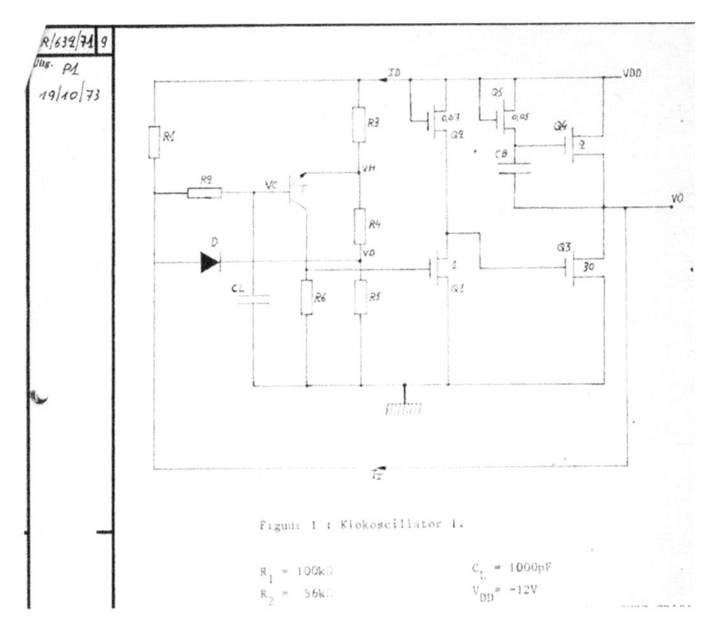

Figure 29.3 Clock generator of PABX system in pMOST technology (1973).

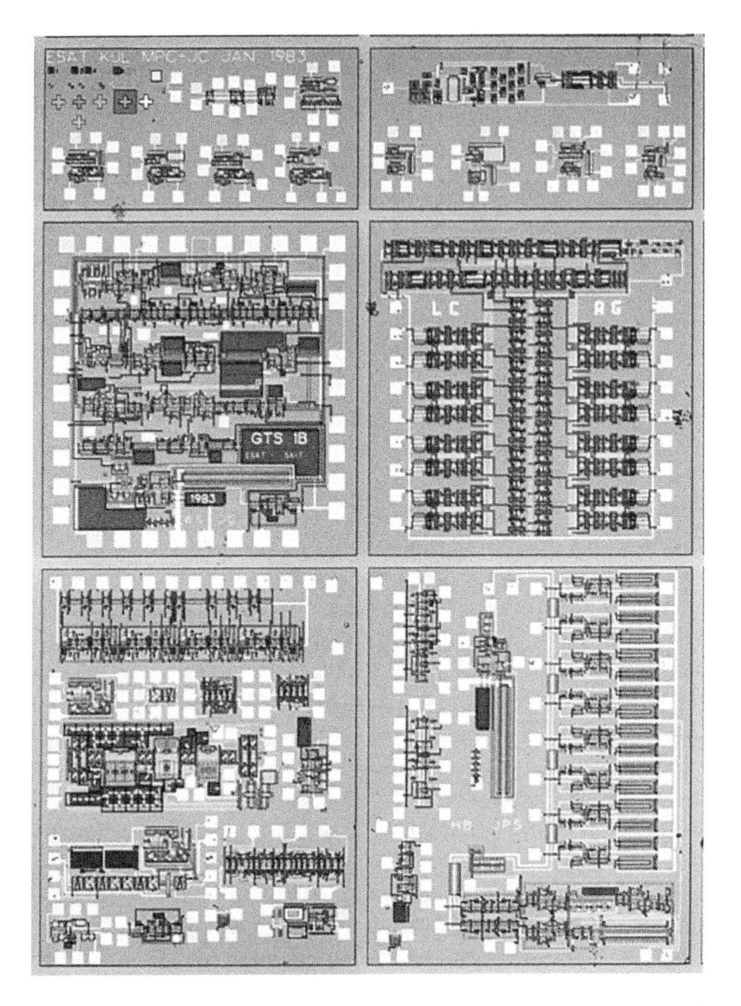

Figure 29.4 MPC-1: 5 μm p-well CMOS; 8.5 × 5.6 mm; January 1983.

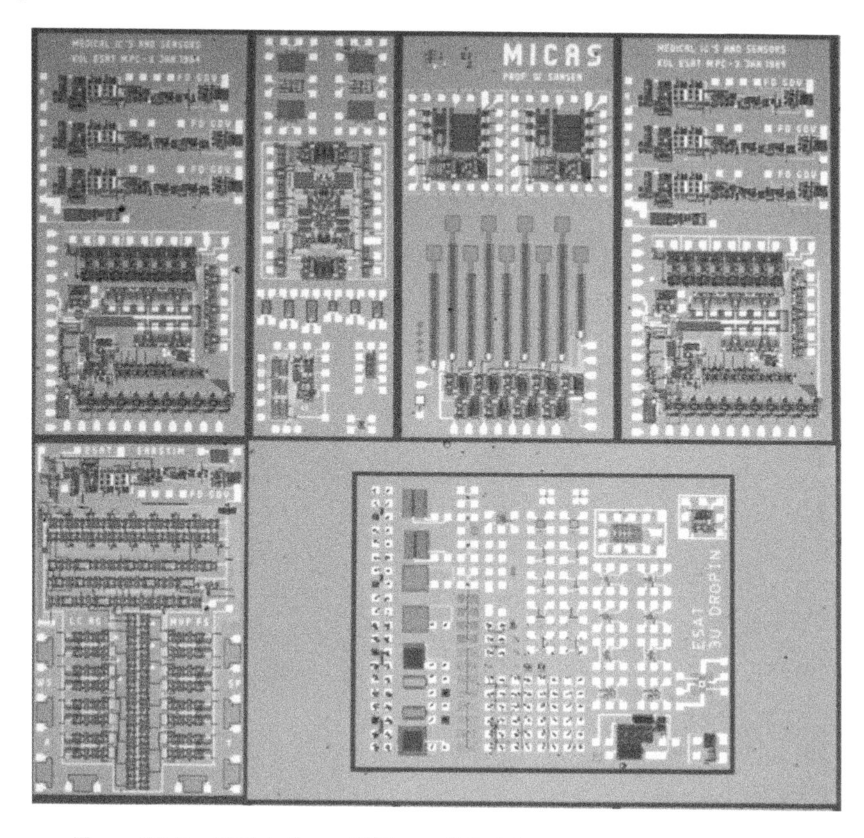

Figure 29.5 MPC-3: 5 μm JFET-p-well CMOS; 12 × 11 mm; January 1984.

a drastic change to CMOS technologies only. On this MPC-1, some transistors are clearly visible, and the opamps on the top row. Also some larger communication systems are easy to see. Our most famous early MPC was the third one however, put together in January 1984 (see Figure 29.5) [4]. It used the same technology as MPC-1 but also included implanted JFETs with low $1/f$ noise. The chemical-sensor heads are clearly visible. The cochlear implant chip is on the bottom row on the left. In that same year, IMEC (Interuniversity Microelectronics Centre), Belgium was created. It made an n-well CMOS process available with 3 μm linewidths at their own facilities. All later MPC's, starting with MPC-10 (in February 1987), have used this technology as abundantly presented on the later ISSCC's and described in the Journal of Solid-State Circuits.

References

[1] Solomon, J.E. and Wilson, G.R. (1966). A highly desensitized, wide-band monolithic amplifier. *IEEE Journal of Solid-State Circuits* 1 (1): 19–28. https://doi.org/10.1109/JSSC.1966.

[2] Sansen, W.M.C. and Meyer, R.G. (1974). An integrated wide-band variable-gain amplifier with maximum dynamic range. *IEEE Journal of Solid-State Circuits* 9 (4): 159–166. https://doi.org/10.1109/JSSC.1974.1050490.

[3] Gilbert, B. (1968). A new wide-band amplifier technique. *IEEE Journal of Solid-State Circuits* 3 (4): 353–365. https://doi.org/10.1109/JSSC.1968.1049924.

[4] Sansen, W., Das, C., and Callewaert, L. (1984). A Monolithic Impedance Buffer, with a Compatible JFET-CMOS Technology. *ESSCIRC '84: Tenth European Solid-State Circuits Conference*, Edinburgh, UK (1984 September), pp. 63–66.

Chapter 30

A Path to the One-Chip Mixed-Signal SoC for Digital Video Systems

Akira Matsuzawa[1,2]

[1] Tokyo Institute of Technology, Tokyo, Japan
[2] Tech Idea Co., Ltd., Kawasaki, Japan

30.1 Introduction

The past 50 years have been an era in which analog electronics have been replaced by digital electronics including Vinyl. Records to Compact Disk, NTSC (national television system committee) and PAL (Phase alternation by line) analog televisions to High-Definition digital television (TV), analog handy camcorder to digital handy camcorder, and Video Home System (VHS) video to Digital Video Disc (DVD). These electronic devices are largely owing to the progress of CMOS integrated circuits with technology scaling, which dramatically improved the performance of digital signal processing (DSP) and memory and reduced power and cost. But the development of analog-to-digital converter (ADC) and digital-to-analog converter (DAC) also contributed greatly. Since these require high accuracy, special circuit techniques different from digital circuits are indispensable to master the device. Based on the author's expertise in the development of ADCs for digital video equipment, this paper presents the development of video-rate ADCs using bipolar, bipolar-complementary metal-oxide-semiconductor (Bi-CMOS), and complementary metal-oxide-semiconductor (CMOS) technology, and the path to the mixed-signal system-on-a-chip (SoC) that can integrate full DVD systems including analog parts as the final goal of integrated circuit (IC) technology.

75th Anniversary of the Transistor, First Edition. Edited by Arokia Nathan, Samar K. Saha, and Ravi M. Todi.

30.2 Bipolar ADCs at Early Development Stage of Digital TVs

I graduated with a master's degree in electronics at the Graduate School of Engineering, Tohoku University in March 1978, and joined Matsushita Electric Industrial Co., Ltd. (currently Panasonic) in April of the same year. After one year of training for new employees, I was assigned to the semiconductor development department division of the Central Research Lab.

In April 1979, I started to work as an IC designer. My first job was designing a video-rate 8-bit ADC for digital TVs and digital video systems. At that time, Matsushita developed six hours-play in the VHS video recorder. The highest level of analog technology was used, such as advanced electro-mechanic technology, material technology, magnetic tape and magnetic head, motor servo, and analog video signal processing circuit technology. When analog technology was at its peak, technological development aimed at digitization of TV and video systems was quietly started. For the digitization of these video systems, the ADC was a major development problem together with the image processing DSP.

Figure 30.1 shows a 10-bit video-rate analog-to-digital (A/D) converter board that was available at the time. Monolithic IC did not exist, and the A/D converter board made by ADI Co. was marketed, but the power consumption was large at 20 W, and the price was as much as 10,000 dollars, which was almost the same price as a small car. Therefore, for the digitization of the video equipment, the ADC development was indispensable to reduce the price and power consumption such that it can finally be used for consumer equipment.

In 1982, we developed a monolithic integrated video-rate 10-bit ADC [1] shown in Figure 30.2. Flash conversion architecture was used and integrated 1023 comparators that used analog bipolar transistors of which mismatch voltage is less than 0.1 mV. This ADC contributed to the development of the digital video switcher, 256 Quadrature Amplitude Modulation (QAM) microwave transmission, and the early-stage High-Definition TV (HDTV) system (MUSE Hi-Vision), etc. It also contributed to the HDTV relaying for the 1988 Seoul Olympics.

In the digitalization of the camera signal for the HDTV, an ADC with conversion speed of over 75 megasamples per second (MS s^{-1}) was necessary, but since such high-speed ADC was not developed at that time, it was difficult to digitize the high-definition camera signal. In 1984, we developed an 8-bit ADC with conversion speed exceeding 100 MS s^{-1} [2]. This ADC was used for the digitizing camera signal for the HDTV system "MUSE (Multiple sub-Nyquist Sampling Encoding)" developed mainly by NHK (Nippon Hoso Kyokai), Japan Broadcasting Corporation, and also for the digital oscilloscopes.

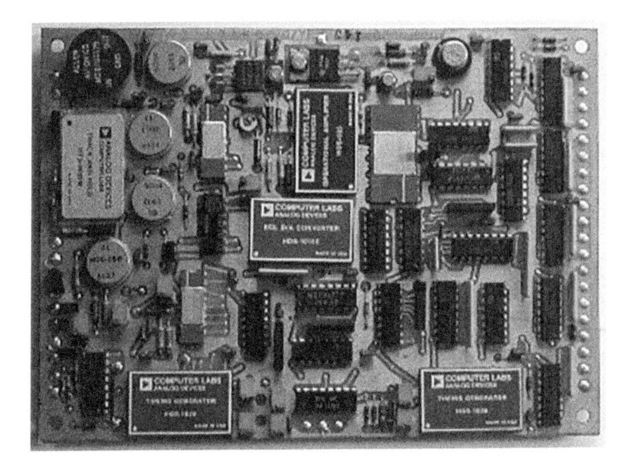

Figure 30.1 10-Bit video-rate A/D converter board.

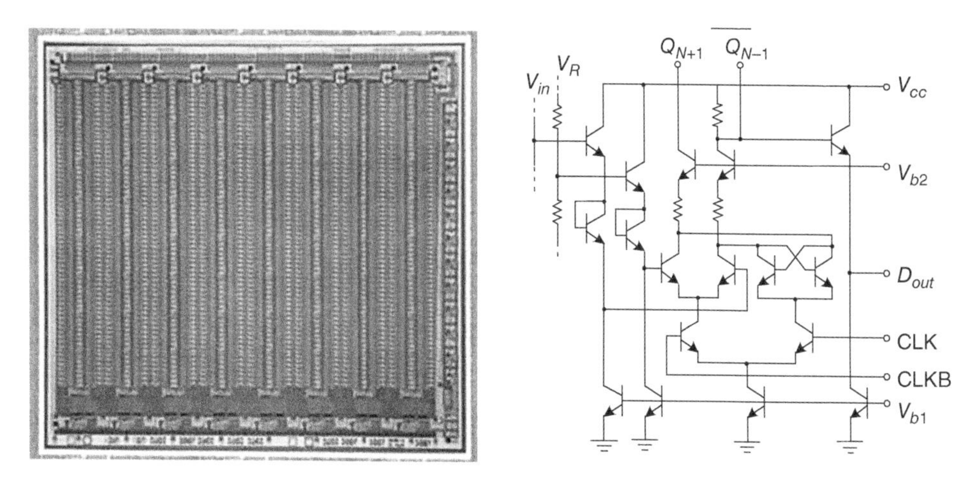

Figure 30.2 Monolithic 10-bit video-rate ADC and comparator.

A flash ADC architecture is not suitable for high-resolution ADC because the circuit size increases in proportion to 2^N for resolution N. In addition to the area, power consumption also increases exponentially. Therefore, if A/D conversion can be performed in two steps, an upper stage and a lower stage, the number of comparators and power consumption can be greatly reduced. If the overall resolution is N and the resolution of the sub-conversion is M, the number of comparators n in the ideal condition is

$$n = 2^M + 2^{N-M} \rightarrow 2^{\left(1+\frac{N}{2}\right)} @ M = \frac{N}{2} \tag{30.1}$$

When the resolution N is 10, the number of comparators can be greatly reduced to 1/16 from 1024 to 64. However, it requires a sample and hold circuit to hold the input signal during two conversion periods and it is almost impossible to realize it with bipolar circuit since the base current of bipolar transistor changes the hold voltage of the sampling capacitor. MOS input circuits are needed. In 1989, we could use Bi-CMOS technology and developed a two-step flash ADC for the home use HDTV receiver. Figure 30.3 shows our developed Bi-CMOS ADC (AN8130K) [3] and HDTV receiver board. Figure 30.4 shows the Bi-CMOS sample and hold circuit.

Figure 30.3 10-Bit video-rate Bi-CMOS ADC and a HDTV receiver board.

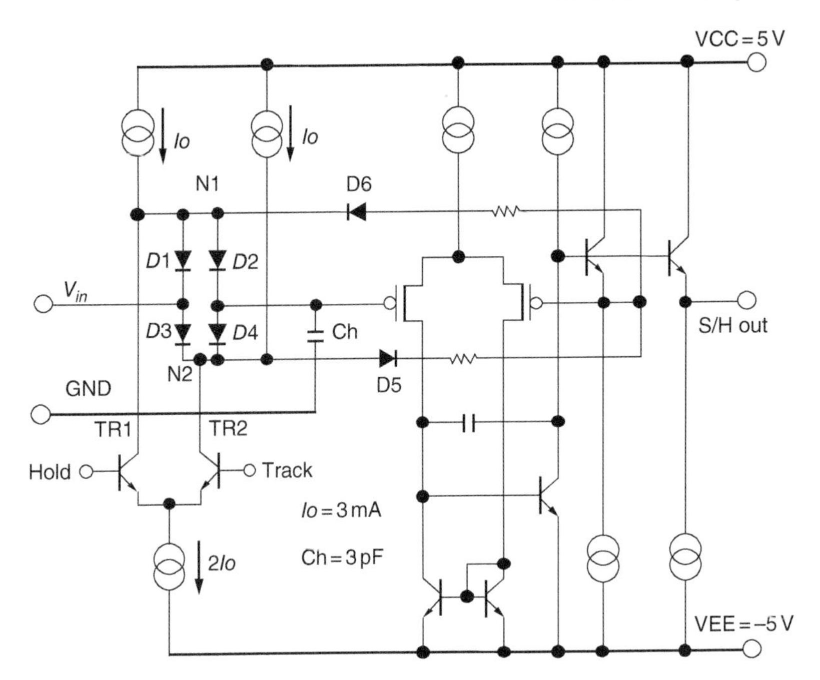

Figure 30.4 Bi-CMOS sample and hold circuit.

Another problem in a two-step flash ADC is a large conversion error occurs at the switching point of the upper ADC due to the deviation of the conversion range between the upper ADC and the lower ADC. Figure 30.5a shows the state of conversion when a voltage shift of ΔV occurs in the upper and lower conversion ranges in the conventional two-step flash A/D conversion. The conventional two-step A/D transformation has a fixed lower A/D conversion range. In this example, the lower part does not have a value over three or more. When there is a voltage shift of ΔV in the conversion range, a large conversion error occurs at the switching point of the upper ADC, and the differential nonlinearity (DNL) deteriorates significantly. In contrast, if lower-conversion is performed using interpolation as shown in Figure 30.5b, the voltage between the top two points is divided equally so as to the spring coil, and the conversion error is distributed and does not concentrate on a specific code. It can achieve a smooth A/D conversion without large DNL degradation. Certainly, this method cannot improve integral nonlinearity (INL), but INL degradation is not important for A/D conversion for video signals. The DNL that is directly linked to image smoothness is important; therefore, interpolation is an effective technique for video-rate ADCs.

Figure 30.6 shows the configuration of the two-step flash A/D converter using resistive interpolation and the output and interpolated voltages of the differential amplifier. A pair of differential amplifiers is selected by the upper conversion result, and the output signal is transferred to the interpolation resistance and the interpolated signals input to the lower comparators.

In 1992, we developed 10-bit 300 MS s^{-1} ADC [4] shown in Figure 30.7 for more accurate and faster A/D conversions that support wider bandwidth and higher dynamic range of HDTV signals. Interpolation circuits using distributed preamplifier and resistor are formed. Although it is a flash type, the offset voltage mismatch of the transistor is effectively relaxed by about 8 times. At that time, as a 10-bit ADC, the ADC enabled conversion speed about 4 times faster than competing products. This ADC was used for digital optical transmission of HDTV signals. This ADC was awarded the R&D100 prize in 1994 as an ADC that expanded the availability of the application of DSP.

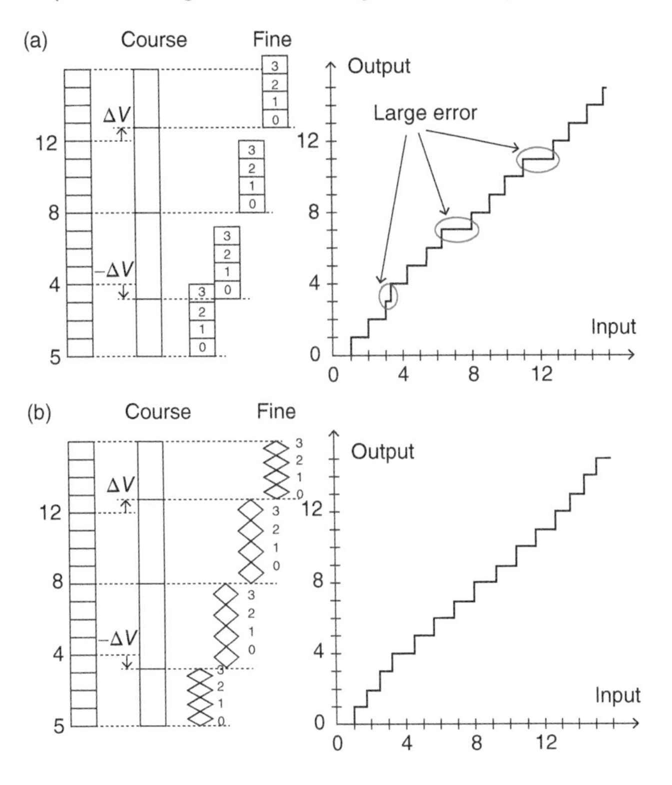

Figure 30.5 Characteristics of two-step flash A/D conversion. (a) Conventional A/D conversion. (b) A/D conversion using interpolation.

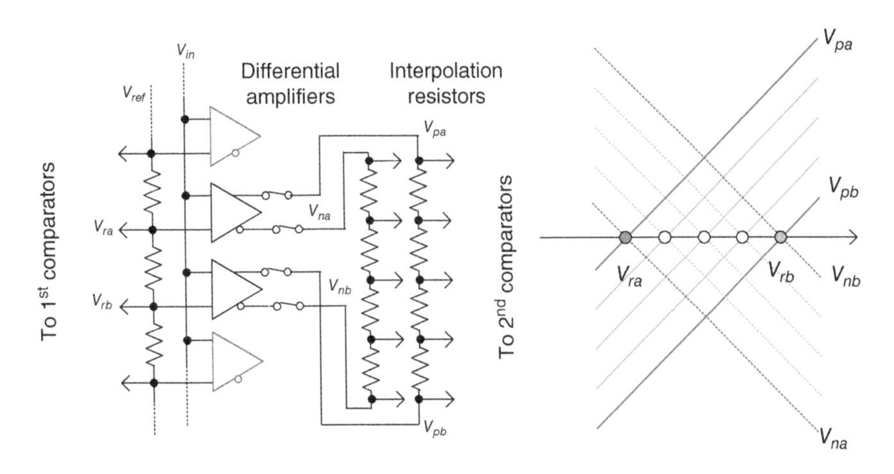

Figure 30.6 Two-step flash video rate A/D converter using resistive interpolation.

Figure 30.8 shows resistive interpolation in flash ADC, which reduces the number of amplifiers and increased the reference voltage between the amplifiers to mV_q. The output voltage of the amplifier is connected by a circuit in which multiple resistors are connected in series, and the A/D conversion value is obtained by comparing the tap voltage of the resistor with a latch. The voltage of the resistor tap is the voltage that interpolates the output voltage of a pair of amplifiers.

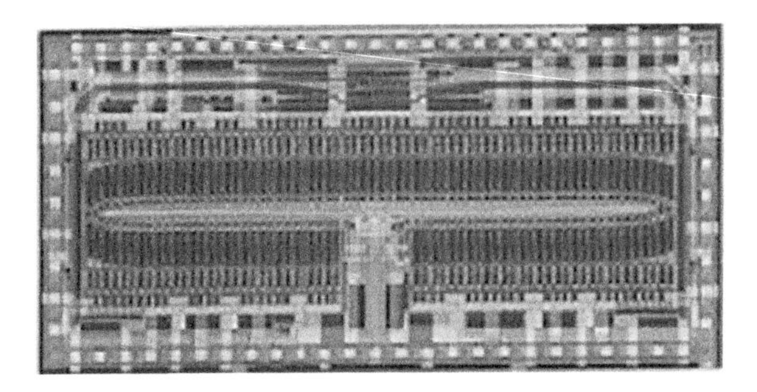

Figure 30.7 10-Bit 300 MS s⁻¹ ADC using resistive interpolation.

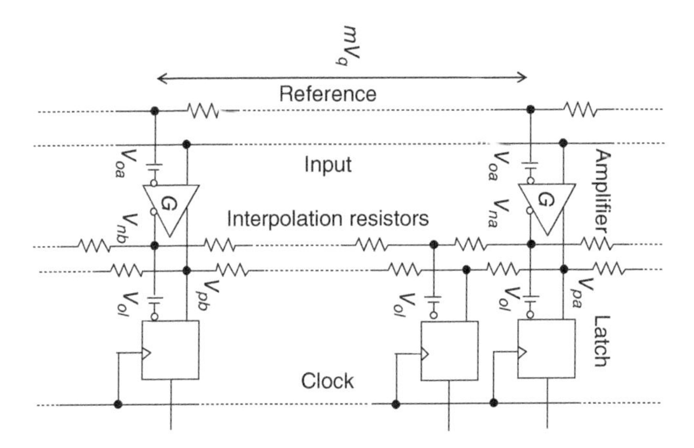

Figure 30.8 Resistive interpolation in flash ADC.

The effect of each offset voltage on the DNL V_{oe} is expressed below. Here, m is the reference voltage difference between amplifiers normalized by the quantization voltage V_q.

$$V_{oe} = \left\{ \left(\frac{V_{oa}}{m} \right)^2 + \left(\frac{V_{ol}}{G} \right)^2 \right\}^{0.5} \tag{30.2}$$

Therefore, the voltage distribution of the offset voltage V_{ol} of the latch remains as V_{ol}/G at the input terminal, but the effect of the voltage distribution of the offset voltage V_{oa} of the amplifier on the DNL is reduced by $1/m$. Therefore, by using this method, the effect of the offset voltage V_{oa} of the amplifier on the DNL can be suppressed, so that high-precision flash ADC can be realized even with small transistors that have a large offset voltage distribution. Not only the number of amplifiers with large power consumption can be reduced to $1/m$, but also the use of small transistors is possible, resulting in significant lower power consumption.

30.3 A CMOS ADC for Digital Handy Camcorder

Early video ADCs from the 1980s to the 1990s used bipolar or Bi-CMOS technologies. At that time, bipolar technology was much better than CMOS in accuracy, speed, and power dissipation. However, for the integration of the system, a CMOS ADC that can realize mixed signal circuits has been demanded strongly. Therefore, we developed a CMOS 10-bit 20 MS s⁻¹ 30 mW ADC [5] shown in Figure 30.9 and announced in 1993.

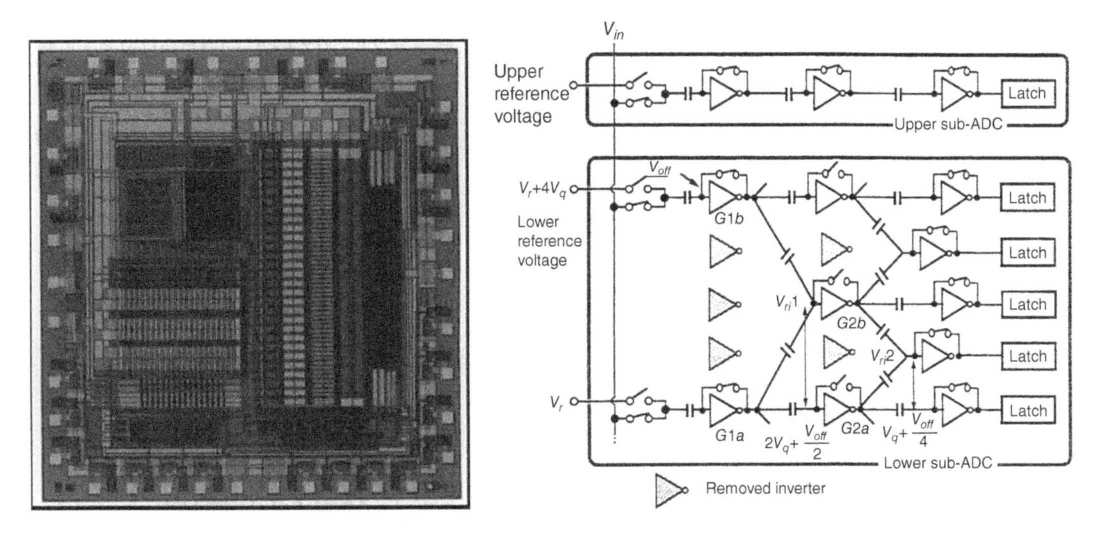

Figure 30.9 10-Bit 20 MS s⁻¹ 30 mW CMOS ADC.

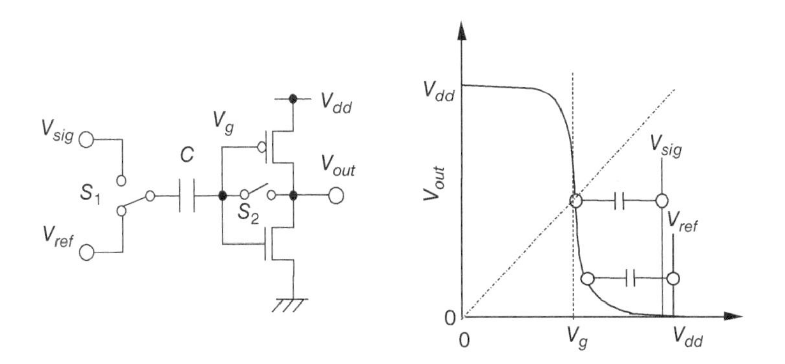

Figure 30.10 Chopper inverter comparator.

The issue of CMOS transistor of insufficient accuracy was addressed by using chopper technology shown in Figure 30.10.

In a sampling period, S_1 selects the input signal and S_2 is closed. The V_g that gives the same voltage to the gate and drain terminals appears at the gate. In the comparison period, S_1 selects reference voltage and S_2 is opened. At this change, the input signal V_{sig} is sampled, and the inverter acts as an amplifier. If V_{sig} is higher than V_{ref}, the output voltage goes down and if V_{sig} is lower than V_{ref}, the output voltage goes up. Thus, accurate comparison can be expected by storing threshold fluctuation in the capacitor and canceling it, but the accuracy of 10-bit required for the video-rate ADC was difficult. Then, we invented the interpolation circuit using the capacitor connected between inverters shown in Figure 30.9.

Furthermore, power consumption was reduced by the reduction of the number of inverters. Figure 30.11 shows the transition of the conversion energy of 10-bit video-rate ADC. Until then, the conversion energy of bipolar ADC and CMOS ADC was antagonistic, and CMOS was not necessarily low power. However, this ADC achieved a decisive low power of 1/8 compared with the other competitive ADCs. Afterward, all video-rate ADCs began using CMOS technology.

When ADC became possible with CMOS, IC technology evolved in the direction of realizing low power and low cost of the whole system. In the evolution of the so-called mixed-signal SoC.

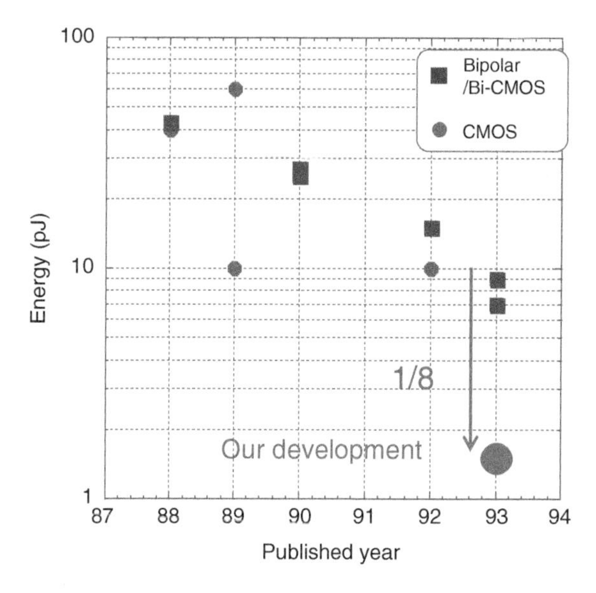

Figure 30.11 Conversion energy of video-rate 10-bit ADCs.

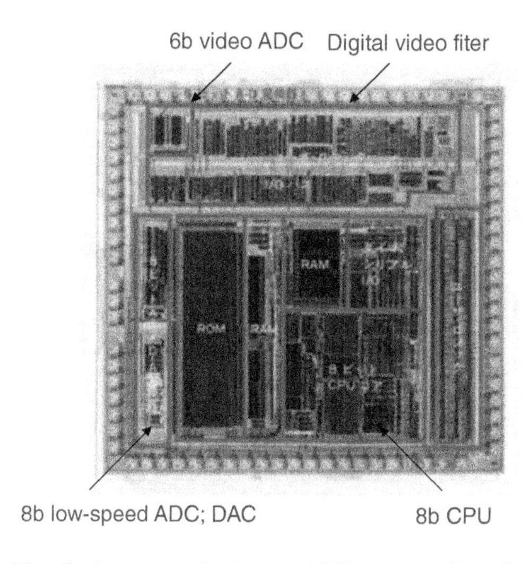

Figure 30.12 Early-stage mixed-signal LSI for digital handy camcorder.

Figure 30.12 shows an early-stage mixed-signal LSI for digital handy camcorder. It has been developed for handy camcorders using DSP shown in Figure 30.13. An 8-bit CPU for controlling the system, low speed ADC and DAC, digital filter for image processing, and video-rate ADC and DAC are integrated in the chip [6]. Digital circuits are used for image stabilization, color correction, and autofocus. Thus, low power operation and cost reduction of integrated circuits including ADC are required for portable equipment. CMOS ADC and low-power mixed-signal LSI became the key technologies for portable video equipment.

At present, the importance of low-power technology has gained common recognition in the world, but it was not so until the middle of the 1990s. The first priority for the IC technology was high performance, especially high speed, with little concern on the low power consumption. However, low power technology is important for the realization of consumer portable equipment and long battery life.

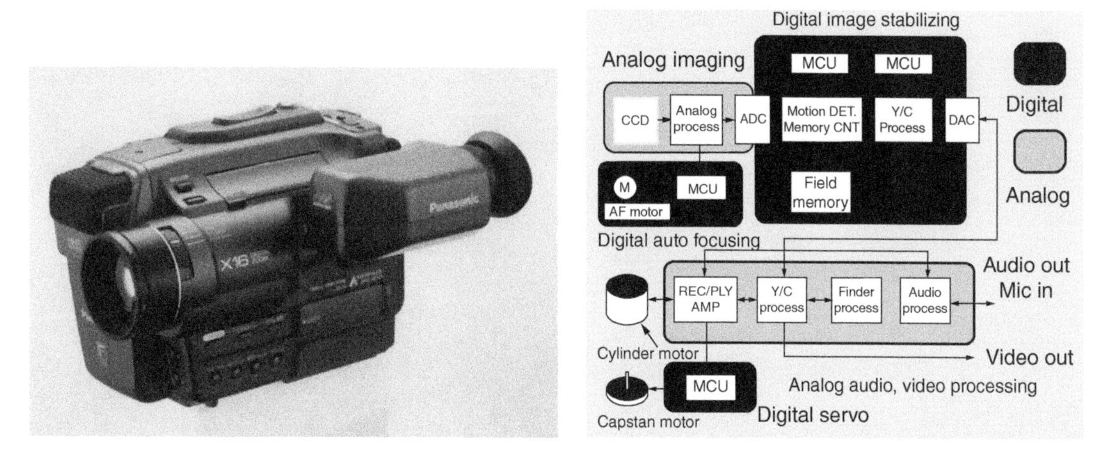

Figure 30.13 Digital handy camcorder and signal processing circuits.

In 1994, I gave an invited talk on low-power IC design at the 1994 IEEE VLSI Circuits Symposium [7] and the invited paper was published in IEEE Journal of Solid-State Circuits in 1994 [6]. On the low power of the consumer portable equipment, not only the digital circuit but also the analog circuit was mentioned, and it was pointed out that the low power of the whole system is important.

30.4 One-Chip Mixed-Signal SoC for DVD

The digitization of the videotape recorder resulted in the emergency of DVD. The problem of the DVD recorder is that the strength of the pickup signal is weak and contains many errors, because the multi-value recording is used as shown in Figure 30.14. Therefore, a digital equalizer and a digital error correction were required, and for this reason, high-speed ADC of the same level as the measuring instrument of 7-bit 400 MS s^{-1} was also required even though it was a consumer equipment.

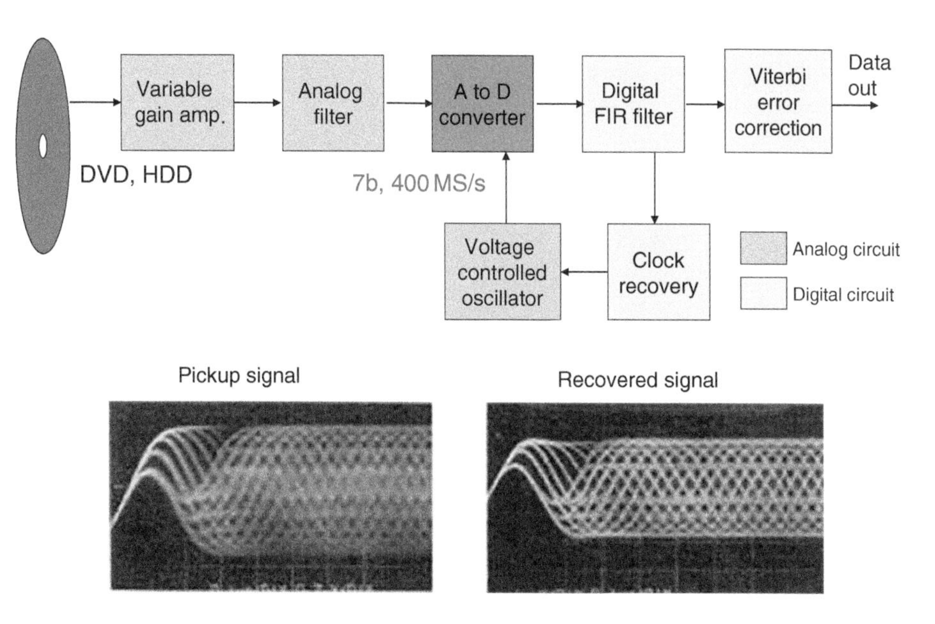

Figure 30.14 DVD recorder system and pickup and recovered signals.

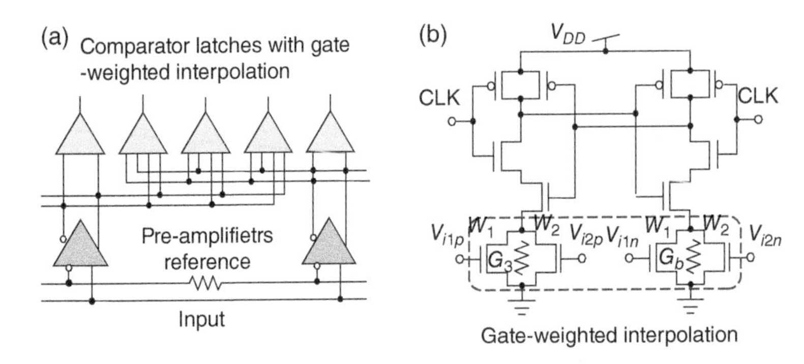

Figure 30.15 ADC and latch circuit with gate-weighted MOS. (a) ADC. (b) Latch circuit using gate-weighted interpolation.

We developed a 6-bit ADC with conversion speeds above 1-gigasamples per second (GS s^{-1}) with bipolar technology in 1991 [8]. In 2000, we developed a 6-bit 800 MS s^{-1} ADC [9] with CMOS technology, but the power consumption was large at 400 mW, and it could not be used for consumer products. In 2002, we succeeded to develop a 7-bit 400 MS s^{-1} CMOS ADC with a low power consumption of 50 mW [10] and opened the way for applications to consumer products.

Figure 30.15 a,b show the conversion architecture, and a unit comparator of the ADC. Basically, it is the same as parallel interpolation [4], but a low-power CMOS dynamic comparator is used, and the interpolation is realized by weighting the gate-width of the input transistors.

Then, the interpolation circuit without the resistance was devised.

Comparisons can be made on the magnitude of the conductors in the MOS linear regions of the left and right transistors. The composed conductors of each transistor are

$$\left.\begin{aligned} G_a &= \frac{\mu C_{ox}}{L}\left[W_1\left(V_{i1p}-V_T\right)+W_2\left(V_{i2p}-V_T\right)\right] \\ G_b &= \frac{\mu C_{ox}}{L}\left[W_1\left(V_{i1n}-V_T\right)+W_2\left(V_{i2n}-V_T\right)\right] \end{aligned}\right\} \tag{30.3}$$

where μ is mobility, C_{ox} is a unit gate capacitance, V_T is the threshold voltage, and L is channel length. The difference in composed conductors is

$$G_a - G_b = \frac{\mu C_{ox}}{L}\left[W_1\left(V_{i1p}-V_{i1n}\right)+W_2\left(V_{i2p}-V_{i2n}\right)\right] \tag{30.4}$$

if the ratio of gate width W_1 to W_2 is

$$W_1 : W_2 = \frac{m-k}{m} : \frac{k}{m} \quad k < m \tag{30.5}$$

according to the principle of the internal division, it is possible to compare the reference voltages by m equality.

This latch has a large offset voltage, but the offset voltage is reduced to $1/G$ at the input end with the gain of the preamplifier as G, and the contribution of the preamplifier offset voltage to DNL is reduced to $1/m$, which in this case is 1/8. The circuit in Figure 30.15b is a dynamic circuit in which no through current flows, and since the interpolation resistor which consumes power is not used, the power consumption of the entire ADC is small. Therefore, this ADC has a power consumption of 40 mW at 400 MS s^{-1} which is 10 times lower than other competitive ADCs and realizes low power consumption that can be embedded on the SoC for the DVD recorder.

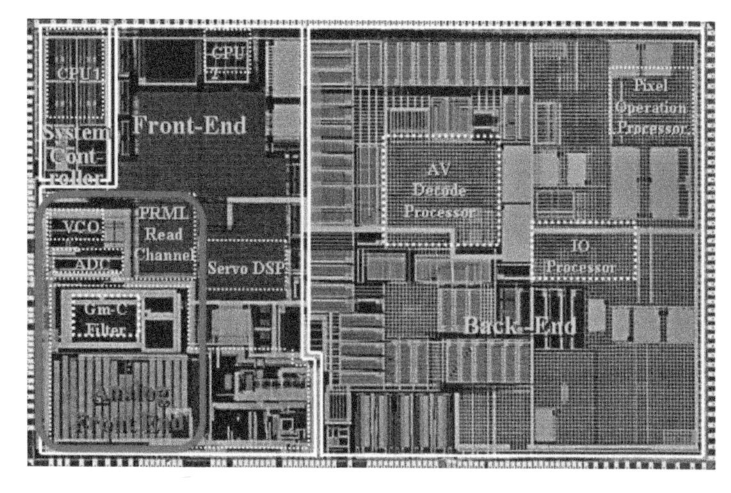

Figure 30.16 Fully one-chip mixed-signal SoC for DVD systems. Source: Photo courtesy of A. Matsuzawa.

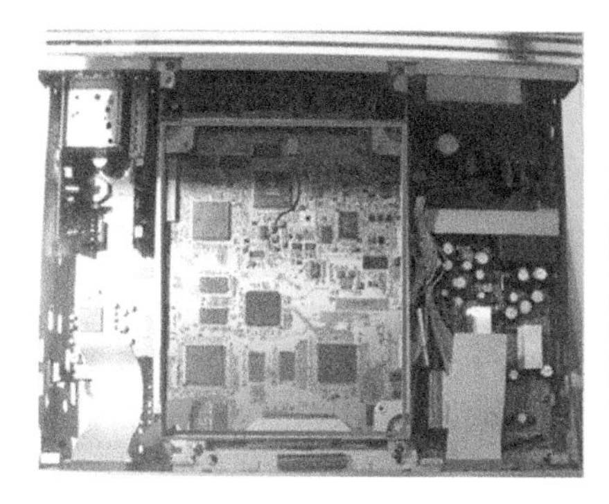

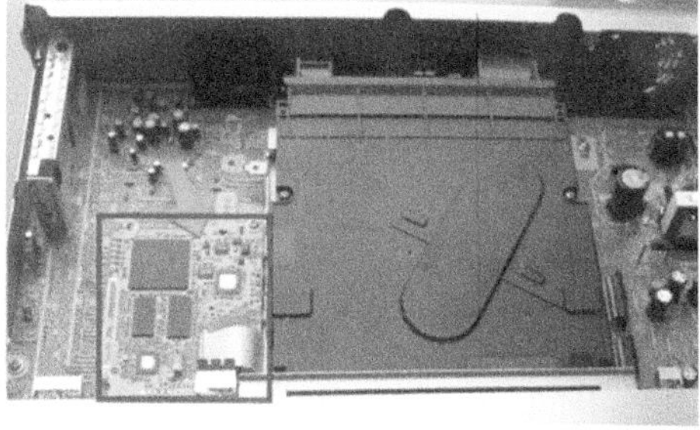

Figure 30.17 DVD recorders. Source: Panasonic Life Solutions India Pvt. Ltd.

The ADC was embedded in a fully one-chip mixed-signal SoC for DVD system shown in Figure 30.16. All DVD functions were integrated into a single IC [11]. It was the beginning of an era of mixed-signal SoC [12, 13].

Figure 30.17 shows the DVD Recorder board before SoC development and after SoC development. The SoC eliminates many ICs and components, lowers manufacturing costs and power consumption, and increases reliability. The smaller form factor also allows DVD to be installed in note PCs. This series of SoC for DVD shipped a total of 520M chips and contributed to the IC sales of 2.5B USD over the course of nine years. A mixed signal SoC that can integrate all circuits is a goal of IC technology.

Appendix: Interpolation Techniques in ADCs

Many ADCs we have developed for digital video systems use interpolation techniques. I will explain this technique briefly.

A flash ADC uses many monotonically shifted reference voltages to simultaneously compare input signals to obtain an AD conversion value, as shown in Figure 30.18a. A major problem is the distribution of the offset voltage of the comparator. In the comparison, a latch circuit that performs polarity

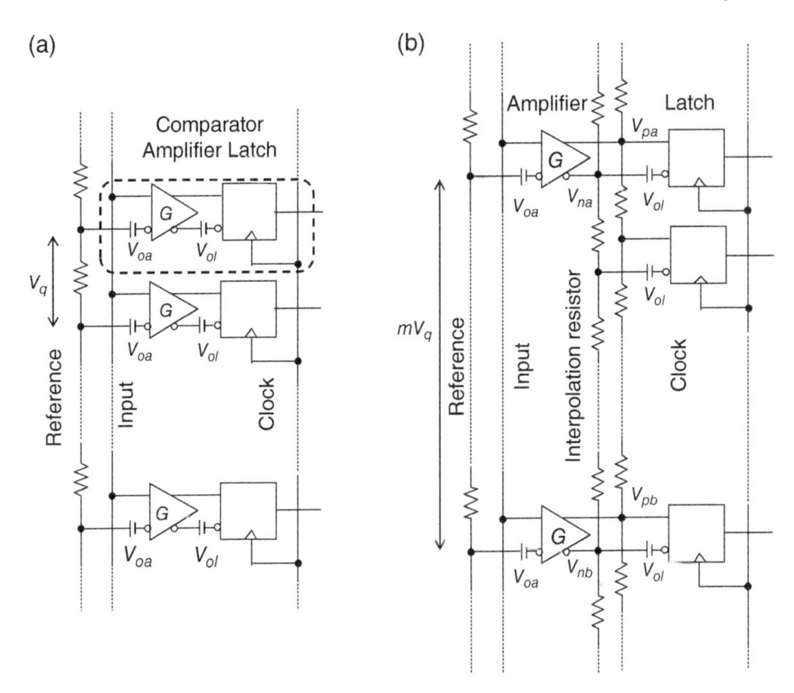

Figure 30.18 (a) Flash ADC. (b) Flash ADC using interpolation technique.

determination in synchronization with the clock is used. However, since the latch circuit is a digital circuit and the voltage distribution of the offset voltage V_{ol} is large, an amplifier is often used in the front stage of the latch. When the gain of the amplifier is G, the voltage distribution of the offset voltage V_{ol} decreases to V_{ol}/G at the input terminal. However, since the voltage distribution of the offset voltage V_{oa} exists in the amplifier, the linearity of ADC deteriorates by this voltage distribution. To obtain a DNL below 1/4 least significant bit (LSB) for the quantization voltage V_{q}, V_{oa} of $V_q/6$ or less is required. For the 10-bit flash ADC at a full-scale voltage of 1.0V, V_{oa} requires a standard deviation of 160 µV or less and this value is challenging for conventional devices.

To solve this problem, Figure 30.18b was configured [4], which reduced the number of amplifiers and increased the reference voltage between the amplifiers to mV_q. The output voltage of the amplifier is connected by a circuit in which multiple resistors are connected in series, and the A/D conversion value is obtained by comparing the tap voltage of the resistor with a latch. Figure 30.19 shows the output voltages of the amplifiers and the voltages of the resistance tap for the input voltage. The voltage of the resistor tap is the voltage that interpolates the output voltage of a pair of amplifiers. The polarity of the output voltage of a pair of amplifiers is crossed by the reference voltages V_{ra} and V_{rb} of the amplifier, and the interpolation voltages V_{ra} and V_{rb} cross at the input voltage that is divided equally between them. Then A/D conversion is possible in the configuration of Figure 30.18b.

The effect of each offset voltage on the DNL V_{oe} is expressed below. Here, m is the reference voltage difference between amplifiers normalized by the quantization voltage V_q.

$$V_{oe} = \left\{ \left(\frac{V_{oa}}{m} \right)^2 + \left(\frac{V_{ol}}{G} \right)^2 \right\}^{0.5} \tag{30.6}$$

Therefore, the voltage distribution of the offset voltage V_{ol} of the latch remains as V_{ol}/G at the input terminal, but the effect of the voltage distribution of the offset voltage V_{oa} of the amplifier on the DNL

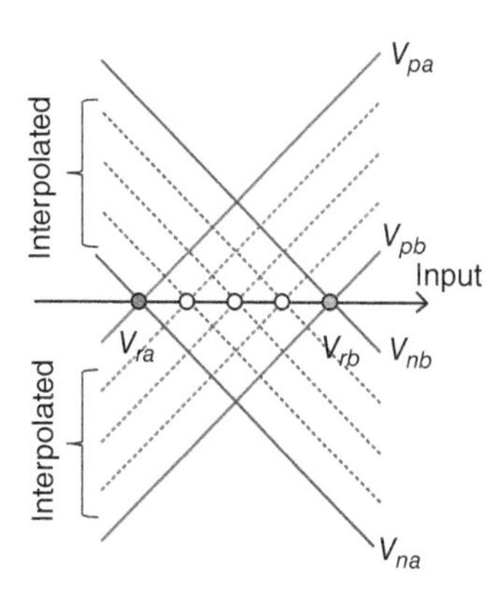

Figure 30.19 Amplifier output voltages and interpolated voltages for input voltage.

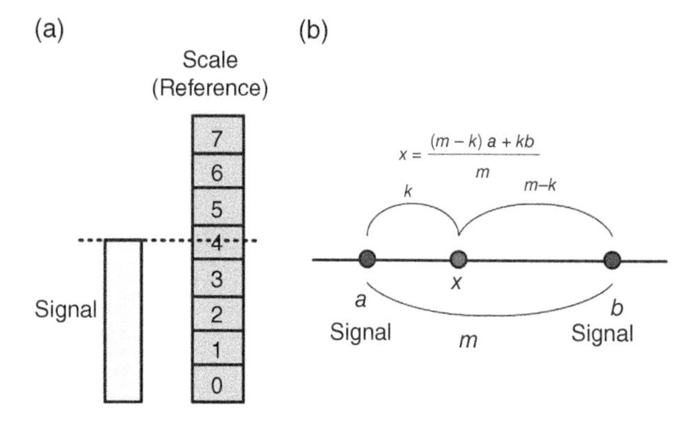

Figure 30.20 A/D conversion; conventional and using interpolation technique. (a) Conventional A/D conversion. (b) A/D conversion using interpolation techniques.

is reduced by $1/m$. Therefore, by using this method, the effect of the offset voltage V_{oa} of the amplifier on the DNL can be suppressed, so that high-precision flash ADC can be realized even with fine transistors that have a large offset voltage distribution. Not only the number of amplifiers with large power consumption can be reduced to $1/m$, but also the use of fine transistors is possible, resulting in significantly lower power consumption. The ADC using this method is 10-bit 300 MS s^{-1} ADC [4] using bipolar transistors.

The conventional A/D conversion is performed by comparing the strength of one signal with the reference voltage as shown in Figure 30.20a.

On the other hand, A/D conversion using interpolation is performed by dividing the two signals a and b equally into m pieces (Figure 30.20b). The intermediate voltage x between signal a and signal b based on the principle of internal division is

$$x = \frac{(m-k)a + kb}{m} \tag{30.7}$$

where m is the number of divisions and k is the weight. That is, the intermediate voltage x is obtained by multiplying the weight of $(m - k)$ on the signal a, adding the weight k to the signal b, and normalizing it with m.

References

[1] Takemoto, T., Inoue, M., Sadamatsu, H. et al. (1982). A fully parallel 10-Bit A/D converter with video speed. *IEEE Journal of Solid-State Circuits* SC-17 (6): 1133–1138.

[2] Inoue, M., Sadamatsu, H., Matsuzawa, A. et al. (1984). A monolithic 8-bit A/D converter with 120 MHz conversion rate. *IEEE Journal of Solid-State Circuits* SC-19 (6): 837–841.

[3] Matsuzawa, A., Kagawa, M., Kanoh, M. et al. (1990). A 10 b 30 MHz two-step parallel Bi-CMOS ADC with internal S/H. *1990 37th IEEE International Solid-State Circuits Conference*, San Francisco, CA, USA (14–16 February 1990), pp. 162–163. https://doi.org/10.1109/ISSCC.1990.110177.

[4] Kimura, H., Matsuzawa, A., Nakamura, T., and Sawada, S. (1993). A 10-b 300-MHz Interpolated-parallel A/D converter. *IEEE Journal of Solid-State Circuits* 28 (4): 438–446.

[5] Kusumoto, K., Matsuzawa, A., and Murata, K. (1993). A 10-b 20-MHz 30 mW pipelined interpolating CMOS ADC. *IEEE Journal of Solid-State Circuits* 28 (12): 1200–1206.

[6] Matsuzawa, A. (1993). Low-voltage and low-power circuit design for mixed analog/digital systems in portable equipment. *IEEE Journal of Solid-State Circuits* 29 (4): 470–480.

[7] Matsuzawa, A. (1993). Low voltage mixed analog/digital circuit design for portable equipment. *1993 Symposium on VLSI Circuits*, Kyoto, Japan (19–21 May 1993), pp. 49–54. https://doi.org/10.1109/VLSIC.1993.920534.

[8] Matsuzawa, A., Nakashima, S., Hidaka, I. et al. (1991). A 6b 1GHz dual-parallel A/D converter. *1991 IEEE International Solid-State Circuits Conference*. Digest of Technical Papers, San Francisco, CA, USA (13–15 February 1991), pp. 174–178. https://doi.org/10.1109/ISSCC.1991.689115.

[9] Sushihara, K., Kimura, H., Okamoto, Y. et al. (2000). A 6 b 800 M Sample/s CMOS A/D converter. *IEEE International Solid-State Circuits Conference*. Digest of Technical Papers (Cat. No.00CH37056), San Francisco, CA, USA (9 February 2000), pp. 428–429. https://doi.org/10.1109/ISSCC.2000.839845.

[10] Sushihara, K. and Matsuzawa, A. (2002). A 7b 450MSPS 50mW CMOS ADC in 0.3 mm^2. *IEEE International Solid-State Circuits Conference*. Digest of Technical Papers (Cat. No.02CH37315), San Francisco, CA, USA (07 February 2002), 1: 170–171. https://doi.org/10.1109/ISSCC.2002.992990.

[11] Okamoto, K., Morie, T., Yamamoto, A. et al. (2003). A fully integrated 0.13-μm CMOS mixed-signal SoC for DVD player applications. *IEEE Journal of Solid-State Circuits* 38 (11): 1981–1991.

[12] Matsuzawa, A. (2003). Mixed signal SoC: a new technology driver in LSI industry. *ISCAS 2003*, Bangkok (May 2003).

[13] Matsuzawa, A. (2004). Mixed signal SoC era. *IEICE Transactions on Electronics* E87-C (6): 867–877.

Chapter 31

Historical Perspective of the Nonvolatile Memory and Emerging Computing Paradigms

Ming Liu

State Key Laboratory of Integrated Chip and Systems, Frontier Institute of Chip and System, Fudan University, Shanghai, China

31.1 Introduction

STORAGE equipment witnesses the development of human civilization. In the era when characters and words were not developed, knotted ropes and cave walls served as the very first storage equipment, on which information was preserved for thousands of years before and finally discovered by archaeologists. Then, compact storage equipment such as oracle bones and clay plates were adapted to record information. Specialized characters, now known as pictographs, are used to encode these information with much more details. With the invention of parchment and papers, the commonly used lightweight storage equipment enables the efficient cultural exchange breaking through spatial limits.

The storage equipment in electrical systems follows the same roadmap as in the history of civilization. In 1936, Alan Turing proposed an unprecedented machine, which operated on an infinite *memory tape* [1]. Since then, memory has been playing one of the most significant parts of computing systems. Electronic engineers are committed to developing memory devices with larger capacity, smaller size, and faster speed, as the storage equipment in human history. Today, we generally divide memory into two categories, volatile memory and nonvolatile memory (NVM), according to whether data can be maintained after the power is off. NVM is generally used for large-volume data storage, while volatile memory is more likely for high-speed data communication. In 1956, IBM introduced the hard disk drive (HDD) NVM, an electromechanical device that stores and retrieves data on rotating platters coated with magnetic material. HDD features extremely low cost but has the disadvantage of low read/write speed because its access order has to follow the physical location.

75th Anniversary of the Transistor, First Edition. Edited by Arokia Nathan, Samar K. Saha, and Ravi M. Todi.

With the ever-expanding requirement of memory speed and cell density, the transistor-based solid-state NVM, such as flash memory and emerging nonvolatile random access memory (RAM), gradually takes the place of HDD. In addition, new integration technologies, like three-dimensional stacking, can further improve the performance of NVMs. At the same time, data explosion drives the communication between memory and processor, a major challenge in today's computers. New computing paradigms are revolutionizing the classical architecture of memory-processing separate systems.

This paper reviews the development of NVM in the past 70 years and looks forward to how solid-state NVM will promote emerging computing paradigms in the future. The remainder of this paper is organized as follows. In Section 31.2, the development history of the most common solid-state NVM is reviewed. How these NVM are applied to today's computer architecture is discussed in Section 31.3, where memory hierarchy and memory walls are emphasized. The exploration of new computing paradigms based on emerging embedded NVM is described in Section 31.4. Finally, Section 31.5 concludes this article.

31.2 Rise of Solid-State Nonvolatile Memory

The history of NVM technology can be traced back to 1952, when Dudley Buck at MIT created the first semiconductor NVM with ferroelectric crystals [2]. Afterward, Dawon Kahng and Simon Sze in Bell Laboratories first proposed floating-gate devices to realize NVM in 1967 [3], based on which electrical programmable read-only memory (EPROM) was invented in 1971. However, these NVM trailblazers were not compact and fast enough, restricted to niche applications before the 1980s. Then comes the major progresses of NVMs, such as flash memory, resistive random access memory (ReRAM), magnetic random access memory (MRAM), phase-change random access memory (PCRAM), ferroelectric field effect transistor (FeFET) and random access memory (FeRAM), which will be detailed as follows (Figure 31.1 and Table 31.1).

31.2.1 Flash Memory

After the floating-gate devices were invented in Bell Lab, Eli Harari invented a new type of thin-oxide-based floating-gate device, achieving the first electrically erasable PROM (EEPROM) cell, eliminating the need for external ultraviolet erasure [4]. In 1980, Fujio Masuoka from Toshiba modified the EEPROM architecture to increase the speed of programming and allow entire memory erasure in a single shot. It was coined as "flash" memory because the rapid erasure process of memory contents is like a flash of the camera. Toshiba announced the first NOR-based flash [5] and NAND-based [6] flash in IEDM 1984 and 1988, respectively. In 1988, Intel launched the first commercial flash memory chip, focusing on computer storage software, which successfully replaced EPROM products. In 1991, Toshiba launched a 4 Mb NAND flash chip. Afterward, flash memory usage grew rapidly and gradually applied to all types of electronic equipment, including cameras, laptops, and mobile phones.

Recently, flash memory has occupied most of the solid-state NVM market due to its high storage density and low cost. However, with process scaling, the development of flash memory is confronted

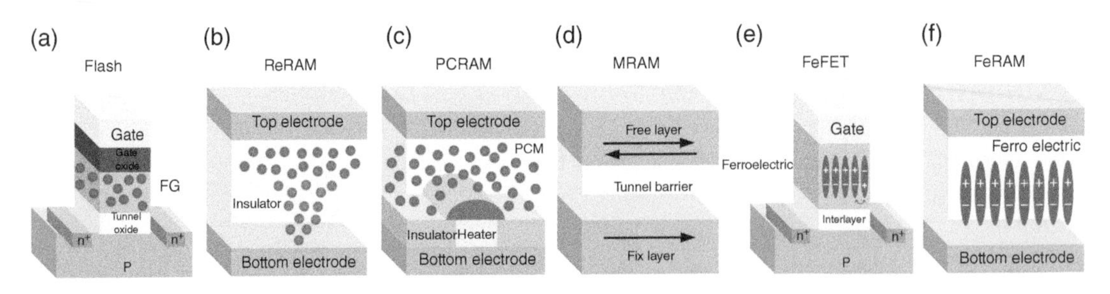

Figure 31.1 Common solid-state NVMs: (a) Flash, (b) ReRAM, (c) MRAM, (d) PCRAM, (e) FeFET, and (f) FeRAM.

Table 31.1 Performance comparison among different NVMs.

	FLASH	MRAM	ReRAM	PCRAM	FeFET	FeRAM
Cell structure	1T	1T1MTJ	1T1R	1T1R	1T	1T1C
Multi-bit	3–4	1	2–3	2–3	2–3	1
Write voltage	~10 V	<1.5 V	<3 V	<3 V	<4 V	<2 V
Write energy	~100 pJ	~1 pJ	~1 pJ	~10 pJ	~0.1 pJ	~0.1 pJ
Write speed	0.1–1 ms	~5 ns	~10 ns	10 ns	~10 ns	<10 ns
Read speed	~10 ns	~5 ns	~10 ns	10 ns	~10 ns	<10 ns
Endurance	10^4–10^6	10^{15}	$>10^7$	$>10^{12}$	$>10^5$	10^9-10^{14}
Latest tech node	40 nm	22 nm	14 nm	22 nm	10 nm	130 nm
Retention	>10 y	>10 y	>10 y	>10 y	>1 y	>10 y

by new challenges, such as lower operation voltage in advanced technology and more complicated process matches. At present, the most serious challenge is how to trade off the performance and reliability of flash memory for sub-40 nm technology. Due to the smaller size, the reduction of the number of electrons in floating gates leads to a smaller variability of the cell current. Therefore, the operating frequency is becoming harder to meet the development requirements of various computing. As for the reliability, with process scaling and increasing number of layers, the trapped charge shift, cross-temperature phenomenon, and layer-to-layer variability become worse, which impacts the endurance and retention of flash memory. An alternative to increase cell density without scaling down is 3D stacking. In 2012, Samsung launched the first 32-layer Vertical-NAND solid-state flash chip, which was the first generation of the 3D NAND flash memory chip. Today, the stacking layer number has approached 176, achieving the storage density of 14.8 Gb mm^{-2} [7].

31.2.2 Resistive Random Access Memory

In contrast with the charge-based floating-gate devices, ReRAM stores data by utilizing the resistive switching (RS) phenomenon of dielectric materials, offering a new storage mechanism and device that can replace traditional memory. The RS phenomena were first reported by Hickmott in 1962, when he discovered a hysteretic resistance change in Al/Al_2O_3/Al metal-oxide-metal structures by applying voltage pulses [8]. However, this RS effect has not gained much attention at that time. In 2000, the electrical-pulse-induced reversible resistance change effect was discovered by Liu et al. in the colossal magnetoresistive (CMR) thin-film materials at room temperature [9]. Following that, Zhuang et al. from Sharp Lab fabricated the 1T1R CMR-based device and a 64-bit test memory array [10]. Their demonstration proved that resistive memory could also work with the reasonable current and voltage, which attracted interest in the RRAM study. In 2004, Baek et al. proposed their resistive memory device called OxRRAM, which was built in transition metal oxide (TMO). This design demonstrated RRAM's high compatibility with CMOS technology and its high performance when used as NVM [11].

After that, ReRAM began to be regarded as a potential NVM candidate in addition to flash memory for its low switching latency, high density, low power dissipation, and high CMOS compatibility. Furthermore, the key advantage of ReRAM is that it can be fabricated after the MOS transistor manufacturing. In other words, only back-end-of-line (BEOL) process is involved, decoupling with transistor scaling and providing larger integration flexibility. The OxRRAM device was first integrated based on CMOS 180 nm technology [10]. During 2017–2020, TSMC, IMECAS, and Intel successfully implemented their ReRAM designs at 40, 28, and 22 nm technology, respectively [12–14]. In 2021, Yang et al. from IMECAS first applied RRAM to the latest 14 nm FinFET process, which further verified the feasibility of ReRAM application in the advanced technology nodes [15].

Recently, ReRAM designs have been approaching industrial-scale production. The aforementioned 1 Mb embedded 1T1R RRAM on 14 nm-FinFET technology [15] achieves the smallest published bit-cell size of 0.022 μm^2 and features fast access speed and high reliability. In 2020, Chou et al. from TSMC also demonstrates a 22 nm-based 96K × 144 ReRAM macro with a stable 10 ns access time after 10k endurance test and wide operation voltage range, which proves the possibility of ReRAM mass production in close future [16]. System-on-Chips (SoC) with embedded RRAM are regarded as a main technical path, when flash memories are hard to be integrated on advanced technologies.

31.2.3 Magnetoresistive Random Access Memory

MRAM is another type of NVM that stores data by utilizing the polarized magnetic field. The first magnetic memory was constructed by Arthur Pohm and Jim Daughton at Honeywell in 1984. One MRAM cell contains two ferromagnetic plates, each of which can hold a magnetization, separated by a thin insulating layer. One plate is set to a particular polarity permanently, while the other plate's magnetization can be changed to match that of an external field to store memory, known as a magnetic tunnel junction. Freescale began to sell 256 Kb and 1 Mb MRAM chips in 2001 and 2002, and Freescale became Everspin in 2008, which still offers mounts of MRAM products. These MRAM devices is realized using toggle memory switching (Toggle MRAM), in which a magnetic field is used to change the electron spin While this type of MRAM is easier for implementation, further application development has been limited by high operation current and large cell size.

Recently, Spin Transfer Torque MRAM (STT-MRAM) is widely adopted. STT-MRAM is an advanced type of magnetoresistive random access memory that uses the magnetism of electron spin to provide nonvolatile properties. Compared with Toggle MRAMs, the large current required to generate the magnetic field during write operation is eliminated. Thereby reducing the power consumption and avoiding the write disturb problem caused by magnetic field induction. STT-MRAM is highly scalable, enabling higher density memory products and also features fast switching speed and high endurance. Samsung and Intel announced they embedded STT-MRAM under 28/22 nm technology in 2018 [17]. In the future, Spin-orbit Torque MRAM (SOT-MRAM) has regarded as a more advanced type of MRAM. There is a good chance that SOT will outperform STT in terms of speed and power consumption. TSMC had demonstrated an 8 Kb SOT-MRAM with high field-free switching speed (1 ns), and low switching voltage (1.5V) [18], which shows the potential to be the energy-efficient high-speed memory application in advanced technologies.

31.2.4 Phase-Change Random Access Memory

PCRAM stores the data through the variable states of chalcogenide materials. The research of PCRAM can be dated back to the 1960s when discovering that chalcogenides can be used to store data [19]. In 2003, the Ovonyx-Intel demonstrated a 4Mb PCRAM chip. The first mass production of PCRAM was a commercial 128 Mb PCRAM chip by Samsung in 2008. PCRAM features higher density than DRAM and is faster than NAND flash; therefore, regarded as a good intermediate before DRAM and SSD in servers. In 2015, Intel and Micron announced the 3D Cross-Point PCRAM using an amorphous selector and initiated a new market under the brand name Optane (Intel) in April 2017. Unfortunately, the Intel-Micron joint development deal was terminated five years later due to business and marketing issues.

PCRAM has the advantages of high on/off ratio, good process compatibility, and scalability. Today, PCRAM is restricted by two challenges. First, PCRAM requires large programming currents to generate high temperatures and change the crystalline state. Second, the resistance value and threshold voltage would drift with time, which affects the reliability of the device and the expansion of the multi-value.

31.2.5 Ferroelectric Field-Effect Transistors and Random Access Memory

The idea of Ferroelectric random access memory (FeRAM) was first proposed by Dudley Allen Buck in 1952 [2]. The FeRAM development began in the 1980s with the semiconductor technology reaching a certain level, and it reached the market in the 1990s and developed fast later. In 2000, Sony used the FeRAM in its commercial product. However, conventional ferroelectrics suffer from various problems such as poor CMOS-compatibility and large physical thickness, which hindered its application at advanced nodes. The discovery of ferroelectricity in doped-HfO_2 in 2011 started a research wave on ferroelectric devices. It has been confirmed that ferroelectric HfO_2 can maintain its planar Pr value when integrated as a 10 nm thin layer into an array of 1.6 μm deep trenches with an aspect ratio of 13 : 1 [20], which indicates that it can achieve storage density similar to that of advanced node DRAM. The density of 1T1C FeRAM is limited by the capacitor and a restoring pulse is required after a destructive readout. FeFET can realize nondestructive read and 3D vertical stack for higher density. In 2016, FeFET was firstly embedded in the 28 nm CMOS process [21]. In 2017, GlobalFoundries has demonstrated a 32 MB FeFET memory based on the 22 nm process with 10 ns writing latency and 1.5 V memory window. Luo et al. from IMECAS manufactured FinFET-based FeFET in the 10 nm node with large on/off current ratio ($>10^4$) [22]. Ferroelectric memories have the advantages of high speed and low writing energy consumption. FeFET even has the potential for high-density memory application such as 3D-NAND [23]. The reliability of FeRAM with 1T1C structure is very close to commercial [24]. The main challenges of FeFET are endurance, retention, and large operating voltage in highly scaled devices.

31.3 NVM in Classical Computer Architectures

In the Turing machine, NVM was only a tape containing dedicated symbols. As the frequency of processors continues to rise, the mechanical-based access method became the weakness of the computing system. Therefore, modern computer architectures have evolved to reduce performance degradation due to slow NVM access. This section reviews the evolution of computing architectures and how NVM interact with them.

31.3.1 Computers Using Von Neumann Architectures

In 1944, Dr. Von Neumann joined the EDVAC project, which provided computing power for the Manhattan Project. In 1945, on the basis of communication and discussion among colleagues, he drafted a 101-page summary report under the title "First Draft of a Report on the EDVAC" [25]. In the report, a complete electronic computer should contain the following five components: A processing unit with both an arithmetic logic unit and processor registers, a control unit that includes an instruction register and a program counter, a memory that stores data and instructions, input device, and output device. The first Von Neumann computer was built by the University of Manchester in 1948, known as Manchester Mark I. However, the high cost and low reliability of electron tubes and mercury delay line memories are annoying. After solid-state transistors were invented in Bell Lab, IBM replaced all the electronic tubes of the IBM 709 computer with transistors in 1959, which became the IBM 7090. The speed was six times faster, and the price was only half. As a result, a new era of modern computers was born. At this stage, the main form of NVM was punched tape and cards.

In the early days of computer technology, memory access was only slightly slower than register access in processors. However, since the 1980s, the performance gap between processors and HDD memory has been widening. The limited throughput between the central processing unit (CPU) and memory leads to a communication bottleneck. As the growth rate of CPU speed and memory size is far faster than the throughput between them, the bottleneck has become a more serious problem. Today, we call this problem *memory wall* [26] or Von Neumann bottleneck.

31.3.2 Memory Hierarchy

Ideally, the access speed of all memories is desired to be as fast as the CPU. However, it is mission impossible for NVMs. In practice, a more economical and feasible way is likely employed, using large capacity low-speed NVM as main storage equipment and introducing multilevel volatile memory to buffer data, thus alleviating the performance gap. Such multilevel architecture is now known as memory hierarchy, from the smallest registers to the largest cloud storage, illustrated in Figure 31.2. The following sections briefly review the development of the cache as the representative of the memory hierarchy.

The concept of cache was first proposed by Wilkes of Cambridge University in 1965 [27]. Cache was called slave memory at that time. After this paper, IBM launched the IBM System/360 Model 85 in January 1968, which was the first commercial computer to use cache. The cache of IBM 360/85 is a 16 KB integrated storage capable of operating every processor cycle. Cache appears relatively late on personal computers. It was until 80386 that the CPU's performance could not be satisfied by the DRAM access speed at that time. In order to speed up memory access, chipsets were used to support cache, and the earliest cache was not placed on the CPU module but on the motherboard. An 8 KB on-chip cache was first integrated into Intel's 80486 CPU, and the previous cache on the motherboard became a level-2 (L2) cache. Since the 1990s, the concept of multilevel cache has become popular as a new and better cache model. Intel's Pentium 4 integrates an L2 cache, Itanium integrates an L3 cache for the first time, and Broadwell architecture even places an embedded DRAM on-chip, which was regarded as an L4 cache.

Recently, the most common local storage is NAND-flash-based SSD. As a result, emerging NVMs can be extra level cache to fill the gaps in Figure 31.2, as long as they are faster than NAND flash and denser than SRAM. As mentioned above, PCRAM-based Intel's OPTANE memory was such a practice. It fills the gap between high-performance DRAM and lower-speed but affordable NAND flash. Cloud servers often use the OPTANE memory to handle *working data*, large-size data that needs to be close to the CPU for rapid access. Alzate et al. from Intel also proposed an STT-MRAM-based L4 cache to replace SRAM [28]. The embedded NVM can achieve 20 ns/4 ns write/read time and endurance of 10^{12} cycles, satisfying the on-chip L4 cache requirement with higher density and lower power consumption than SRAM.

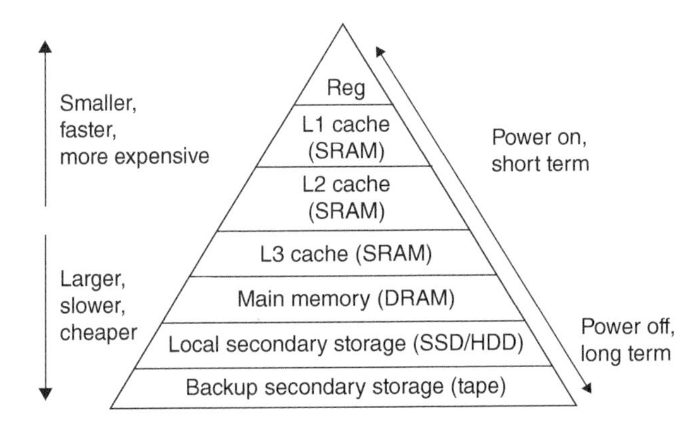

Figure 31.2 Multilevel memory hierarchy structure for modern computers.

31.4 NVM-Driven New Computing Paradigm

Excepting from the relentless pursuit of inserting more memory hierarchy levels to bridge the performance gap between memory and processor, new computing paradigms are being intensively studied to reduce the data movement in between. Computing-in-memory (CIM) is a representative scheme that directly performs logic or arithmetic operations between the input and stored data inside the memory array, as shown in Figure 31.3. One of the earliest demonstrations of CIM can be traced back to 1988. Howard from Bell Laboratories presented their work on processing perceptron models using a 12×12 cross-bar array based on thin film resistors with predefined fixed conductance [29]. It demonstrates the promise of CIM in terms of computing parallelism, energy- and area-efficiency, yet suffers from poor flexibility due to the non-programmable cell conductance and limited functions because of the small scale. Meanwhile, digital processors have been prevailing because of the rapid development of VLSI design methodology and the continuous performance gain brought by Moore's Law.

Recently, the CIM paradigm has redrawn intensive research attention and become the innovation frontier ranging from semiconductor sciences to computer architectures. We speculate the underlying reasons as follows. Firstly, the performance gain due to scaling down gradually gets saturated. Secondly, the explosive growth of data-centric computing tasks further underscores the memory wall issues. Lastly, the rapid development of emerging memory technologies has offered much more competent platforms for CIM design than in the past.

Flash-based CIM has been investigated for decades. In 1996, C.A. Mead et al. [30] demonstrated the concept using a 2×2 array of floating-gate devices for low power and parallel subthreshold analog computing. Since then, it has attracted increasing research attention to achieve CIM chips based on commodity flash memory, which are already productized nowadays. Recently, L. Fick et al. [31] from Mythic have demonstrated NOR flash-based CIM accelerators having 80M flash cell pairs. It shows one order of magnitude improvement in energy efficiency compared to the mainstream parallel computing hardware. Besides, researchers from Macronix [32] have demonstrated 512Gb CIM based on 3D NAND flash memory, which reveals the possibility of computing in large-scale SSD.

Emerging memory-based CIM has been drawing continually growing attention in recent years. Because of its simple process and cost-effectiveness, ReRAM has become one of the most used platforms to implement CIM chips. In 2005, D.B. Strukov et al. [33] demonstrated a 12×12 ReRAM cross-bar array for CIM to process perceptron models. Since then, multilayer fully connected, convolution, and recurrent neural networks have been demonstrated with Kb-level ReRAM arrays [34]. Moreover, based on foundry-developed ReRAM technology, several ReRAM CIM macros [35] and systems [36]

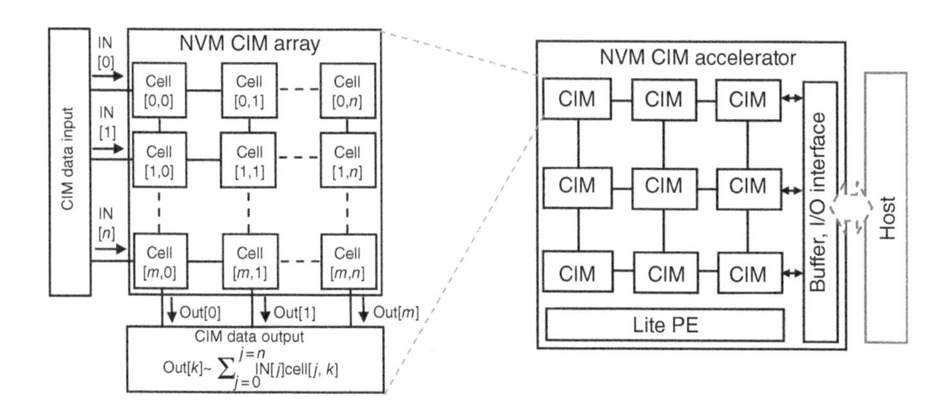

Figure 31.3 NVM-based computing-in-memory paradigm, where tensor computing circuits are redistributed inside CIM macros. The complete accelerator/processor architecture contains multiple CIM macros.

integrating Mb-level ReRAM cells have been demonstrated for deep neural networks processing. More recently, an 8-layer 3D ReRAM CIM chip has been demonstrated [37], indicating its promise of extending scale as well as density.

Except for ReRAM, other types of emerging memory are also under intensive studying. Researchers for IBM [38, 39] carry out a systematical study on PCM-based CIMs from array-to chip-level demonstration for linear equations solving and DNN processing stressing the multilevel cell (MLC) properties of PCRAM. On the other hand, because of the small memory window, MRAM is usually proposed for near-memory computing. More recently, researchers from Samsung [40] verified the possibility of realizing MRAM CIM using a customized array and computing approach.

To date, CIM has been almost verified in all types of commercial or emerging NVMs, and hence enables to perform computing in different layers of memory hierarchy from cache, embedded, and persistent to storage memory. It, therefore, provides more opportunities for design-technology cooptimization, architecture- and application-level explorations to overcome the Von Neumann bottleneck than ever before.

31.5 Conclusion

This paper reviews and discusses the emergence and development of NVM and the research of emerging computing paradigm from a historical perspective. We discussed the development and evolution of five solid-state NVM driven by demand and technology. Then, we review the development of memory hierarchy computer architecture and how NVM is applied in these architectures. Finally, we discuss the emerging CIM paradigm, especially in NVM, where the data movement and unnecessary data copy in caches can be eliminated.

References

[1] Turing, A.M. (1936). On computable numbers, with an application to the entscheidungsproblem. *Journal of Mathematics* 58 (345–363): 5.

[2] Buck, D.A. (1952). *Ferroelectrics for Digital Information Storage and Switching*. Massachusetts Inst Of Tech Cambridge Digital Computer Lab, Tech. Rep.

[3] Kahng, D. and Sze, S.M. (1967). A floating gate and its application to memory devices. *The Bell System Technical Journal* 46 (6): 1288–1295.

[4] Harari, E. (1978). Electrically erasable non-volatile semiconductor memory. US Patent 4,115,914.

[5] Masuoka, F., Asano, M., Iwahashi, H. et al. (1984). A new flash e^{2}prom cell using triple polysilicon technology. *International Electron Devices Meeting (IEDM)*, San Francisco, CA (9–12 December 1984), pp. 464–467. IEEE.

[6] Masuoka, F., Momodomi, M., Iwata, Y. et al. (1987). New ultra high density eprom and flash eeprom with nand structure cell. *International Electron Devices Meeting (IEDM)*, Washington, DC (6–9 December 1987), pp. 552–555. IEEE.

[7] Cho, W., Jung, J., Kim, J. et al. (2022). A 1-tb, 4b/cell, 176-stacked-wl 3d-nand flash memory with improved read latency and a 14.8gb/mm^2 density. *IEEE International Solid- State Circuits Conference (ISSCC)*, San Francisco, CA (20–26 February), 65: 134–135. IEEE.

[8] Hickmott, T. (1962). Low-frequency negative resistance in thin anodic oxide films. *Journal of Applied Physics* 33 (9): 2669–2682.

[9] Liu, S., Wu, N., and Ignatiev, A. (2000). Electric-pulse-induced reversible resistance change effect in magnetoresistive films. *Applied Physics Letters* 76 (19): 2749–2751.

[10] Zhuang, W., Pan, W., Ulrich, B.D. et al. (2002). Novel colossal magnetoresistive thin film nonvolatile resistance random access memory (rram). *Digest. of IEEE International Electron Devices Meeting(IEDM)*, San Francisco, CA (8–11 December 2002), pp. 193–196. IEEE.

[11] Baek, I., Lee, M.S., Seo, S. et al. (2004). Highly scalable nonvolatile resistive memory using simple binary oxide driven by asymmetric unipolar voltage pulses. *Digest. of IEEE International Electron Devices Meeting (IEDM)*, San Francisco, CA (13–15 December 2004), pp. 587–590. IEEE.

[12] Chou, C.-C., Lin, Z.-J., Tseng, P.-L. et al. (2018). An n40 256k×44 embedded rram macro with sl-precharge sa and low-voltage current limiter to improve read and write performance. *IEEE International Solid-State Circuits Conference (ISSCC)*, San Francisco, CA (11–15 December 2018), pp. 478–480. IEEE.

[13] Lv, H., Xu, X., Yuan, P. et al. (2017). Beol based rram with one extra-mask for low cost, highly reliable embedded application in 28 nm node and beyond. *IEEE International Electron Devices Meeting (IEDM)*, San Francisco, CA (2–6 December 2017), pp. 2.4.1–2.4.4. IEEE.

[14] Jain, P., Arslan, U., Sekhar, M. et al. (2019). A 3.6mb 10.1mb/mm2 embedded non-volatile reram macro in 22nm finfet technology with adaptive forming/set/reset schemes yielding down to 0.5v with sensing time of 5ns at 0.7v. *IEEE International Solid-State Circuits Conference (ISSCC)*, San Francisco, CA (17–21 February 2019), pp. 212–214. IEEE.

[15] Yang, J., Xue, X., Xu, X. et al. (2021). 24.2 a 14nm-finfet 1mb embedded 1t1r rram with a 0.022μm2 cell size using self-adaptive delayed termination and multi-cell reference. *IEEE International Solid- State Circuits Conference (ISSCC)*, San Francisco, CA (13–22 February), 64: 336–338. IEEE.

[16] Chou, C.-C., Lin, Z.J., Lai, C.A. et al. (2020). A 22nm 96kx144 rram macro with a self-tracking reference and a low ripple charge pump to achieve a configurable read window and a wide operating voltage range. *IEEE Symposium on VLSI Circuits*, Honolulu, HI (16–19 June 2020), pp. 1–2. IEEE.

[17] Wei, L., Alzate, J.G., Arslan, U. et al. (2019). A 7mb stt-mram in 22ffl finfet technology with 4ns read sensing time at 0.9v using write-verify-write scheme and offset-cancellation sensing technique. *IEEE International Solid- State Circuits Conference – (ISSCC)*, San Francisco, CA (17–21 February 2019), pp. 214–216. IEEE.

[18] Song, M.Y., Lee, C.M., Yang, S.Y. et al. (2022). High speed (1ns) and low voltage (1.5v) demonstration of 8kb sot-mram array. *IEEE Symposium on VLSI Technology and Circuits (VLSI Technology and Circuits)*, Honolulu, HI (12–17 June 2022), pp. 377–378. IEEE.

[19] Ovshinsky, S.R. (1968). Reversible electrical switching phenomena in disordered structures. *Physical Review Letters* 21 (20): 1450.

[20] Müller, J., Böscke, T.S., Müller, S. et al. (2013). Ferroelectric hafnium oxide: a cmos-compatible and highly scalable approach to future ferroelectric memories. *Digest of IEEE International Electron Devices Meeting (IEDM)*, Washington, DC (9–11 December 2013), pp. 10.8.1–10.8.4. IEEE.

[21] Trentzsch, M., Flachowsky, S., Richter, Ret al. (2016). A 28nm hkmg super low power embedded nvm technology based on ferroelectric fets. *IEEE International Electron Devices Meeting (IEDM)*, San Francisco, CA (3–7 December 2016), pp. 11.5.1–11.5.4. IEEE.

[22] Luo, Q., Gong, T., Cheng, Y. et al. (2018). Hybrid 1t e-dram and e-nvm realized in one 10 nm node ferro finfet device with charge trapping and domain switching effects. *IEEE International Electron Devices Meeting (IEDM)*, San Francisco, CA (1–5 December 2018), pp. 2.6.1–2.6.4. IEEE.

[23] Florent, K., Lavizzari, S., Di Piazza, L. et al. (2017). First demonstration of vertically stacked ferroelectric al doped hfo_2 devices for nand applications. *Symposium on VLSI Technology*, Kyoto (5–8 June 2017), pp. T158–T159. IEEE.

[24] Okuno, J., Kunihiro, T., Konishi, K. et al. (2020). Soc compatible 1t1c feram memory array based on ferroelectric $hf_{0.5}zr_{0.5}o_{2.}$ *IEEE Symposium on VLSI Technology*, Honolulu, HI (16–19 June 2020), pp. 1–2. IEEE.

[25] Von Neumann, J. (1993). First draft of a report on the edvac. *IEEE Annals of the History of Computing* 15 (4): 27–75.

[26] Wulf, W.A. and McKee, S.A. (1995). Hitting the memory wall: implications of the obvious. *ACM SIGARCH Computer Architecture News* 23 (1): 20–24. https://doi.org/10.1145/216585.216588.

[27] Wilkes, M.V. (1965). Slave memories and dynamic storage allocation. *IEEE Transactions on Electronic Computers* EC-14 (2): 270–271.

[28] Alzate, J.G., Arslan, U., Bai, P. et al. (2019). 2 mb array-level demonstration of stt-mram process and performance towards l4 cache applications. *IEEE International Electron Devices Meeting (IEDM)*, San Francisco, CA (7–11 December 2019), pp. 2.4.1–2.4.4. IEEE.

[29] Howard, R.E., Jackel, L.D., and Graf, H.P. (1988). Electronic neural networks. *AT&T Technical Journal* 67 (1): 58–64.

[30] Diorio, C., Hasler, P., Minch, A., and Mead, C. (1996). A single-transistor silicon synapse. *IEEE Transactions on Electron Devices* 43 (11): 1972–1980.

[31] Fick, L., Skrzyniarz, S., Parikh, M. et al. (2022). Analog matrix processor for edge ai real-time video analytics. *2022 IEEE International Solid-State Circuits Conference (ISSCC)*, San Francisco, CA (20–26 February), 65: 260–262. IEEE.

[32] Hu, H.-W., Wang, W.C., Chen, C.K. et al. (2022). A 512gb in-memory-computing 3d-nand flash supporting similar-vector-matching operations on edge-ai devices. *IEEE International Solid-State Circuits Conference (ISSCC)*, San Francisco, CA (20–26 February), 65: 138–140. IEEE.

[33] Prezioso, M., Merrikh-Bayat, F., Hoskins, B.D. et al. (2015). Training and operation of an integrated neuromorphic network based on metal-oxide memristors. *Nature* 521 (7550): 61–64.

[34] Wang, Z., Wu, H., Burr, G.W. et al. (2020). Resistive switching materials for information processing. *Nature Reviews Materials* 5 (3): 173–195.

[35] Chang, M.-F., Hung, J.M., Chen, P.C. et al. (2022). Reliable computing of reram based compute-in-memory circuits for ai edge devices. *Proceedings of IEEE/ACM International Conference on Computer-Aided Design (ICCAD)*, San Diego, CA (29 October 2022 – 3 November 2022), p. 6. IEEE.

[36] Yao, P., Wu, H., Gao, B. et al. (2020). Fully hardware-implemented memristor convolutional neural network. *Nature* 577 (7792): 641–646.

[37] Huo, Q., Yang, Y., Wang, Y. et al. (2022). A computing-in-memory macro based on three-dimensional resistive random-access memory. *Nature Electronics* 5 (7): 469–477.

[38] Ambrogio, S., Narayanan, P., Tsai, H. et al. (2018). Equivalent-accuracy accelerated neural-network training using analogue memory. *Nature* 558 (7708): 60–67.

[39] Khaddam Aljameh, R., Stanisavljevic, M., Mas, J.F. et al. (2022). Hermes-core – a 1.59-tops/mm^2 pcm on 14-nm cmos in-memory compute core using 300-ps/lsb linearized cco-based adcs. *IEEE Journal of Solid-State Circuits* 57 (4): 1027–1038.

[40] Jung, S., Lee, H., Myung, S. et al. (2022). A crossbar array of magnetoresistive memory devices for in-memory computing. *Nature* 601 (7892): 211–216.

Chapter 32

CMOS Enabling Quantum Computing

Edoardo Charbon

Faculty of Engineering, EPFL, Lausanne, Switzerland

32.1 Why Cryogenic Electronics?

Upon the introduction of integrated circuits in the late 1950s and of the microprocessor in the early 1970s, it became clear that integrated transistors would dominate the computing scene, being ideally suited to the binary logic at the core of all processing paradigms. In the later 1980s, neuromorphic computing based on analog versions of neurons briefly captured the attention of the academic and industrial world but we need to wait until the mid-2000s to see the emergence of neuromorphic computing, based however on digital logic, due to the power advantage of multicore processors over intrinsically analog circuits.

Quantum computing was born conceptually in the 1980s with the famous Feynman proposal of using quantum physics, in particular employing superposition, entanglement, and quantum interference, to compute [1, 2]. Relegated to the realm of theory for many years, Shor's prime-factorization algorithm being an example of it [3], quantum computing became feasible only with the introduction of solid-state quantum bits (qubits) and ion traps in the mid-2000s [4, 5]. However, it became obvious early on that the control of qubits, i.e. their interface to the classical world, required advanced circuits and systems. The first choice fell on multicore processors but it soon became clear that high levels of programmability were not necessary, while simple, massively-parallel processors could be preferable. The choice of field-programmable gate-arrays (FPGAs) became obvious, whereas operation at room temperature (RT) immediately posed a bottleneck to the expansion of quantum processors comprising large qubit arrays.

75th Anniversary of the Transistor, First Edition. Edited by Arokia Nathan, Samar K. Saha, and Ravi M. Todi.

A solution to this limitation was proposed in 2016 [6], where the control of qubits was to be performed at cryogenic temperature (CT), not far from millikelvin temperatures, which qubits needs to be operated at. While transmons are a very mature technology, paving the way to large qubit counts [7], spin qubits [8] are very interesting thanks to their most popular implementation that appears to be closer to modern CMOS transistors [9–11].

32.2 The Quantum Stack

The control of a qubit starts from a quantum algorithm, described in a certain high-level language that is compiled into quantum circuits, i.e. a sequence of operations to be applied on a set of qubits. Quantum arithmetic is also used in this context, along with compilation of certain functions into a quantum instruction set used for quantum execution (QEX) and error correction (QEC). Finally, the QEX and QEC are implemented using the quantum-classical interface, which operates directly on the qubits. This process is depicted in Figure 32.1 and is referred to as the quantum stack.

At the bottom of the quantum stack, the qubit controller is located in close proximity to the qubits and thus operating at CT. Qubit control is dominated by transistors and thus, the need emerged immediately to model cryogenic transistors and to design complex transistor-based systems. Nevertheless, the control of both types of qubits is surprisingly similar, involving single-sideband modulation of carriers with frequencies ranging from a few GHz to 20 GHz. Thus, a generic architecture soon emerged similar to that of a transceiver, as shown in Figure 32.2.

32.3 Modeling Cryo-CMOS Devices

CMOS technologies operating at deep-cryogenic temperatures are known collectively as Cryo-CMOS integrated circuits. The specifications for each component can be derived from the overall fidelity using a number of methods. Recently, simulation tools based on SPICE and Q-TIP have emerged [12, 13], which could be used to extract constraints on the input-referred noise of the low-noise amplifier (LNA), for instance, and on the other components of Figure 32.2, such as the carrier frequency inaccuracy and phase noise of the carrier generator, the pulse width of the arbitrary waveform generator (AWG), and the signal-to-distortion ratio of the mixers, to name a few.

Whether or not these specifications can be met is dependent on the performance of each single transistor and thus the selection of an appropriate CMOS technology is critical to the success of the approach. While, in principle most transistors can operate at CT, it is a question of their transconductance, maximum frequency, and on the overall model that will make it possible to meet the component

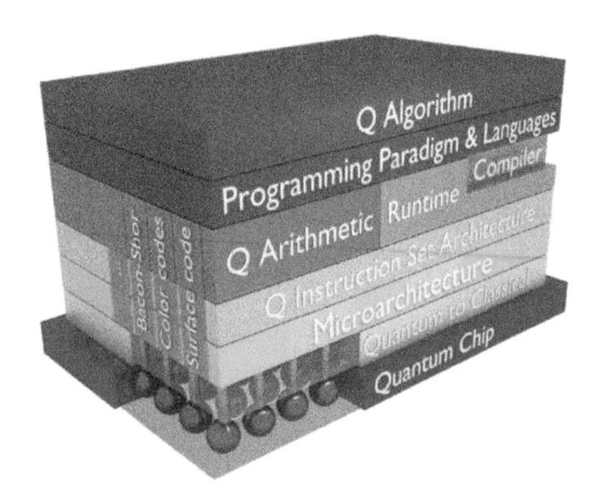

Figure 32.1 The quantum stack from the public talk in [6]. Source: Charbon et al. [6]/IEEE.

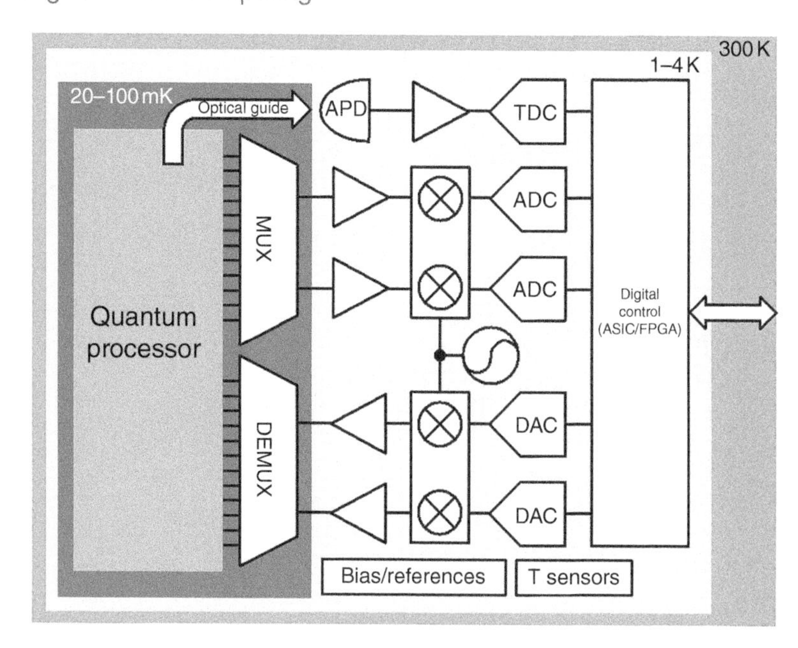

Figure 32.2 Generic architecture of quantum-classical interface for the control of qubits. Source: Charbon et al. [6]/IEEE.

specifications. Indeed, due to different levels of doping in the drain and source regions of MOSFETs, the channel may or may not achieve freeze-out conditions causing the transistor not to conduct at low enough temperature. For technologies with feature size below 0.5 µm, the condition of freeze-out is almost never reached at and around 3 K. However, non-idealities, such as a "kink" at high source–drain voltages may be observed. For instance, this behavior was observed in 160-nm bulk CMOS but not in 40-nm bulk CMOS transistors of both N and P type [14, 15]. In addition, increases in threshold voltage and in transconductance were measured in both technologies. Both physics-based (PSP) and BSIM4 models were modified to account for these changes and to enable the design of circuits with accurate prediction of performance.

An example from [14, 15] is shown in Figure 32.3, where the $I_d - V_{ds}$ and $I_d - V_{gs}$ characteristics are plotted for a 0.16-µm and 40-nm CMOS transistor. Measurements at 4 K (solid lines) and 300 K (dashed lines) show a general increase of about 2× in mobility in NMOS and a 30% increase of threshold voltage. Mobility increase is due to an overall decrease in electron scattering at low temperatures. Threshold increase is caused by an increase in ionization energy [15]. Another difference at 4 K is the reduction of velocity saturation, leading to I_d curves that saturate at a lower V_{ds} compared to those at 300 K. The plots show good matching of simulated characteristics (solid lines) using with the extended models compared to measurements (dotted lines) for both 1 K and 100 mK.

From 300 to 4 K the subthreshold slope (*SS*) is shown in Figure 32.4. The discrepancy from the theory can be explained by the incomplete ionization of impurity atoms at CT. This leads to an increase of the non-ideality factor $n(T)$. Other physical phenomena, such as carrier freeze-out of the substrate, can lead to a "kink" effect in the $I_d - V_{ds}$ characteristic above 2 V in some older processes, starting with 0.16 µm. This is due to the increase of the substrate potential caused by impact ionization at the drain, generating electron–hole pairs with holes flowing through the substrate raising its potential, thus forward biasing the source–substrate junction.

Processes with feature sizes below 90 nm do not exhibit this problem, generally, due to higher overall levels of doping. However, tubs and wells need to be properly contacted and biased to avoid latch-up.

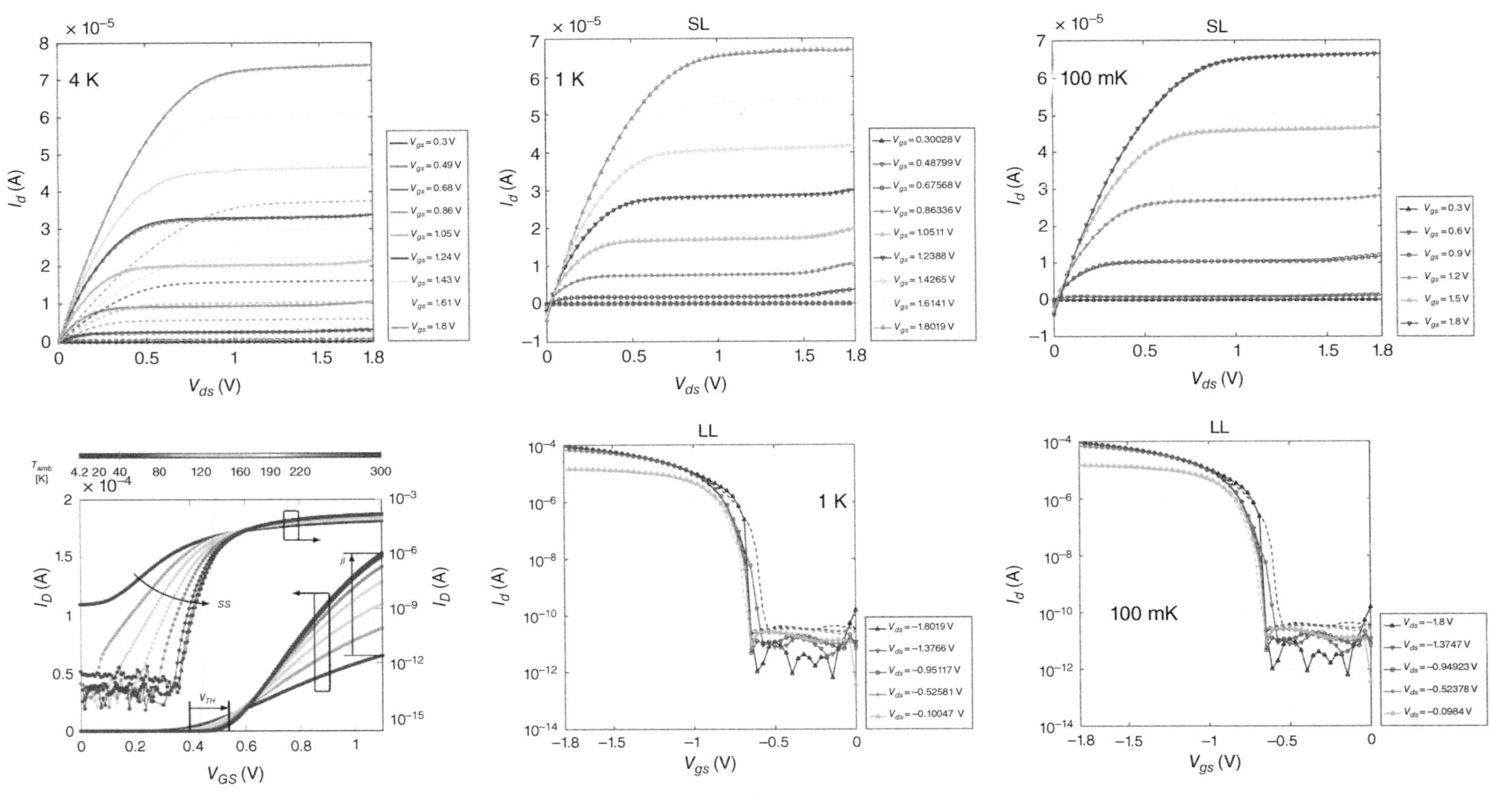

Figure 32.3 Top: 0.16 μm long/narrow NMOS $I_d - V_{ds}$ as a function of V_{gs}. Bottom: 40 nm large PMOS $I_d - V_{gs}$ as a function of V_{ds}.

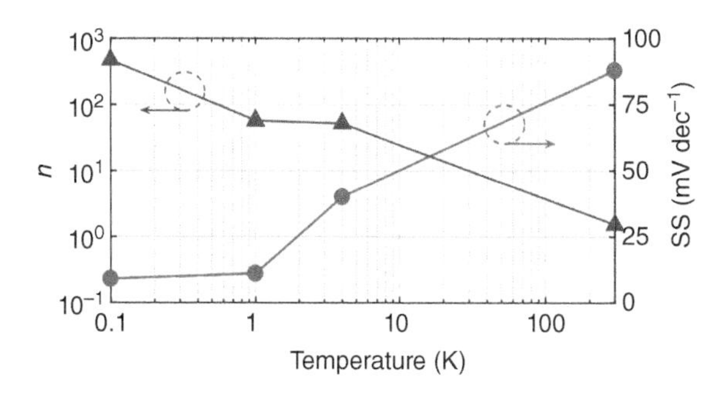

Figure 32.4 Subthreshold slope (*SS*) and non-ideality factor (*n*) temperature dependence.

This was shown in [16], where a standard cell library was redesigned, see Figure 32.5, with that in mind and tested in the design of a RISC-V processor, which operated normally and at full clock frequency at CT.

32.4 Specific Effects in Cryo-CMOS Transistors

At cryogenic temperatures, discontinuous or bumpy behavior of the subthreshold current is observed in MOS transistors. It is the explained by the incomplete ionization of dopants and Coulomb barrier. Fully Depleted Silicon-on-Insulator (FD-SOI) MOSFETs and FinFETs show different behavior at 4 K than a standard bulk CMOS process. All MOSFETs in state-of-the-art technologies, however, show correct transistor operation in moderate and strong inversion.

Other effects, such as device variability, mismatch, and self-heating, are generally higher at CT. Additionally, while thermal noise is lower, other types of noise, such as flicker noise can be significant, thus especially impacting analog and mixed-signal circuits. Transistor matching worsens at CT as shown by many authors [17–22]. Self-heating is also a major consideration, which is currently being researched by many authors. Indeed, the temperature of active devices may be higher than the ambient temperature due to Joule heating. This is a particularly critical issue as the power budget for the entire CMOS quantum controller is limited to 1 W. Especially in technologies like SOI and FD-SOI, that suffer from a limited thermal contact to the surrounding substrate due to the buried oxide, this effect can be very pronounced, while bulk devices are considered to be far less susceptible to self-heating, since these devices are in direct contact with the surrounding silicon [23–25].

32.5 Perspectives and Trends

There has been a renewed interest in BJTs and MOS transistors for applications involving the use of cryogenic temperatures. Although, active devices and, in particular, these types of transistors have been characterized at low temperatures for many years, it is now apparent that a new generation of cryo-CMOS circuits and systems based on these devices are emerging, requiring robust and continuous operation to guarantee appropriate interfacing between classical and quantum circuits. This trend is particularly important when scalable quantum computers are the target [26–31]. In Figure 32.6, an example of such a circuit is shown from [26]. We outlined some of the challenges in designing integrated circuits to operate at 4 K and below, based on complete characterizations of transistors at these temperatures. Well-known and new effects are currently under study, whereas more work is required to improve our knowledge of cryo-CMOS electronics. Moreover, well-known circuit design styles are more used and are optimized for sub-Kelvin operation, such as traveling-wave amplifiers and parametric amplifiers. This will certainly spur new activities and an increased interest in cryogenic electronic design in the future.

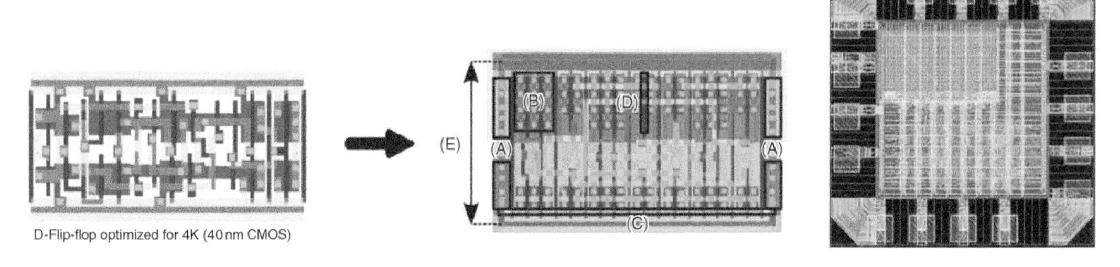

Figure 32.5 Left: redesign of a D-flip-flop in a 40-nm CMOS technology. Right: fabricated RISC-V microprocessor. The letters in the figure relate to specific rules outlined in [16]. Source: Adapted from Schriek et al. [16]/IEEE/CC BY 4.0.

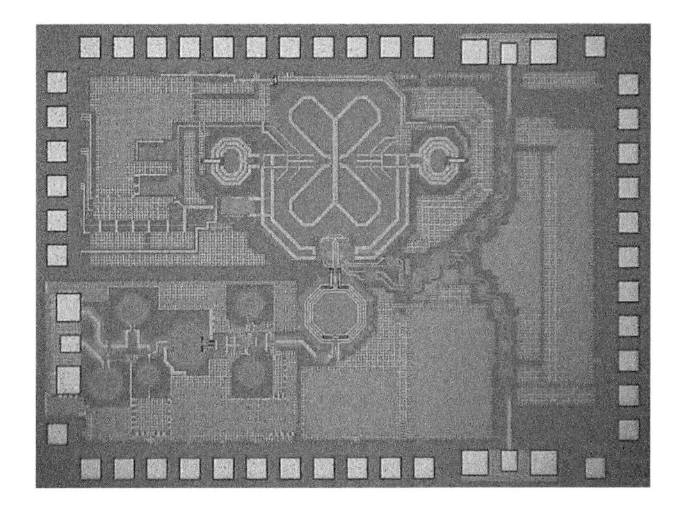

Figure 32.6 Example of quantum-classical interface implemented in 40-nm CMOS technology. Source: Peng et al. [26]/IEEE/CC BY 4.0.

References

[1] Feynman, R.P. (1982). Simulating physics with computers. *International Journal of Theoretical Physics* 21: 67.

[2] Leighton, R. and Feynman, R. (1985). *Surely, You're Joking, Mr. Feynman.* W.W. Norton.

[3] Shor, P. W. (1994). Algorithms for quantum computation: discrete logarithms and factoring. In: *Proc. 35th Annual Symposium on Foundations of Computer Science.* IEEE Computer Society Press. Santa FE, New Mexico, pp. 124–134, https://doi.10.1109/SFCS.1994.365700.

[4] Koch, J., Yu, T.M., Gambetta, J. et al. (2007). Charge-insensitive qubit design derived from the Cooper pair box. *Physical Review A* 76 (4): 042319.

[5] Debnath, S. (2016). *A Programmable Five Qubit Quantum Computer Using Trapped Atomic Ions.* University of Maryland.

[6] Charbon, E., Sebastiano, F., Vladimirescu, A. et al. (2016). Cryo-CMOS for quantum computing. *IEEE IEDM* 1351–1354.

[7] Gambetta, J., Javadi-Abhari, A., Johnson, B. et al. (2021). Driving quantum performance: more qubits, higher Quantum Volume, and now, a proper measure for speed. *IBM Research.*

[8] Loss, D. and DiVincenzo, D.P. (1998). Quantum computation with quantum dots. *Physical Review A* 57: 120.

[9] Nowack, K.C., Koppens, F., Nazarov, Y.V., and Vandersypen, L. (2007). Coherent control of a single electron spin with electric fields. *Science* 318: 1430.

[10] Veldhorst, M., Yabg, C.H., Hwang, J.C.C. et al. (2015). A two-qubit logic gate in silicon. *Nature* 526: 410.

[11] Xue, X., Russ, M., Samkharadze, N. et al. (2022). Quantum logic with spin qubits crossing the surface code threshold. *Nature* 601: 343.

[12] van Dijk, J.P.G. (2021). Designing the electronic interface for qubit control. PhD thesis. TU Delft.

[13] van Dijk, J., Vladimirescu, A., Babaie, M. et al. (2018). A co-design methodology for scalable quantum processors and their classical electronic interface. *Proc. DATE*. Dresden, Germany (19–23 March 2018).

[14] Incandela, R., Song, L., Homulle, H.A.R. et al. (2017). Nanometer CMOS characterization and compact modeling at deep cryogenic temperatures. *Proc. ESSDERC*. Leuven, Belgium (11–14 September 2017).

[15] Incandela, R.M., Song, L., Homulle, H.A.R. et al. (2018). Characterization and compact modeling of nanometer CMOS transistors at deep-cryogenic temperatures. *IEEE Journal of the Electron Devices Society* 6: 996–1006.

[16] Schriek, E., Sebastiano, F., and Charbon, E. (2020). A cryo-CMOS digital cell library for quantum computing applications. *IEEE Solid-State Circuits Letters* 3: 310–313.

[17] Galy, P., Lemyre, J.C., Lemieux, P. et al. (2018). Cryogenic temperature characterization of a 28-nm FD-SOI dedicated structure for advanced CMOS and quantum technologies co-integration. *IEEE Journal of the Electron Devices Society* 6: 594–600.

[18] t'Hart, P.A., Babaie, M., Charbon, E. et al. (2020). Characterization and modeling of mismatch in cryo-CMOS. *IEEE Journal of the Electron Devices Society* 8: 263–273.

[19] t'Hart, P.A., Babaie, M., Charbon, E. et al. (2020). Subthreshold mismatch in nanometer CMOS at cryogenic temperatures. *IEEE Journal of the Electron Devices Society* 8: 797–806.

[20] Croon, J.A., Rosmeulen, M., Decoutere, S. et al. (2002). An easy-to-use mismatch model for the MOS transistor. *IEEE Journal of Solid-State Circuits* 37 (8): 1056–1064.

[21] Jomaah, J., Ghibaudo, G., and Balestra, F. (1995). Analysis and modeling of self heating effects in thin-film SOI MOSFETs as a function of temperature. *Solid-State Electronics* 38 (3): 615–618.

[22] Triantopoulos, K., Cassé, M., Barraud, S. et al. (2019). Self-heating effect in fdsoi transistors down to cryogenic operation at 4.2 K. *IEEE Transactions on Electron Devices* 66 (8): 3498–3505.

[23] t'Hart, P.A., Babaie, M., Vladimirescu, A., and Sebastiano, F. (2021). Characterization and modeling of self-heating in nanometer bulk-CMOS at cryogenic temperatures. *IEEE Journal of the Electron Devices Society* 9: 891–901.

[24] Artanov, A., Cabrera-Galicia, A., Kruth, A. et al. (2021). Self-heating effect in 65nm CMOS technology. *Workshop on Low-Temperature Electronics*.

[25] t'Hart, P.A., Huizinga, T., Babaie, M. et al. (2022). Integrated cryo-CMOS temperature sensors for quantum control ICs. *Workshop on Low Temperature Electronics*. Matera, Italy (6–9 June 2022).

[26] Peng, Y., Ruffino, A., Yang, T.Y. et al. (2022). A cryo-CMOS wideband quadrature receiver with frequency synthesizer for scalable multiplexed readout of silicon spin qubits. *IEEE Journal of Solid-State Circuits* 57 (8): 2374–2389.

[27] Van Dijk, J.P.G., Patra, B., Subramanian, S. et al. (2020). A scalable cryo-CMOS controller for the wideband frequency-multiplexed control of spin qubits and transmons. *IEEE Journal of the Solid-State Circuits* 55 (11): 2930–2946.

[28] Bardin, J.C., Jeffrey, E., Lucero, E. et al. (2019). Design and characterization of a 28-nm bulk-CMOS cryogenic quantum controller dissipating less than 2 mW at 3 K. *IEEE Journal of the Solid-State Circuits* 54 (11): 3043–3060.

[29] Le Guevel, L., Billiot, G., Jehl, X. et al. (2020). A 110mK 295μW 28nm FDSOI CMOS quantum integrated circuit with a 2.8GHz excitation and nA current sensing of an on-chip double quantum dot. *IEEE International Solid-State Circuits Conference*, pp. 306–308. San Francisco (16–20 February 2020).

[30] Bardin, J. (2022). Beyond-classical computing using superconducting quantum processors. *IEEE International Solid-State Circuits Conference*, pp. 422–424. San Francisco (20–26 February 2022).

[31] Patra, B., Incandela, R.M., Van Dijk, J.P.G. et al. (2018). Cryo-CMOS circuits and systems for quantum computing applications. *IEEE Journal of the Solid-State Circuits* 53 (1): 309–321.

Chapter 33

Materials and Interfaces

How They Contributed to Transistor Development

Bruce Gnade

Materials Science and Engineering, University of Texas at Dallas, Richardson, TX, USA

33.1 Introduction

The importance of materials on the advancement of transistors can be summed up simply by looking at the number of elements and compounds present in a silicon transistor in the early days vs the number of elements and compounds present in today's transistors. In the early days, there were only a few elements that were used in the fabrication process. For instance, in an early silicon-based transistor, there was the silicon substrate, a few dopant materials – boron and phosphorous, SiO_2 as a dielectric, and aluminum as a contact and wiring material. Today, a significant portion of the periodic table is used to fabricate an integrated circuit. While silicon is the still the workhouse substrate material, transistors are now fabricated in volume from materials ranging from ZnO-based materials for thin-film transistors used in active matrix backplanes for liquid crystal and organic light emitting diode displays (OLEDS) to GaN and SiC for high voltage, high power devices, to GaAs and InP-based devices for high-speed transistors. For this article, I will mainly focus on the role of materials and interfaces for silicon-based transistors because of the dominance of silicon-based technology. As time goes on, III–V- and II–VI-based transistors will continue to gain importance, primarily based on advances in materials. By no means is this meant to be a rigorous, comprehensive review of all of the great materials work that has been carried out over the last 75 years, but instead an example of a few highlights where materials science has helped advance silicon-based transistor technology. This article is divided into four general topics: (i) back-end-of-line materials, (ii) channel materials, (iii) gate stack, and (iv) contacts. I will try to highlight a few of the areas where materials allowed silicon-based transistors to continue to advance.

75th Anniversary of the Transistor, First Edition. Edited by Arokia Nathan, Samar K. Saha, and Ravi M. Todi.

33.2 Back-End-of-Line

One of the biggest differences between scaled silicon technology and compound semiconductor technologies are the advances that have been made in back-end-of-line (BEOL) processing for scaled silicon. BEOL processing refers to all the processes that take place after the gate stack is formed during the integrated circuit fabrication process. While the transistor is the heart of integrated circuits, the BEOL processing has allowed transistors to scale to sub-10 nm gate lengths, by providing high-performance local and global interconnects. There have been very many materials advances that were needed to maintain circuit performance as transistors scaled below 100 nm. While BEOL materials and processes are not the focus of this article, I will mention a few that have allowed continued scaling. Probably the single biggest change in BEOL was the introduction of copper metallization. Until the late 1990s, aluminum was the metallization of choice. While aluminum was adequate for the first 40 years of transistor development, there were two primary issues that limited its continued use. The electrical resistivity of aluminum (2.65 μΩcm) [1] limited circuit performance due to RC delay, and electromigration in aluminum, even when the aluminum is doped with copper, limited long-term reliability as the width of aluminum wires continued to shrink, increasing the current density to $>10^6$ A cm^{-2}. While copper would appear to be a good choice to replace aluminum, with low resistivity (1.72 μΩcm) [1] and high electromigration resistance [2], integrating copper into a silicon process flow presented many challenges. Because there were no good precursors for the chemical vapor deposition (CVD) of copper, nor chemistries for the dry etching of copper, a totally new process technology was developed by IBM to implement copper [3]. The new Damascene copper electroplating process involved depositing a seed layer of metal, electroplating copper and then using chemical mechanical polishing to remove the excess copper. This was a dramatic change for the industry, where introducing copper into a silicon wafer fab was unthinkable. Who would have thought that you could plate copper on a silicon wafer, and then "sand off" the extra to make it smooth? An example of the results of the Damascene copper process is shown in Figure 33.1. The introduction of copper also necessitated the development of tungsten plugs to fill vias between metal 0 and metal 1, as well as barrier layers to minimize diffusion of copper between layers and into the surrounding dielectric layers [5]. The introduction of CVD tungsten started in 4-Mbit DRAMS, and became pervasive in the industry [6]. Current ICs have >15 layers of interconnect metal.

Below the 0.25 μm node devices, circuit performance was beginning to be limited by the RC delay between the transistors, rather than the speed of the transistors themselves. While copper reduced the resistance, the capacitance of the interlevel dielectric also needed to be reduced, as both the thickness

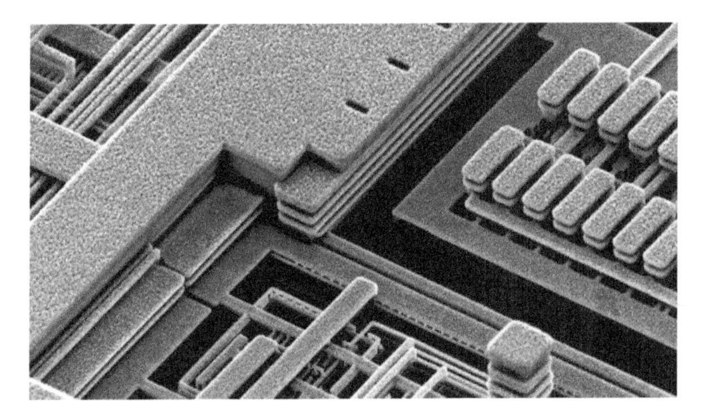

Figure 33.1 An example of the IBM copper-Damascene metallization left after the interlevel dielectric is removed [4]. Source: Reprint Courtesy of IBM Corporation ©.

of the dielectric layer and the distance between lines were shrinking. As with aluminum, for the first 40 years of transistor development, interlevel dielectrics were dominated by SiO_2, deposited by various techniques such as spin-on glass and plasma enhanced chemical vapor deposition (PECVD). The dielectric constant of bulk SiO_2 is 3.9. While there were several variations of dense SiO_2 implemented to incrementally reduce the dielectric constant, the eventual solution was to make the interlevel dielectric porous, reducing the dielectric constant by incorporating air in the film [7]. The introduction of low-κ dielectrics and the continued scaling of the metal lines have made the introduction of new barrier materials such as Co/Ru and Co/Ru/TaN necessary [8].

33.3 Channel Materials

Silicon has played a major role as the channel material in transistors for many reasons, but one of the most important reasons is the silicon surface. From early days it was understood that the quality and cleanliness of the silicon surface was important. One of the early, detailed studies of cleaning silicon surfaces was done by Kern at RCA labs in 1970 [9]. The wet chemical cleaning procedure that was developed by Kern used two chemical processes referred to as RCA-1 and RCA-2, and is still used today in some fashion. What makes this work even more impressive is that these studies were done before the development of surface analysis tools such as X-ray photoelectron spectroscopy (XPS) and time-of-flight secondary ion mass spectroscopy (TOF-SIMS). Kern relied on using radioactive tracers to determine the deposition and removal of metals from silicon surfaces. Kern's early work helped define the limits for how pure the chemicals, such as acids and bases, had to be to be used in semiconductor processing. Another important step in controlling the silicon surface was determining the importance of hydrogen passivation of a clean silicon surface. Higashi et al. [10, 11] at Bell Labs used infrared absorption spectroscopy to show that basic solutions of HF with a pH between 9 and 10 could produce an ideally terminated Si(111) surface, with silicon monohydride passivating the surface, with a very low defect density. While it was thought for many years that HF left a fluorine-terminated surface, this work showed that the silicon–hydrogen bond was stable and could protect the silicon surface before subsequent processing. An alternative to the RCA cleanup was demonstrated by Ohmi [12] in 1996, where the initial acid cleanup used in the RCA process was replaced with ozonized DI water to oxidize the contaminants on the surface. The process demonstrated by Ohmi also produced a hydrogen-terminated surface at room temperature.

The stability of the hydrogen-terminated surface was demonstrated by the high temperature required to thermally desorb the monohydride at approximately 795 K [13]. Using ultra-high vacuum (UHV) surface analysis techniques, the group at the University of Pittsburg was able to show that the mono-hydride Si(100) surface was stable in low oxygen and water environments to approximately 500 °C, which opened the door for UHV processing at relatively high temperatures. The stability of the hydrogen-terminated silicon surface allowed for the development of UVH CVD processes that provided the ability to deposit epitaxial films at relatively low temperatures. Epitaxial growth requires a highly perfect, chemically pure surface to grow on. In the early days of silicon epitaxy, this clean, perfect silicon surface was obtained by a high temperature, 1100 °C pre-clean to remove contaminants such as carbon and oxygen [14]. UHV CVD was the first growth technique to rely on the hydrogen-passivated surface to eliminate the high-temperature cleaning process. The reduced thermal budget not only allowed for lower temperature growth, but it also reduced dopant diffusion, which was required for smaller transistor dimensions. The UHV CVD process allowed the low-temperature epitaxial deposition of SiGe, developed by Myerson at IBM [15], which was instrumental in demonstrating alternate channel materials for silicon-based transistors. The use of UHV deposition processes became a standard in the industry. The use of strained silicon, SiGe, and germanium channels in MOSFETs provide higher transistor speeds because of the higher carrier mobility in the strained layers [16].

33.4 Gate Stack

Another important property of the silicon surface is the near-perfect interface that forms between SiO_2 and silicon during thermal oxidation. For the first 50 years of silicon transistors, thermal SiO_2 was the default gate dielectric. As transistor gate lengths continued to shrink, the thickness of the SiO_2 gate dielectric had to continue to be reduced to increase the gate capacitance. As the SiO_2 thickness was reduced below 1.5 nm, leakage current through the gate dielectric became prohibitively large. Also, even though operating voltages were being reduced, the electric field across the gate dielectric was increasing because the thickness was going down faster than the gate voltage. This increased electric field was impacting long-term reliability of the gate dielectric [17]. It was found that incorporating nitrogen in the SiO_2 could reduce the number of defects and improve the reliability [18]. The ultimate limit to reducing SiO_2 thickness is leakage current through the SiO_2 due to electron tunneling. In order to overcome this limitation, the SiO_2 can be replaced with a dielectric with a higher dielectric constant, allowing the use of a thicker dielectric. While there are many materials that meet the criteria of having a dielectric constant higher than SiO_2, when the requirement of integrating the material into a process flow, as well as providing a low defect, stable interface is taken into account, the number of choices is quickly reduced [19]. One of the primary considerations is whether or not the material is thermodynamically stable in contact with silicon. There was early work on higher κ gate dielectrics that could be epitaxially deposited on silicon, such as CaF_2 [20] and CeO_2 [21] but these were deemed impractical because it was difficult to consistently get high-quality interfaces, and the process flow was difficult to implement. There were a number of oxide materials that were evaluated, including Ta_2O_5 [22], TiO_2 [23], and Al_2O_3 [24]. They failed for various reasons, ranging from being thermodynamically unstable next to silicon to low crystallization temperature. $HfSi_xO_y$ ended up being one of the materials that met the requirements needed to be an advanced gate dielectric [25]. Research in advanced gate dielectrics is an ongoing active area of research. Because of the introduction of FinFETS to increase gate area and provide better gate control, it became necessary to deposit highly conformal films for the gate dielectric [26]. The deposition technique of choice was atomic layer deposition (ALD). ALD was first developed to deposit epitaxial films used in electroluminescent displays [27], but it became a widely used process in the semiconductor industry in the 2000s [28]. ALD provides thickness control at the atomic level, and in principle can provide perfect conformality and compositional control. There is ongoing research in the development of precursors for ALD for depositing a wide range of material. What started out as a clever demonstration has turned into a mainstream materials deposition technique [29].

Just as SiO_2 was the gate dielectric of choice for the first 40 years, polysilicon was the gate contact of choice for the first 40 years. As the industry transitioned from SiO_2 to high-κ gate dielectrics, there were several challenges associated with using polysilicon that became apparent. One issue was the thermodynamic stability of polysilicon in contact with high κ dielectrics during the high-temperature activation anneal, often done above 1000 °C. It was also discovered that the effective work function of the polysilicon gate could not be tuned simply by changing the doping concentration. The effective work function is fixed near the silicon conduction band edge due to Fermi-level pinning [30]. The solution to the problem was the introduction of metal gates, where the work function could be tuned by the composition of the metal, and materials that were thermodynamically stable next to different high κ materials could be chosen. There have been a wide range of materials and interfaces studied as potential metal gate candidates [31, 32]. The introduction of metal gates has not been without challenges though. In addition to process integration challenges, reliability has also been an issue due to effects such as negative bias temperature instability [33]. The integration of metal gates with high-κ gate dielectrics continues to be an active area of research [34].

33.5 Contacts

While transistor performance and reliability are driven by the channel materials and gate stack, low-resistance source–drain contacts are a critical component of the transistor. Early silicon transistors used aluminum as the metal interconnect and the contact metal. It was discovered that silicon dissolution into aluminum contacts caused a major reliability problem. Early attempts to overcome this problem included using a barrier layer such as Ti/W to minimize diffusion between the aluminum metallization and the silicon substrate [35]. One of the next steps in contact development was the introduction of silicides [36]. These silicides can be formed by codepositing a metal and silicon. Silicides can also be formed by a solid-state reaction of a metal and silicon. A wide range of materials including NiSi, $TiSi_2$, $CoSi_2$, and WSi_2 have been investigated. Each of these materials has limitations [37]. As the transistor structures become more complex, contact materials and processes become more complicated [38].

33.6 Summary

The advancement of silicon-based transistors over the last 75 years has been nothing short of phenomenal. Early transistors cost approximately \$1 each in 1960 (\$10 each in today's dollars). Transistors in today's high-performance integrated circuits cost on the order of a nanocent, a $>10^{10}$ reduction in cost. It is hard to imagine any other technology that has had this level of advancement in a mere 75 years. Advances in transistor technology have been made possible by coordinated research and development by the industry and their university partners, guided by long-range roadmaps, which define technology needs >10 years out. This has been made possible by \$100s of billions of dollars of investment by the semiconductor industry. Many of the advances have been enabled by engineering and technological advances, but this level of improvement would not have been possible without significant materials advances. Materials will play an even bigger role over the next 75 years as more and more functionality is added to silicon integrated circuits. While advancements over the past 75 years have been phenomenal, it is impossible to even image what will happen over the next 75 years [39].

References

[1] Weast, R.C. (ed.) (1987). *CRC Handbook of Chemistry and Physics*, 68e. Boca Raton, FL: CRC Press, Inc.

[2] Ohmi, T., Hoshi, T., Sakai, S. et al. (1993). Evaluating the large electromigration resistance of copper interconnects employing a newly developed accelerated life-test method. *J. Electrochem. Soc.* 140: 1131.

[3] Andricacos, P.C., Uzoh, C., Dukovic, J.O. et al. (1998). Damascene copper electroplating for chip interconnections. *IBM J. Res. Develop.* 42: 567.

[4] http://www-03.ibm.com/ibm/history/ibm100/images/icp/A270169B56060G37/us__en_us__ibm100__copper_interconnects__copper_in_chips__620x350.jpg.

[5] Nicolet, M.A. (1995). Ternary amorphous metallic thin films as diffusion barriers for Cu metallization. *Appl. Surf. Sci.* 91: 269.

[6] Lee, P.I., Cronin, J., and Kaanta, C. (1989). Chemical vapor deposition of tungsten (CVD W) as submicron interconnection and via stud. *J. Electrochem. Soc.* 136: 2106.

[7] C.C. Cho, Gnade, B.E., and Smith, D.M. (1996). Porous dielectric material with improved pore surface properties for electronics applications. US Patent 5,504,042.

[8] Kim, H.-W. (2022). Recent trends in copper metallization. *Electronics* 11: 2914.

[9] Kern, W. (1970). Radiochemical study of semiconductor surface contamination. *RCA Rev.* 31: 234.

[10] Higashi, G.S., Chabal, Y.J., Tucks, G.W., and Raghavachari, K. (1990). Ideal hydrogen termination of the Si(111) surface. *Appl. Phys. Lett.* 56: 656.

[11] Chabal, Y.J., Higashi, G.S., and Raghavachari, K. (1989). Infrared spectroscopy of Si(111) and Si(100) surfaces after HF treatment: hydrogen termination and surface morphology. *J. Vac. Sci. Technol. A* 7: 2104.

[12] Ohmi, T. (1996). Total room temperature wet cleaning for Si substrate surface. *J. Electrochem. Soc.* 143: 2957.

[13] Sinniah, K., Sherman, M.G., Lewism, L.B. et al. (1990). Hydrogen desorption from the monohydride phase on Si(100). *J. Chem. Phys.* 92: 5700.

[14] Joyce, B.A. and Bradley, R.R. (1963). Epitaxial growth of silicon from the pyrolysis of monosilane on silicon substrates. *J. Electrochem. Soc.* 110: 1235.

[15] Meyerson, B.S. (1992). UHV/CVD growth of Si and Si:Ge alloys: chemistry, physics, and device applications. *Proc. IEEE* 80: 1592.

[16] Lee, M.L., Fitzgerald, E.A., Bulsara, M.T. et al. (2005). Strained Si, SiGe, and Ge channels for high-mobility metal-oxide-semiconductor field-effect transistors. *J. Appl. Phys.* 97: 011101.

[17] Buchanan, D.A. (1999). Scaling the gate dielectric: materials, integration and reliability. *IBM J. Res. Develop.* 43: 245.

[18] Lucovsky, G., Yasuda, T., Ma, Y. et al. (1994). Control of Si-SiO_2 interface properties in MOS devices prepared by plasma-assisted and rapid thermal processes. *Mater. Res. Soc. Symp. Proc.* 318: 81.

[19] Wilk, G.D., Wallace, R.M., and Anthony, J.M. (2001). High-k gate dielectrics: current status and materials properties considerations. *J. Appl. Phys.* 89: 5243.

[20] Schowalter, L.J., Fathauer, R.W., Goehner, R.P. et al. (1985). Epitaxial growth and characterization of CaF_2 on Si. *J. Appl. Phys.* 58: 302.

[21] Tye, L., El-Masry, N.A., Chikyow, T. et al. (1994). Electrical characteristics of epitaxial CeO_2 on Si(111). *Appl. Phys. Lett.* 65: 3081.

[22] Autran, J.L., Devine, R., Chaneliere, C., and Balland, B. (1997). Fabrication and characterization of Si-MOSFET's with PECVD amorphous Ta_2O_5 gate insulator. *IEEE Electron Device Lett.* 18: 447.

[23] Campbell, S.A., Gilmer, D.C., Wang, X.C. et al. (1997). MOSFET transistors fabricated with high permitivity TiO_2 dielectrics. *IEEE Trans. Electron Devices* 44: 104.

[24] George, S.M., Snch, O., Way, J.D., and Appl. (1994). Atomic layer controlled deposition of SiO_2, and Al_2O_3 using ABAB . . . binary reaction sequence chemistry. *Surf. Sci.* 82/83: 460.

[25] Wilk, G.D., Wallace, R.M., and Anthony, J.M. (2000). Hafnium and zirconium silicates for advanced gate dielectrics. *J. Appl. Phys.* 87: 484.

[26] van Dal, M.J.H., Vellianitis, G., Duffy, R. et al. (2008). Material aspects and challenges for SOI FinFET integration. *ECS Trans.* 13: 223.

[27] Suntola, T. (1992). Atomic layer epitaxy. *Thin Solid Films* 216: 84.

[28] George, S.M. (2010). Atomic layer deposition: an overview. *Chem. Rev.* 110: 111.

[29] Oviroh, P.O., Akbarzadeh, R., Pan, D., and Coetzee, R.A.M. (2019). New development of atomic layer deposition: processes, methods and applications. *Sci. Technol. Adv. Mats.* 20: 465.

[30] Lee, B.H., Oh, J., Tseng, S.H. et al. (2006). Gate stack technology for nanoscale devices. *Mater. Today* 9: 32.

[31] Gusev, E.P., Narayanan, V., and Frank, M.M. (2006). Advanced high-κ dielectric stacks with poly-Si and metal gates: recent progress and current challenges. *IBM J. Res. Dev.* 50: 387.

[32] Alshareef, H.N., Quevedo-Lopez, M., Wen, H.C. et al. (2006). Work function engineering using lanthanum oxide interfacial layers. *Appl. Phys. Lett.* 89: 232103.

[33] O'Sullivan, B.J., Ritzenthaler, R., Simoen, E. et al. (2017). *IEEE International Reliaibility Physics Symposium*, DG-8.1, Monterey CA (2–6 April 2017).

[34] Robertson, J. and Wallace, R.M. (2015). High-K materials and metal gates for CMOS application. *Mats. Sci. Eng. Rep.* 88: 1.

[35] Chang, P.-H., Hawkins, R., Bonifield, T.D., and Melton, L.A. (1988). Aluminum spiking at contact windows in Al/Ti-W/Si. *Appl. Phys. Lett.* 52: 272.

[36] Chen, L., Lur, W., Chen, J. et al. (1994). Silicide formation by rapid thermal processing. *MRS Online Proc. Lib.* 342: 99.

[37] Gambino, J.P. and Golgan, E.G. (1998). Silicides and ohmic contacts. *Mats. Chem. Phys.* 52: 99.

[38] Koh, S.-M., Kong, E.Y.J., Liu, B. et al. (2011). Contact-resistance reduction for strained n-FinFETs with silicon–carbon source/drain and platinum-based silicide contacts featuring tellurium implantation and segregation. *IEEE Trans. Electron Devices* 58: 3852.

[39] Doering, R. and Nishi, Y. (2008). *Handbook of Semiconductor Manufacturing Technology*, 2e. CRC Press – Taylor and Francis Group.

Chapter 34

The Magic of MOSFET Manufacturing

Kelin J. Kuhn

Department of Materials Science and Engineering, Cornell University, USA

34.1 Introduction

In 1965, in one of the most famous papers in electronics [1], Gordon Moore made several statements predicting where integrated circuit (IC) technology would be in 1975. One of these statements was, "This allows at least 500 components per linear inch or a quarter million per square inch." Today, the Apple M2 has approximately 20 billion transistors in ~155 mm^2 (0.24 in.^{2}) or about 83 billion transistors in a square inch. Just to appreciate the difference between 1965 and 2022, Figure 34.1 compares the number of zeros between 250,000 (speculated in 1965 for 1975) versus 83,000,000,000 (achieved in 2022) for components per square inch.

Just pause for a moment and think about this. An IC chip like the Apple M2 has (literally) tens of billions of transistors in it, AND they all need to work correctly. This is beyond remarkable. How could normal human beings (who struggle to write tweets without spelling errors!) build something roughly half the size of a penny with 20 billion working transistors in it? Even more startling, the cost to build such a chip is roughly the price of taxi fare from the San Francisco Airport to Silicon Valley.

Well, the answer is simple. It is magic!

My goal in the remainder of this paper is to describe how the magic works – or more precisely, how the transistor architecture interlaces with the semiconductor process flow to create these remarkable chips. Moreover, I am going to speculate a bit on where this may take us in the next 75 years.

75th Anniversary of the Transistor, First Edition. Edited by Arokia Nathan, Samar K. Saha, and Ravi M. Todi.

250,000 per square inch (Predicted by Moore in 1965 for 1975)
83,000,000,000 per square inch (Achieved by Apple M2 in 2022)

Figure 34.1 Appreciating the number of zeros!

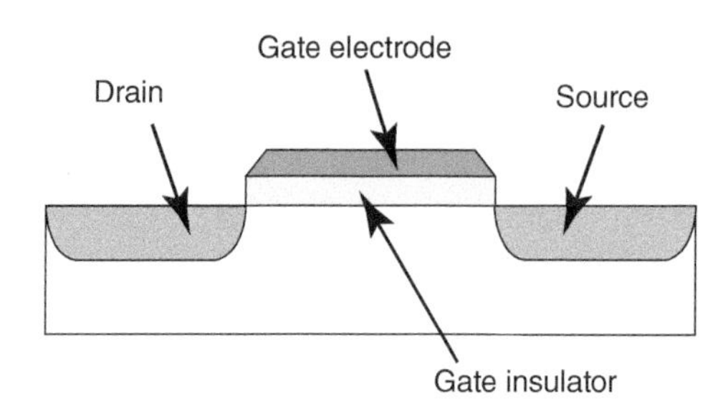

Figure 34.2 The simplicity of MOS.

34.2 The Magic of MOS

Metal-oxide semiconductor (MOS) devices are ideally (perhaps even magically!) suited for use in ICs. The most important reasons are discussed below.

34.2.1 Simplicity

Perhaps the most important property of a MOS device is its simplicity (see Figure 34.2). There are only three important structures in an ideal MOS device: (i) the source/drain regions (the device is symmetric, so the source and drain are identical), (ii) the gate electrode, and (iii) the gate insulator. Carrier flow from the source to the drain is controlled by the electric field generated from the gate electrode. The gate electrode is isolated from the channel by the gate insulator, so minimal current flows through the gate (all the work is done by the capacitor formed by the gate electrode and the gate insulator).

The simplicity arises because two of these structures (the source/drain regions and the gate electrode) are fundamentally straightforward to fabricate. Source/drain regions can be as simple as heavily doped "n" (or "p") regions in the substrate. The gate electrode is just a conductive material, and can be formed of materials ranging from doped polysilicon to metal. The only fundamentally difficult structure to fabricate in the MOS device is the gate insulator. The gate insulator needs to be both very thin AND exceptionally free of physical, chemical, and electronic defects.

34.2.2 Smaller is Better

Another magical property of a MOS device is the smaller you make it, the better it works. Most things in life do not work that way (and a lot of transistors do not either!) but MOS devices do. This property of MOS devices is usually characterized by the term "Scaling Laws" and the seminal description of these laws is table I of Dennard's 1974 paper [2] and reproduced as Table 34.1 below.

Applying Table 34.1 to real life suggests shrinking all the relevant dimensions of an ideal MOS device (length L, width W, and insulator thickness t_{ox}) by a factor of 0.7 (and appropriately modifying the doping and voltage by the same factor) will result in a delay time (VC/I) improvement of

Table 34.1 Dennard scaling.

Device or circuit parameter	Scaling factor
Scaling results for circuit performance	
Device Dimension t_{ox}, L,W	$1/\kappa$
Doping Concentration N_A	κ
Voltage V	$1/\kappa$
Current I	$1/\kappa$
Capacitance $\varepsilon A/t$	$1/\kappa$
Delay time/circuit VC/I	$1/\kappa$
Power dissipation/circuit VI	$1/\kappa^2$
Power density VI/A	$1/\kappa$

Source: Adapted from Dennard et al. [2].

1/0.7 or about 1.4×. Remarkably, from around 1975 to about 1995 (ending around the 0.25 μm generation) Dennard-style scaling worked (in production even!) with minimal modifications. From around 1995 to about 2005, Dennard-style scaling mostly worked, but the gate length was overscaled to stay on the performance targets. Overall, Dennard-style scaling was astonishingly successful for around 30 years.

34.2.3 Compatible with a Flat Wafer

Yet another important property of MOS devices is the structure lends itself to a flat architecture where the current flow is parallel to the substrate. IC manufacturing is similar to silk screen printing in that most IC manufacturing steps create or alter thin uniform layers of material on the wafer. A flat transistor architecture permits more layers and more accurate control of those layers. Even modern "2D" transistor architectures (such as the FinFET, described later) still endeavor to restore the wafer to a flat geometry after the transistor is built. This is in contrast to many types of transistors, for example, bipolar junction transistors, BJTs (which possess an inherently vertical architecture, and need to be distorted to work well in ICs) and true vertical transistors (such as power transistors, which also do not integrate well into ICs).

34.2.4 Upgradeable

The ideal MOS device is remarkably easy to upgrade. Performance improvements (in some cases dramatic performance improvements!) are possible by simply upgrading the basic elements of the MOS device with more sophisticated elements.

The first round of upgrades (done between ~1965 and ~2010) were material replacements. For example, the original silicon dioxide gate insulator was first upgraded with a silicon oxy-nitride gate insulator, and then further upgraded with a high-κ bilayer formed of a silicon oxy-nitride interface layer and a high-κ insulator (the most common high-κ insulator being hafnium dioxide). The original metal gate electrode would first be upgraded with a dual workfunction doped polysilicon gate electrode, and (in an interesting turn of history!) would return to a metal gate electrode (but this time a dual workfunction approach fabricated from more sophisticated material systems). The original straightforward implanted source drain region would be upgraded with multiple overlapping implanted regions including the source–drain itself, a tip (also called a source–drain extension, or SDE), and a halo (also called a pocket). The implanted source drain would, in turn, be upgraded with an epitaxial (and in some cases epitaxial and strained!) raised source–drain (Figure 34.3).

The second round of updates (beginning in about 2011) were structural. The most dramatic of these is the FinFET, which conceptually folds the transistor along the width direction, creating a pillar or "fin" that protrudes from the wafer. Among the many benefits of a FinFET is an improvement in "short-channel control." Saying this another way, the FinFET geometry permits a dramatically smaller device (shorter gate length) than the conventional geometry (Figure 34.4).

Of course, once the FinFET was proven successful, many of the prior upgrades (high-κ gate insulator, dual workfunction metal gate, and epitaxial source drain) were immediately applied to the FinFET (Figure 34.5).

However, as different as Figure 34.5 appears from Figure 34.2, all of these changes can still be understood as upgrades from the basic MOS device. Carriers still flow parallel to the substrate, from the source to the drain, controlled by the electric field generated from the gate electrode. The gate electrode is still isolated from the channel by an insulator, and the electric field formation is still controlled by the gate capacitor.

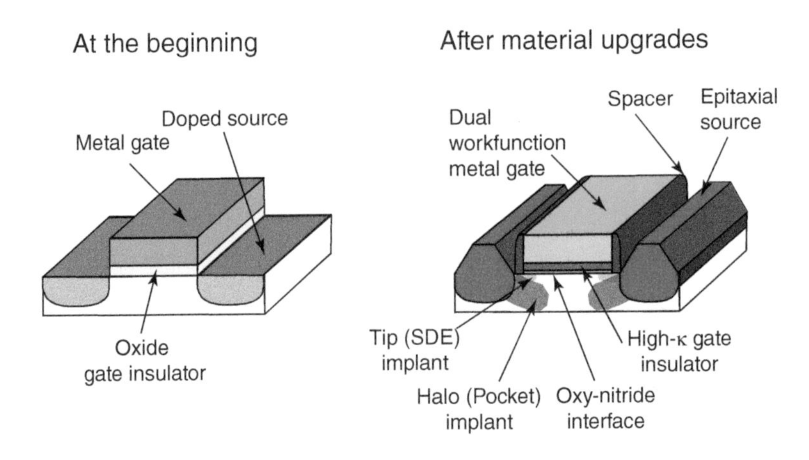

Figure 34.3 The first round of upgrades – material replacements.

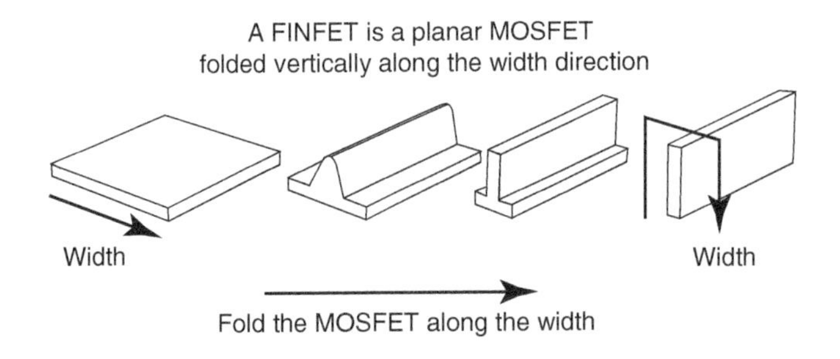

Figure 34.4 A FinFET is a planar MOS device folded along the width direction.

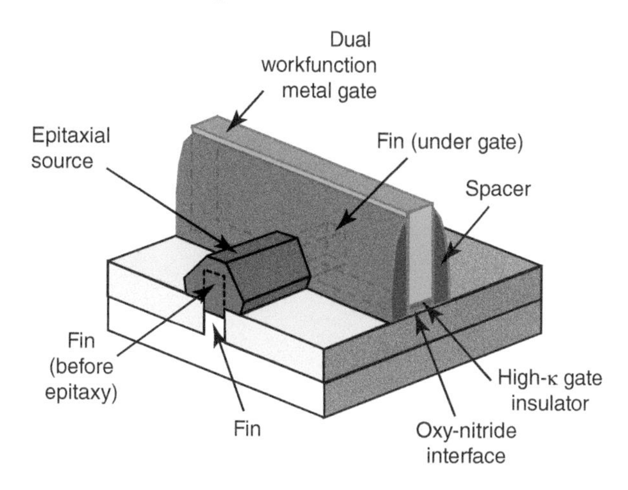

Figure 34.5 A modern FinFET with both structural and material upgrades.

34.3 The Magic of Self-alignment

Self-alignment is an immensely powerful process tool. While self-alignment does not receive a lot of press, it is a major part of the magic behind the success of integrated circuits.

I am not sure there is a standard definition of self-alignment, but I will offer one. My definition is: *Self-alignment is a process technique where pre-existing structures on the wafer are used to set the boundaries of subsequent structures.* The traditional example of self-alignment is the use of a pre-existing transistor gate electrode to set the edge of an implanted source/drain region as shown in Figure 34.6.

The importance of self-alignment in this architecture is it removes misalignment variations between the gate and the source/drain. Saying this another way, in a self-aligned process, the source/drain regions cannot "wiggle" with respect to the gate – they can only "wiggle together" (see Figure 34.7).

Self-alignment alone is a very powerful process technique. However, it can be combined with another process element, called a sidewall spacer, to do truly magical things.

Sidewall spacers are formed by (i) depositing a conformal film over a vertical structure (for example, a pillar) on a wafer, (ii) performing a highly directional etch (anisotropic etch) over the entire wafer, and (iii) setting the depth of the etch equal to the thickness of the conformal film. The result is a sidewall spacer (see Figure 34.8, the purple "bookend" on each side of the pillar is the sidewall spacer).

Sidewall spacers are relatively straightforward to make and remove. Moreover, they can be stacked on top of each other (using combinations of different materials and their associated selective etches, for example oxides, nitrides, carbides, and the like). As such, they are immensely powerful tools to

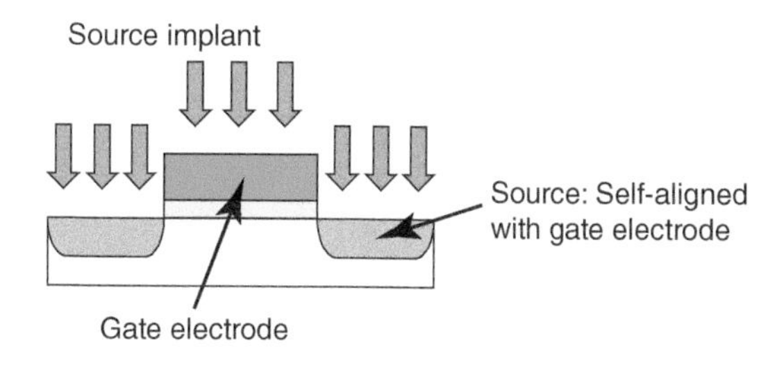

Figure 34.6 Self-alignment of the source using the pre-existing gate electrode.

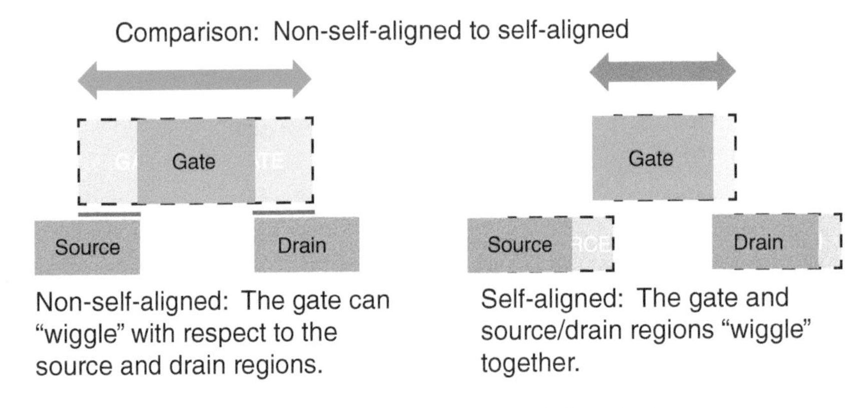

Figure 34.7 Why self-alignment is important.

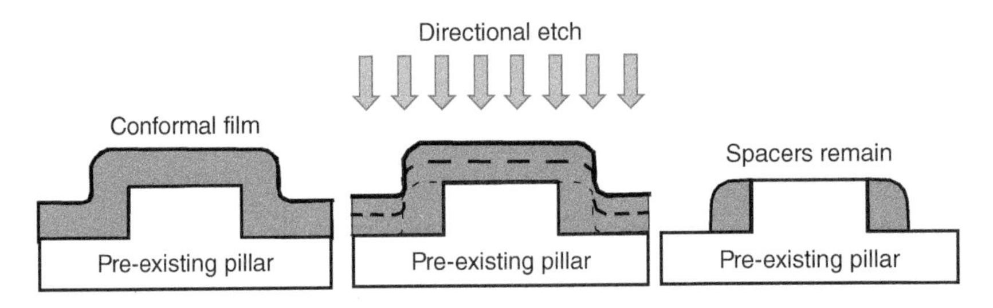

Figure 34.8 Fabrication of sidewall spacers.

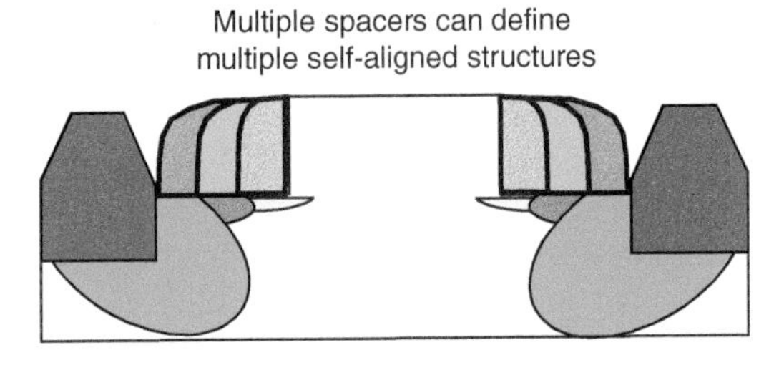

Figure 34.9 Multiple spacers can define multiple self-aligned structures (tips, halo, epitaxial source/drain).

architect subtle overlapping features of the MOS transistor (for example, tips, halos, and the like) as shown in Figure 34.9. Moreover, spacers are not limited to the source/drain sides of the gate electrode. It is also possible to use spacers in the orthogonal direction, for example in self-aligned gate endcap architectures. Furthermore, spacers are not limited to the transistor, as they can be applied to a variety of other self-aligned structures within the process.

34.4 The Magic of Semiconductor Manufacturing

In 1979, Robert Barrow (27th Commandant of the Marine Corps) made the comment that, "Amateurs talk about strategy and tactics. Professionals talk about logistics and sustainability.…" It would be great if there were some semiconductor manufacturing version of the same quote (by Craig Barrett, for example!) but unfortunately, there is not. However, if Craig HAD made such a comment, I suspect it would have been something like "Amateurs talk about paradigms and synergy. Professionals talk about quality control and automation."

34.4.1 Quality Control

The most critical aspect of establishing and maintaining a healthy semiconductor manufacturing process is quality control. The fundamental rule is "When something breaks, fix it so it cannot happen again." In a healthy process flow, failures due to a single incident, as well as failures due to two simultaneous incidents, should be vanishingly rare. These are relatively high probability events and protections against such failures should have long ago been embedded in the quality system. As an example, the first time a quality issue occurred because a bar-code was misread, some type of quality fix would have been added so the process does not fail on a single misread. The first time a bar-code was misread AND the process recipe misloaded; again, a quality fix would have been added to prevent the two simultaneous failures. This type of quality control is sometimes termed "Continual Improvement." Unfortunately, the name is not very descriptive of the technique (among other things, it is applied to a lot of different techniques!) but I suspect that "Systematic Elimination of Multiple-Failure-Mode Low-Probability Events," just did not stick (it does not make a good acronym either, SEMFMLPE?)

If this quality methodology is rigorously followed over time, most quality issues will be extremely low probability multiple-simultaneous-failure events (the bar-code was misread AND the process recipe misloaded AND the station controller was running an old version of software AND and the purge valve was replaced by a seemingly-correct but software-incompatible model). Perhaps more than any other factor, it is this type of quality methodology that enables (literally) billions of working transistors to be manufactured at a very modest cost.

34.4.2 Automation

Automation is essential to managing the complexity of a modern semiconductor process. Automation has been continually evolving throughout the lifetime of IC manufacturing. In some sense the old adage of "needing computers to make computers" is precisely correct.

Automation serves a number of useful functions and I discuss a few below.

Automation permits manufacturing processes more complex than unaided humans can support. Humans simply are not very good at repetitive routine detailed operations. Automated systems are. However, automated systems can also do inane things (nothing quite like watching a robot at a cleans station methodically break every wafer in a lot – each one broken in precisely the same way). Manufacturing operations work best when the routine work is automated and the humans spend their time debugging and fixing the intermittent failures.

Automation permits process operations which humans find uncomfortable, unhealthy, or dangerous. One of the biggest automation transitions in semiconductor manufacturing occurred between 200 mm (8-in.) and 300 mm (12-in.) factories. The 200 mm factories required humans to wear the uncomfortable "spacemen" bunny suits (with the bubble masks and air filters) as the humans had to manually move wafer boxes between process steps. The 300 mm factories stored the wafers in clean self-contained modules called FOUPs and automation handled the wafer movement. As a consequence, humans in the factory could wear substantially less restrictive clean-room clothing and injuries (typically ergonomic) associated with wafer handling were eliminated.

Last, but certainly not least, automation allows embedding quality checks within the automation for low probability multiple-failure events (automation and quality control are intrinsically coupled). Humans are very unlikely to remember the time the bar-code was misread AND the process recipe misloaded AND the station controller was running an old version of software AND the purge valve was replaced by a seemingly correct but software-incompatible model – but an automated system can!

34.5 Transistor Magic for the NEXT 75 Years?

I am going to group my predictions into three categories (i) more of the same, but better-smaller-faster, (ii) harnessing our ability to manage complexity by introducing more (useful) complexity into the entire semiconductor manufacturing process, and (iii) harnessing our ability to manage complexity to enable more complex transistor logic systems.

34.5.1 Better-Smaller-Faster

There are an enormous number of transistor architectures presently being researched or developed with the intent of high-volume manufacturing [3]. These include continued upgrades to MOS transistors such as nanoribbons (structural upgrade), negative capacitance (gate dielectric upgrade), ferroelectric (gate electric upgrade), metal source–drain (source/drain upgrade), and the like. There are also transistors under discussion that use beyond-MOS technologies (tunnel FET, electrochemical devices, impact ionization devices, nano-switches, and so on). Finally, there are transistors that use spin, not charge, a fascinating and rapidly developing field typically called spintronics. HOWEVER, while all these new developments are immensely exciting, my personal sense is that only the ones that can effectively "mimic" upgrades to MOS will make it into manufacturing.

34.5.2 Further Evolution of Complexity

The last 75 years of transistor development has taught us more than simply how to make nifty transistors. It has also taught us how to manage indescribably complex systems at high quality. This leads to the critical question of "What else can we do with our ability to manage complexity?" This opens a number of possibilities. For example, there is amazing new magic possible by having all parts of the process exchanging information in real-time (feed-forward systems, feed-back systems, and the like). There is the option of increased customization (custom processing on small lots, custom processing on single wafers, custom products at the die level, and so on). Moreover, the sophisticated techniques developed for manufacturing transistors on chips can also be extended across the entire semiconductor manufacturing environment (chip to package, package to board, board to system). These may be evolutionary changes from the perspective of manufacturing, but could be revolutionary changes for the culture (product customization for each individual chip, for example!)

34.5.3 Is It Time to Explore Other Logic Systems?

For the last 75 years, we have relied on transistors enabling binary logic (i.e. where ON is "1" and OFF is "0"). Binary logic is not the only way to design a logical system. Multivalued logic (where the transistor has more than two states) is another alternative. Multi-value memory (memory with more states than just "0" or "1") has recently seen increased interest, primarily in the form of flash to support solid-state drives (SSDs). (As one recent example, Samsung's 870 QVO SATA SSD (introduced in 2020) is a quad (4-bit) V-NAND flash memory.) The next step after multivalued memory is multivalued logic. However, unlike with memory, the implementation of transistor-level multivalued logic (i.e. logic systems that take advantage of a multiplicity of states in each transistor) is an enormously complex transition. Although it has been known for a number of years that multi-value logic has the potential to achieve significant improvements due to the reduction in the number of elements and connections, it has also been known to be immensely complex [4]. However, unlike 75-years ago when this journey began, today we have the expertise to manage that complexity!

References

[1] Moore, G.E. (1965). Cramming more components onto integrated circuits. *Electronics* 38, 19 (8): 114–118.

[2] Dennard, R.H., Gaensslen, F.H., Yu, H.-N. et al. (1999). Design of ion-implanted MOSFET's with very small physical dimensions. *Proceedings of the IEEE* 87 (4): 668–678.

[3] Liu, T.-J.K. and Kuhn, K. (2015). *CMOS and Beyond: Logic Switches for Terascale Integrated Circuits*, 1e. Cambridge University Press.

[4] Hurst, S.L. (1984). Multiple-valued logic – its status and its future. *IEEE Transactions on Computers* C-33 (12): 1160–1179.

Chapter 35

Materials Innovation

Key to Past and Future Transistor Scaling

Tsu-Jae King Liu and Lars Prospero Tatum

Department of Electrical Engineering and Computer Sciences, University of California Berkeley, Berkeley, CA, USA

35.1 Introduction

Steady advancement in semiconductor process technology over the past 60+ years made microelectronics ever more affordable and functional, resulting in the proliferation of smart devices, information processing, and communication systems – and the digital transformation of every aspect of modern society. The key to continual improvements in computing cost and performance has been transistor miniaturization: the smaller a transistor is, the more compactly a microprocessor chip comprising interconnected transistors can be implemented and hence the more chips that can be yielded from a single silicon wafer substrate of fixed size (e.g. 300 mm diameter in state-of-the-art semiconductor manufacturing facilities), resulting in lower manufacturing cost per chip; alternatively, more transistors can be integrated onto a chip of fixed size for increased chip functionality. Furthermore, smaller transistors and shorter interconnecting wires (called interconnects) between them have smaller associated capacitance so that voltage signals propagate more quickly through them, i.e. the integrated circuit (IC) can be operated with a higher clock frequency.

Advancements in IC manufacturing technology have enabled ever finer control of the thickness of thin (below 1 μm in thickness) films that are used to form the transistors and interconnects, as well as ever finer resolution of the lithographic process that is used to pattern the thin films. Accordingly, the minimum size of transistors has shrunk over time so that the number of transistors incorporated on a

75th Anniversary of the Transistor, First Edition. Edited by Arokia Nathan, Samar K. Saha, and Ravi M. Todi.

single IC chip has roughly doubled every two years according to Moore's Law [1]. Today, the most advanced transistors have features with minimum dimension below 10 nm, so that the most advanced microprocessors pack more than 100 million transistors within an area of one square millimeter [2].

In very-large-scale integrated (VLSI) circuits used for digital computing, transistors function very simply as electronic switches. Due to its superior scalability to nanometer dimensions, the MOSFET is the transistor design of choice for microprocessors and microcontrollers. However, MOSFETs are imperfect switching devices because they can leak current when they are turned off. This off-state leakage current (I_{OFF}) worsens as the transistor channel length is scaled down. Materials innovations have been key to suppressing the short-channel effect – as well as to enhancing on-state current for faster IC operation – and hence for sustaining Moore's Law. This chapter first reviews the basics of MOSFET operation and complementary MOSFET (CMOS) digital ICs. It then discusses nonideal MOSFET characteristics that pose challenges for device miniaturization, motivating materials innovations that have helped to address these challenges and thereby have enabled continued transistor scaling to ever-smaller dimensions. It concludes with an outlook for additional materials innovations to enable MOSFET scaling to 1 nm gate length.

35.2 MOSFET Basics

35.2.1 Transistor Structure and Operation

Figure 35.1 illustrates a conventional planar bulk-silicon MOSFET structure. The Source and Drain (S/D) semiconductor regions have high concentrations of impurities (dopants) that make these regions electrically conductive, either with mobile *positive* charge (p-type conductivity) or with mobile *negative* charge (n-type conductivity). Typically, the semiconductor channel region beneath the Gate electrode is doped to be of opposite conductivity type than the S/D regions, so that there naturally exists a built-in electric potential barrier to diffusion of mobile charges from the Source region into the channel region. The S/D regions are formed using a fabrication process sequence that naturally aligns them to the edges of the Gate electrode, so that the channel length (L) automatically scales down with Gate length (L_G). This self-aligned process provides for minimum unnecessary overlap between the Gate and S/D regions, which is desirable for minimizing parasitic Gate capacitance.

When the voltage (V_{GS}) applied between the Gate and Source is greater than a threshold voltage (V_T), the semiconductor beneath the Gate becomes depleted of its original mobile charges and a channel "inversion layer" of opposite conductivity type (matching that of the S/D regions) is formed at the semiconductor surface so that current (I_{DS}) can readily flow between the Source and Drain regions if their electric potentials are unequal, i.e. if there is a nonzero voltage difference (V_{DS}) that induces

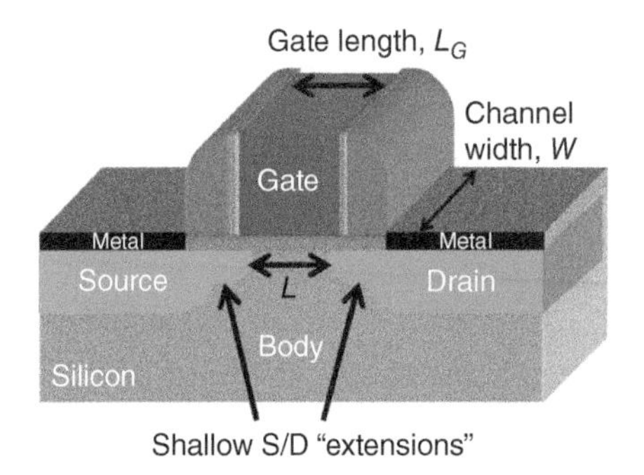

Figure 35.1 Schematic illustration of a planar bulk-silicon metal-oxide-semiconductor field-effect transistor (MOSFET) structure.

mobile charges to drift between these regions. Note that the Gate electrode (shown in red) is electrically insulated from the semiconductor channel region by a thin oxide layer (shown in light gray) to prevent direct current flow between the Gate and semiconductor.

In the on state, the current flowing between the Source and the Drain increases super-linearly with increasing gate overdrive voltage $|V_{GS} - V_T|$, which has a maximum magnitude of $|V_{DD} - V_T|$, where V_{DD} is the IC power supply voltage. More specifically, I_{DS} is proportional to the width of the channel region (W), to the mobile charge drift velocity (v_{drift}) and to the amount of mobile charge in the inversion layer (Q_{inv}):

$$I_{DS} = W \times v_{drift} \times Q_{inv} \tag{35.1}$$

Below the saturation velocity limit, the drift velocity is proportional to the lateral electric field E within the inversion layer:

$$v_{drift} = \mu_{eff} \times E \tag{35.2}$$

where μ_{eff} is the effective mobility of the mobile charge. (Note: In sub-50 nm gate length MOSFETs, ballistic transport becomes significant; however, the "source injection velocity" of mobile charge is still proportional to the effective mobility [3].) Since a minimum electric potential difference (equal to V_T) between the Gate and semiconductor is needed to form the inversion layer, the maximum possible electric potential of the inversion layer at the Drain end of the inversion layer is $|V_{GS} - V_T|$, so E is limited to be no greater than $|V_{GS} - V_T|/L$.

The inversion-layer charge density Q_{inv} is proportional to gate overdrive voltage:

$$Q_{inv} = C_{ox} \times |V_{GS} - V_T| = (\varepsilon_{ox} / t_{ox}) \times |V_{GS} - V_T| \tag{35.3}$$

where C_{ox} is the Gate capacitance per unit area, and ε_{ox} and t_{ox} are the permittivity and thickness of the gate-insulating oxide layer, respectively.

35.2.2 Subthreshold and Off-State Leakage Current

The magnitude of the aforementioned potential barrier that impedes the flow of mobile charges from the Source into the channel region of a MOSFET depends on V_{GS}. Specifically, it decreases with increasing $|V_{GS}|$, by a proportionality factor that depends on capacitive coupling between the Gate and the channel region near the Source, relative to the capacitive couplings between this region and the Drain (C_D) and the charge-neutral semiconductor Body region beneath the channel region (C_{dep}). This proportionality factor is C_{ox}/C_{total}, where $C_{total} = C_{ox} + C_{dep} + C_D$. A change in V_{GS} (ΔV_{GS}) thus results in a change in potential barrier height equal to $\Delta V_{GS} \times (C_{ox}/C_{total})$.

The mobile charges in the Source have kinetic energy with an exponential probability distribution: their population decreases exponentially with linearly increasing kinetic energy relative to kT, where k is the Boltzmann constant and T is the absolute temperature. As a result, the number of mobile charges in the Source region with sufficient kinetic energy to surmount the potential barrier (to diffuse into the channel region and subsequently drift to the Drain) increases exponentially as the barrier height decreases. Hence, as $|V_{GS}|$ increases, lowering the potential barrier, $|I_{DS}|$ increases exponentially, as illustrated in Figure 35.2.

When the MOSFET is in the on state (i.e. when $|V_{GS}| > V_T$), the potential barrier between the Source region and channel region is very small so that I_{DS} is limited not by the rate of mobile charge diffusion into the channel region but by the rate at which mobile charges drift under the influence of a lateral electric field induced by the applied voltage V_{DS} (Eq. 35.1). The maximum transistor "drive" current (I_{ON}) is equal to I_{DS} with $V_{GS} = V_{DS} = V_{DD}$.

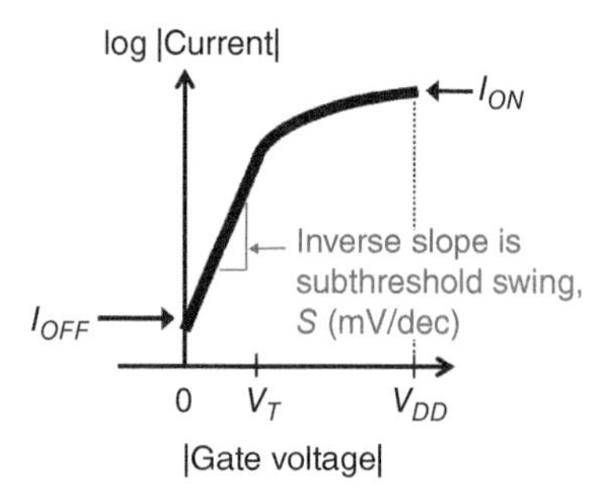

Figure 35.2 Semi-log plot of MOSFET output (drain) current vs. input (gate) voltage. Below threshold (i.e. in the subthreshold region), the current is an exponential function of the gate voltage.

The inverse slope of the log I_{DS} vs. V_{GS} plot in the subthreshold region (i.e. for $|V_{GS}| < V_T$) is referred to as the subthreshold swing (S), and represents the amount of change in V_{GS} needed to effect 10× (one decade) change in I_{DS}:

$$S = (kT/q)\ (\ln 10)\ (C_{total}/C_{ox}) \tag{35.4}$$

where q is the electronic charge and kT/q is the thermal voltage (26 mV at room temperature).

From Figure 35.2 it can be seen that I_{OFF} is proportional to $10^{-|V_T/S|}$. Therefore, to minimize I_{OFF} for a fixed value of V_T, S should be minimized. From Eq. (35.4) it can be deduced that the minimum value of S is $(kT/q)(\ln 10)$, which is approximately 60 mV decade^{-1} at room temperature (300 K). This minimum value is achieved when the Gate is capacitively coupled to the channel region much more strongly than either the Drain or the Body so that $C_{total}/C_{ox} \cong 1$. Such a device is considered to have superior "electrostatic integrity."

35.2.3 Threshold Voltage Adjustment

The absolute value of V_T depends on various material parameters and is determined during the IC manufacturing process:

$$|V_T| = V_{FB} + |V_S| + |V_{ox}| \tag{35.5}$$

where V_{FB} is the flat-band voltage, corresponding to the difference in work functions of the gate material and the semiconductor channel region; V_S is the voltage that is vertically dropped across the semiconductor depletion region (i.e. from the semiconductor surface to the charge-neutral Body region) in order to invert the semiconductor conductivity type at the surface, which increases with (approximately the square root of) the dopant concentration within this region; and V_{ox} is the voltage dropped across the gate-insulating oxide layer, which according to Gauss's Law is proportional to the total charge within the semiconductor beneath the Gate at the threshold condition ($Q_{inv} = 0$):

$$|V_{ox}| = |Q_{dep}|/C_{ox} \tag{35.6}$$

where Q_{dep} is the total space charge per unit area in the depletion region beneath the Gate.

Typically, the gate-insulating oxide thickness (t_{ox}) is set to the minimum value allowed by specifications for Gate leakage current (which is due to quantum-mechanical tunneling through the physically thin oxide), to provide for maximum C_{ox}, hence Q_{inv} and on-state current. Conventionally Q_{dep} has been adjusted by selective (lithographically masked) ion implantation of additional dopants during the IC fabrication process, to adjust V_T to the desired value.

35.3 Complementary MOS (CMOS) Technology

As described above, when a MOSFET is in the on state, an electrically conductive channel comprising mobile electronic charges of the same type as in the S/D regions is formed at the surface of the semiconductor beneath the Gate electrode, electrically connecting the Source and Drain. If the channel comprises *negatively charged* electrons, induced by applying a positive Gate voltage relative to the source voltage ($V_{GS} > V_T$), then the MOSFET is an *n-channel* transistor and has n-type S/D regions. If the channel instead comprises *positively charged* holes (locally missing valence electrons), induced by applying a negative Gate voltage relative to the Source voltage ($V_{GS} < -|V_T|$), then the MOSFET is a *p-channel* transistor and has p-type S/D regions.

35.3.1 CMOS Inverter Circuit

If the Gate electrodes of a pair of n-channel and p-channel MOSFETs are connected together to form an input node and their source electrodes are biased as shown in Figure 35.3, complementary switching behavior is achieved, i.e. only one transistor is turned on at a time when the input node is high (biased at V_{DD}) or low (biased at 0V). With their Drains connected together to form the output node, the n-channel device is used to connect and discharge ("pull down") the output node to 0V when the input node is high, whereas the p-channel device is used to connect and charge ("pull up") the output node to V_{DD} when the input node is low; therefore, this circuit performs a simple inverting function. The larger the transistor on-state current, the faster the output node voltage is charged or discharged and hence the faster the circuit operates. Note that the capacitance of the output node comprises not only the inverter transistors' Drain junction capacitances but also wiring capacitance as well as the input capacitance of the circuit that is presumably being driven by this inverter circuit. Also note that, in either inverter state, the MOSFET that is off sustains a large drain-to-source voltage difference $|V_{DS}| = V_{DD}$. The static power dissipation of this CMOS inverter circuit is equal to $I_{OFF} \times V_{DD}$.

35.3.2 CMOS Digital Logic and Memory Circuits

Digital computers use a binary number system: a binary digit (bit) value of "0" is represented by a low voltage, while a bit value of "1" is represented by a high voltage. Computation is implemented with circuits called logic gates, which perform Boolean logic functions. Any logic gate can be constructed with a combination of n-channel pull-down devices and a complementary set of p-channel pull-up devices. (Each input voltage signal drives the gates of a pair of complementary n-channel and p-channel MOSFETs.) For example, a circuit that performs the NOT-AND (NAND) logic function is shown in

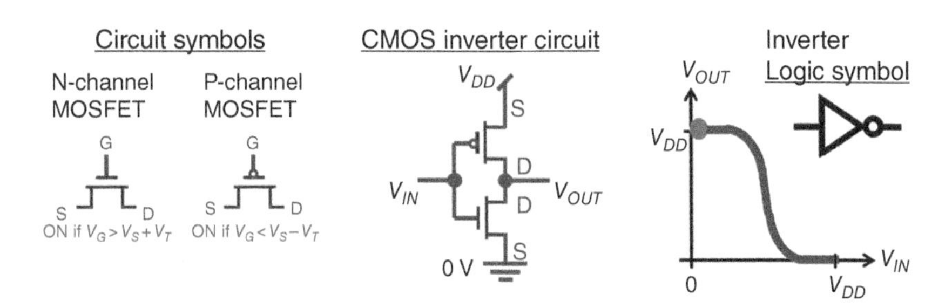

Figure 35.3 (left) Circuit symbols for n-channel and p-channel MOSFETs, (center) schematic circuit diagram for a CMOS inverter circuit, and (right) voltage transfer curve and logic symbol for the inverter.

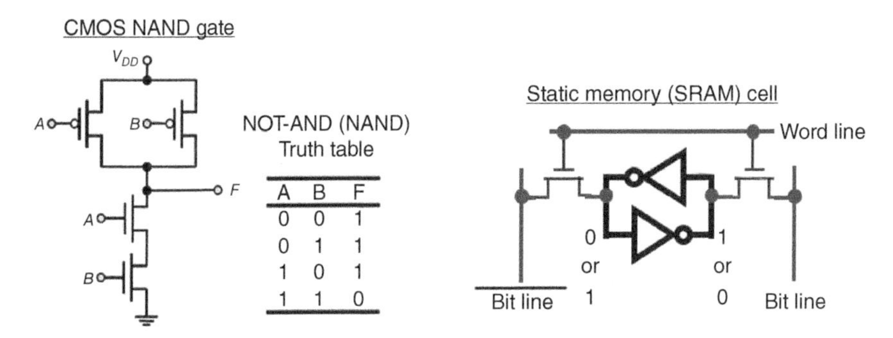

Figure 35.4 (left) Circuit diagram and truth table for a 2-input NAND logic gate; (right) circuit schematic for a 6-transistor static memory (SRAM) cell comprising cross-coupled CMOS inverters forming a latch.

Figure 35.4 (left). A microprocessor typically comprises many combinational (time-independent) logic circuits, clocked sequential logic circuits (whose output values depend not only on the present values but also on the past values of the input signals), and memory cells for temporary storage of data. Figure 35.4 (right) shows the circuit schematic for a static random access memory (SRAM) cell, which has two stable states so that it can store one bit of information.

The total power (P_{total}) dissipated by a CMOS IC comprises two major components: dynamic dissipation associated with charging/discharging of nodal capacitances, which is proportional to the average capacitance charged every clock cycle (C_{eff}), the clock frequency (f), and V_{DD} squared; and static dissipation due to transistor off-state leakage in all of the logic gates:

$$P_{total} = f \times C_{eff} \times (V_{DD})^2 + I_{OFF,all} \times V_{DD} \tag{35.7}$$

35.3.3 Threshold Voltage Design Trade-off

A large value of $|V_T|$ is desirable for lower I_{OFF}, for lower static power dissipation, whereas a smaller value of $|V_T|$ is desirable for higher I_{ON}, for faster circuit operation; thus, there is a fundamental design trade-off for V_T. Modern IC manufacturing processes offer multiple (at least 3) choices of V_T, for both n-channel and p-channel MOSFETs, to give chip designers the ability to separately optimize the power-performance trade-off for different block of circuitry on the chip.

35.4 MOSFET Scaling Challenges

35.4.1 Short-Channel Effects

As the gate length L_G is scaled down, the Drain is physically closer to the Source, causing V_T to be reduced because the built-in electric field of the Drain p–n junction helps to deplete a larger proportion of the semiconductor beneath the Gate. This reduction in V_T with decreasing L_G is referred to as the "short-channel effect" and it is undesirable because unavoidable process-induced variations in L_G (from device to device on a chip, across dies on a wafer, from wafer to wafer, and from lot to lot) result in significant variations in V_T and hence in I_{OFF} and I_{ON}, leading to undesired variations in IC static power consumption and performance, respectively. L_G scaling also worsens the undesirable drain-induced barrier lowering (DIBL) effect of a dynamic reduction in $|V_T|$ increasing I_{OFF} with increasing $|V_{DS}|$. Finally, increased C_D relative to C_{ox} with decreasing L_G degrades the subthreshold swing, resulting in undesirably larger I_{OFF}.

To help suppress the aforementioned short-channel effects, the depth of the S/D regions must be scaled down together with the channel length. Shallow S/D extensions (cf. Figure 35.1) were introduced to facilitate L_G scaling below 50 nm while maintaining reasonably low S/D series resistance [4]. Also, the dopant concentration in the semiconductor channel region can be increased to reduce the extent of drain junction depletion into the channel region [5]. However, the higher channel-region dopant concentration results in degraded μ_{eff} (due to increased ionized impurity scattering as well as a larger transverse electric field within the inversion layer); also a concomitant reduction in drain junction depletion length results increased band-to-band tunneling across the reverse-biased Drain-Channel/Body p–n junction resulting in higher off-state I_{DS}, limiting the efficacy of this approach.

35.4.2 Random Dopant Fluctuation Effect

As the MOSFET gate length L_G and channel width W are proportionately scaled down, and the channel-region dopant concentration is increased, the volume of the depleted semiconductor region shrinks, so that the number of ionized dopants (N_{dep}) in this region becomes statistically small, less than ~150 for minimally sized transistors (e.g. used in compact SRAM cells) with sub-50 nm L_G. Since the standard deviation in the number of dopant atoms increases with the inverse of the square root of N_{dep}, random variation in V_T due to Q_{dep} variation increases rapidly with bulk-silicon MOSFET scaling below 50 nm [6]. This issue is exacerbated by the fact that V_T depends more so on the number and location of dopant atoms in the channel region near to the Source p–n junction.

35.4.3 Parasitic Resistance

For a MOSFET in the on state, current flows through the metal–semiconductor (S/D) contacts and the doped S/D regions to "access" the channel that is modulated by the applied Gate voltage. If the contacts and/or the S/D regions connected in series with the channel present non-negligible "parasitic" resistance to current flow then they sustain non-negligible voltage drops in the on state, effectively reducing the voltages (V_{GS} and V_{DS}) applied to the intrinsic transistor, thereby undesirably decreasing I_{DS}, resulting in slower IC operation.

The resistance of each of the shallow S/D extension regions of a MOSFET is given by:

$$R_{SDE} = \rho_{SDE} \times L_{SDE} / \left(W \times x_j\right) \tag{35.8}$$

where ρ_{SDE} is the resistivity (inversely proportional to the dopant concentration) in these regions, and L_{SDE} and x_j are the length and depth of these regions, respectively. As x_j is scaled down with L to suppress short-channel effects, R_{SDE} increases. It should be noted that L_{SDE} can be reduced to compensate for this, but at a trade-off of increasing parasitic capacitance between the Gate and metal contact hole "plugs" which results in slower circuit operation and higher power consumption (cf. Eq. 35.7).

The resistance of an ohmic metal–semiconductor contact is given by:

$$R_C = \rho_C / A_C \tag{35.9}$$

where ρ_C is the specific contact resistivity (units: Ω-cm^2) and A_C is the contact area (units: cm^2). With transistor miniaturization, the length and width of the S/D regions are scaled down, so that A_C is geometrically reduced, resulting in increased R_C unless ρ_C can be commensurately reduced.

35.4.4 Thin-Body MOSFET Structures

To more effectively suppress short-channel effects and to circumvent RDF-induced variations, thin-body MOSFET structures (with semiconductor body thickness smaller than L_G) were proposed [7, 8], developed [9, 10], and eventually adopted in high-volume manufacturing to enable gate-length scaling below 25 nm. In 2011, Intel Corp. was the first to adopt "three-dimensional" transistor structures called FinFETs in which the Gate/gate-oxide stack wraps around three sides of a fin-shaped channel region for improved Gate control, in high-volume manufacturing for its 22 nm technology node [11].

A technological challenge introduced by thin-body MOSFET technology is that Q_{dep} is small and has negligible impact on V_T, i.e. $|V_{ox}|$ in Eq. (35.5) is negligible and $|V_S|$ also is small. Therefore, different gate materials with work functions closer to the mid-gap (rather than conduction- and valence-band edges) of silicon are required to achieve the desired values of V_T.

35.5 MOSFET Materials Innovations

35.5.1 Improving Intrinsic Transistor Performance

Silicon-germanium ($Si_{1-x}Ge_x$) S/D regions were originally proposed for reduced parasitic resistance [12] and subsequently shown to be beneficial for enhancing effective hole mobility in p-channel MOSFETs because they induced uniaxial compressive strain in the channel region [13]. Eventually, Si:C was used to induce uniaxial tensile strain and thereby enhance effective electron mobility in n-channel MOSFETs [14]. Stressed SiN_x contact-etch-stop liner (CESL) layers deposited over the transistor structures have been used as an additional method for inducing strain in the channel region [15].

Due to the need to avoid increasing short-channel effects and RDF-induced V_T variations, planar bulk-silicon MOSFET scaling proceeded without L_G scaling for multiple CMOS process technology generations (90, 65, 45, and 32 nm) [16–19]. To achieve improvements in I_{ON}, alternative approaches to L reduction were developed: strained silicon for increased μ_{eff} starting for 90 nm technology [16]; and high-permittivity (high-κ) metal gate stacks for increased C_{ox} starting for 45 nm technology [18, 19].

Although thin-body MOSFET structures are effective for suppressing short-channel and RDF effects, they present new challenges for continued L_G scaling toward atomic dimensions. The effective mobilities of electrons and holes degrade with decreasing silicon body thickness below 5 nm due to surface roughness scattering [20]. Therefore, alternative semiconductor materials that can provide for higher mobilities, such as $Si_{1-x}Ge_x$ for p-channel FinFETs [21], recently have been adopted.

35.5.2 Improving Threshold Voltage Control

To achieve multiple V_T values without the need to adjust channel/Body doping, heavily p-type doped (p+) polycrystalline silicon-germanium (poly-$Si_{1-x}Ge_x$) was the first proposed tunable work function gate material [22]. It was used as a near-midgap gate material to achieve a reasonable value of V_T in the first p-channel FinFETs [23]. The Ni–Ti metal interdiffusion technique was introduced to achieve different gate work function values for thin-body MOSFETs in an integrated CMOS fabrication process flow [24]; since then, a TiN/Al-based tunable gate work function technology has been developed [25]. It should be noted here that gate work function variability (due to grain orientation dependency of metal work function) is the largest source of random variation in V_T for thin-body MOSFETs [26]. Metal-gate/oxide dipole formation is a relatively new technique for adjusting V_T [27] that may be advantageous for "gate-all-around" (GAA) MOSFETs (which Samsung introduced in 2022 for its 3 nm technology node [28]) which can achieve even better electrostatic integrity than the FinFET but cannot as easily incorporate thick gate material stacks.

35.5.3 Reducing Parasitic Resistance

Materials innovations have enabled S/D series resistance and specific contact resistivity to be reduced with advances in IC manufacturing. Titanium was the first metal used to convert the top portion of the S/D contact regions into silicide, i.e. $TiSi_2$ (which is much more electrically conductive than heavily doped silicon) by metal deposition and subsequent thermal annealing to react the metal (cf. Figure 35.1) with silicon [29], to effectively increase A_C. At the time, heavily doped polycrystalline silicon (poly-Si) was the standard material used for MOSFET Gate electrodes, so the top surface of the Gate electrode was also silicided in this process; the Gate-sidewall spacers (colored light purple in Figure 35.1) allowed the Gate and S/D regions to be silicided in a self-aligned manner. Over the years, with transistor miniaturization, alternative metals (cobalt, nickel) were used to form self-aligned silicide ("salicide"): due to difficulty of forming low-resistivity-phase $TiSi_2$ in small features (i.e. the Gate electrodes), $CoSi_2$ started to be used in MOSFETs with gate lengths below approximately 0.25 µm [30]; to mitigate issues with silicon consumption during the silicidation process – which becomes more problematic with S/D junction depth scaling because it can result in silicide spiking through the junction resulting in undesirable Drain–Body current conduction – NiSi started to be used in MOSFETs with gate lengths below approximately 100 nm [31]. For sub-10 nm CMOS technology nodes, nickel-platinum silicide (NiPtSi) can be used to contact the $Si_{1-x}Ge_x$ S/D regions with ultralow ρ_C (~10^{-9} Ω-cm^2) [32], and titanium silicidation can be used to lower the contact resistance to a similarly low level [33]; therefore, Ti and NiPt silicides are the semiconductor-contacting materials of choice for n-channel and p-channel FinFETs, respectively [34]. In the future, further advancements in low-temperature epitaxial growth of $Si_{1-x}Ge_x$ to achieve higher dopant concentration and new contact materials (e.g. scandium) for n-type silicon to provide for lower Schottky barrier height [35] will be needed to achieve even lower values of ρ_C to compensate for continued scaling of S/D contact area.

35.6 Outlook for Continued Transistor Scaling

Quantum mechanical tunneling through the Source potential barrier sets a fundamental scaling limit for MOSFET channel length scaling [36]. Alternative semiconductor materials that provide for a larger potential barrier (i.e. with a larger bandgap energy than silicon) are needed to overcome this scaling limit. Wider-bandgap two-dimensional (2D) semiconductor materials such as transition metal dichalcogenides (TMDCs) can be used to realize ultrathin and atomically smooth semiconductor channels for superior suppression of short-channel effects and surface-roughness scattering, to facilitate gate-length scaling to 1 nm [37]. Therefore, transistor miniaturization is projected to continue – enabled by materials innovations – well into the future.

Three-dimensional stacking of transistors, e.g. n-channel MOSFETs located in a separate device layer above p-channel MOSFETs, have been proposed as a parallel technological pathway for continuing to improve transistor density (i.e. the number of transistors per unit area) [38]. Thin-film semiconductor materials that can be formed at relatively low process temperatures (<400 °C), such as TMDCs, oxide semiconductors (e.g. indium gallium zinc oxide, IGZO), can facilitate this pathway while providing for improved current per unit layout area.

A fundamental challenge for increasing transistor density is increasing power density. CMOS energy efficiency is fundamentally limited by transistor off-state leakage current, which limits the benefits of voltage (V_{DD}) scaling [39]: if V_{DD} is lowered (to reduce dynamic power dissipation, cf. Eq. (35.7)) to be below V_T, then it takes exponentially longer for a digital IC to perform its function; the resultant exponential increase in energy consumed due to static power dissipation results in increased total energy consumed per digital operation. Material innovations such as an ultrathin ferroelectric gate-insulating film that has the effect of negative gate capacitance [40], or a tunneling semiconductor heterojunction material system that provides for high on/off ratio [41], which enable sub-60 mV dec^{-1} subthreshold swing – hence, lower V_T for a fixed I_{OFF} specification – will be needed to address this challenge and fully unlock the benefits of continued transistor scaling.

References

[1] Moore, G.E. Cramming more components onto integrated circuits, Reprinted from Electronics, volume 38, number 8 April 19, 1965, pp. 114 ff. *IEEE Solid-State Circuits Society Newsletter* 11 (3): 33–35. https://doi.org/10.1109/N-SSC.2006.4785860.

[2] Auth, C., Aliyarukunju, A., Asoro, M. et al. (2017). A 10nm high performance and low-power CMOS technology featuring 3rd generation FinFET transistors, self-aligned quad patterning, contact over active gate and cobalt local interconnects. *2017 IEEE International Electron Devices Meeting (IEDM)*, San Francisco, CA, USA (December 2017), pp. 29.1.1–29.1.4. http://doi.org/10.1109/IEDM.2017.8268472.

[3] Lundstrom, M. and Ren, Z. (2002). Essential physics of carrier transport in nanoscale MOSFETs. *IEEE Transactions on Electron Devices* 49 (1): 133–141. https://doi.org/10.1109/16.974760.

[4] Ono, M., Saito, M., Yoshitomi, T. et al. (1993). Sub-50 nm gate length n-MOSFETs with 10 nm phosphorus source and drain junctions. *Proceedings of IEEE International Electron Devices Meeting*, Washington, DC, USA (December 1993), pp. 119–122. http://doi.org/10.1109/IEDM.1993.347385.

[5] Bin, Y., Wann, C.H.J., Nowak, E.D. et al. (1997). Short-channel effect improved by lateral channel-engineering in deep-submicronmeter MOSFET's. *IEEE Transactions on Electron Devices* 44 (4): 627–634. https://doi.org/10.1109/16.563368.

[6] Asenov, A., Brown, A.R., Davies, J.H. et al. (2003). Simulation of intrinsic parameter fluctuations in decananometer and nanometer-scale MOSFETs. *IEEE Transactions on Electron Devices* 50 (9): 1837–1852. https://doi.org/10.1109/TED.2003.815862.

[7] Yu, B., Tung, Y.-J., Tang, S. et al. (1997). Ultra-thin-body silicon-on-insulator MOSFET's for terabit-scale integration. *Proc. Int. Semiconductor Device Research Symp.*, Charlottesville, VA, USA (December 1997), pp. 623–626.

[8] Chang, L., Tang, S., King, T.-J. et al. (2000). Gate length scaling and threshold voltage control of double-gate MOSFETs. *International Electron Devices Meeting. Technical digest. IEDM (cat. No.00CH37138)*, San Francisco, CA, USA (December 2000), pp. 719–722. http://doi.org/10.1109/IEDM.2000.904419.

[9] Hisamoto, D., Lee, W.-C., Kedzierski, J. et al. (1998). A folded-channel MOSFET for deep-sub-tenth micron era. *International Electron Devices Meeting. Technical Digest (Cat. No.98CH36217)*, San Francisco, CA, USA (January 1998), pp. 1032–1034. http://doi.org/10.1109/IEDM.1998.746531.

[10] Yu, B., Chang, L., Ahmed, S. et al. (2002). FinFET scaling to 10 nm gate length. *Digest. International Electron Devices Meeting*, San Francisco, CA, USA (December 2002), pp. 251–254. http://doi.org/10.1109/IEDM.2002.1175825.

[11] Auth, C., Allen, C., Blattner, A. et al. (2012). A 22nm high performance and low-power CMOS technology featuring fully-depleted tri-gate transistors, self-aligned contacts and high density MIM capacitors. *Symposium on VLSI Technology (VLSIT)*, Honolulu, HI, USA (June 2012), pp. 131–132. http://doi.org/10.1109/VLSIT.2012.6242496.

[12] Takeuchi, H., Lee, W.-C., Ranade, P., and King, T.-J. (1999). Improved PMOSFET short-channel performance using ultra-shallow Si0.8Ge0.2 source/drain extensions. *International Electron Devices Meeting. Technical digest (cat. No.99CH36318)*, Washington, DC, USA (December 1999), pp. 501–504. http://doi.org/10.1109/IEDM.1999.824202.

[13] Thompson, S.E., Armstrong, M., Auth, C. et al. (2004). A logic nanotechnology featuring strained-silicon. *IEEE Electron Device Letters* 25 (4): 191–193. https://doi.org/10.1109/LED.2004.825195.

[14] Yeo, Y.-C. (2006). Enhancing CMOS transistor performance using lattice-mismatched materials in source/drain regions. *International SiGe Technology and Device Meeting*, Princeton, NJ, USA (May 2006), pp. 1–2. http://doi.org/10.1109/ISTDM.2006.246557.

[15] Yang, H.S., Malik, R., Narasimha, S. et al. (2004). Dual stress liner for high performance sub-45nm gate length SOI CMOS manufacturing. *IEDM Technical Digest. IEEE International Electron Devices Meeting*, San Francisco, CA, USA (December 2004), pp. 1075–1077. http://doi.org/10.1109/IEDM.2004.1419385.

[16] Thompson, S., Anand, N., Armstrong, M. et al. (2002). A 90 nm logic technology featuring 50 nm strained silicon channel transistors, 7 layers of Cu interconnects, low k ILD, and 1um^2 SRAM cell. *Digest. International Electron Devices Meeting*, San Francisco, CA, USA (December 2002), pp. 61–64. http://doi.org/10.1109/IEDM.2002.1175779.

[17] Tyagi, S., Auth, C., Bai, P. et al. (2005). An advanced low power, high performance, strained channel 65nm technology. *IEEE International Electron Devices Meeting. IEDM Technical Digest.*, Washington, DC, USA (December 2005), pp. 245–247. http://doi.org/10.1109/IEDM.2005.1609318.

[18] Mistry, K., Allen, C., Auth, C. et al. (2007). A 45nm logic technology with high-k+metal gate transistors, strained silicon, 9 Cu interconnect layers, 193nm dry patterning, and 100% Pb-free packaging. *IEEE International Electron Devices Meeting*, Washington, DC, USA (December 2007), pp. 247–250. http://doi.org/10.1109/IEDM.2007.4418914.

[19] Packan, P., Akbar, S., Armstrong, M. et al. (2009). High performance 32nm logic technology featuring 2nd generation high-k + metal gate transistors. *IEEE international electron devices meeting (IEDM)*, Baltimore, MD, USA (December 2009), pp. 1–4. http://doi.org/10.1109/IEDM.2009.5424253.

[20] Uchida, K., Watanabe, H., Kinoshita, A. (2002). Experimental study on carrier transport mechanism in ultrathin-body SOI nand p-MOSFETs with SOI thickness less than 5 nm. *Digest. International Electron Devices Meeting*, San Francisco, CA, USA (February 2002), pp. 47–50. http://doi.org/10.1109/IEDM.2002.1175776.

[21] Xie, R., Montanini, P., Akarvardar, K. et al. (2016). A 7nm FinFET technology featuring EUV patterning and dual strained high mobility channels. *2016 IEEE International Electron Devices Meeting (IEDM)*, San Francisco, CA, USA (December 2016), pp. 2.7.1–2.7.4. http://doi.org/10.1109/IEDM.2016.7838334.

[22] King, T.-J., Pfiester, J.R., Shott, J.D. et al. (1990). A polycrystalline-Si1-xGex-gate CMOS technology. *International Technical Digest on Electron Devices*, San Francisco, CA, USA (December 1990), pp. 253–256. http://doi.org/10.1109/IEDM.1990.237181.

[23] Huang, X., Lee, W.-C., Kuo, C. et al. (1999). Sub 50-nm FinFET: PMOS. *International Electron Devices Meeting. Technical Digest (Cat. No.99CH36318)*, Washington, DC, USA (December 1999), pp. 67–70. http://doi.org/10.1109/IEDM.1999.823848.

[24] Polishchuk, I., Ranade, P., King, T.-J., and Hu, C. (2001). Dual work function metal gate CMOS technology using metal interdiffusion. *IEEE Electron Device Letters* 22 (9): 444–446. https://doi.org/10.1109/55.944334.

[25] Lima, L.P.B., Dekkers, H.F.W., Lisoni, J.G. et al. (2014). Metal gate work function tuning by Al incorporation in TiN. *Journal of Applied Physics* 115: 074504. https://doi.org/10.1063/1.4866323.

[26] Nawaz, S.M., Dutta, S., Chattopadhyay, A., and Mallik, A. (2014). Comparison of random dopant and gate-metal workfunction variability between junctionless and conventional FinFETs. *IEEE Electron Device Letters* 35 (6): 663–665. https://doi.org/10.1109/LED.2014.2313916.

[27] Bao, R., Zhou, H., Wan, M. et al. (2018). Extendable and manufacturable volume-less multi-Vt solution for 7nm technology node and beyond. *2018 IEEE International Electron Devices Meeting (IEDM)*, San Francisco, CA, USA (December 2018), pp. 28.5.1–28.5.4. http://doi.org/10.1109/IEDM.2018.8614518.

[28] Bae, G., Bae, D.-I., Kang, M. et al. (2018). 3nm GAA technology featuring multi-Bridge-Channel FET for low power and high performance applications. *2018 IEEE International Electron Devices Meeting (IEDM)*, San Francisco, CA, USA (December 2018), pp. 28.7.1–28.7.4. http://doi.org/10.1109/IEDM.2018.8614629.

[29] Alperin, M.E., Hollaway, T.C., Haken, R.A. et al. (1985). Development of the self-aligned titanium silicide process for VLSI applications. *IEEE Journal of Solid-State Circuits* 20 (1): 61–69. https://doi.org/10.1109/JSSC.1985.1052277.

[30] Zhang, S.-L. and Smith, U. (2004). Self-aligned silicides for Ohmic contacts in complementary metal–oxide–semiconductor technology: TiSi2, CoSi2, and NiSi. *Journal of Vacuum Science & Technology A: Vacuum, Surfaces, and Films* 22 (4): 1361–1370. https://doi.org/10.1116/1.1688364.

[31] Iwai, H., Ohguro, T., and Ohmi, S. (2002). NiSi salicide technology for scaled CMOS. *Microelectronic Engineering* 60 (1): 157–169. https://doi.org/10.1016/S0167-9317(01)00684-0.

[32] Zhang, Z., Koswatta, S.O., Bedel, S.W. et al. (2013). Ultra low contact resistivities for CMOS beyond 10-nm node. *IEEE Electron Device Letters* 34 (6): 723–725. https://doi.org/10.1109/LED.2013.2257664.

[33] Yu, H., Schaekers, M., Rosseel, E. et al. (2015). 1.5×10–9 Ωcm^2 Contact resistivity on highly doped Si:P using Ge pre-amorphization and Ti silicidation. *2015 IEEE International Electron Devices Meeting (IEDM)*, Washington, DC, USA (December 2015), pp. 21.7.1–21.7.4. http://doi.org/10.1109/IEDM.2015.7409753.

[34] Adusumilli, P., Alptekin, E., Raymond, M. et al. (2016). Ti and NiPt/Ti liner silicide contacts for advanced technologies. *IEEE symposium on VLSI Technology*, Honolulu, HI, USA (June 2016), pp. 1–2. http://doi.org/10.1109/VLSIT.2016.7573382.

[35] Porret, C., Everaert, J.-L., Schaekers, M. et al. (2022). Low temperature source/drain epitaxy and functional silicides: essentials for ultimate contact scaling. *2022 International Electron Devices Meeting (IEDM)*, San Francisco, CA, USA (December 2022), pp. 34.1.1–34.1.4. http://doi.org/10.1109/IEDM45625.2022.10019501.

[36] Wang, J. and Lundstrom, M. (2002). Does source-to-drain tunneling limit the ultimate scaling of MOSFETs? *Digest. International Electron Devices Meeting*, San Francisco, CA, USA (December 2002), pp. 707–710. http://doi.org/10.1109/IEDM.2002.1175936.

[37] Desai, S.B., Madhvapathy, S.R., Sachid, A.B. et al. (2016). MoS2 transistors with 1-nanometer gate lengths. *Science* 354 (6308): 99–102.

[38] Thean, A., Tsai, S.-H., Chen, C.-K. et al. (2022). Low-thermal-budget BEOL-compatible beyond-silicon transistor technologies for future monolithic-3D compute and memory applications. *2022 International Electron Devices Meeting (IEDM)*, San Francisco, CA, USA (December 2022), pp. 12.2.1–12.2.4. http://doi.org/10.1109/IEDM45625.2022.10019511.

[39] Calhoun, B.H., Wang, A., and Chandrakasan, A. (2005). Modeling and sizing for minimum energy operation in subthreshold circuits. *IEEE Journal of Solid-State Circuits* 40 (9): 1778–1786. https://doi.org/10.1109/JSSC.2005.852162.

[40] Salahuddin, S. and Datta, S. (2008). Use of negative capacitance to provide voltage amplification for low power nanoscale devices. *Nano Letters* 8 (2): 405–410. https://doi.org/10.1021/nl071804g.

[41] Choi, W.Y., Park, B.-G., Lee, J.D., and Liu, T.-J.K. (2007). Tunneling field-effect transistors (TFETs) with subthreshold swing (SS) less than 60 mV/dec. *IEEE Electron Device Letters* 28 (8): 743–745. https://doi.org/10.1109/LED.2007.901273.

Chapter 36

Germanium

Back to the Future

Krishna C. Saraswat

Department of Electrical Engineering, Stanford University, Stanford, CA, USA

36.1 Introduction

The transistor has been one of the most important innovations of the past century and has revolutionized the world by ushering in the era of digital electronics. In 1947, Walter Brattain and John Bardeen demonstrated the point-contact transistor amplifier and in 1948, William Shockley invented the junction transistor semiconductor amplifier. The trio were awarded the Nobel Prize in physics in 1956 for their inventions [1]. Soon after, the inventions of the integrated circuit (IC) [2, 3] and the silicon metal-oxide-semiconductor field-effect transistor (MOSFET) [4] led to the birth of modern-day electronics as we know it. In 1965, Gordon Moore observed that the number of transistors on integrated circuits was doubling every two years [5]. This trend came to be known as Moore's Law and has been the driving force behind the meteoric rise of the semiconductor industry over the past half century. This growth has largely been powered by the scaling of transistors to smaller dimensions, enabling the exponential increase in transistor density.

Figure 36.1 shows technology progression through disruptive innovations. For many decades this was done by simply shrinking transistors dimensions, in what is known as classical scaling or Dennard scaling [6]. However, around the early 2000s, the benefits gained by classical scaling began to diminish once transistor gate lengths shrunk below 100 nm. This ushered in the equivalent scaling era, where new innovations had to be made to continue increasing transistor and circuit performance. Some of these major breakthroughs that enabled the continuation of Moore's Law were the introduction of silicon germanium (SiGe) source/drain at the 90 nm node, high-dielectric constant (high-κ) gate dielectric

75th Anniversary of the Transistor, First Edition. Edited by Arokia Nathan, Samar K. Saha, and Ravi M. Todi.

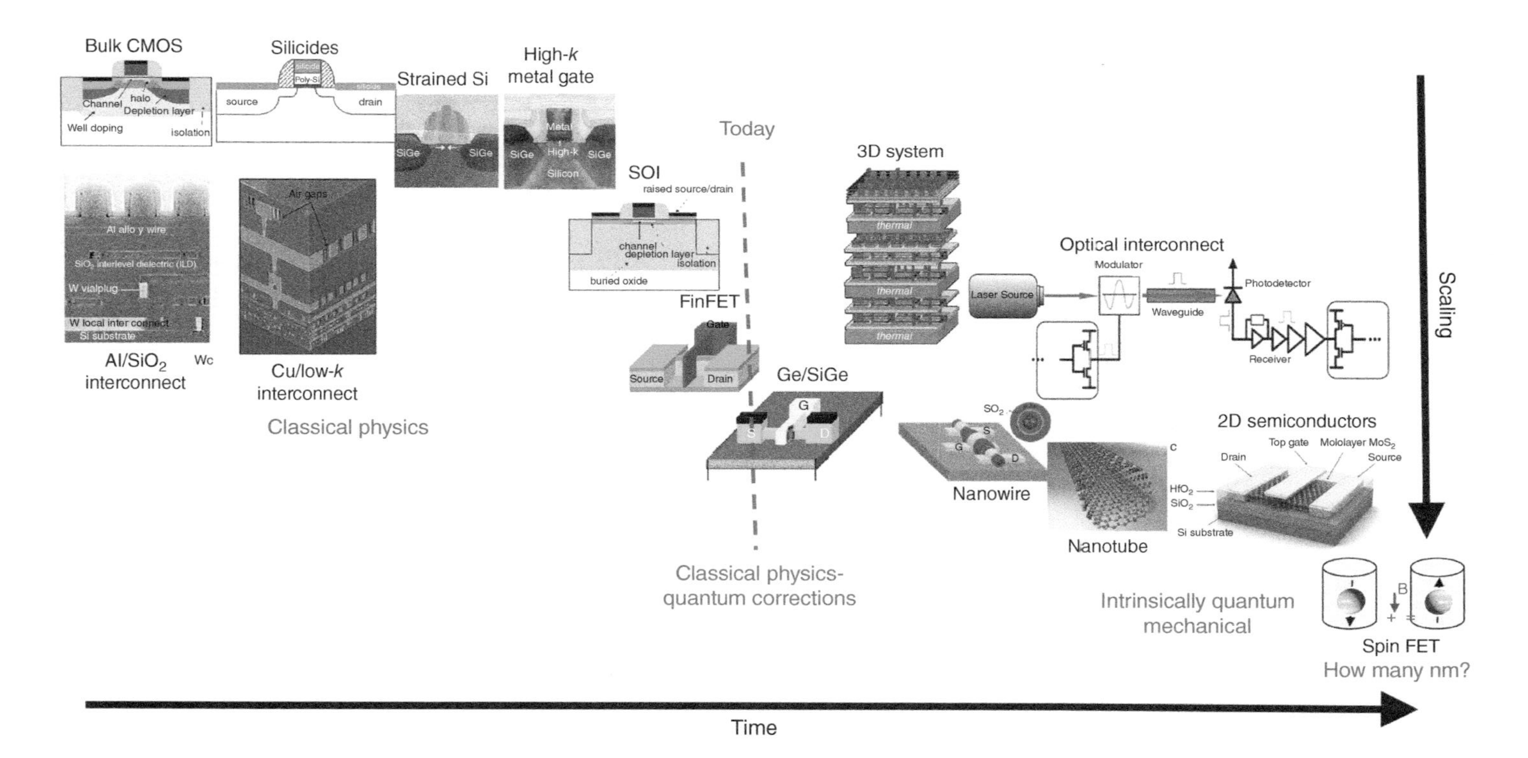

Figure 36.1 For many decades transistor performance was improved by shrinking transistor dimensions, in what is known as Dennard scaling era. The subsequent equivalent scaling era saw the introduction of strained channel, high-κ metal gate, and FinFET architectures. Future improvements will require innovations in materials, devices, and heterogeneous integration technologies.

and metal gate at the 45 nm node, and the transition from planar transistors to the FinFET architecture at the 22 nm node [7]. It is important to note that in the equivalent scaling era the technology node does not correspond to the minimum transistor feature size as it did previously and is primarily used as a marketing term to indicate the advancement of system performance.

Throughout all of these changes, the one constant has been the use of Si as the transistor material since the 1960s. Interestingly, Brattain and Bardeen's first point-contact transistor in 1947 as well as Shockley's bipolar junction transistors (BJTs) in 1948 and Jack Kilby's hybrid IC in 1958 were made using Ge as the transistor material. However, Si soon replaced Ge and became the semiconductor of choice due in large part to the high-quality interface it forms with silicon dioxide, which enabled the formation of high-quality gate stacks. Even after this advantage was removed with the introduction of high-κ/metal gate, Si channel transistors continued to yield high performance because of the advancements discussed earlier. However, as we advance beyond the so-called 5 nm node, Si is beginning to reach its limits, and alternative channel materials are being heavily researched to potentially augment Si. In fact, Taiwan Semiconductor Manufacturing Company (TSMC) introduced SiGe as the channel material in their p-channel metal-oxide-semiconductor (PMOS) transistors at the 5 nm node [8], marking the first time that a transistor without a pure Si channel has been used in mass manufacturing. Use of the SiGe channel led to an average performance enhancement of 18% over the Si channel transistors.

36.2 Need for High Mobility Material for MOS Channel

The primary motivation for introducing a new channel material is to improve the transistor drive current and switching speed and reduce power drain. In devices where the channel length is long, charge carriers undergo many scattering events as they travel from the source to the drain and reach velocity saturation. However, at channel lengths less than 100 nm, injected carriers can traverse the channel before reaching saturation velocity, as shown in Figure 36.2. In this regime, carrier transport can be considered to be quasi-ballistic and the drive current (*IDsat*) can be described by [9, 10]:

$$I_{Dsat} = qN_{source}v_{inj}\left(\frac{1-r}{1+r}\right)$$

where q is the elementary charge, N_{source} is the charge carrier density at the source, v_{inj} is the carrier injection velocity at the source, and r is the back scattering coefficient. We see that a higher injection velocity and lower back scattering rate are critical for increasing drive current in the quasi-ballistic regime. v_{inj} can be increased by using a channel material with a lower carrier effective mass along the channel direction. The mobility in a material, μ, is inversely related to both the carrier effective mass and the back scattering rate, meaning that the low-field mobility can still be a good indicator for

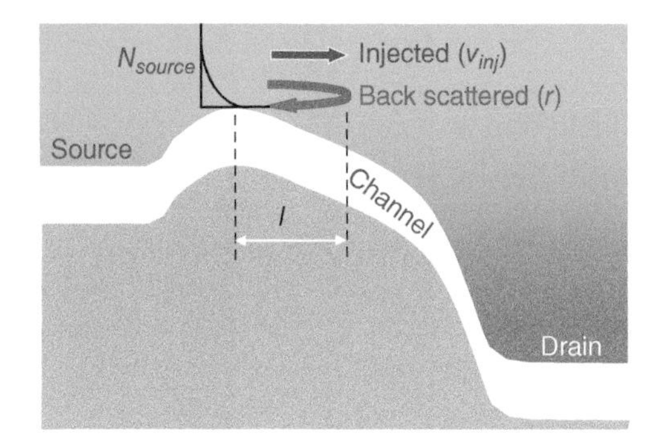

Figure 36.2 Carrier transport in a MOSFET.

Table 36.1 Properties of bulk, undoped Si, Ge, InAs, and GaSb at room temperature.

Material property	Si	Ge	InAs	GaSb	Importance
μ_n ($cm^2 V^{-1} s^{-1}$)	1600	3900	40,000	6000	I_{ON}
μ_p ($cm^2 V^{-1} s^{-1}$)	430	1900	500	1000	I_{ON}
Bandgap E_G (eV)	1.12	0.66	0.36	0.72	I_{OFF}
Dielectric constant	11.6	16	14.8	15.7	SS
Electron barrier (eV)	$\sim 2E_G/3$	0.6	−0.2	0.7	ρ_c N^+ S/D
Hole barrier (eV)	$\sim E_G/3$	0.06	0.38	0	ρ_c P^+ S/D

predicting current drive under quasi-ballistic transport. Therefore, replacing Si with a higher mobility material can increase the transistor drive current.

Table 36.1 lists the bulk electron and hole mobilities for several undoped semiconductors. Ge has the highest hole mobility (μ_p) of all the candidate materials shown here by a significant margin. The electron mobility (μn) of Ge is also 2.5× higher than Si. The relatively large and symmetric electron and hole mobilities suggest that Ge can be used as the channel material for both high-performance PMOS and NMOS transistors, making it a particularly attractive candidate to replace Si. Also, Ge is compatible with current Si processing techniques, making it easier to be adopted by industry. This is evidenced by the use of SiGe in the source/drain and in the channel for PMOS as mentioned before.

Note that GaAs and other III–Vs have very high μn, but their density of states is low due to low effective mass, meaning they are not suitable for achieving high electron carrier concentrations and drive currents [13]. They also suffer from high off-state leakage currents.

Significant research has been done over the past few decades to address some of the major challenges facing Ge transistor technology, including defect passivation at the gate oxide interface [14–16] and reduction of leakage current caused by its small bandgap [17]. High-performance Ge PMOS has been demonstrated, but it has been very difficult to achieve high drive currents in Ge NMOS. This is one of the last major hurdles that needs to be overcome for realization of Ge CMOS technology.

In this paper, we will describe major advances made during the last couple of decades in Ge technology.

36.3 Surface Passivation of Ge-Based MOSFETs

Ge is a promising MOSFET channel material candidate with numerous advantages over Si; however, unlike Si, it lacks a stable native oxide for MOSFET gate insulation and IC field isolation. For instance, a mixture of Ge oxides (GeO_x and GeO_2) would form on the Ge surface upon air exposure with the former desorbs at moderate temperatures while the latter dissolves in water. The water-soluble nature of the Ge oxide was one of the key properties that led to the success of the point-contact transistor, yet it cannot become a high-quality gate dielectric for Ge MOSFET applications. The first Ge n-MOSFET demonstration archived in literature could be dated back in 1965 with pyrolytically decomposed silicon dioxide (SiO_2) gate dielectric and antimony (Sb) out-diffused junctions [9]. In contrast, the first Ge p-MOSFET was realized in 1975 using chemical vapor deposited (CVD) SiO_2 gate dielectric and boron ion-implanted source and drain [10]. Subsequently, several Ge n-MOSFETs and p-MOSFETs were fabricated using various MOS gate dielectric and junction formation technologies [13–17]. Even so, none of those technologies would be applicable for deeply scaled MOSFETs. This mandated the development and evaluation of more advanced Ge CMOS technologies.

Inspired by the success of the high-κ gate dielectric techniques on Si, in 2002, we investigated the possibility of applying high-κ dielectrics to Ge. In that work, we focused on the electrical characteristics of the MOS capacitors using ZrO_2 as the gate dielectrics (Figure 36.3a) [18] followed by demonstrated Ge MOSFETs using HfO_2 as the gate dielectrics [14]. In our subsequent work, Ge passivation with its native oxynitride (GeO_xN_y) and HfO_2 or ZrO_2 deposited in an atomic-layer deposition (ALD) system was studied. The optimum dielectric stack could be attained by rapid thermal nitridation (RTN) of Ge in ammonia to form GeO_xN_y followed by ALD of the high-κ film. Excellent electrical characteristics were obtained from MOSCAPs with low leakage, good *C*–*V* characteristics, and reasonably low interface state density (Figure 36.3b, c) [15]. The RTN technique was also employed to passivate the Ge surface prior to the deposition of SiO_2 for field isolation. During the ensuing years, surface passivation of Ge has been extensively investigated by many researchers with high-κ metal oxides of Zr, Hf, Al, La, and Er. In another work Ge oxidation in ozone at 400 °C to grow GeO_2 followed by high-κ ALD gave very low values of $D_{it} \sim 2\times10^{11}\,cm^{-2}V^{-1}$ [19]. However, these techniques were not suitable for obtaining ultralow gate dielectric thickness. In a later work <1 nm EOT gate dielectric with low D_{it} was demonstrated by first depositing a high-κ dielectric on Ge and then oxidize it either using oxidation by ozone [16] or oxygen plasma [20].

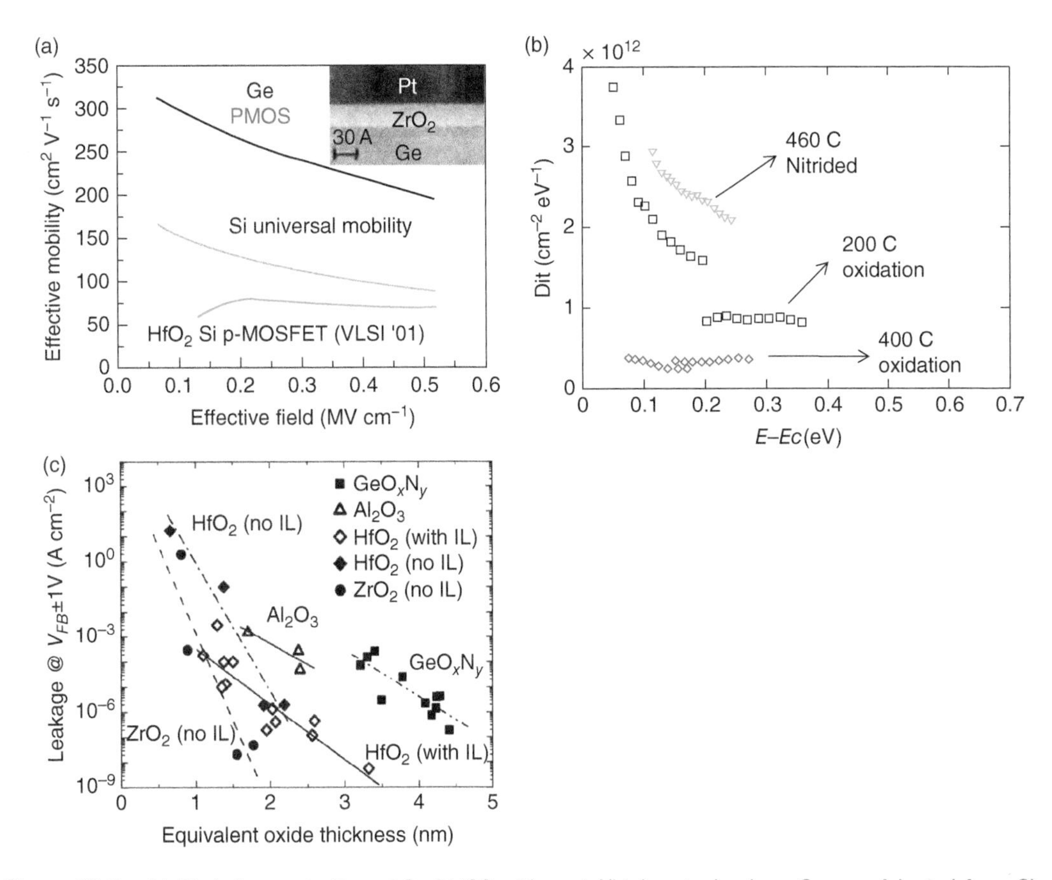

Figure 36.3 (a) First demonstration of Ge PMOS with metal/high-κ technology. Source: Adapted from Chui et al. [18]. (b) Optimization if interface state density in Ge MOS systems by oxidation. Source: Adapted from Kuzum et al. [12]. (c) Gate leakage in various Ge gate dielectrics. Source: Chui et al. [15]/IEEE.

36.4 Low Resistance Contacts to Ge

As device scaling continues, parasitic resistance largely dominated by contact resistance, limits the device performance. Specific contact resistivity, ρ_c, of a metal–semiconductor (M/S) contact is dependent on the Schottky barrier height, Φ_B, and the electrically active dopant density N at that interface. In M/S contacts the metal Fermi level is pinned at the charge neutrality level (CNL), ECNL due to metal-induced gap states (MIGS), resulting in fixed electron and hole Schottky barrier heights (Figure 36.4a). Historically, the method to reduce ρ_c is by increasing N to $>1E20\,\text{cm}^{-3}$ thereby thinning the barrier (Figure 36.4b), thus allowing more tunneling current. This method works well for p-Ge which can be doped heavily approaching $10^{21}\,\text{cm}^{-3}$. Furthermore, p-Ge with low barrier to holes also gives low ρ_c combined with heavy doping.

High Φ_B and low N results in high ρ_c, eclipsing the promise of intrinsic performance of scaled devices. Large parasitic contact resistance to n-Ge exists due to metal Fermi-level pinning near the

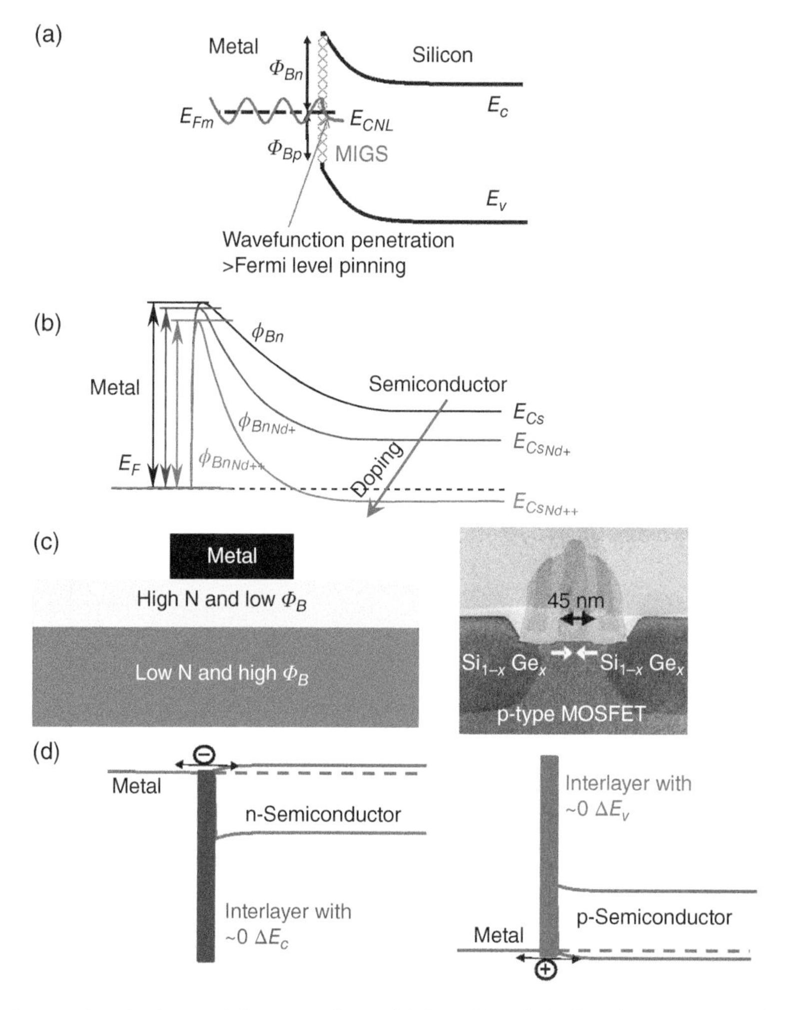

Figure 36.4 Approaches for low resistance contacts. (a) Fermi level pinning due to metal-induced gap states (MIGS). (b) Decrease barrier thickness by increasing doping. Best for n-Si, p-Ge, and n-InAs. (c) Heterostructures, e.g. p-Ge on p-Si n-InGaAs on n-Ge. Source: S.E. Thompson et al/IEEE. (d) De-pinning Fermi level by metal-interlayer-semiconductor (MIS) ZnO for n-contacts NiO for p-contacts.

valence band resulting in high Φ_B. This provides a large tunnel barrier to the electrons in n-MOS. Furthermore, it is difficult to obtain heavy electrically active n-type doping in Ge.

One method for increasing active dopant concentration is by laser annealing. Laser annealing has been used to obtain heavy n-type doping and low ρ_c to n-Ge. Highly doped Si:P epitaxial growth was performed in contacts resulting in heavy chemical doping of phosphorus $>10^{21}\,cm^{-3}$, followed by millisecond laser anneal to achieve high activation of electrons. Low contact resistivity of 2E-9 $\Omega\,cm^2$ was demonstrated in n-Ge [21]. However, thermal stability of laser annealed n-Ge is problematic as during subsequent processing at higher temperatures the dopant may deactivate increasing contact resistance.

Another method is interlayer placement between metal and Ge to reduce effective barrier height. We have experimentally demonstrated low contact resistivities of III–V/Ge heterocontacts to n-Ge [22]. III–V is favorable in that the CNL is located near the conduction band as well as its low electron effective mass, beneficial for high electron transmission. 60× improvement is achieved in the hetero-contact at N_d of $3\times10^{19}\,cm^{-3}$. With increase in electron density, ρ_c can be further decreased (Figure 36.5).

A variation of this method is by inserting a wider bandgap metal oxide interlayer between the metal and semiconductor (MIS) with the aim of de-pinning the metal Fermi level and thus reduce the Schottky barrier height (Figure 36.4c) [23]. In accordance with the MIGS theory, the ultrathin wider bandgap interlayer physically separates the two materials, allowing the metal electron wavefunction to be attenuated in the interlayer prior to penetrating the semiconductor. The attenuated metal states result in fewer charges available to drive E_F toward E_{CNL}. After de-pinning the Fermi level, the metal workfunction Φ_M can then be used to tune the effective barrier height. Metal workfunction now pins at the interlayer. By choosing a metal with proper workfunction and by choosing an interlayer with E_{CNL} close to the band edge and larger pinning factor, the overall barrier between metal and semiconductor can be minimized. For n-type ohmic contacts, metals with a low Φ_M near the semiconductor conduction band should result in a near-zero barrier height.

Simulations in [24] have shown that ZnO, ITO, and TiO_2 are excellent candidates for this as they have near-zero difference between the conduction bands of the metal and many semiconductors including Ge. Furthermore, they can be doped very heavily and thus the barrier between metals and these oxides becomes very thin. Of these, heavily doped ZnO and ITO appear to be better candidates as transmission coefficient of electrons through them is much higher than TiO_2 [24]. Experimentally, several demonstrations have shown the use of ZnO [25, 26] and ITO [25] to reduce Φ_{BN} and ρ_c in the case of n-Ge.

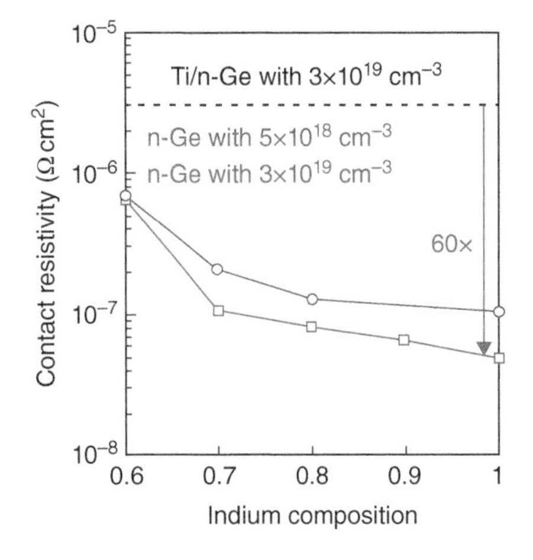

Figure 36.5 ρ_c vs. In composition with different N_d. The blue line and the red one correspond to Ge with electron density of $5\times10^{18}\,cm^{-3}$ and $3\times10^{19}\,cm^{-3}$, respectively. Source: Suh et al. [22]/arXiv.

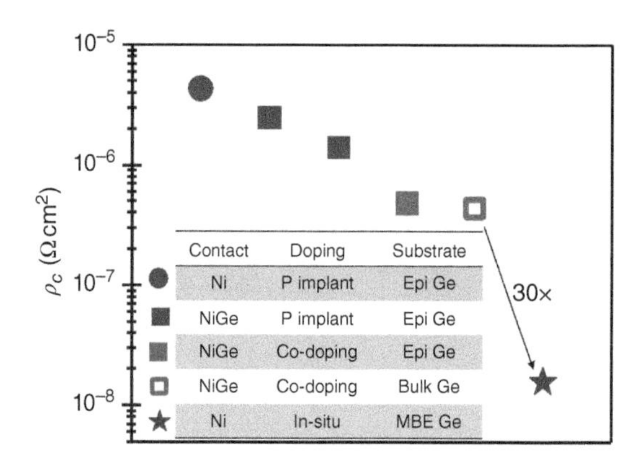

Figure 36.6 Comparison of contact schemes to n-type Ge. NiGe formation and co-doping both improve ρ_c. Nonthermal equilibrium low-temperature MBE of Ge with in-situ Sb doping gives best results. Source: Adapted from Ramesh [27].

Low temperature nonthermal equilibrium molecular beam epitaxy (MBE) of Ge with in-situ Sb doping can be used to realize heavy n-type electrically active doping $>10^{20}$ cm^{-3}. We fabricated nickel (Ni) contacts on n-Ge substrate to achieve an ultralow ρ_c of $1.6 \times 10^{-8}\,\Omega$cm^2, which is among the lowest reported values to date for contacts to n-type Ge (Figure 36.6). This work demonstrates that low resistance contacts to n-type Ge can be achieved and provides a pathway toward the realization of Ge NMOS technology [27, 28].

36.5 Heteroepitaxial Growth of Ge on Si

In order for Ge to make inroads into semiconductor products, it must be integrated onto Si. Thus, it is critical to develop new methods for heteroepitaxial Ge technology because its growth on Si is hampered by the large lattice mismatch (4.2%). This mismatch results in growth that is dominated by "islanding" and misfit dislocations that are formed at the Si substrate/Ge film interface terminating at the film surface as threading dislocations, thus degrading device performance.

We invented a novel technique for growing thick films of relaxed high quality heteroepitaxial Ge layers on Si. The technique involves CVD growth of a thin film of Ge on Si at a low temperature (~400 °C), followed by in-situ H_2 annealing at a high temperature (~825 °C) and hence the name Multiple Hydrogen Annealing for Heteroepitaxy (MHAH) [29]. Following the first Ge growth and H_2 annealing, the Ge surface roughness is reduced by 90% to 2.5 nm rms. Multiple MHAH cycles are done to obtain desired thickness. Using this process, we have demonstrated the growth of heteroepitaxial-Ge on silicon, with defects confined near the Si/Ge interface (Figure 36.7b). The results achieved are smooth single-crystal Ge layers on Si with threading dislocations as low as 3×10^6 cm^{-2} and surface roughness of 0.7 nm rms. We have also demonstrated selective heteroepitaxial deposition of high-quality thick Ge layers through SiO_2 windows on Si by the MHAH technique (Figure 36.8) [30]. The Ge layer is grown only on Si, using the selectivity between the Si and SiO_2. This technique can also be used to grow Ge films laterally over the SiO_2 surface. Therefore, Ge on insulator (GOI) structure is achieved by using over-lateral growth. The Ge layer on SiO_2 window has a low threading dislocation density count of ~10^6 cm^{-2}. This technique has been used in achieving heterogeneous integration of a high mobility pure Ge channel MOSFETs [30], optical detectors [31, 35], modulators [32], and light emitters [33, 42] directly on Si for future technology nodes.

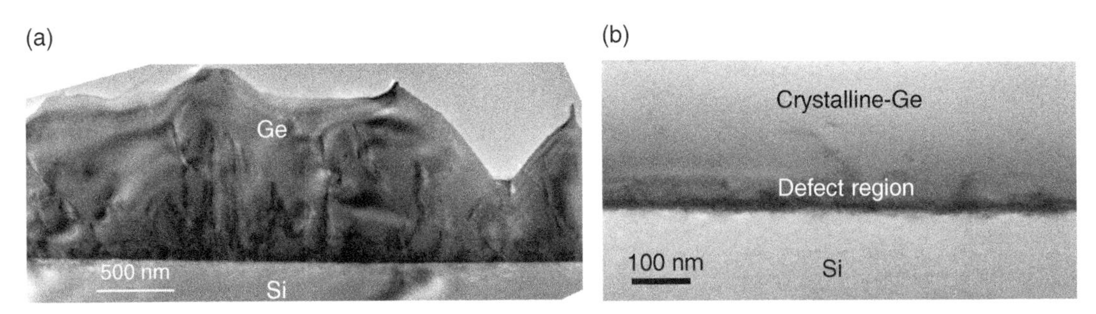

Figure 36.7 Cross-sectional TEM image of a heteroepitaxial-Ge layer on Si grown by (a) conventional method and (b) MHAH method on bare Si. Source: Nayfeh et al. [29]/AIP Publishing LLC.

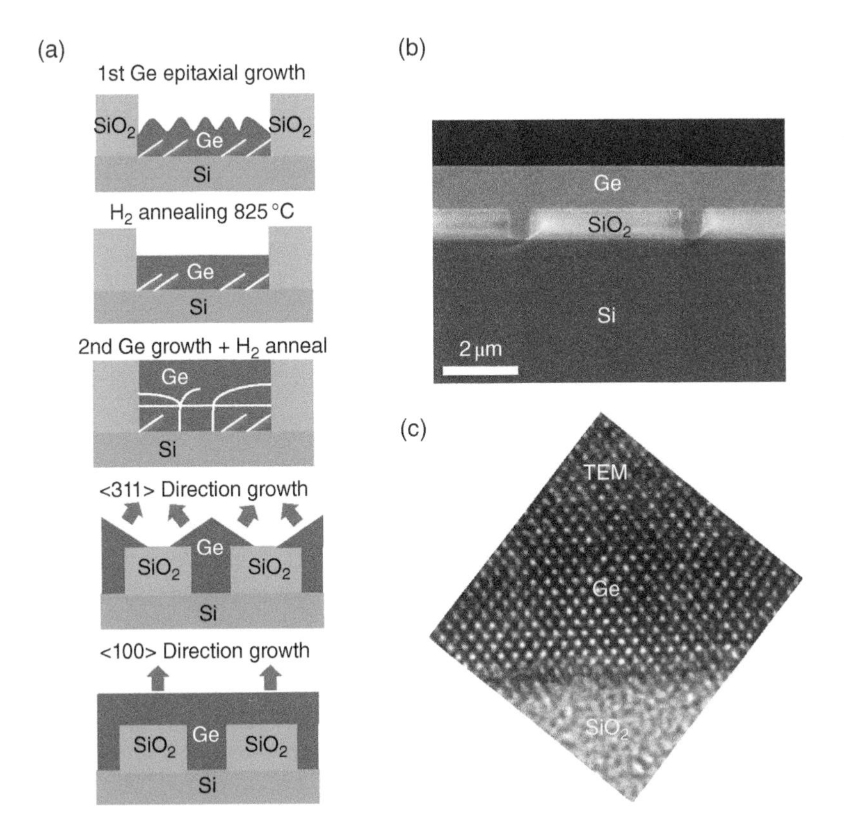

Figure 36.8 (a) MHAH selective Ge growth on Si, (b) cross-sectional SEM, and (c) TEM. Source: Adapted from Yu et al. [30]/IEEE.

36.6 Strained Ge and Heterostructure FETs

High mobility materials like Ge and $Si_xGe_{(1-x)}$ are very promising channel materials for nanoscale MOSFETs. Straining Ge and Si_xGe_{1-x} can significantly increase carrier mobility. Hole mobility increases with increasing strain and increasing Ge concentration because of a reduction in effective mass. The materials, such as strained Si and many III–V materials, have larger carrier mobility than unstrained Si, but the enhanced leakage because of their smaller bandgap or direct bandgap may limit

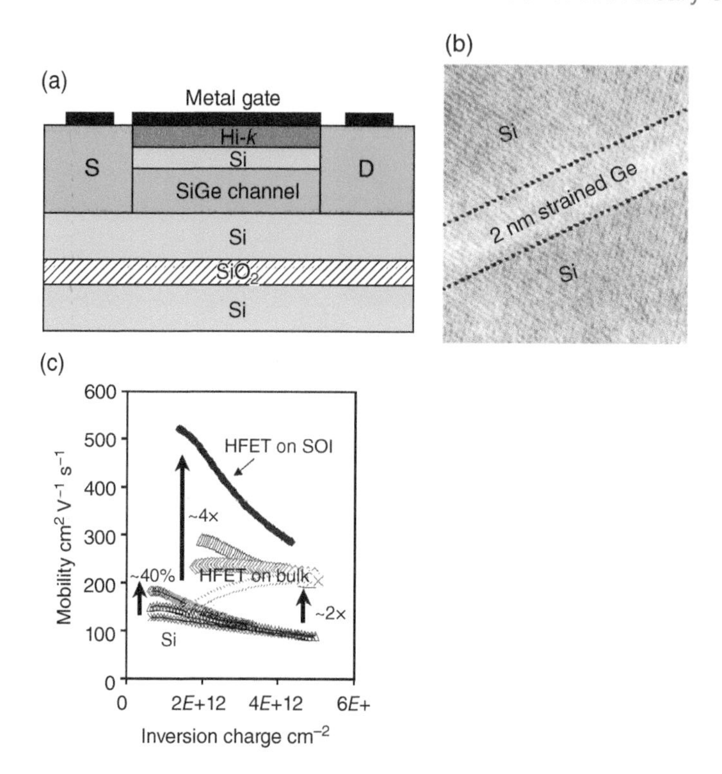

Figure 36.9 (a) Schematic and (b) cross-section TEM of Si/s-Ge/Si hetero-structure FET (H-FET). (c) Mobility of Si and Ge PMOS FETs. Source: Adapted from Krishnamohan et al. [17]/IOP Publishing.

their scalability. In a nanoscale transistor, generally the minimum standby off-state current (IOFF, min) is determined by the band to band tunneling (BTBT) leakage, IBTBT.

Below a critical thickness, around a few nm, defect-free Ge films can be grown on Si; however, they are strained due to the lattice mismatch. Above the critical thickness, the layer will have many misfit dislocations making it unusable for any practical applications. Below the critical thickness, the material is defect free. Strain in general results in reduction in the EG and hence enhanced IBTBT. Quantum mechanical confinement on the other hand results in increased EG and hence reduced IBTBT. Figure 36.9 shows the structure of a novel device to combine strain and quantum mechanical confinement to obtain desired transport properties with reduced off-state leakage. In this structure, the transport can be confined to the center of the channel in a high mobility material, Ge, flanked by a high EG material, Si. The mobility is further enhanced due to strain, reduced electric field in the center of the double gate structure due to symmetry, and the channel being away from the dielectric interface. The bandgap of the center channel is increased due confinement by keeping it very thin.

We have demonstrated [17] a novel Si/s-Ge/Si hetero-structure FET (H-FET), in which the transport occurs in high μ_p strained Ge and leakage in wider EG Si (Figure 36.9). The confinement of thin Ge between Si results in an increase in the EG and hence reduction in IBTBT, while strain keeps μ_p high. Experimentally, the resulting optimal structure obtained was an ultrathin, low defect, 2 nm fully strained Ge epi channel on relaxed Si, shown in Figure 36.9. H-FETs on bulk Si show a ~2× μ_p enhancement over Si, while H-FETs on SOI show even higher μ enhancements of >4× over Si devices. Both types of H-FETs show reduction in IOFF compared to bulk Ge devices. In particular, H-FETs on SOI show significant reduction in IOFF due to reduced E-field in Ge and E_g increase due to strain. Ultimately, device performance is determined by intrinsic gate delay (CV/I) and the minimum IOFF achievable.

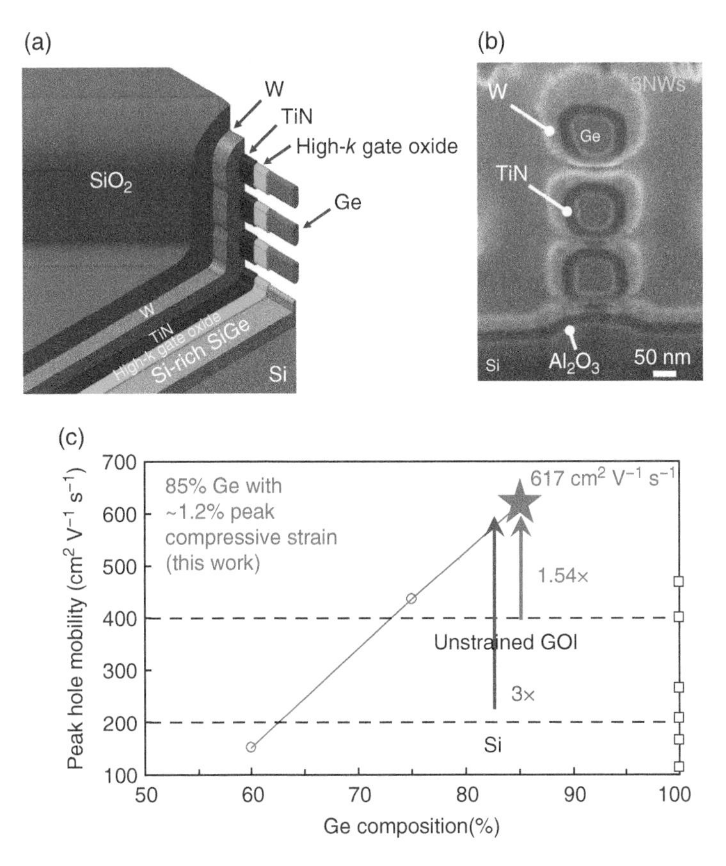

Figure 36.10 (a) Schematic and (b) cross-section TEM (c) hole mobility of nanoscale SiGe/Ge PMOS FETs fabricated by Gate-All-Around (GAA) 3D Ge condensation technique. Source: Ohmi et al. [34]/IEEE.

36.7 Nanoscale Ge FETs

Incorporating high mobility material in channel and 3D stacking channels in a multigate/surround gate device can enable next-generation high-performance CMOS technology. Ge is a potential channel material due to its superior μ and light effective mass. We demonstrated a novel method of fabricating 3D-stacked strained SiGe/Ge Gate-All-Around (GAA) by 3D Ge condensation [34]. The method allows Ge-rich SiGe and/or Ge to be incorporated in channel with a large uniaxial compressive strain. Either nanosheet (NS) or nanowire (NW) channels GAA device architecture can be fabricated. Through the geometric optimization of GAA channels, the large uniaxial compressive strain of ~2.5% and the excellent electrical characteristics of the pFETs with high hole mobility of 617 $cm^2\ V^{-1}\ s^{-1}$ were demonstrated (Figure 36.10) [34].

36.8 Ge NMOSFETs

Ge NMOSFETs have in the past exhibited poor drive current. Low n-type dopant activation and Fermi level pinning near the valance band (E_v) give high contact resistance to n-Ge, problematic for NMOS. Ge surface passivation with GeO_2 shows low D_{it} near E_v but high near the conduction band (E_c), another severe problem for NMOS. Low and symmetric D_{it} through sulfur passivation followed by ALD of Al_2O_3 and then annealing in ozone was demonstrated. With these two treatments, a record-low D_{it} (<1E11 cm^{-2}eV) at both band edges was achieved [15, 16]. We also demonstrated Fermi level de-pinning in metal/Ge Schottky junctions by inserting an ultrathin interfacial TiO_2 layer [35]. This technique has been demonstrated to lower contact resistance to n-Ge. Additionally, use of co-doping of Ge with Sb and P has been shown to increase dopant activation to $10^{20}\,cm^{-3}$ [36]. Similar activation has been achieved by laser annealing [21].

Semiconducting germanium tin (GeSn) alloy has recently emerged as a promising candidate for high-performance CMOS. However, to enable complimentary logic on GeSn, there is a need for high mobility n-channel devices on GeSn. A detailed theoretical analysis shows that alloying Ge with Sn drastically modifies the Ge conduction band structure without significantly affecting the valence band [37]. With increasing Sn content, Γ valley comes down with respect to the indirect valleys, increasing population of electrons in this low electron mass valley which boosts mobility and thus V_{inj}. High-quality GeSn films have been realized on Ge-on-Si using a CVD process. A novel surface passivation scheme using ozone oxidation of thin Ge cap achieve record low trap densities at high-κ/GeSn interface. With a low thermal budget Si-compatible device fabrication process, n-channel MOSFETs on GeSn with channel as high as 8.5% have been demonstrated with excellent characteristics [37].

These techniques will allow high-performance nanoscale NMOS. However, additional work is needed in this area.

36.9 Ge-Based Novel Devices for Optical Interconnects

Si CMOS dominated the microelectronics in the past; however, scaling to future nodes is reaching practical and fundamental limits. To go beyond these limits, novel materials and device structures are being pursued to enhance performance, the opposite is true for the interconnects [38]. Scaled metal interconnects have several limitations, including excessive power dissipation, insufficient communication bandwidth, and signal latency [39, 40]. Many of these obstacles stem from the physical limitation, in particular, the increase in resistivity, as wire dimensions and grain size become comparable to the bulk mean free path of electrons in copper. This is compounded by the fact that the highly resistive diffusion barrier is not scalable making it appreciable fraction of the interconnect.

Optical interconnects (Figure 36.11) use photons instead of electrons for communication and thus offer larger bandwidth at lower power consumption and lower delay [40, 41]. While III–V-based optical links are routinely used for long-distance communication, for on-chip and off-chip communication in conventional machines, all the optical components need to be integrated seamlessly on a Si platform.

Simulations show that as compared to Cu, carbon nanotubes (CNTs) are advantageous for local interconnects, while optical interconnects are better suited for global, semi-global, and chip to chip interconnects [40, 41]. For optical interconnect to be accepted in manufacturing, the technology must be compatible with Si platform and superior to electrical wires in terms of delay, energy/bit, and bandwidth. Ge and its alloys (GeSn and SiGeSn) are emerging as viable candidates for Si-compatible integration of monolithic optical components: laser [33, 42], detector [35, 43, 44], and modulator [47]. While Ge growth on Si, surface passivation, and low resistance contacts were discussed in earlier sections, additional techniques needed for Ge-based optical components are briefly discussed here.

We have demonstrated p–i–n [45] and junction-less metal-semiconductor-metal (MSM) photodetectors [44] in Ge grown on Si with excellent quantum efficiency and responsivities at a wavelength of 1.55 µm. Dark current, a concern for Ge photodetectors due to its small bandgap, has been mitigated by using asymmetric workfunction electrodes in MSM detectors [35]. By inserting a thin TiO_2 electron-selective hole blocking Fermi level de-pinning layer, the junction-less photodiode shows a dark current reduction by a factor of 3×10^3, with an extended absorption spectrum due to built-in strain in Ge [35].

Modulators act as optical switches which control the flow of photons. Modulators exploiting the free carrier dispersion effect in Si have been demonstrated in Mach–Zehnder interferometer and ring resonator configurations but have not proven to be accepted for on-chip interconnects. We have demonstrated

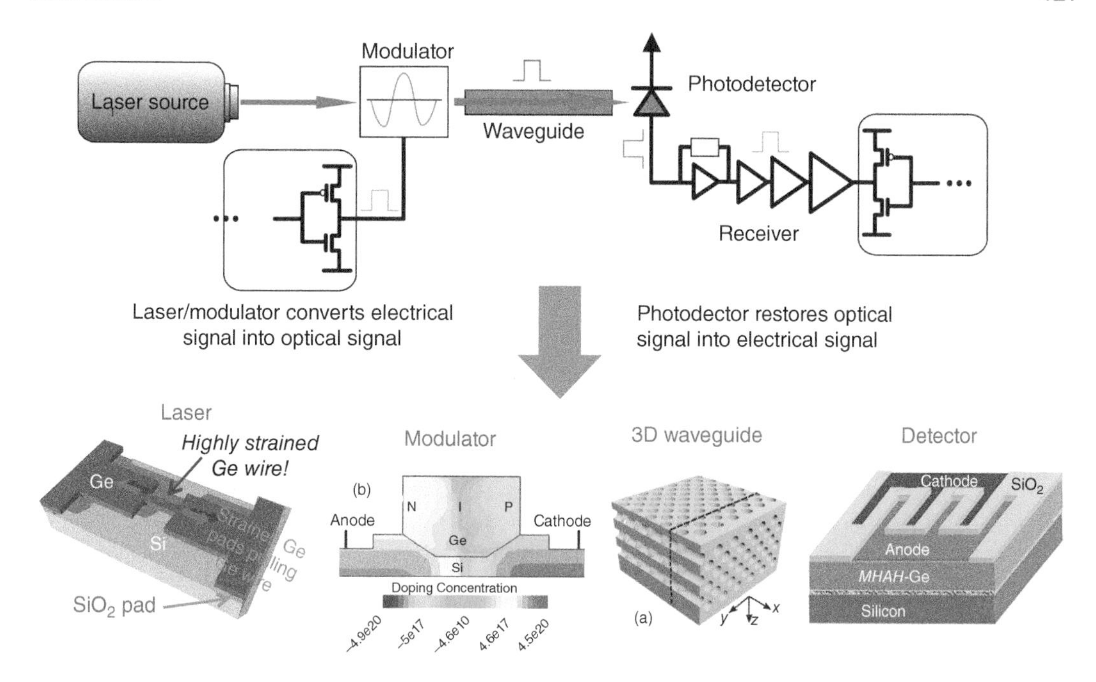

Figure 36.11 Schematic illustration of an optical interconnect system and Si-compatible photonic interconnect using Ge-based components integrated on Si.

Si-compatible electro-absorption modulators with the best reported energy-delay product based on the Franz–Keldysh effect (FKE) or the quantum-confined Stark effect in GeSi or Ge epitaxially grown on Si [32]. The FKE is known to be a sub-picosecond phenomenon enabling high-speed modulation, higher efficiency, and lower capacitance due to strong confinement of optical and electrical field enabled by submicron Ge/Si waveguide platform and as such can be considered as a potential candidate to meet low energy, high bandwidth targets for Si-based modulators.

A Ge laser, however, remains particularly challenging due to its indirect bandgap. However, the energy difference between the direct Γ valley and the indirect L valley is only 136 meV and this difference can be reduced further by introducing tensile strain in Ge and adding Sn [33, 42], resulting in more efficient light emission We have demonstrated methods to fabricate thin Ge membranes integrated on a Si substrate and induce sustainable and large tensile strain to make it a direct bandgap material and improve its optical properties [42]. Recent work has demonstrated that addition of >8% tin (Sn) to Ge makes it a direct bandgap semiconductor suitable for light emission and strain-tunable Ge or GeSn membranes can ultimately be utilized for high-efficiency near-infrared lasers, which are essential to realize on-chip optical interconnects [45].

36.10 Summary

In this paper, we have summarized the historical progress in the area of Ge MOSFET and briefly mentioned the progress in the area of Ge-based devices for optical interconnects.

Acknowledgment

I would like to acknowledge my colleagues and students who have been the main contributors in this work.

References

[1] Garrett, A.B. (1963). The discovery of the transistor: W. Shockley, J. Bardeen, and W. Brattain. *J. Chem. Educ.* 40 (6): 302–303. https://doi.org/10.1021/ed040p302.

[2] Kilby, J.S. (1959). Miniaturized electronic circuits. US Patent 3,138,743, 3,138,743. http://doi.org/10.1109/N-SSC.2007.4785580.

[3] Noyce, R.N. (1959). Semiconductor device-and-lead structure. US Patent 2,981,877, 2,981,877. http://doi.org/10.1109/N-SSC.2007.4785577.

[4] Kahng, D. (1960). Electric field controlled semiconductor device. US Patent 3,102,230, 3,102,230.

[5] Moore, G.E. (1965). Cramming more components onto integrated circuits. *Electronics* 38 (8): 114–117. https://doi.org/10.1109/N-SSC.2006.4785860.

[6] Dennard, R.H., Gaensslen, F.H., Yu, H.-N. et al. (1974). Design of ion-implanted MOSFET's with very small physical dimensions. *IEEE J. Solid State Circuits* 9 (5): 256–268. https://doi.org/10.1109/JSSC.1974.1050511.

[7] Bohr, M. (2011). The evolution of scaling from the homogeneous era to the heterogeneous era. *International Electron Devices Meeting (IEDM)*, Washington, DC, USA (December 2011), pp. 1.1.1–1.1.6. http://doi.org/10.1109/IEDM.2011.6131469.

[8] TSMC logic: high mobility channel. https://research.tsmc.com/english/research/logic/high-mobility-channel/publish-time-1.html (accessed 21 August 2021).

[9] Natori, K. (1994). Ballistic metal-oxide-semiconductor field effect transistor. *J. Appl. Phys.* 76 (8): 4879–4890. https://doi.org/10.1063/1.357263.

[10] Lundstrom, M. and Ren, Z. (2002). Essential physics of carrier transport in nanoscale MOSFETs. *IEEE Trans. Electron Devices* 49 (1): 133–141.

[11] Chui, C.O. and Saraswat, K.C. (2004). Advanced germanium MOSFET technologies with high-κ gate dielectrics and shallow junctions. *International Conference on Integrated Circuit Design and Technology*, Austin, Texas (17–20 May 2004), vol. 650, pp. 245–252. http://doi.org/10.1109/ICICDT.2004.1309955.

[12] Kuzum, D., Krishnamohan, T., Pethe, A.J. et al. (2008). Ge-interface engineering with ozone oxidation for low interface-state density. *IEEE Electron Device Lett.* 29 (4): 328–330. https://doi.org/10.1109/LED.2008.918272.

[13] Krishnamohan, T., Kim, D., and Saraswat, K.C. (2010). Properties and trade-offs of compound semiconductor MOSFETs. In: *Fundamentals of III–V Semiconductor MOSFETs* (ed. P. Serge and L. Ye), 7–27. Boston, MA: Springer US. https://doi.org/10.1007/978-1-4419-1547-4_2.

[14] Chui, C.O., Kim, H., McIntyre, P.C., and Saraswat, K.C.A (2003). Germanium NMOSFET process integrating metal gate and improved Hi-k dielectrics. *IEEE International Electron Devices Meeting (IEDM) Tech. Digest,* Washington, DC, (7–10 December 2003), pp. 437–440.

[15] Chui, C.O., Kim, H., Chi, D. et al. (2006). Nanoscale germanium MOS dielectrics - part II: high-k gate dielectrics. *IEEE Trans. Electron Devices* 53 (7): 1509–1516.

[16] Yang, B., Gupta, S., and McVittie, J.P. (2012) High quality germanium gate stack by sulfur passivation and novel ozone oxidation. *IEEE Semiconductor Interface Specialists Conference (SISC)*, San Diego, CA, (6–8 December 2012), pp. 32.

[17] Krishnamohan, T., Kim, D., Jungemann, C. et al. (2006). High performance, ultra-thin, strained-Ge, heterostructure FETs with high mobility and low leakage. *ECS Trans.* 3 (7): 687–695. https://doi.org/10.1149/1.2355864.

[18] Chui, C.O., Kim, H.S., David Chi, B. et al. (2002). A sub-400'C germanium MOSFET technology with high-k dielectric and metal gate. *IEEE Int. Electron Dev. Meet.*, San Francisco, CA (December 2002).

[19] Kuzum, D., Pethe, A.J., Krishnamohan, T. et al. (2007). Interface-engineered Ge (100) and (111), N- and P-FETs with high mobility. *Tech Digest of IEEE IEDM* Washington, DC (December 2007), pp. 723–726.

[20] Zhang, R., Huang, P.-C., Lin, J.-C. et al. (2012). Physical mechanism determining Ge p- and n-MOSFETs mobility in high ns region and mobility improvement by atomically flat GeOx/Ge interfaces. *IEEE IEDM.* San Francisco, CA (December 2012).

[21] Ni, C.-N., Li, X., Sharma, S. et al. (2015). Ultra-low contact resistivity with highly doped Si:P contact for nMOSFET, *Symp. VLSI Tech.* Kyoto, T118–T119. https://doi.org/10.1109/VLSIT.2015.7223711.

[22] Suh, J., Ramesh, P., Meng, K. et al. (2018). Low resistance III–V heterocontacts to N-Ge. *Int. Conf. on Solid State Dev. and Mat. (SSDM)*, Tokyo (13 September 2018). pp. 229–230.

[23] Connelly, D., Faulkner, C., Grupp, D.E., and Harris, J.S. (2004). A new route to zero-barrier metal source/drain MOSFETs. *IEEE Trans. Nanotech.* 3 (1): 98–104.
[24] Shine, G. and Saraswat, K.C. (2017). Analysis of atomistic dopant variation and Fermi level Depinning in nanoscale contacts. *IEEE Trans Electron. Devices* 64 (9): 3768–3774.
[25] Manik, P.P. and Lodha, S. (2015). Contacts on n-type germanium using variably doped zinc oxide and highly doped indium tin oxide interfacial layers. *Appl. Phys. Express* 8: 051302.
[26] Kim, J.-K., Kim, G.-S., Shin, C. et al. (2014). Analytical study of interfacial layer doping effect on contact resistivity in metal-interfacial layer-Ge structure. *IEEE Electron Device Lett.* 35 (7): 705–707.
[27] Ramesh, P. (2021). Approaching the limits of low resistance contacts to n-type germanium. Ph.D. dissertation. Stanford Univ. (August 2021).
[28] Jeon, J., Suzuki, A., Nakatsuka, O., and Zaima, S. (2018). Formation of ultra-low resistance contact with nickel stanogermanide/heavily doped n+ Ge1-xSnx structure. *Semicond. Sci. Technol.* 33 (12): 124001. https://doi.org/10.1088/1361-6641/aae624.
[29] Nayfeh, A., Chui, C.O., Saraswat, K.C., and Yonehara, T. (2004). Effects of hydrogen annealing on heteroepitaxial-Ge layers on Si: Surface roughness and electrical quality. *Appl. Phys. Lett.* 85 (14): 2815–2817.
[30] Yu, H.-Y., Park, J.-H., Okyay, A.K., and Saraswat, K.C. (2012). Selective-area high-quality germanium growth for monolithic integrated optoelectronics. *IEEE Electron Device Lett.* 33 (4): 579–581.
[31] Okyay, A.K., Chui, C.O., and Saraswat, K.C. (2006). Leakage suppression by asymmetric area electrodes in metal-semiconductor-metal photodetectors. *Appl. Phys. Lett.* 88: 063506.
[32] Gupta, S., Srinivasan, S.A., Pantouvaki, M. et al. 50GHz Ge waveguide electro-absorption modulator integrated in a 1615nm Si photonics platform. *OFC, Paper #Tu2A.4*, Los Angeles (March 2015).
[33] Gupta, S., Nam, D., Vuckovic, J., and Saraswat, K. (2018). Room temperature lasing unraveled by a strong resonance between gain and parasitic absorption in uniaxially strained germanium. *Phys. Rev. B* 97: 155127.
[34] Suh, J., Meng, A.C., Kim, T.R. et al. (2019). 3D-stacked highly strained SiGe/Ge Gate-All-Around (GAA) pFETs fabricated by 3D Ge condensation. *2019 Int. Conf. on Solid State Dev. and Mat. (SSDM)*, Nagoya, Japan (2–5 September 2019), pp. 571–572.
[35] Nam, J.H., Afshinmanesh, F., Nam, D. et al. (2015). Monolithic integration of high performance germanium-on-insulator p–i–n photodetector on silicon. *Opt. Express* 23 (12): 15816–15823.
[36] Baik, S., Kwon, H., Paeng, C. et al. (2019). Boosting n-type doping levels of Ge with co-doping by integrating plasma-assisted atomic layer deposition and flash annealing process. *IEEE Electron Device Lett.* 40 (9): 1507.
[37] Suyog, G., Vincent, B., Yang, B. et al. (2012). Towards high mobility GeSn channel nMOSFETs: improved surface passivation using novel ozone oxidation method. *Tech Abstracts IEEE IEDM*, San Francisco, pp. 16.2.1–16.2.4 (December 2012).
[38] Saraswat, K.C. and Mohammadi, F. (1982). Effect of interconnection scaling on time delay of VLSI circuits. *IEEE Trans. Electron Devices* ED-29: 645.
[39] Kapur, P., Chandra, G., McVittie, J.P., and Saraswat, K.C. (2002). Technology and reliability constrained future copper interconnects – part II: performance implications. *IEEE Trans. Electron Devices* 49 (4): 598.
[40] Koo, K.-H., Kapur, P., and Saraswat, K.C. (2007). Compact performance models and comparisons for Gigascale on-Chip global interconnect technologies. *IEEE Trans. Electron Devices* 56 (9): 1787.
[41] Miller, D.A.B. (2017). Attojoule optoelectronics for low-energy information processing and communications – a tutorial review. *J. Lightwave Technol.* 35 (3).
[42] Bao, S., Kim, D., Onwukaeme, C. et al. (2017). Low-threshold optically pumped lasing in highly strained Ge nanowires. *Nature Communications* 8 (1): 1845.
[43] Yu, H.-Y., Kim, D., Ren, S. et al. (2009). Effect of uniaxial-strain on Ge p-i-n photodiodes integrated on Si. *Appl. Phys. Lett.* 95: 161106.
[44] Okyay, A.K., Chui, C.O., and Saraswat, K.C. (2006). Leakage suppression by asymmetric area electrodes in metal-semiconductor-metal photodetectors. *Appl. Phys. Lett.* 88: 063506.
[45] Sukhdeo, D.S., Kim, Y., Gupta, S. et al. (2015). Theoretical modeling for the interaction of tin alloying with N-type doping and tensile strain for GeSn laser. *IEEE Electron Device Lett.* 36 (8): 1307–1310.

Index

75th Anniversary of the Transistor, First Edition. Edited by Arokia Nathan, Samar K. Saha, and Ravi M. Todi.